FLOW PHENOMENA
IN POROUS MEDIA

ENERGY, POWER, AND ENVIRONMENT

A Series of Reference Books and Textbooks

Editor

PHILIP N. POWERS
Professor Emeritus of Nuclear Engineering
Purdue University
West Lafayette, Indiana

Consulting Editor
Energy Management and Conservation
PROFESSOR WILBUR MEIER, JR.
Dean, College of Engineering
The Pennsylvania State University
University Park, Pennsylvania

1. Heavy Oil Gasification, *edited by Arnold H. Pelofsky*

2. Energy Systems: An Analysis for Engineers and Policy Makers, *edited by James E. Bailey*

3. Nuclear Power, *by James J. Duderstadt*

4. Industrial Energy Conservation, *by Melvin H. Chiogioji*

5. Biomass: Applications, Technology, and Production, *by Nicholas P. Cheremisinoff, Paul N. Cheremisinoff, and Fred Ellerbusch*

6. Solar Energy Technology Handbook: Part A, Engineering Fundamentals; Part B, Applications, Systems Design, and Economics, *edited by William C. Dickenson and Paul N. Cheremisinoff*

7. Solar Voltaic Cells, *by W. D. Johnston, Jr.*

8. Energy Management: Theory and Practice, *by Harold W. Henry, Frederick W. Symonds, Robert A. Bohm, John H. Gibbons, John R. Moore, and William I. Snyder*

9. Utilization of Reject Heat, *edited by Mitchell Olszewski*

10. Waste Energy Utilization Technology, *by Yen-Hsiung Kiang*

11. Coal Handbook, *edited by Robert A. Meyers*

12. Energy Systems in the United States, *by Asad T. Amr, Jack Golden, Robert P. Ouellette, and Paul N. Cheremisinoff*

13. Atmospheric Effects of Waste Heat Discharges, *by Chandrakant M. Bhumralkar and Jill Williams*

14. Energy Conservation in Commercial and Residential Buildings, *by Melvin H. Chiogioji and Eleanor N. Oura*

15. Physical Cleaning of Coal: Present and Developing Methods, *edited by Y. A. Liu*

16. Flow Phenomena in Porous Media: Fundamentals and Applications in Petroleum, Water, and Food Production, *by Robert A. Greenkorn*

Additional Volumes in Preparation

FLOW PHENOMENA IN POROUS MEDIA

Fundamentals and Applications in Petroleum, Water, and Food Production

Robert A. Greenkorn
Purdue University
West Lafayette, Indiana

MARCEL DEKKER, INC.   New York and Basel

Library of Congress Cataloging in Publication Data

Greenkorn, Robert Albert, [date]
  Flow phenomena in porous media.

  (Energy, power, and environment ; 16)
  Includes bibliographies and index.
  1. Groundwater flow. 2. Oil reservoir engineering.
3. Soil moisture. I. Title. II. Series.
  GБ1197.7.G73 1983   532'.051   82-22119
  ISBN 0-8247-1861-5

MARCEL DEKKER, INC.
270 Madison Avenue, New York, New York 10016

Current printing (last digit):
10 9 8 7 6 5 4 3 2 1

PRINTED IN THE UNITED STATES OF AMERICA

To my family

Rosemary, David, Eileen, Susan, and Nancy

Preface

The increasing demand for oil, water, and food produced in an environmentally sound manner has placed emphasis on petroleum production, water production, and food production. A major part of this production concerns the movement of fluids in porous media. Oil occurs in porous sandstone and limestone; its effective production, especially by waterflood and enhanced recovery mechanisms, requires a knowledge of flow phenomena in porous media. The movement of water and its recovery from underground reservoirs for drinking and for irrigation requires a knowledge of flow phenomena in porous media. The movement of moisture, nutrients, and pollutants in soils affects crop growth and requires a knowledge of flow phenomena in porous media.

This book is meant to provide the fundamentals of flow phenomena through porous media, including a detailed discussion of the pseudotransport coefficients, permeability, capillary pressure, and dispersion and their relation to the geometry of porous media. Single-fluid flow, multifluid immiscible flow, multifluid miscible flow, including the effects of heterogeneity of the media, phase change, and adsorption, are discussed. Scaling, analogs, and numerical techniques are used in the prediction of flow phenomena in porous media. Specific applications to petroleum engineering, hydrology, and soils science are investigated.

The point of view of the topic in this book is different from that in other books on this subject. The book is written from the point of view of a chemical engineer—in terms of process simulation. As a chemical engineer, I am concerned with mechanisms; with a model of how things happen and the mathematical description of the mechanisms. My experience working in oil production research especially concerned with recovery processes and simulation of these processes impressed upon me the importance of describing the porous medium system under consideration. Reservoirs are heterogeneous, nonuniform, and anisotropic and the success or failure of many recovery processes eventually rests on being able to describe these nonidealities in a manner adequate for realistic simulation. This point of view—concern with the interrelation of the microscopic and macroscopic description of the porous media and the parameters in the mechanistic description of the flow phenomena—is integrated throughout the book.

The book is organized into two parts. The first six chapters are concerned with the description of flow phenomena in porous media by describing the media and modeling phenomena mathematically, physically, or numerically. The second part (Chapters 7, 8, and 9) are concerned with a sequence of specific applications of the material in the first six chapters to problems in reservoir engineering, hydrology, and soils science.

Although the book was not written as a textbook or a reference book, it can be used for both. As with other authors who have attempted to encompass this subject area, I must include a disclaimer that the topic is so large that I have not attempted to cover all its aspects. For example, the entire area of thermal effects, which are of great importance in reservoir and geothermal engineering, is completely ignored.

I wrote most of the book while a visiting professor in Petroleum Engineering at Stanford University during the winter and spring quarters of 1978. It was a terrific place for this purpose. The people were stimulating, the library exceptional, and the weather relative to the Midwest that year was outstanding. I especially want to thank Hank Ramey for inviting me and to thank Hank and Bill Brigham for putting up with my questions and giving me the answers to many of my problems. My officemate while at Stanford, "Muz" Standing, taught me all about relative permeability and flow with phase change—topics I thought I understood until I was fortunate enough to be with him.

Robert A. Greenkorn

Contents

Contents   ix

FLOW PHENOMENA
IN POROUS MEDIA

1

Introduction

Flow in porous media requires a description of both the media and the flow. A porous medium generally is an extremely complicated network of channels and obstructions. It is usually described statistically and in some average sense. Microscopically the distribution of particles or channels is described by assumed statistical descriptions.

Darcy's law, the phenomenological law, describing flow rate as a function of potential drop is used to calculate flow in porous media. Darcy's law, permeability, and porosity are used to describe porous media in terms of the flow conductance and the connected void space.

1.1 DEFINITION OF A POROUS MEDIUM

A porous medium is a solid with holes in it. In the broadest sense an open pipe is a limit case of a porous medium. Usually the number of holes or pores is sufficiently large that a volume average is needed to calculate pertinent properties. Pores which occupy some definite fraction of the bulk volume form a complex network of voids. In describing flow phenomena in porous media we are most interested in interconnected pores since these are the ones that affect flow. Dead-end pores, that is, pores with only one entry (or exit), must also be considered in certain flow phenomena, especially those where mass transfer takes place. Unconnected pores do not affect flow directly but may affect the compressibility of the matrix. In addition, there are other properties that create flow channels such as fractures and openings along bedding planes. Normally, we consider these channels as part of the total pore volume of the medium.

The matrix of a porous medium is the material in which the holes or pores are imbedded. The manner in which the holes are imbedded, how they are interconnected, and the description of their location, size, shape, and interconnection characterize the porous medium. The porosity is a quantitative property that describes the fraction of the medium that is voids. When we are concerned with flow, the pores or fraction of the medium that contributes to flow constitutes the effective porosity. Recognize that in defining porosity we replace the complex network of voids with a single number that represents an average property. A porous medium

1

of a given porosity can be extremely different from another porous medium that has the same porosity.

There are many methods of determining porosity. The common methods involve measuring the liquid saturation where the medium is completely dried, weighed, and then filled with a liquid and weighed again. One can determine the porosity by determining the volume of the liquid that is needed to fill the pores taking into account any necessary corrections for temperatures, pressures, and any fittings that might be involved. Porosity is the ratio of the pore volume, the volume of the fluid used to fill the medium, to the total volume of the porous medium.

In the gas expansion method, a medium of known bulk volume and a given amount of air or gas are placed in a container of known volume under pressure. This container is then connected with an evacuated container of known volume. The new pressure of the system is read, and this allows one to calculate from the gas laws the volume of gas that was originally in the porous medium.

Other methods are used, such as measuring the bulk volume of a piece of medium and then compacting it by destroying the voids and measuring the difference in the volume. One can look at thin sections of a porous medium and estimate porosity by determining statistically the number or the size of the pores. One can displace a fluid of known concentration from a porous medium with a fluid of different concentration and determine the porosity from the mass balance of the different concentrations.

The methods of forcing a fluid such as water or mercury into the porous medium and determining the amount of invading fluid or the gas expansion methods are the most popular means of obtaining porosity. (See standard API porosity measurements.)

Since our main interest is the passage of materials into and out of the pore space, we characterize or describe porous media in terms of the properties of the matrix that affect flow. The flow properties which describe the matrix from the point of view of the contained fluid are pseudotransport properties: permeability, the conductance of the medium; dispersion, the mixing caused by the tortuous paths in the medium; capillary pressure, the interfacial force due to constrictions. These properties depend on the structure of the matrix. These properties are bulk properties and have meaning only when applied to a medium having some minimum number of pores—a piece large enough to be volume averaged. These properties are normally only useful when used in equations that are at the same level of description as the properties themselves. In other words, to try to describe the permeability of a small number of pores where Darcy's law does not apply is meaningless. As with the porosity, whenever we replace the medium with an average number such as permeability or capillary pressure or dispersion, we lose information concerning the microscopic description since media of greatly different microscopic character may have similar permeabilities, capillary pressures, and dispersion. The problem of characterizing a porous medium is one of relating the geometrical properties of the matrix and the pseudotransport properties.

There is an extremely large array of materials that can be classified as porous media. Broadly we classify media as unconsolidated and consolidated; as ordered and random. Examples of unconsolidated media are beach sand, glass beads, catalyst pellets, column packings, soil, gravel, and packing, such as charcoal. Examples of consolidated media are most of the naturally occurring rocks such as sandstones and limestones. In addition, concrete, cement, bricks, paper, cloth, etc., are manmade consolidated media. Wood can be considered a consolidated medium, as can the human lung. Ordered media are regular packing of various types of media such as spheres, column packings, wood, etc. Random media are media that are without any particular correlating factor. They are hard to find, since if one looks carefully at almost any media one can correlate some factor. Perhaps media that come the closest to being random are beach or river sands that were not sorted when they were laid down. Bread seems to be random since the pores result from random expansion of gases inside.

1.2 DARCY'S LAW

Since we are concerned with the movement of fluids through porous media, the equation describing motion (the momentum balance) is of central importance. Darcy's law (Darcy, 1856) relates the volumetric flow rate Q of a fluid flowing linearly through a porous medium directly to the energy loss, inversely to the length of the medium, and proportional to a factor called the hydraulic conductivity, K. Figure 1.1 is a rough sketch of the kind of experiment Darcy performed on a sandbed. In this figure fluid enters a packed bed of length $\Delta \ell$ between points 1 and 2, at the top above the bed at volumetric flow rate Q, and passes through the bed and comes out of the bottom of the bed at volumetric flow rate Q. One can measure the difference in the hydraulic head between points 1 and 2 with a water manometer from some datum point z = 0. The difference between the height of the fluid at points 1 and 2 is the hydraulic head or the pressure drop and takes place over the length 1 to 2. In the figure it is indicated by $h_1 - h_2$.

Darcy's law is expressed as

$$Q = \frac{KA(h_1 - h_2)}{\Delta \ell} \qquad (1.1)$$

where

$$h = z + \frac{p}{\rho} + \text{constant} \qquad (1.2)$$

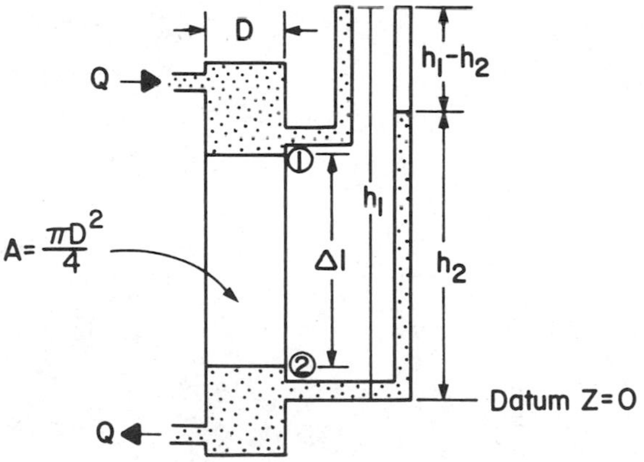

FIG. 1.1 Darcy's experiment.

Darcy's law is empirical in that it is not derived from first principles; rather it is the result of experimental observation.

Darcy's law is usually considered valid for creeping flow where the Reynolds number defined for a porous medium is less than one. The Reynolds number in open conduit flow is defined as the ratio of inertial to viscous forces and in terms of a characteristic length perpendicular to flow for the system. In order to define the Reynolds number for flow in porous media, we use the hydraulic radius concept. We define the hydraulic radius as the void volume of a porous medium divided by the surface area of the medium. The value of the hydraulic radius for a medium of spherical particles is

$$R_H = \frac{\text{void volume of porous medium}}{\text{surface area of porous medium}} \qquad (1.3)$$

If the particle volume is V_p and the surface area of the particles is S_p, then the specific surface for spherical particles is

$$S = \frac{6}{D} \quad \left(= \frac{S_p}{V_p} \right) \qquad (1.4)$$

and ϕ is the porosity of the medium

$$\phi = \frac{\text{void volume}}{\text{bulk volume}} (100) \qquad (1.5)$$

and the void volume of the porous medium is

$$\text{Void volume} = \frac{\phi}{1 - \phi} \text{ (volume of particles)} \qquad (1.6)$$

We can write the hydraulic radius in terms of the porosity, the number of particles, the volume of the particles, and the surface area of a particle in the following way:

$$R_H = \frac{\phi V_p N/(1 - \phi)}{S_p N} = \frac{V_p \phi}{S_p (1 - \phi)} \qquad (1.7)$$

For nonspherical particles, the effective particle diameter can be defined by

$$D_p = \frac{6}{S} \qquad (1.8)$$

The velocity of approach and the interstitial velocity are related by a mass balance as in the following equation.

$$v_\infty A_{tot} \rho = v A_{void} \rho \qquad (1.9)$$

As long as the porosity in terms of the area ratio is the same as the volumetric porosity, we write, as an approximation,

$$v_\infty = \phi v \tag{1.10}$$

The Reynolds number, which is the ratio of viscous to inertial forces, is written for flow in a pipe as

$$Re = \frac{vD\rho}{\mu} \tag{1.11}$$

In this case, if we substitute for velocity and diameter by using the velocity of approach divided by porosity and $4R_h$, we obtain a Reynolds number for the porous medium as follows:

$$Re = \frac{4R_h v_\infty \rho}{\phi \mu} = \frac{2}{3(1-\phi)} \frac{D_p v_\infty \rho}{\mu} \tag{1.12}$$

Usually the numerical constants are dropped and a Reynolds number is defined without numerical constants

$$Re_p = \frac{D_p v_\infty \rho}{\mu(1-\phi)} \tag{1.13}$$

In some correlations the $1 - \phi$ is also dropped. Darcy's law is usually considered valid in the creeping flow regime where Re_p is less than 1.

The hydraulic conductivity K, defined by Darcy's law Eq. (1.1), is dependent on the properties of the fluid as well as the pore structure of the medium. Hydraulic conductivity can be written more specifically in terms of the intrinsic permeability and the properties of the fluids

$$K = \frac{k\rho g}{\mu} \tag{1.14}$$

where k is the intrinsic permeability of the porous medium and in principle is only a function of the pore structure. The hydraulic conductivity is temperature dependent since the properties of the fluid density and viscosity are temperature dependent. The intrinsic permeability is not temperature dependent. Darcy's law is often written in differential form, so that in one dimension

$$\frac{Q}{A} = q = -\frac{k}{\mu}\frac{dp}{dx} \tag{1.15}$$

The minus sign results from the definition of Δp which is equal to $p_2 - p_1$, a negative quantity. The term q is called the seepage velocity and is equivalent to the velocity of approach v_∞ which was used in the definition of the Reynolds number.

It is usually assumed that Darcy's law is valid in three dimensions and that the permeability k is a second-order tensor dependent on the directional properties of the pore structures and $\underline{q}$ is a vector given by

$$\underline{q} = -\frac{\underline{\underline{k}}}{\mu} \underline{\nabla}p \tag{1.16}$$

In the discussion that follows we assume the porous medium is ideal, that is, homogeneous, uniform, and isotropic. These terms are defined in more detail in Chapter 2. If the ratio of permeability to viscosity k/μ is constant, then

$$\underline{q} = \underline{\nabla}\Phi \tag{1.17}$$

We can determine superficial potentials and streamlines from Eq. (1.17) using the real and imaginary parts of a complex function $w = \Phi + i\Psi$. The use of potential theory and the theory for ideal fluids is valid for many two-dimensional problems in flow through porous media, since in the creeping flow regime one does not have to concern oneself with the viscosity of the fluid. The creeping flow or potential flow solutions are valid on the average. From Darcy's law the potential must be equal to the conductivity times the thickness of the medium and

$$\Phi = Kh = K\left(\frac{p}{\rho g} + z\right) + \text{constant} \tag{1.18}$$

Although Darcy's law is empirical, DeWiest (1965) showed heuristically that Darcy's law is the empirical equivalent of the Navier-Stokes equations. We substitute v_x/ϕ and v_z/ϕ into the Navier-Stokes equations for two dimensions using x and z, where z is the dimension in the direction of gravity.

$$\frac{1}{\phi}\frac{\partial v_x}{\partial t} + \frac{v_x}{\phi^2}\frac{\partial v_x}{\partial x} + \frac{v_z}{\phi^2}\frac{\partial v_x}{\partial z} = -\frac{1}{\rho}\frac{\partial p}{\partial x} + \frac{\mu}{\rho\phi}\nabla^2 v_x \tag{1.19}$$

$$\frac{1}{\phi}\frac{\partial v_z}{\partial t} + \frac{v_x}{\phi^2}\frac{\partial v_z}{\partial x} + \frac{v_z}{\phi^2}\frac{\partial v_z}{\partial z} = -\frac{1}{\rho}\frac{\partial p}{\partial z} + \frac{\mu}{\rho\phi}\nabla^2 v_z - g \tag{1.20}$$

We assume statistical averages yield the following:

$$\nabla^2 v_x = \frac{1}{c}\frac{v_x}{d^2} \tag{1.21}$$

$$\nabla^2 v_z = \frac{1}{c}\frac{v_z}{d^2} \tag{1.22}$$

where d is an average or characteristic pore size and c is a dimensionless shape parameter. Assuming the properties of the fluid and the medium are constant, we write the velocities in terms of velocity potentials Φ so that

$$v_x = -\frac{\partial\Phi}{\partial x} \tag{1.23}$$

$$v_z = -\frac{\partial\Phi}{\partial z} \tag{1.24}$$

Substituting Eqs. (1.21) through (1.24) into Eqs. (1.19) and (1.20)

$$-\frac{1}{\phi}\frac{\partial}{\partial x}\left(\frac{\partial\phi}{\partial t}\right) + \frac{1}{\phi^2}\frac{\partial}{\partial x}\left[\frac{1}{2}\left(\frac{\partial\phi}{\partial x}\right)^2 + \frac{1}{2}\left(\frac{\partial\phi}{\partial z}\right)^2\right] = -\frac{1}{\rho}\frac{\partial p}{\partial x} + \frac{\mu}{c\rho d^2\phi}\frac{\partial\phi}{\partial x} \tag{1.25}$$

$$-\frac{1}{\phi}\frac{\partial}{\partial z}\left(\frac{\partial\phi}{\partial t}\right) + \frac{1}{\phi^2}\frac{\partial}{\partial z}\left[\frac{1}{2}\left(\frac{\partial\phi}{\partial x}\right)^2 + \frac{1}{2}\left(\frac{\partial\phi}{\partial z}\right)^2\right] = -\frac{1}{\rho}\frac{\partial p}{\partial z} + \frac{\mu}{c\rho d^2\phi}\frac{\partial\phi}{\partial z} - g \qquad (1.26)$$

Integrate both equations, assuming that μ and ρ are constant, to obtain

$$-\frac{1}{\phi}\frac{\partial\phi}{\partial t} + \phi^2\frac{1}{2}\left[\left(\frac{\partial\phi}{\partial x}\right)^2 + \left(\frac{\partial\phi}{\partial z}\right)^2\right] + \frac{p}{\rho} - \frac{\mu\phi}{c\rho d^2\phi} + gz = F(t) \qquad (1.27)$$

Finally, assuming steady flow and neglecting the inertial terms in Eq. (1.27), we find

$$\phi = \frac{k\rho g}{\mu}\left(\frac{p}{\rho} + z\right) + \text{constant} \qquad (1.28)$$

where

$$k = cd^2\phi \qquad (1.29)$$

Equation (1.28) is equivalent to Eq. (1.18) and we have shown that Darcy's law is the empirical equivalent of the Navier-Stokes equations.

The integrated forms of Darcy's law for an incompressible fluid are, for the linear case,

$$Q = \frac{kA(p_1 - p_2)}{\mu L} \qquad (1.30)$$

and for the radial case:

$$Q = \frac{2\pi kh(p_1 - p_2)}{\mu \ln r_1/r_2} \qquad (1.31)$$

For a compressible fluid, the volumetric flow rate Q varies with the pressure change. It is usually assumed that

$$pQ = p_m Q_m = \text{constant} \qquad (1.32)$$

where p_m is the arithmetic average of p_1 and p_2, and Q_m is the volumetric flow rate at p_m. Using Q_m, the linear and radial forms are the same as Eqs. (1.30) and (1.31) with Q_m replacing Q.

Permeability is normally determined using either Eq. (1.30) or (1.31) either in the incompressible or the compressible form, depending on whether a liquid or gas is used as the flowing fluid. Normally linear flow is used and most often liquid is used as the flowing fluid since one does not have to correct for compressible effects or slip flow. The volumetric flow rate Q (or Q_m) is determined at several pressure drops. Q (or Q_m) is plotted versus the average pressure p_m. The slope of this line will yield the fluid conductivity K or knowing the fluid density and viscosity, the intrinsic permeability k. For gases, the fluid conductivity apparently depends on pressure so that

$$\tilde{K} = K\left(1 + \frac{b}{p}\right) \tag{1.33}$$

where b is a parameter dependent on the fluid and the porous medium. Under such circumstances a straight line results as with a liquid, but it does not go through the origin, rather it has a slope of bK and intercept K. This difference between the liquid and gas flow was pointed out by Klinkenberg (1941). The explanation for this phenomenon is that gases do not stick to the walls of the porous medium and slip occurs. This slip shows up as an apparent dependence of the permeability on pressure. (See API permeability measurements.)

1.3 SCOPE OF FLOW PHENOMENA IN POROUS MEDIA

Flow phenomena in porous media can be classified as single-fluid and multifluid flow. In a limiting sense single-fluid flow need not be considered alone. However modeling of single-fluid flow is simpler if we consider it by itself rather than as a limiting case. For single-fluid or homogeneous flow there is one pseudotransport coefficient, the intrinsic permeability k.

For multifluid flow there are two major subdivisions: immiscible and miscible flow. For immiscible flow the capillary pressure (or relative permeability) is the additional pseudotransport coefficient. It is recognized that part of the capillary pressure is an equilibrium property, the interfacial tension. For miscible flow the mixing is described by another pseudotransport coefficient, the dispersion coefficient.

Multifluid flow may be stable or unstable since the densities, viscosities, and interfacial tensions of the fluids may differ. Instability can occur in both immiscible and miscible flows due to density and viscosity differences.

Wave phenomena and effects of overburden pressure must be considered when looking at regimes which include compressibility of the matrix and inertial effects.

We will consider only the preceding phenomena. Thermal effects, reactions other than adsorption, and turbulence are outside the scope of this book.

REFERENCES

Darcy, H. (1856): Les Fontaines Publiques de la Ville de Dijon, Dalmont, Paris.

DeWiest, R. J. M. (1965): Geohydrology, Wiley, New York.

Klinkenberg, L. J. (1941): API Drill. Prod. Pract. 200.

SUGGESTED READING

Bear, J., Dynamics of Fluids in Porous Media, Elsevier, New York, 1972.

Calhoun, J. C., Jr., Fundamentals of Reservoir Engineering, University of Oklahoma Press, Norman, 1953.

Collins, R. E., Flow of Fluids through Porous Materials, Van Nostrand, 1961.

Churchill, R. V., <u>Complex Variables and Applications</u>, McGraw-Hill, New York, 1960.

DeWiest, R. J. M. (ed.), <u>Flow through Porous Media</u>, Academic, New York, 1969.

Greenkorn, R. A., <u>Matrix Properties of Porous Media</u>, Proc. Second Symp. IAHR-ISSS, Fundamentals of Transport Phenomena in Porous Media, Guelph, 1972.

Greenkorn, R. A., and D. P. Kessler, <u>Transfer Operations</u>, McGraw-Hill, New York, 1972.

Muskat, M., <u>The Flow of Homogeneous Fluids through Porous Media</u>, McGraw-Hill, New York, 1937.

Poluborinova-Kochina, P. Ya., <u>Theory of Ground Water Movement</u>, Princeton University Press, Princeton, N. J., 1962.

Richardson, J. G., Flow through Porous Media, Sec. 16, in <u>Handbook of Fluid Dynamics</u>, V. I. Streeter (ed.), McGraw-Hill, New York, 1961.

Scheidegger, A. E., <u>The Physics of Flow through Porous Media</u>, Macmillan, New York, 1957.

2
Description of Porous Media

Our main interest in describing porous media is to understand and to predict the passage of materials into and out of porous media. It is possible to break down the description or characterization of porous media in terms of geometrical or structural properties of the matrix that affect flow and in terms of the flow properties which describe the matrix from the point of view of the contained fluid. Our problem of description of a porous medium is one of describing the geometrical or structural properties in some average fashion and interrelating these average structural properties with the flow properties.

There are two levels of description. At the microscopic level, we describe pore structure in some way. Usually, this "some way" is nebulous and amounts to a statistical description of a "pore size" distribution. The reason for calling it nebulous is that the description of pore size is nebulous—it depends on how one decides to describe such a distribution. The use of the term <u>microscopic</u> here is not meant to infer molecular, since we would have a similar problem if we went to a molecular level in trying to describe a property. Microscopic means in terms of pores or in terms of a distribution of pores. This is opposed to a macroscopic description, which is a description of the media in terms of the average or bulk properties and their variation at sizes or scales much larger than pores. The macroscopic level may be based on Darcy's law, in other words, for a piece of media large enough to have meaning in some volume-averaged sense. We use also statistical descriptions or distributions of the volume-averaged properties, just as we use statistical descriptions of pores at the microscopic level.

In addition, we consider the mechanical properties of fluids and media since the mechanical properties are important when compressibility and elasticity enter the problem, when we are concerned with the effects of overburden pressure on a reservoir, or when we are concerned with wave phenomena.

2.1 MICROSCOPIC DESCRIPTION

Since our main interest in description of porous media is calculating and/or predicting the passage of material into and out of pore spaces, a microscopic description is one that characterizes the structure of the pores. The objective of pore

structure analysis is to provide a description that can be related to the macroscopic
or bulk flow properties. The bulk properties that we want to relate to pore description are porosity, permeability, dispersion, tortuosity, capillarity, connectivity,
relative permeability, adsorption, and wettability. We discuss these properties in
the next several chapters in relation to whether they are structure properties,
volume properties, flow properties, mixing properties, etc. By pore structure
analysis we mean a description of the size, shape, orientation, and manner of interconnection of the pores. When one examines samples of the same kind of media,
such as a sandstone, it is apparent that the number of pore sizes, shapes, orienta-
tions, and interconnections is enormous. Because of this complexity, pore structure
description is most often attempted in terms of a statistical distribution of apparent
pore sizes. This distribution is apparent because to convert measurements to pore
sizes we resort to models which provide "average" or model pore sizes. As we
shall see, often we are describing a distribution of what seems to be pore diameters;
however, in consolidated media, often the pore lengths and pore diameters in the
direction of flow and perpendicular flow are significantly different, and it sometimes becomes necessary to describe both.

Pore Size Distributions

There are many ways of attempting a description of pore size. The most frequently
used methods are ones associated with measurements and models used to interpret
the measurements. One way is to define a pore diameter at a point as the diameter
of the largest sphere which contains the point and fills the pore space. A pore size
distribution is defined statistically as that fraction of the total pore space that has
pore diameter in the range δ and $\delta + d\delta$. Delta is assumed to follow a probability
density function.

$$\int_0^\infty f(\delta) \, d\delta = 1 \tag{2.1}$$

The corresponding probability distribution function is

$$F(\delta) = \int_\delta^\infty f(\delta) \, d\delta \tag{2.2}$$

If we fit a given distribution function to a pore size distribution, then we describe
the pore size distribution with the parameters of that distribution function, normally
the mean and the variance. Thus we have carried our description of the structure
of a porous medium through an assumed pore size measurement, such as the enclosed sphere, to a distribution of this measurement. Further, we reduce the information to two numbers, the average and the variance of the distribution function.

Another way to define a pore size distribution is to model the porous medium
as a bundle of straight cylindrical capillaries. The diameters of the model capillaries are distributed according to some distribution function. Perhaps a more
realistic model, especially for consolidated media, would have a distribution function for the lengths of the capillaries as well as the diameters. The capillaries in
the usual model, however, are given a length to match the pore volume based on
the average diameter of this distribution of capillaries. This model is used when
interpreting capillary pressure measurements in terms of pore size.

When interpreting photomicrographs, another definition is to use the mean
intercept length to characterize pore size. One measures the mean intercept length

by taking the arithmetic average of all the chords obtained by intersecting the pores in a photomicrograph with parallel straight lines in all directions. Dullien and Mehta (1972) and Dullien and Dhawan (1973) discuss the problem of interpreting pore sizes from photomicrographs in detail. With photomicrographic interpretation, the additional problem of interpreting a two-dimensional surface in three dimensions exists. For a bundle of capillaries, the mean intercept length is equal to four times the ratio of the volume to the surface area. This ratio is four times the hydraulic radius. Photomicrographs may be interpreted many ways, using various size grids, successive circles, and so on, most often depending on the equipment available for measurement. Most of these various approaches of using chord lengths, sized grids, successive circles, and so on, will give the same answer as long as the number of samples measured and the number of measurements made is large.

A common method of obtaining a pore size distribution is from capillary pressure measurements or mercury porosimetry. The capillary pressure is a multifluid property, related to the specific free energies of the interface between fluids, and fluids and pore walls. The capillary pressure is an equilibrium property, directly related to the interfacial tension (Morrow, 1970). At equilibrium, the surface free energy between the fluids is a minimum. The equilibrium condition is expressed by the Laplace equation

$$p_c = \gamma_{12}\left(\frac{1}{r_1} + \frac{1}{r_2}\right) \tag{2.3}$$

where p_c is the capillary pressure, γ_{12} is the specific free energy interface between fluids 1 and 2 and is equivalent to σ the interfacial tension between the two fluids, r_1 and r_2, are the two principal radii of curvature of the interface at any point.

Consider a straight cylindrical capillary. The radius of curvature for such a capillary is the harmonic average of the two principal radii, which is the term in parentheses in Eq. (2.3); therefore

$$p_c = \frac{2\gamma_{12}}{r} \tag{2.4}$$

For two immiscible fluids in contact with the solid walls of the capillary, the fluid-fluid interface intersects the solid surface at an angle

$$\cos \theta = \frac{\gamma_{s1} - \gamma_{s2}}{\gamma_{12}} \tag{2.5}$$

where γ_{12} is the specific free energy of the interface between fluids 1 and 2, γ_{s1} is the specific free energy of the interface between the solid and fluid 1, and γ_{s2} is the specific free energy between the solid and fluid 2. Since the angle θ may be different for an advancing or receding interface, there may be a hysteresis in the capillary pressure depending on the direction of motion of the interface—whether it is advancing into the pores or receding out of the pores. The radius of curvature of the meniscus is

$$r = \frac{\delta}{2 \cos \theta} \tag{2.6}$$

and

$$p_c = \frac{\gamma_{12} \cos \theta}{\delta} = \frac{2\sigma \cos \theta}{r} \qquad (2.7)$$

where δ is the capillary diameter. Equation (2.7) is for a capillary tube where we usually assume $\gamma_{12} = \sigma$. For fluid surrounding a bundle of rods $\gamma_{s2} = 0$ and $\gamma_{s1} = \gamma_{12} = \sigma$, then $\cos \theta = 1$ and

$$p_c = \frac{4\sigma}{\delta} = \frac{2\sigma}{r} \qquad (2.8)$$

Imagine a porous medium to be a fundle of capillaries; the combined pressure in all the capillaries is the capillary pressure of such a medium. If we apply a given pressure to a fluid-filled porous medium, then the saturation of the mdedium will be a function of the applied pressure. The relation between saturation and capillary pressure if a given pressure p_c is applied is

$$S = \int_{\delta_c}^{\infty} f(\delta) \, d\delta \qquad (2.9)$$

Figure 2.1 is a typical capillary pressure-wetting fluid saturation curve showing hysteresis. These curves were determined in a sandstone where the fluid system was oil and water. The drainage curve, curve 1, begins with the sample saturated with the water. The imbibition curve, curve 2, begins with the sample saturated with the nonwetting fluid, oil. The hysteresis in the capillary behavior may be due to several mechanisms.

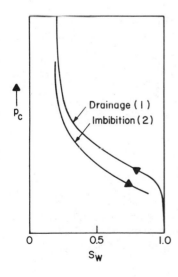

FIG. 2.1   Capillary pressure-wetting fluid saturation for a porous rock. (From Corey, 1977.)

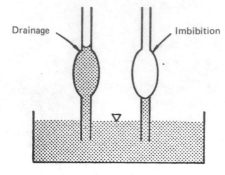

FIG. 2.2 "Ink bottle" mechanism for
capillary hesteresis. (From Corey,
1977.)

1. A difference in the advancing and receding contact angles in the fluid in
   contact with the solid surfaces.
2. An ink bottle effect. This effect is sketched in Fig. 2.2.

If one assumes a pore shape which contains two equal small radii on the drainage
curve the pore will empty to the upper radius. Thus the fluid saturation will be
higher than with the imbibition curve. The fluid saturation will be lower on imbibition since the fluid fills to the lower radius. In a real situation it is obviously much
more complicated since there will be several radii and many different connections.

   If the pores are not straight capillaries, Eq. (2.3) still applies. To find the
actual expression between saturation and capillary pressure for a porous medium,
we must know the average interfacial curvature as a function of saturation. Since
this is difficult to determine, we generally use the capillary model. The capillary
model does not have to be circular, nor the capillaries be of equal lengths. Although
normally we assume them to be of equal length.

   A straight cylindrical capillary model is often used to interpret capillary
pressure data. To obtain a pore size distribution from capillary pressure data, we

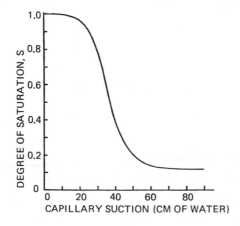

FIG. 2.3 Moisture characteristic of
Oso Flaco sand. (From Day and Luthin,
1956, The Williams and Wilkins Co.,
Baltimore.)

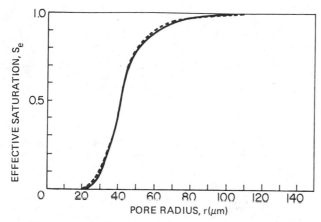

FIG. 2.4 Comparison between experimentally obtained effective pore size distribution (1) and calculated theoretical functions for $S_r = 0.119$. (From Brutsaert. © 1966, The Williams and Wilkins Co., Baltimore.)

measure the capillary pressure as a function of saturation by injecting a nonwetting fluid into a saturated porous media (air-water, mercury-air, etc.). Figure 2.3 is a soil moisture suction curve where air displaces water on Oso Flaco sand (Day and Luthin, 1956). The data in Fig. 2.4 are then integrated in terms of Eqs. (2.7) and (2.2) to obtain a cumulative pore size distribution as a function of the effective radius (diameter) of the capillary model and as a function of the volume of fluid penetrating the sample. Figure 2.4 is the pore size distribution for the data of Fig. 2.3 (Brutsaert, 1966), where

$$S_e = \frac{S - S_r}{1 - S_r} \qquad (2.10)$$

At small radii the assumption for the capillary model breaks down, and the shape of the curve at the low saturation end is difficult to determine. This difficulty is obviated by introducing the transformation of Eq. (2.10), where S_r is the residual or irreducible saturation. It is the saturation at which the drainage curve begins to parallel the ordinate. Now we take the derivative of the curve in Fig. 2.4 to obtain the effective pore size density function. Figure 2.5 is this result, and this curve is a pore size distribution.

The pore size distribution determined from capillary pressure-saturation data shown in Fig. 2.5 detects only the necks of the pores. Such a distribution does not detect the voids because these voids are only accessible through the necks, which control the pressure or flow. If one uses a capillary pressure description, the voids are assigned the dimensions of the necks, and the lengths, if one could calculate them, would be greater to account for the effective porosity or pore volume. In effect, the pore size distribution will have a smaller variance than actual, and it will be skewed toward the smaller pore diameters. This is generally satisfactory when considering flow since it is the necks which cause the most resistance to the flow. On the other hand, if the voids are much larger than the necks, it is possible

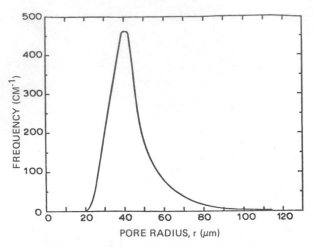

FIG. 2.5 Effective pore size density for $S_r = 0.119$ obtained from the experimental effective pore size distribution. (From Brutsaert. © 1966, The Williams and Wilkins Co., Baltimore.)

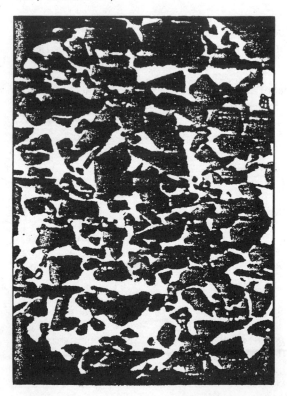

FIG. 2.6 Photomicrograph of section of salt pack saturated with Wood's metal after salt has been dissolved (pores are black). (From Dullien and Mehta, 1971–1972.)

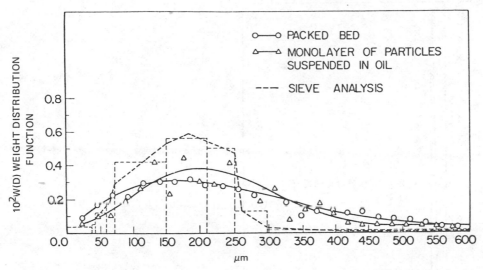

FIG. 2.7 Comparison of results of photomicrographic work with sieve analysis.
(From Dullien and Mehta, 1971-1972.)

that under certain circumstances errors in pore volume will create errors in estimation of flow properties.

Another method of obtaining a pore size distribution is to make a photomicrograph of a thin section of a porous medium and infer the pore sizes by direct measurement of the pores on a set of such photomicrographs. By a set we mean a set
large enough so that the number of measurements give a good statistical average.
To use this approach, we must use a method of "measuring" pores that is stereo-
logically sound (i.e., it correctly infers the three-dimensional geometry from an
interpretation of the two-dimensional photograph). Further, it must be correct for
known pore sizes. If we use a mean intercept length (or a grid or a sequence of
circles), the approach should correctly infer known geometries. The approach used
to interpret the photomicrographs most often depends on the equipment available to
interpret them. Dullien and co-workers (1970, 1971-1972, 1973, 1974) have discussed photomicrographic methods in detail.

Dullien and Mehta (1971-1972) compared the results of photomicrographic
analysis, sieve analysis, and mercury porosimetry for beds of packed salt particles. Figure 2.6 is a reproduction of a photomicrograph of a salt pack saturated
with Woods metal after the salt is dissolved. Figure 2.7 is a comparison of the
photomicrographic analysis with sieve analysis. Figure 2.8 is a comparison of the
results from mercury porosimetry and the photomicrographic method. Study of
these figures shows that the pore size distribution from the photomicrograph and
the sieve analysis are in agreement. There is a marked difference in the results
by photomicrographic analysis and mercury porosimetry. Mercury porosimetry,
which is a capillary pressure method, is measuring the pore necks. This distribution has most meaning in a flow situation. The photomicrographic analysis, on the
other hand, is measuring mostly voids, since they contribute or are highly weighted
in the averaging process. They have more meaning in porosity calculation. It is

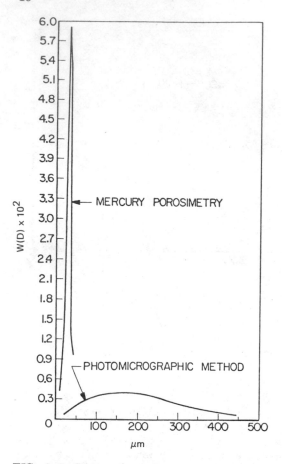

FIG. 2.8 Comparison of results obtained by mercury porosimetry and photomicrographic method. (From Dullien and Mehta, 1971-1972.)

interesting that the sieve analysis and the photomicrographic method give similar
results and as pointed out earlier, these results tend to agree with the hydraulic
radius concept.

The use of sorption isotherms to determine pore size distributions is a
method used for small pores of the order of 10 to 1000 Å. The surface area and
pore size distribution are inferred from adsorption (or desorption) of gases such
as nitrogen, oxygen, argon, or helium. The method assumes that in the pores the
saturation vapor pressure is reduced from that in an open space. This situation is
represented by the Kelvin equation using a model of circular capillaries

$$\ln \frac{p^*}{p^\circ} = -\frac{4\sigma}{\delta}\frac{\tilde{V}}{RT} \qquad\qquad (2.11)$$

where p° is the vapor pressure of a liquid, p^* is the vapor pressure of the same
liquid in pores of diameter δ, $\tilde{V}$ is volume per mole, R is the gas constant, and T

is absolute temperature. This equation is used to infer a pore size distribution from vapor pressure or adsorption pressure measurements. One measures the volume of gas condensed in the pores as a function of pressure. A sample of a porous medium is equilibrated in an atmosphere of fixed vapor pressure. Then the liquid saturation of the sample is measured. These methods are used by kineticists to determine surface area of catalysts and by soil scientists to determine pore size distributions from moisture suction curves. Figure 2.9 shows typical vapor pressure versus saturation curves.

Sorption curves exhibit hysteresis similar to that discussed earlier with capillary pressure measurements. One obtains a different curve depending on whether gas is adsorbed or desorbed. When a fluid is condensed in the pores (adsorption), as pressure is increased the pores of larger radii are filled. When a fluid is desorbed, the pores of largest radii will empty first, but the pressure at which they empty is slower than that at which the same pores filled. In addition to the mechanisms mentioned earlier with capillary pressure measurements (i.e., difference in advancing and receding contact angles and the ink bottle effect), for the kind of pores used with adsorption measurements, that is, small pores, the presence of a liquid film in the pore changes the effective pore size. If the hysteresis loop can be repeated, then the hysteresis is generally considered to be caused by the pore shape.

The review of Dullien and Batra (1970) discusses in great detail pore size determinations by the methods described here as well as by other methods including particle size and packing.

Particle Size and Packing

Pore structure for unconsolidated media is inferred from a particle size distribution, the geometry of the particles, and the manner in which the particles are packed. The theory of packing has been determined for symmetrical shapes, especially spheres. A knowledge of particle size, their symmetry, and the theory of packing allows one to establish relationships between pore size distributions and particle size distributions.

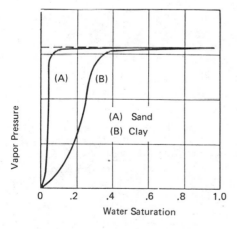

FIG. 2.9 Vapor pressure versus wetting fluid saturation. (From Collins, 1961.)

With ideal models based on symmetrical shapes, we are normally not interested in using such models to predict flow in real media. These ideal models, which
are homologs of one possible real porous medium, have size, shape, distribution,
and packing that can be precisely described, at least in a statistical sense. It is
possible to use these ideal models to investigate the forms of the flow equations
needed to predict and investigate details of the flow. Since these models, even though
ideal, do represent a porous medium, any relationship that we use to understand
mechanism or predict flow must work for these models. Therefore they are useful
when studying the mechanisms of flow or the meaning or connection of flow parameters to structure.

The common regular packing of identical spheres is described by the geometry
of the packing. The packing of spheres has two extreme cases (Graton and Fraser,
1935). A cubic packing is one limit, and its pore space is shown in Fig. 2.10. This
packing has a calculated porosity of 0.4764. The other extreme is a rhombohedral
packing, and its pore space is shown in Fig. 2.11. It has a calculated porosity of
0.2595. The model of Mayer and Stowe (1965) is defined in terms of a single angle σ
which is determined by the edges of the rhombohedron formed by connecting the
centers of a set of spheres as sketched in Fig. 2.12. This model may be utilized
to calculate porosities of intermediate packings of uniform spheres.

When spheres of identical size are randomly packed, the porosity will depend
on the method of packing (Haughey and Beveridge, 1969). Very loosely packed
spheres result from spheres settling in a liquid filled system. Such a system will
have a porosity of about 0.44. A loose random pack is made by letting spheres roll
in place one at a time over the previously packed bed of spheres and will have a
porosity of about 0.42. A poured random pack, resulting from pouring spheres in
place continually, will have a porosity of about 0.38. A close random pack, which
results from pouring spheres and at the same time vibrating the bed, will have a
porosity of about 0.36. When spheres of varying size are packed, the porosity also
depends on the size range. Usually the porosity of a close random pack of spheres
of varying sizes will be in the range of 0.32 to 0.35. The reason for this lower

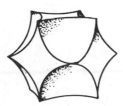

FIG. 2.10 Packing and pore space of
uniform spheres in a cubic arrangement. (From S. C. Graton and H. S.
Fraser, Systematic packing of spheres
with particular reference to porosity
and permeability, J. Geol. 43(8), 810.
© 1935 by the University of Chicago
Press.)

FIG. 2.11 Packing and pore space of uniform spheres in a rhombohedral arrangement. (From S. C. Graton and H. S. Fraser, Systematic packing of spheres with particular reference to porosity and permeability, J. Geol. 43(8), 785. © 1935 by the University of Chicago Press.)

porosity is that the smaller spheres will take up the spaces in between the larger spheres resulting in lower pore volume.

Haughey and Beveridge (1966) used the porosity to model the coordination number, i.e., the number of points of adjacent contacts for a single imbedded sphere. The coordination number n describes the packing. For example, n is 6 for cubic packing, 8 for orthorhombic, 10 for tetragonal-sphenoidal, 12 for rhombohedral. For random packing,

$$n = 22.47 - 39.39\phi \qquad 0.254 \leq \phi \leq 0.5 \tag{2.12}$$

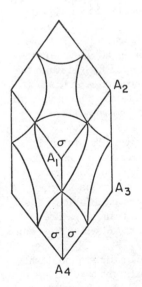

FIG. 2.12 Unit cell of packed spheres. (From Mayer and Stowe, 1965.)

With nonuniform angular particles such as unconsolidated sand and soils, random packs usually have porosities in the range of 0.3 to 0.45 depending on how they are packed. When the particles have different sizes, the smaller particles will fill in the interstices between the larger particles, giving a smaller or lower pore size than a packing with one size of particle. When packing particles of different sizes, it is possible to get grading of particles and pore sizes. In small diameter beds of packing materials it may be necessary to consider wall effects since there may be a significantly different porosity along the wall than in the main part of the bed.

With unconsolidated random packing of various size particles, the particle size distribution will correlate with the pore size distribution and with the porosity and permeability. Many correlations of flow properties in packed beds are correlated in terms of particle size distributions or in terms of the distribution of equivalent spheres. It is well to remember that this entire discussion concerning pore size distributions is at best approximate. Real pores are very complex.

2.2 A STATISTICAL MODEL OF A POROUS MEDIUM

This discussion is to provide a model pore structure so we can relate the microscopic and macroscopic descriptions of porous media. We will use this model to discuss relations between the pore structure of the model and macroscopic properties in the following chapters.

With a statistical model structure we relate model pore areas (or radii) and model pore lengths to distribution functions for areas and lengths. Anisotropy can be built into the model by relating anisotropy with the orientation of the model pores. Statistical models are not used to predict flow in real media; rather they are used to understand or to postulate relations between pore structure and the form of the macroscopic pseudotransport coefficients. Although these models are limited in relation to real-world homologs, they are well-described models of a possible porous medium. Any relationships we postulate relating to macroscopic flow properties, whether as to form or their relation to pore structure, must work for the statistical model as well as for real porous media. Likewise, any insights into the form of a flow property or its relation to structure that we find for the statistical model will be useful in discussing real porous media flow relations. The model will allow us to define the macroscopic terms nonuniformity and anisotropy more precisely.

The model is imagined as a collection of capillary tubes with radii distributed according to a distribution function and length distributed according to some other distribution function. You might imagine the model as resulting from taking a bunch of capillary tubes of different sizes and busting them into various lengths and stuffing them into a box and letting the flow pass through the capillaries. The model capillaries may be oriented (anisotropic) or nonoriented (random). The capillaries are assumed to intersect randomly to form a network in which flow can take place.

Guin and co-workers (1970, 1971a, 1971b) discuss a generalized model where the model pore space is a network of randomly intersecting straight capillaries. These capillaries do not have to be circular but can have any shape. The elemental

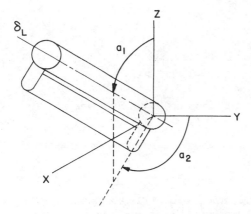

FIG. 2.13 An elemental pore. (From Guin et al., 1971b.)

pore is length ℓ, area A, and is oriented with two angles α_1 and α_2, shown in Fig. 2.13. An ensemble of elemental pores are assumed distributed according to a function $f(A, \ell, \alpha_1, \alpha_2)$. The function $f(A, \ell, \alpha_1, \alpha_2)$ is a probability density function and is the number of pores having area A to A + dA, length ℓ to ℓ + dℓ, orientation α_1 to α_1 + dα_1, and α_2 to α_2 + dα_2. The function f is the number of pores per unit volume.

Several properties may be computed for the model. The volumetric porosity of the model is

$$\psi = \int_0^{2\pi} \int_0^{\pi/2} \int_0^{\infty} \int_0^{\infty} A\ell f(A, \ell, \alpha_1, \alpha_2) \, dA \, d\ell \, d\alpha_1 \, d\alpha_2 \qquad (2.13)$$

The Darcy velocity is an area average defined by

$$\underline{q} = \frac{1}{S} \int_S \underline{v} \, dS \qquad (2.14)$$

where $\underline{v}$ is the interstitial velocity and S is an area element of a porous medium intersecting many pores. The number average pore velocity is

$$\langle \underline{v} \rangle = \frac{\int_D \underline{v} f(A, \ell, \alpha_1, \alpha_2) \, dD}{\int_D f(A, \ell, \alpha_1, \alpha_2) \, dD} \qquad (2.15)$$

where $\langle \underline{v} \rangle$ is the average velocity in an individual capillary (dD $\equiv$ dA dℓ dα_1 dα_2).
The Dupuit-Forchheimer relation defines the average interstitial velocity as

$$\langle \underline{v}_{DF} \rangle = \frac{\underline{q}}{\phi} \qquad (2.16)$$

It is usually assumed that $\langle \underline{v} \rangle$ is equal to $\langle \underline{v}_{DF} \rangle$; however we shall see this relationship does not hold for a medium with nonuniform pores. Let $\underline{n}\ell$ be a unit vector along the axis of a model pore. There are $\ell(\underline{n}_\ell \cdot \underline{n}) f(A, \ell, \alpha_1, \alpha_2) \, dA \, d\ell \, d\alpha_1 \, d\alpha_2$ model pores intersecting a unit area with normal $\underline{n}$ and $da = dS(\underline{n}_\ell \cdot \underline{n})$. Equation (2.14) may be written

$$\underline{q} = \int_D \underline{v} A \ell \, f(A, \ell, \alpha_1, \alpha_2) \, dD \qquad (2.17)$$

Combine Eqs. (2.13), (2.16), and (2.17) to find the Dupuit-Forchheimer velocity of the model

$$\langle \underline{v}_{DF} \rangle \frac{\langle \underline{v} A \ell \rangle}{\langle A \ell \rangle} \qquad (2.18)$$

The assumption that $\langle \underline{v} \rangle = \langle v_{DF} \rangle$ is true only when A and ℓ are constant, in other words, for a uniform medium. However, for a larger number of pores $\langle \underline{v} \rangle \doteq \langle v_{DF} \rangle$.
    The permeability of the model is

$$\underline{\underline{k}} = c \int_D \underline{\delta}_\ell \underline{\delta}_\ell A^2 \, \ell f(A, \ell, \alpha_1 \alpha_2) \, dD \qquad (2.19)$$

The permeability is a symmetric tensor for an arbitrary distribution function f. If the pores are not geometrically similar, the shape factor c must be included inside the integral. Preferential orientation of pores does not affect the symmetry of the permeability tensor. The permeability is nonuniform if the model pores are nonuniform. For a homogeneous medium the permeability has a single value rather than a distribution of values.
    The ergodic assumption—that spatial averages coincide with mathematical expectation—is implied in many statistical models of porous media (Matheron, 1967; Scheidegger, 1957). It is assumed that $\underline{q}$ is the mathematical expectation of $\langle \underline{v} \rangle$.

$$\hat{\underline{q}} = E\{\alpha(\delta)\underline{v}\} \qquad (2.20)$$

where a caret is added to differentiate it from $\underline{q}$ of Eq. (2.17). The permeability of this model using Eq. (2.20) is

$$\hat{\underline{\underline{k}}} = \frac{\phi}{8} E\{A\alpha_1\alpha_2\} \qquad (2.21)$$

The value of $\underline{q}$ for the model using Eq. (2.17) is

$$\underline{q} = E\{\underline{v} A \ell\} \qquad (2.22)$$

and

$$\underline{\underline{k}} = \frac{\phi}{8} \frac{E\{A^2 \alpha_1 \alpha_2\}}{E\{A\}} \qquad (2.23)$$

Therefore, $\underline{\underline{k}}$ is not equal to $\hat{\underline{\underline{k}}}$ and $\underline{q}$ is not equal to $\hat{\underline{q}}$ unless the model has uniform pores. (For a large number of pores $\underline{q} \doteq \hat{\underline{q}}$.)

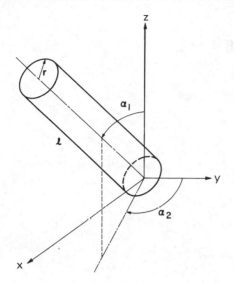

FIG. 2.14 Description of an elemental
pore space. (From Haring and Greenkorn, 1970.)

To directly relate the parameters of the model pore structure with the microscopic properties of a porous medium, we use a density function for the size of the
elemental pores. The statistical model of Haring and Greenkorn (1970) uses a
parametric density function, the beta function, to describe the nonuniformity of
circulary capillaries. The elemental pore for this model is shown in Fig. 2.14.
The pore space is approximated by a large number of randomly oriented straight
cylindrical pores. We assume that all directions for a pore are equally likely. The
domain for the model is $0 \leq r \leq R$, $0 \leq \ell \geq L$, $0 \leq \alpha \leq \pi/2$, $0 \leq \alpha_2 \leq 2\pi$, where R is
the largest pore in the ensemble and L is the longest pore in the ensemble. We de-
fine dimensionless length and radius as

$$\ell^* = \frac{\ell}{L} \tag{2.24}$$

$$r^* = \frac{r}{R} \tag{2.25}$$

The model is made nonuniform by distributing the dimensionless length ℓ^* and the
dimensionless radius r^*, according to the beta function. The beta function is a
parametric distribution where the random variable has a range of 0 to 1 and the
function gives a variety of skew and symmetric shapes.

$$B(x; p_1, p_2) = \frac{(p_1 + p_2 + 1)!}{p_1! \, p_2!} (x)^{p_1} (1 - x)^{p_2} \tag{2.26}$$

Several examples of the distribution of x according to Eq. (2.26) for various values
in p_1 and p_2 are shown in Fig. 2.15.

We assume length, radius, and orientation of the elemental pores are independent. In a uniform isotropic model, length and radius are dependent through

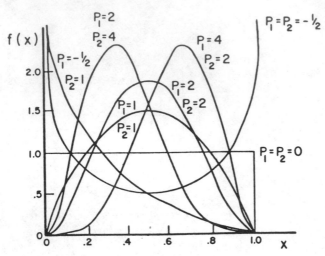

FIG. 2.15 Beta distribution. (From Haring and Greenkorn, 1970.)

pore volume. In a nonuniform model, length and radius are not dependent. The probability density functions for length and radius are

$$f(\ell^*) = \frac{(a + b + 1)!}{a! \, b!} \, (\ell^*)^a (1 - \ell^*)^b \qquad (2.27)$$

$$g(r^*) = \frac{(\alpha + \beta + 1)!}{\alpha! \, \beta!} \, (r^*)^\alpha (1 - r^*)^\beta \qquad (2.28)$$

The average values of ℓ^* and r^* are determined from the first moment of the probability density function.

$$\langle \ell^* \rangle = \frac{a + 1}{a + b + 2} = \frac{\langle \ell \rangle}{L} \qquad (2.29)$$

$$\langle r^* \rangle = \frac{\alpha + 1}{\alpha + \beta + 2} = \frac{\langle r \rangle}{R} \qquad (2.30)$$

The model is imagined as a large number of randomly intersecting elemental pores. The probability a given pore exists with size and orientation in the range ℓ^* to $\ell^* + d\ell^*$, r^* to $r^* + dr^*$, α_1 to $\alpha_1 + d\alpha_1$, and α_2 to $\alpha_2 + d\alpha_2$ is given by the normalized product of the independent probabilities so that

$$dE = \frac{1}{2\pi} \left[\frac{(a+b+1)!}{a! \, b!} (\ell^*)^a (1 - \ell^*)^b \, d\ell^* \right] \left[\frac{(\alpha+\beta+1)!}{\alpha! \, \beta!} (r^*)^\alpha (1 - r^*)^\beta \, dr^* \right] \alpha_1 \, d\alpha_1 \, d\alpha_2 \qquad (2.31)$$

The capillary pressure in a saturated elemental pore is given by

$$p_c = \frac{2\sigma \cos \theta}{r} \qquad (2.32)$$

Since the capillary pressure of the model pore depends on the radius, the capillary pressure of an ensemble of pores depends on the distribution of pore radii. The differential volume of fluid invading the pore space increased by a fractional saturation dS is V_p dS.

$$V_p \, dS = -N\pi R^2 L r^{*2} \langle \ell \rangle g(r^*) \, dr^* \tag{2.33}$$

where N is the number of pores.

Equation (2.33) may be written:

$$dS = -\frac{r^{*2}}{\langle r^{*2} \rangle} g(r^*) \, dr^* \tag{2.34}$$

Since ℓ^* and r^* are independent, we use the average length of the elemental pores in the calculation of capillary pressure. As the radius of a capillary becomes smaller, the capillary pressure and its associated saturation increases. Therefore as the radius varies from zero to one, saturation varies from one to zero. Integrating Eq. (2.34) after substituting Eq. (2.28)

$$1 - S = \int_0^{r^*} \frac{(\alpha + \beta + 3)!}{(\alpha + 2)! \, \beta!} (r^*)^{\alpha+2} (1 - r^*)^\beta \, dr^* \tag{2.35}$$

The invading fluid does not have access to all pores when r^* approaches 1. The error involved diminishes rapidly as S increases. Combining the definition of r^* and the capillary pressure Eq. (2.32), we obtain

$$r^* = \frac{r}{R} = \frac{p_c(lt)}{p_c(r)} = \frac{1}{p_c^*} \tag{2.36}$$

which normalizes p_c with respect to the threshold pressure corresponding to the largest pore. Substituting Eq. (2.36) into Eq. (2.35)

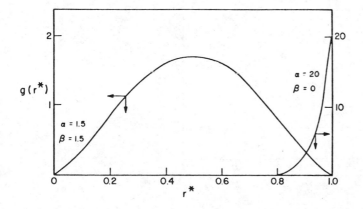

FIG. 2.16 Radius distribution function. (From Haring and Greenkorn, 1970.)

$$1 - S = \int_0^{1/p_c^*} \frac{(\alpha + \beta + 3)!}{(\alpha + 2)!\,\beta!}\left(\frac{1}{p_c^*}\right)^{\alpha+2}\left(1 - \frac{1}{p_c^*}\right)^{\beta} d\left(\frac{1}{p_c^*}\right) \qquad (2.37)$$

The integral of Eq. (2.37) is calculated from tabulated values of the incomplete beta function

$$1 - S = B_I\left(\frac{1}{p_c^*}\,;\,\alpha + 2,\,\beta\right) \qquad (2.38)$$

Figure 2.16 shows two distribution functions for pore radii: a uniform and a wide pore distribution. Figure 2.17 shows the capillary pressure versus saturation for the distributions of Fig. 2.16 calculated from Eq. (2.38).

The measured drainage capillary pressure of a random pack of glass beads with a size range of 590 to 840 μm and a measured porosity of $\phi = 35.4$ percent are plotted as solid dots in Fig. 2.18. The open squares were calculated using Eq. (2.38) where the parameters of the radius distribution function were determined from the threshold pressure.

The permeability of the model with a large number of pores is found by relating the average velocity in the elemental pore to the average velocity in the ensemble. Consider bulk flow through the model in the z direction.

$$v_z = -\frac{R^2}{8\mu} r^{*2} \frac{dp}{dz} \cos \alpha_1 \qquad (2.39)$$

The Dupuit-Forchheimer equation is

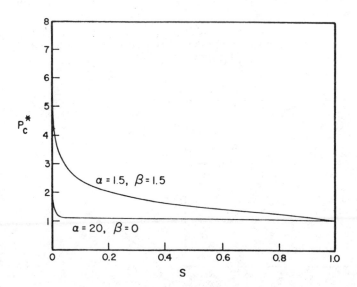

FIG. 2.17 Capillary pressure curves from figure distributions. (From Haring and Greenkorn, 1970.)

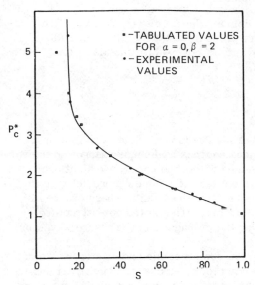

FIG. 2.18 Comparison of measured and calculated capillary pressure. (From Haring and Greenkorn, 1970.)

$$\langle v_z \rangle = \frac{q_z}{\phi} = - \frac{k}{\phi\mu} \frac{dp}{dz} \cos \alpha_1 \qquad (2.40)$$

To find the corresponding velocity for the model, we determine

$$\langle v_z \rangle = \frac{\int r^* \ell^* v_z \cos \alpha_1 \, dE}{\int r^* \ell^* \, dE} \qquad (2.41)$$

integrating

$$\langle v_z \rangle = - \frac{R^2}{24} \frac{dp}{dz} \frac{(\alpha + 2)(\alpha + 1)}{(\alpha + \beta + 3)(\alpha + \beta + 2)} \qquad (2.42)$$

equating Eqs. (2.39) and (2.41)

$$\frac{k}{\phi} = \frac{R^2}{24\mu} \frac{(\alpha + 2)(\alpha + 1)}{(\alpha + \beta + 3)(\alpha + \beta + 2)} \qquad (2.43)$$

or in terms of the average radius

$$\frac{k}{\phi} = \frac{\langle r \rangle^2}{24} \frac{(\alpha + 2)(\alpha + \beta + 2)}{(\alpha + \beta + 3)(\alpha + 1)} \qquad (2.44)$$

This section related model pore structure properties to the macroscopic flow properties of the model. The model can have an explicit pore size distribution. At

the level of the distribution we require both length and radius (or area) to describe
nonuniformity. The anisotropy of the model can also be clearly defined and directly
related to the permeability. The capillary pressure and distribution of radii of the
model are clearly related. The nonuniformity of permeability is also related to the
parameters of the pore size distribution.

2.3 MECHANICAL PROPERTIES

So far, we have treated the matrix of a porous medium as rigid. Changes in fluid
pressure or external pressure from various sources will cause changes in the volume of both the matrix and the fluid contained within the pores. We additionally
characterize porous media to determine the effect of volume change, change in
internal and external forces in the matrix and contained fluid. These changes in
volume will change pore volume and the flow properties in the pore structure.
These effects are important in considering the propagation of pressure waves in
fluid-filled media.

     To describe volume changes due to the compressible and elastic nature of a
porous medium, the volume change is related to the stress-strain characteristics
of the medium. The description of stresses in a porous medium is difficult since
it is not possible to determine such characteristics at every point. Normally the
stresses are developed as if a medium is a macroscopic continuum. The basic
literature associated with the following discussion is due to Biot (1941, 1956, 1962),
Gassman (1951), Geertsma (1957), van der Knapp (1959), Jacob (1940), Paria (1963),
Terzaghi (1951), and Verruijt (1969). Generally the bulk volume and the pore volume
of the medium must be known as a function of the pore fluid tension (negative of the
pore pressure) and the external tension (negative of the external pressure force).
The bulk compressibility can be found if these relations are known.

     We usually assume the stresses causing compression (or expansion) are
elastic. Under certain conditions or with certain media (e.g., clay) flow-plastic
deformation, may occur giving rise to additional complexities in the constitutive
relationships describing the stress. In what follows, we will not consider plastic
deformations we only consider elastic effects.

     The following are several definitions that will be used to discuss elastic
porous media. Matrix compressibility is the fractional change in the volume of the
solid matrix per unit change in uniform pressure.

$$c_s = -\frac{1}{V_s}\frac{\partial V_s}{\partial p} \qquad\qquad (2.45)$$

Bulk compressibility is the fractional change in bulk volume per unit in uniform
pressure.

$$c_b = -\frac{1}{V_b}\frac{\partial V_b}{\partial p} \qquad\qquad (2.46)$$

Pore compressibility is the fractional change in pore volume per unit change in
uniform pressure.

$$c_p = -\frac{1}{V_p}\frac{\partial V_p}{\partial p} \tag{2.47}$$

Figure 2.19 shows pore compressibility as a function of porosity.
The three compressibilities are related by

$$c_b = (1 - \phi)c_s + \phi c_p \tag{2.48}$$

Internal or pore stress variation is the variation of pressure with external stresses constant. External or a bulk stress variation is the variation of stresses on the outer boundaries of the matrix while the fluid pressure in the pores is held constant.

Heuristically, we use a stress tensor for porous media analogous to that for a fluid or a solid. Figure 2.20 shows the stresses on a cube where stresses are positive in the direction of the arrows, the positive directions of the coordinate axes. The stress is expressed

$$\begin{bmatrix} \tau_{xx} & \tau_{xy} & \tau_{xz} \\ \tau_{yx} & \tau_{yy} & \tau_{yz} \\ \tau_{zx} & \tau_{zy} & \tau_{zz} \end{bmatrix} = \begin{bmatrix} p & 0 & 0 \\ 0 & p & 0 \\ 0 & 0 & p \end{bmatrix} \begin{bmatrix} \tau'_{xx} - p & \tau'_{xy} & \tau'_{xz} \\ \tau'_{yx} & \tau'_{yy} - p & \tau'_{yz} \\ \tau'_{zx} & \tau'_{zy} & \tau'_{zz} - p \end{bmatrix} \tag{2.49}$$

where the τ_{ij} are the total stresses, τ'_{ij} are the effective stresses, and p is the

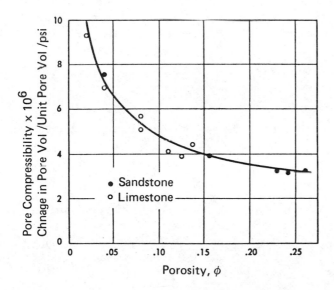

FIG. 2.19 Pore compressibilities of rocks. (From H. N. Hall, Compressibility of reservoir rocks, Pet. Trans. AIME 198, SPEJ 309. © 1953 SPE-AIME.)

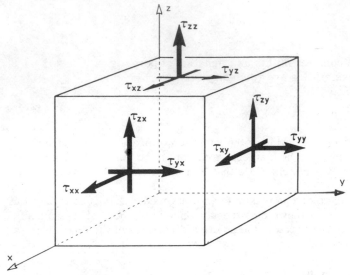

FIG. 2.20 Total stresses on a cube of a porous medium. (From Verruijt, 1969.)

hydrostatic pressure acting equally in the three coordinate directions. At equilibrium, ignoring inertial forces

$$(\underline{\nabla} \cdot \underline{\underline{\tau}}) + \underline{F} = 0 \tag{2.50}$$

and assuming zero moments

$$\tau_{ij} = \tau_{ji} \tag{2.51}$$

in terms of the effective stress and pressure, Eq. (2.50) is

$$(\underline{\nabla} \cdot \underline{\underline{\tau}}') + \underline{F} = \underline{\nabla}p \tag{2.52}$$

Letting effective stresses in terms of initial stresses $\underline{\underline{\tau}}'^{(0)}$ and incremental effective stress $\underline{\underline{\sigma}}'$ be

$$\underline{\underline{\tau}}' = \underline{\underline{\tau}}'^{(0)} + \underline{\underline{\sigma}}' \tag{2.53}$$

and pressure in terms of an initial pressure $p^{(0)}$ and an incremental pressure

$$p = p^{(0)} + \sigma \tag{2.54}$$

assuming body forces (gravity) are independent of time

$$\underline{\nabla} \cdot \underline{\underline{\sigma}}' = \underline{\nabla}\sigma \tag{2.55}$$

As a first approximation, assume the deformations of a porous medium are related to effective stresses by Hooke's law

$$\epsilon_{ii} = \frac{1}{E}[\sigma'_{ii} - \nu(\sigma_{ij} + \sigma'_{jk})] \qquad (2.56)$$

$$\epsilon_{jk} = \frac{\sigma'_{jk}}{\mu} \qquad (2.57)$$

where ϵ_{ij} is incremental strain, E is Young's modulus, ν is Poisson's ratio, and μ is the sheer modulus.

$$\mu = \frac{E}{2(1 + \nu)} \qquad (2.58)$$

where the porous medium is assumed to be homogeneous and uniform. Hooke's law in terms of the stresses is

$$\sigma'_{ii} = 2\mu\epsilon_{ii} + \lambda\epsilon \qquad (2.59)$$

$$\epsilon = \epsilon_{xx} + \epsilon_{yy} + \epsilon_{zz} \qquad (2.60)$$

$$\lambda = \frac{\nu E}{(1 - 2\nu)(1 + \nu)} \qquad (2.61)$$

the μ and λ in these equations are referred to as the Lamé constants. The strains are related to the displacements by

$$\epsilon_{ii} = \frac{\partial u_i}{\partial x_i} \qquad (2.62)$$

$$\epsilon_{jk} = \frac{1}{2}\left(\frac{\partial u_j}{\partial x_k} + \frac{\partial u_k}{\partial x_j}\right) \qquad (2.63)$$

and therefore

$$\epsilon = \underline{\nabla} \cdot \underline{u} \qquad (2.64)$$

The volume strain e is related to the incremental volume strain ϵ and the initial strain $e^{(0)}$ by

$$e = e^{(0)} + \epsilon \qquad (2.65)$$

Substituting Eqs. (2.62) and (2.63) into Eqs. (2.60) and (2.61), we can rewrite Eq. (2.56) as

$$\mu\nabla^2\underline{u} + (\lambda + 2\mu)(\underline{\nabla} \cdot \underline{\epsilon}) = \underline{\nabla} \cdot \underline{\sigma} \qquad (2.66)$$

From Darcy's law and the equation of state of a compressible fluid

$$\frac{k}{\rho g} \nabla^2 p = \frac{\partial \epsilon}{\partial t} + \phi \beta' \frac{\partial \sigma}{\partial t} \tag{2.67}$$

where

$$\rho = \rho_0 e^{\beta p} \tag{2.68}$$

$$\beta' = \beta + \frac{1 - S}{p} \tag{2.69}$$

We can rewrite Eq. (2.67) by differentiating it successively with respect to each coordinate and summing the result to get

$$(\lambda + 2\mu)\epsilon = \nabla^2 \sigma \tag{2.70}$$

If we integrate Eq. (2.70)

$$(\lambda + 2\mu)\epsilon = \sigma + f(x, y, z, t) \tag{2.71}$$

where f must follow Laplace's equation

$$\nabla^2 f = 0 \tag{2.72}$$

Combining Eqs. (2.67) and (2.71) for the case where f equals zero,

$$(\alpha + \phi\beta') \frac{\partial \sigma}{\partial t} = \frac{k}{\rho g} \nabla^2 \sigma \tag{2.73}$$

$$\alpha = \frac{1}{\lambda + 2\mu} \tag{2.74}$$

This situation is possible, that is, f = 0 if the displacements are in one direction only, and the total stresses do not change.

Van der Knapp (1959) considered nonlinear stress-volume relations. To describe these nonlinear volume changes one needs to measure the bulk volume V_b and the pore volume V_p as a function of the pore fluid tension σ and the hydrostatic tension $\bar{\sigma}$. In this instance, finite variations are considered as opposed to the infinitesimal variations discussed above. For an elemental volume large compared to the dimensions of the pores, we expand the bulk volume in a Taylor series around the initial state $(\sigma_i, \bar{\sigma}_i)$.

$$V_b(\sigma_i + \sigma, \bar{\sigma}_i + \sigma) - V_b(\sigma_i, \bar{\sigma}_i) =$$

$$\sigma \frac{\partial V_b}{\partial \sigma} + \bar{\sigma} \frac{\partial V_b}{\partial \bar{\sigma}} + \frac{1}{2!} \left(\sigma^2 \frac{\partial^2 V_b}{\partial \sigma^2} + 2\sigma\bar{\sigma} \frac{\partial^2 V_b}{\partial \sigma \, \partial \bar{\sigma}} + \bar{\sigma}^2 \frac{\partial^2 V_b}{\partial \bar{\sigma}^2} \right) + \cdots \tag{2.75}$$

Starting with an unstressed condition, applying tensions, $\sigma = \bar{\sigma} = \Pi$,

$$V_b(\Pi, \Pi) - V_b(0, 0) = c_s V_b(0, 0) \Pi \tag{2.76}$$

where c_s is matrix compressibility and is assumed constant. Starting with an arbitrary prestressed condition $(\sigma_i, \bar{\sigma}_i)$

$$V_b(\sigma_i + \Pi, \bar{\sigma}_i + \Pi) - V_b(\sigma_i, \bar{\sigma}_i) = c_s V_b(\sigma_i, \bar{\sigma}_i) \Pi \tag{2.77}$$

Substituting Eqs. (2.77) into Eq. (2.75)

$$V_b(\sigma_i + \Pi, \bar{\sigma}_i + \Pi) - V_b(\sigma_i, \bar{\sigma}_i) =$$

$$c_s V_b(\sigma_i, \bar{\sigma}_i) \Pi = \left(\frac{\partial V_b}{\partial \sigma} + \frac{\partial V_b}{\partial \bar{\sigma}} \right) \Pi + \frac{1}{2!} \left(\frac{\partial^2 V_b}{\partial \sigma^2} + 2 \frac{\partial^2 V_b}{\partial \bar{\sigma}^2} \right) \Pi^2 + \cdots \tag{2.78}$$

Since the coefficients of Π are equal,

$$\frac{\partial V_b}{\partial \sigma} + \frac{\partial V_b}{\partial \bar{\sigma}} = c_s V_b \tag{2.79}$$

where the sums of the higher derivatives are zero. Since

$$V_b = V_s + V_p \tag{2.80}$$

$$\frac{\partial V_b}{\partial \sigma} = \frac{\partial V_s}{\partial \sigma} + \frac{\partial V_p}{\partial \sigma} \tag{2.81}$$

$$\frac{\partial V_b}{\partial \bar{\sigma}} = \frac{\partial V_s}{\partial \bar{\sigma}} + \frac{\partial V_p}{\partial \bar{\sigma}} \tag{2.82}$$

The work of expansion of the bulk volume equals the work of contraction of the pores; therefore,

$$\frac{\partial V_b}{\partial \sigma} = - \frac{\partial V_p}{\partial \bar{\sigma}} \tag{2.83}$$

Variation in matrix volume is relatively small compared to bulk and pore volume. Combining Eq. (2.81), (2.82), and (2.83) with Eq. (2.79)

$$V_s(\sigma_i + \sigma, \bar{\sigma}_i + \bar{\sigma}) - V_s(\sigma_i, \bar{\sigma}_i) = -c_s V_{pi}\sigma + c_s V_{bi}\bar{\sigma} \tag{2.84}$$

The following are the expressions for the bulk, pore, and matrix volume for finite systems. The decrease in bulk volume when applying external hydrostatic stress is of the order of 1 percent. Therefore we use V_{b_1} the bulk volume at one atmosphere. The three-dimensional representation of the volume of the nonlinear pore system is a cylindrical surface represented by

$$\frac{V_b - V_{b_1}}{V_{b_1}} = c_s \sigma - f(\sigma - \bar{\sigma}) \tag{2.85}$$

This expression relates the fractional change in bulk volume to the total change in pore tensions σ and the effective frame stress $\sigma - \bar{\sigma}$. From Eq. (2.84) it follows

$$\frac{V_s - V_{s_1}}{V_{s_1}} = c_s \sigma - \frac{1}{1 - \phi_1} c_s (\sigma - \bar{\sigma}) \tag{2.86}$$

and

$$\frac{V_p - V_{p_1}}{V_{p_1}} = c_s \sigma + \frac{1}{\phi_1} c_s (\sigma - \bar{\sigma}) - \frac{1}{\phi_1} f(\sigma - \bar{\sigma}) \tag{2.87}$$

It is possible, then, to characterize changes in volume of a stressed porous medium by measuring c_s, V_b, or V_p as a function of $\bar{\sigma}$, keeping σ constant. Figure 2.21 is a graphical representation of the fractional changes in bulk, pore, and matrix volume as a function of pore and external pressures.

The bulk compressibility is defined by van der Knapp (1959) in terms of the effective hydrostatic tension on the frame and the bulk volume at one atmosphere with σ constant.

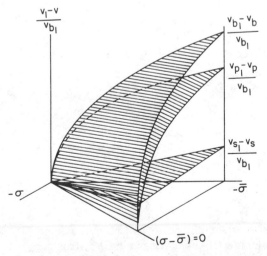

FIG. 2.21 Fractional change in bulk, pore, and rock volume as functions of pore and external pressures. (From W. van der Knapp, Nonlinear behavior of elastic porous media, Pet. Trans. AIME 216, SPEJ 179. © 1959 SPE-AIME.)

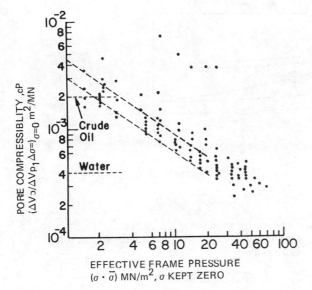

FIG. 2.22 Pore compressibility data of sandstones: 1 MN m^{-2} = 145.04 psi; 1 m^2 MN^{-1} = 6.89 × 10^{-3} psi^{-1}. Most data lie in between $1/V_{p_1} \, \Delta V_p / \Delta \bar{\sigma}$ = +3 = 10$^{-3}(\sigma - \bar{\sigma})^{-07}$ and $1/V_p \, \Delta V_p / \Delta \bar{\sigma}$ = +4.5 = 10$^{-3}(\sigma - \bar{\sigma})^{-07}$. (From W. van der Knapp, Nonlinear behavior of elastic porous media, Pet. Trans. AIME 216, SPEJ 179. © 1959 SPE-AIME.

$$c_b = \frac{1}{V_{b_1}} \frac{\partial V_b}{\partial \bar{\sigma}} \tag{2.88}$$

This is essentially the same equation as Eq. (2.46) since $\bar{\sigma}$ = -p. Pore compressibility is represented by

$$c_p = -\frac{1}{\phi_1} c_s + \frac{1}{\phi_1} \frac{\partial f}{\partial (\sigma - \bar{\sigma})} = \frac{1}{V_{p_i}} \frac{\partial V_p}{\partial \bar{\sigma}} \tag{2.89}$$

The effect of reservoir pressure decline on pore volume can be calculated using Eqs. (2.86) and (2.87). Figure 2.22 shows the pore compressibility of van der Knapp (1959(and Carpenter and Spencer (1940) in the porosity range 15 to 30 percent. The straightline behavior is similar to that for packed spheres (Hara, 1935). Porosity and permeability under reservoir conditions may differ from laboratory measurements due to the overburden pressure (Fatt, 1953) expressed by

$$\frac{\phi}{\phi_1} = \frac{V_p}{V_{p_i}} \frac{V_{b_1}}{V_b} \tag{2.90}$$

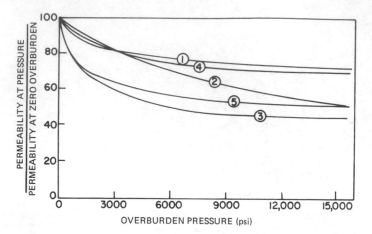

FIG. 2.23 Curves showing the dependence of (specific) permeability on total external stress (i.e., overburden pressure).

Curve	Source of material	Permeability (millidarcys)
1	Colorado	3–86
2	Southern California coast	40–8
3	San Joachin Valley, Calif.	45–0
4	Arizona	4–35
5	Arizona	632

(From I. Fatt and D. H. Davis, Reduction in permeability with overburden pressure, Pet. Trans. AIME 195, SPEJ 329. © 1952 SPE-AIME.)

For sandstones in the porosity range of 15 to 30 percent at pressures of 10 MN m^{-2} the porosity is smaller by about 1 or 2 percent. Limestones change more, relatively, since their initial porosities are usually lower. Figure 2.23 shows the change of permeability with overburden pressure (Fatt and Davis, 1952).

As a result of reservoir pressure decline there may be reservoir subsidence. If we assume all of the bulk volume change is in the vertical direction,

$$\frac{1}{V_b}\frac{dV_b}{d\sigma} = c_s - c_b\left(1 - \frac{d\bar{\sigma}}{d\sigma}\right) = \frac{1}{h}\frac{dh}{d\sigma} \tag{2.91}$$

For a reservoir thickness of 100 m and a fluid pressure decline of 10 MN m^{-2}, dh = 7 cm—not much.

The velocity of compression waves in an isotropic solid is given by

$$v = \frac{3(1 - \nu)}{c\rho(1 + \nu)} \tag{2.92}$$

where ν is Poisson's ratio, c is compressibility, and ρ is density. For an elastic system:

$$c = c_b \frac{\phi(c_\ell - c_s) + c_s(c_b - c_s)}{\phi(c_\ell - c_s) + (c_b - c_s)}$$

(2.93)

Figure 2.24 shows the data of Hicks and Berry (1956) where velocity is plotted versus external pressure. The change in velocity due to external pressure change is significant. If we combine the continuity equation and Darcy's law (see Chap. 3), we obtain the single–phase equation of condition:

$$\phi c_\ell \frac{\partial p}{\partial t} = \frac{k}{\mu} \nabla^2 p$$

(2.94)

Taking the interaction of fluid and matrix into account (Biot, 1941),

$$\left(1 - \frac{c_s}{c_b}\right) \frac{\partial \epsilon}{\partial t} - c_s \left(1 - \phi - \frac{c_s}{c_b} + \phi \frac{c_\ell}{c_s}\right) \frac{\partial p}{\partial t} = -\frac{k}{\mu} \nabla^2 p$$

(2.95)

where ϵ is the bulk volume dilitation. Neglecting the effect of pressure drop on boundary conditions and assuming overburden pressure is constant with no horizontal displacement and further neglecting c_s with respect to c_b. Using $\nu = 1/5$

$$\phi\left(c_\ell + \frac{c_p}{2}\right) \frac{\partial p}{\partial t} = \frac{k}{\mu} \nabla^2 p$$

(2.96)

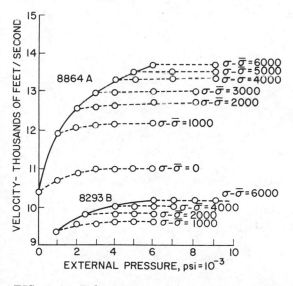

FIG. 2.24 Laboratory measurements on two sandstone cores showing acoustic wave velocity as a function of the difference between external and internal pressure. (From W. G. Hicks and J. E. Berry, 1956.)

2.4 MACROSCOPIC DESCRIPTION

As stated in Section 2.1 a macroscopic description of a porous medium is a description in terms of average or bulk properties at sizes much larger than a single pore.
When we reach the point of characterizing a porous medium macroscopically, we
must cope with the scale of description. The scale we use will depend on the way
we wish to model and the size at which we wish to model the porous medium. Macroscopically, we assume if at all possible that a given medium is "ideal," that is,
homogeneous, uniform, and isotropic. The term homogeneous normally implies an
average property can be used to replace the entire media; for example, a single
value of permeability can be used for a "homogeneous" reservoir. And this value
of permeability will characterize flow in this reservoir. Unfortunately, a medium
which is homogeneous in this sense for one property may not be homogeneous for
another.

The term reservoir description often means description in the sense of
homogeneous as opposed to heterogeneous, as discussed above. Reservoir description means describing the reservoir at a level where a property changes enough
such that more than a single average has to be used to model flow. In this sense a
reservoir composed of a section of coarse gravel and a section of fine sand, where
these two materials are separated and have significantly different permeabilities,
is heterogeneous. Defining the dimensions locating the area, and determining the
average properties of the gravel and the sand is an example of what is meant. This
description is satisfactory for reservoir-level problems; however, if one studies
mechanisms of fluid flow or a process at a given scale and wants to use this data
at a different scale, the effects of nonideal media at a given scale requires more
specific definitions.

Slattery (1972) describes averages to use in volume-averaging the equations
of change for flow in porous media. Equations describing imcompressible creeping
flow of a Newtonian fluid of constant viscosity through a porous medium are
the continuity equation

$$\underline{\nabla} \cdot \underline{v} = 0 \tag{2.97}$$

and the void distribution function

$$\alpha(\delta) = \begin{cases} 1 & \text{if } \delta \text{ is in the pores} \\ \\ 0 & \text{if } \delta \text{ is in the matrix} \end{cases} \tag{2.98}$$

Since $\alpha(\delta)$ is unknown, we volume-average the equation of continuity and equation
of motion. (See Chap. 3.)

We assume for every point in space an average volume which contains the
volume of the fluid v_f. The porosity of this space is

$$\phi = \frac{1}{V} \int_V \alpha(\delta) \, dV \tag{2.99}$$

Assume a microscopic characteristic length d, over which there is significant variation in point velocity. Further assume that a macroscopic characteristic length L

representative of the distance over which significant variation in the average velocity $\langle v \rangle$ takes place. Let d approximate pore diameter or grain size and L approximate the length of the system. The volume average of a point quantity associated
with the fluid is

$$\langle \psi \rangle = \frac{1}{V} \int_V \psi \, dV \qquad (2.100)$$

If we plot $\langle \psi \rangle$ versus V (starting with V in the solid), as V increases, portions of
the fluid are contained within V. The average $\langle \psi \rangle$ passes through several fluctuations until V reaches a size where $\langle \psi \rangle$ smooths. Figure 2.25 is a schematic pic-
ture of $\langle \psi \rangle$ versus V. For values of V larger than at the dashed line on Fig. 2.25
the volume average $\langle \psi \rangle$ is smooth. However, it is not necessarily constant.

If we let ℓ be a characteristic macroscopic length for the average value and
$d \ll \ell$ and $\langle \langle \psi \rangle \rangle = \langle \psi \rangle$, then

$$\langle \langle \psi \rangle \rangle = \frac{1}{V} \int_V \langle \psi \rangle \, dV \qquad (2.101)$$

which leads to $\ell \ll L$. Whitaker (1970) shows for the above restrictions that the
volume and area averages are essentially equivalent.

Bear (1972) uses porosity as a point quantity in a porous medium and defines
a representative elementary volume (REV) to characterize a medium macroscopically. This definition is a specific case of the more general discussion of Whitaker
(1970). The size of the REV around a point P is smaller than the total medium so it
can represent flow at P. There must be enough pores to allow statistical averaging.
If porosity varies, the maximum length is the characteristic length that indicates
the rate of change of porosity. The minimum length is the pore size. Let P be a
point inside a porous medium surrounded by a spherical volume ΔV_i and define a
ratio

$$n_i = \frac{(\Delta V_y)_i}{\Delta V_i} \qquad (2.102)$$

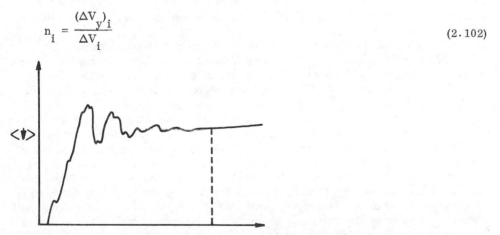

FIG. 2.25 Dependence of average on averaging volume. (From S. Whitaker, Advances in theory of fluid motion in porous media, in Flow through Porous Media,
Amer. Chem. Soc., Washington, D.C., 1970.

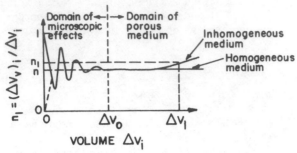

FIG. 2.26 Definition of porosity and representative elementary volume. (From Bear, 1972.)

$(\Delta V_v)_i$ is the volume of void space within volume ΔV_i. Consider a sequence of values of ℓ_i such that $\Delta V_1 > \Delta V_2 > \Delta V_3 > \cdots$. For large values of ΔV_i the ratio n_i may change gradually as ΔV_i gets smaller. As ΔV_i falls below a certain value, there will be large fluctuations in n_i when ΔV_i is approaching the dimension of the pores. When ΔV_i converges to a point, n_i will be 0 or 1 depending on whether the point ends up in the matrix or in a pore. Figure 2.26 shows n_i versus ΔV_i. The volume represented by ΔV_0 is the REV. Below this value the average value for porosity has no meaning. The REV is equivalent to the generalized value defined by Whitaker (1970) as the dashed line in Fig. 2.25, if $\langle \psi \rangle$ is porosity. Bear replaces the porous medium in a "continuum sense" with a model which gives the correct average porosity at any point in the medium.

We should define nonideal media (the terms heterogeneity, nonuniformity, and anisotropy) in the volume-average sense. Since the definitions for nonideal properties are arbitrary it seems reasonable to define them in direct relation to flow, but in the correct volume-average sense. The concepts of heterogeneity, nonuniformity, and anisotropy can be defined at the level of Darcy's law in terms of permeability. Permeability is more sensitive to conductance, mixing, and capillary pressure than porosity.

Greenkorn and Kessler (1970) define heterogeneity, nonuniformity, and anisotropy in reference to permeability as follows. First, by macroscopic we imply averaging over elemental volumes of radius ϵ about a point in the media, where ϵ is large enough that Darcy's law can be applied for appropriate Reynolds numbers. In other words, we are at volumes large with respect to that of a single pore. Further, we intend ϵ to be the minimum radius that satisfied such a condition; otherwise, by making ϵ too large we may obscure certain nonidealities by burying their effects far within the elemental volume. Obviously one can, for all practical purposes, remove certain effects by scale.

For example, consider a single Ping-Pong ball buried in a bed of 0.01-in. glass spheres. An elemental volume of radius 1 in. will certainly not behave in a homogeneous, uniform, isotropic (ideal) fashion if it includes the Ping-Pong ball; however, in an elemental volume of radius 500 ft, one will probably never detect the effects of one lone Ping-Pong ball.

We base our definitions of heterogeneity, nonuniformity, and anisotropy on the probability density distribution of permeability of random macroscopic elemental volumes selected from the medium, the permeability being expressed by the one-dimensional form of Darcy's law

$$q_x = -\frac{k}{\mu}\frac{dp}{dx}$$ (2.103)

By a uniform medium we mean one in which the probability density function for permeability is either a Dirac delta function or a linear combination of N functions that satisfy the relation:

$$f(k) = \sum_{i=1}^{N} \xi_i \delta$$ (2.104)

where

$$\sum_{i=1}^{N} \xi_i = 1$$ (2.105)

(The ξ_i are constants, and the side condition is necessary to ensure that the integral of the density function is equal to one.) Examples of this sort of behavior are shown in Fig. 2.27. By nonuniform we mean a medium in which the probability density function cannot be constructed with a finite number of weighted delta functions (Fig. 2.28).

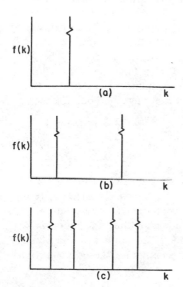

FIG. 2.27 Observed permeability for uniform media (a) homogeneous, (b,c) heterogeneous. (From R. A. Greenkorn and D. P. Kessler, Dispersion in the heterogeneous non-uniform anisotropic porous media, in Flow through Porous Media, Amer. Chem. Soc., Washington, D.C., 1970.)

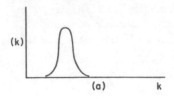

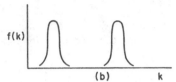

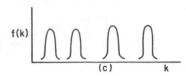

FIG. 2.28 Observed permeability for nonuniform media (a) homogeneous, (b, c) heterogeneous. (From R. A. Greenkorn and D. P. Kessler, Dispersion in the heterogeneous nonuniform anisotropic porous media, in Flow through Porous Media, Amer. Chem. Soc., Washington, D.C., 1970.)

By a heterogeneous medium we mean one in which the permeability distribution is at least bimodal (Fig. 2.27b and c, and Fig. 2.28b and c).

Figure 2.27a is homogeneous and uniform; Fig. 2.28a is homogeneous and nonuniform.

We have drawn the distributions in Fig. 2.28 as if all have the same variance (homoscedastic). We can also speak of a second-order nonideality where the variance of the permeability distribution depends on spatial location or orientation in the medium.

Thus far, we have said nothing about anisotropy. By anisotropy we mean the permeability varies with direction in an elemental volume in the medium.

We now can make a more mathematical statement of our definitions. The probability density function for permeability is, in general, a function of location and orientation. We can describe this function with five independent variables: the rectangular coordinates x_i (i = 1, 2, 3) for location, and the angular coordinates θ, ψ for orientation.

$$P(k_1 \leq k \leq k_2) = \int_{k_1}^{k_2} f(x_i, \theta, \psi) \, dx_i \, d\theta \, d\psi \qquad (2.106)$$

If the probability density distribution is independent of θ and ψ, the medium is isotropic; if the distribution is expressible by a finite linear combination of delta functions, the medium is uniform; if the distribution is monomodal, the medium is homogeneous; and if the variance of the distribution is not constant, the medium has second-order nonidealities. The second-order nonidealities may also be classified with respect to the probability density function for variance of the permeability distribution:

1. Nonuniformity (variance density function not a finite series of delta functions)

2. Heterogeneity (variance density function bimodal)
3. Anisotropy (variance density function a function of orientation), where the variance distribution is

$$P(\text{var}_1 \leq \text{var } k \leq \text{var}_2) = \int_{\text{var}_1}^{\text{var}_2} g(x_i, \theta, \psi) \, dx_i \, d\theta \, d\psi \qquad (2.108)$$

(To get the probability density function for the variance we obviously must consider elemental volumes much larger than for the parent permeability distribution.)

REFERENCES

Bear, J. (1972): Dynamics of Fluids in Porous Media, Elsevier, New York.

Biot, M. A. (1941): General Theory of Three-Dimensional Consolidation, J. Appl. Phys. 12, 155.

Biot, M. A. (1956): Theory of Propagation of Elastic Waves in Fluid-Saturated Porous Solids. I. Low-Frequency Range, J. Acoust. Soc. Amer. 28(3), 168.

Biot, M. A. (1962): Mechanics of Deformation and Acoustic Propagation in Porous Media, J. Appl. Phys. 33(4), 1482.

Brutsaert, W. (1966): Probability Laws for Pore-Size Distributions, Soil Sci. 101(2), 85.

Carpenter, C. B., and G. B. Spencer (1940): Measurements of Compressibility of Consolidated Oil-Bearing Sandstones, RI 3540 U.S.B.M. October.

Collins, R. E. (1961): Flow of Fluids through Porous Materials, Van Nostrand, New York.

Day, P. R., and J. W. Luthin (1956): Numerical Solution of the Differential Equation of Flow for a Vertical Drainage Problem, Soil Sci. Soc. Amer. Proc. 20, 443.

Dullien, F. A. L., and V. K. Batra (1970): Determination of the Structure of Porous Media, IEC 62(10), 25.

Dullien, F. A. L., and P. N. Mehta (1971–1972): Particle Size and Pore (Void) Size Determination by Photomicrographic Methods, Powder Technol. 5, 179.

Dullien, F. A. L., and G. K. Dhawan (1973): Photomicrographic Size Determination of Non-Spherical Objects, Powder Technol. 7, 305.

Dullien, F. A. L., and G. K. Dhawan (1974): Characterization of Pore Structure by a Combination of Quantitative Photomicrography and Mercury Porosimetry, J. Colloid Interface Sci. 47(2), 337.

Fatt, I., and D. H. Davis (1952): Reduction in Permeability with Overburden Pressure, Pet. Trans. AIME 195, 329.

Fatt, I. (1953): Effect of Overburden Pressure on Relative Permeability, Pet. Trans. AIME 198, 325.

Gassman, F. (1951): Uber die Elastizitat Poroser Medien, Viertelj. Natf. Gesdlsch. Zürich 96, 1.

Geertsma, J. (1957): Effect of Fluid Pressure Decline on Volumetric Change of Porous Rocks, Pet. Trans. AIME 210, 331.

Graton, S. C., and H. S. Fraser (1935): Systematic Packing of Spheres with Particular Reference to Porosity and Permeability, J. Geol. 43(8), 785.

Greenkorn, R. A., and D. P. Kessler (1970): Dispersion in Heterogeneous Non-uniform, Anisotropic Porous Media, in Flow through Porous Media, Amer. Chem. Soc., Washington, D.C., Chap. 8, p. 159.

Guin, J. A., D. P. Kessler, and R. A. Greenkorn (1971a): Average Pore Velocities in Porous Media, Phys. Fluids 14(1), 181.

Guin, J. P., D. P. Kessler, and R. A. Greenkorn (1971b): Permeability Tensor for Anisotropic Non-uniform Porous Media, Chem. Eng. Sci. 26, 1475.

Hall, H. N. (1953): Compressibility of Reservoir Rocks, Pet. Technol. Tech. Note 149, January.

Hara, G. (1935): Theorie der Akustischen Schwingunaustrertung in gekornten Substazen und experimentelle Untersuchungen an Kohlepulver, Elektr. Nachr. Techn. 12, 191.

Haring, R. E., and R. A. Greenkorn (1970): Statistical Model of a Porous Medium with Non-uniform Pores, A.I.Ch.E. J. 16(3), 477.

Haughey, D. P., and G. S. G. Beveridge (1966): Local Voidage Variation in a Randomly Packed Bed of Equal-Sized Spheres, Chem. Eng. Sci. 21, 905.

Haughey, D. P., and G. S. G. Beveridge (1969): Structural Properties of Packed Beds: A Review, Can. J. Chem. Eng. 47, 130.

Hicks, W. G., and J. E. Berry (1956): Fluid Saturation of Rocks from Velocity Logs, Geophysics 21, 739.

Jacob, C. E. (1940): Flow of Water in an Elastic Artesian Aquifer, Trans. Amer. Geophys. Union 21, 574.

Matheron, G. (1967): Elements Porous une Theorie des Milleux Poreus, Maison, Paris.

Mayer, R. D., and R. A. Stowe (1965): Mercury Porosimetry: Breakthrough Pressure for Penetration between Packed Spheres, J. Colloid. Sci. 20, 893.

Morrow, N. R. (1970): Physics and Thermodynamics of Capillary Action in Porous Media, in Flow through Porous Media, American Chemical Society, Washington, D.C., Chap. 6, p. 103.

Paria, G. (1963): Flow of Fluids through Deformable Solids, Appl. Mech. Rev. 16, 421.

Scheidegger, A. E. (1957): The Physics of Flow through Porous Media, Macmillan, New York.

Slattery, J. C. (1972): Momentum, Energy and Mass Transfer in Continua, McGraw-Hill, New York.

Terzaghi, A. (1951): Theoretical Soil Mechanics, Chapman and Hall, London.

van der Knapp, W. (1959): Nonlinear Behavior of Elastic Porous Media, Pet. Trans. AIME 216, 179.

Verruijt, A. (1969): Elastic Storage of Aquifers, in Flow through Porous Media, R. J. M. DeWiest (ed.), Academic, New York, Chap. 8, p. 331.

Whitaker, S. (1970): Advances in Theory of Fluid Motion in Porous Media, in Flow through Porous Media, American Chemical Society, Washington, D.C., Chap. 2, p. 31.

SUGGESTED READING

Bear, J., D. Zaslavsky, and S. Irmay (eds.), Physical Principles of Water Percolation and Seepage, UNESCO, Paris, 1968.

Brutsaert, W., On Pore Size Distribution and Relative Permeabilities of Porous Mediums, J. Geophys. Res. 68(8), 2233 (1963).

Corey, A. T., Mechanics of Heterogeneous Fluids in Porous Media, Water Resources Publications, Fort Collins, Colo., 1977.

deBoer, J. H., The Shapes of Capillaries in the Structure and Properties of Porous Materials, D. H. Everett and F. S. Stone (eds.), Butterworth, London, 1958.

Dullien, F. A. L., Porous Media Fluid Transport and Pore Structure, Academic, New York, 1979.

Fara, H. D., and A. E. Scheidegger, Statistical Geometry of Porous Media, J. Geophys. Res. 66(10), 3279 (1961).

Greenkorn, R. A., Matrix Properties of Porous Media, Proc. Second IAHR–ISSS, Symp. Fundamentals of Transport Phenomenon in Porous Media, Guelph, 1972, p. 26.

3

Single-Fluid Flow

The equations of change for single-fluid flow can be obtained as limit cases of those for multifluid flow. However, the inherent simplification of one fluid allows simplification in the flow modeling such that it is useful to treat this limit case (single-fluid flow) separately.

The parameters that relate the media and flow in this case are porosity, permeability, tortuosity, and connectivity. These parameters are usually defined with reference to single-fluid flow. Complex potentials can be used to model single-fluid flow. Complex potentials can be used to model single-fluid flow problems since Laplace's equation results in single-fluid steady flow situations. Also compressible flow can be modeled as steady flow.

Much of the mathematics and techniques for transient flow in multifluid systems is an extension of the single-flow situation. Many multifluid transient situations are modeled as single-fluid flow. Inertial effects (wave propagation) are also modeled with a single-fluid model. The topic of non-Newtonian flow in porous media is introduced as a single-fluid flow problem.

3.1 EQUATIONS OF CHANGE

Darcy's law is an empirically determined relation for single-fluid flow in a linear system

$$q = \frac{Q}{A} = -\frac{k}{\mu}\frac{dp}{dx} \tag{3.1}$$

Slattery (1972) raises three questions concerning Eq. (3.1)—two of them concern us at this point.

1. How should the equation of flow (Darcy's law) be written in other geometries and with other boundary conditions?

Usually the general three-dimensional equation for Darcy's law is written heuristically as

48

$$\underline{q} = -\frac{k}{\mu} \underline{\nabla} p \tag{3.2}$$

2. How should Darcy's law be written if the medium is anisotropic (oriented)?

An oriented porous medium has distinct directional effect on flow, that is, at a "point" the resistance to flow may be a function of direction. It is usually assumed that permeability is a second-order tensor and Darcy's law is written (Ferrandon, 1948; Scheidegger, 1954, 1957).

$$\underline{q} = -\frac{\underline{\underline{k}}}{\mu} \underline{\nabla} p \tag{3.3}$$

Whitaker (1970) poses the questions raised by Slattery (1972) in a slightly different manner. For the creep flow of a constant viscosity single Newtonian fluid in a rigid porous medium, the continuity equation of the fluid is

$$\frac{\partial \rho}{\partial t} + \rho(\underline{\nabla} \cdot \underline{v}) = 0 \tag{3.4}$$

The equation of motion for the fluid is

$$\frac{\partial \rho \underline{v}}{\partial t} + (\underline{\nabla} \cdot \rho \underline{v}\underline{v}) - (\underline{\nabla} \cdot \underline{\underline{\tau}}) - \rho \underline{F} = 0 \tag{3.5}$$

where $\underline{\underline{\tau}}$ are the stresses in the fluid and $\underline{F}$ are body forces. With these two equations we require a void distribution function

$$\alpha(\delta) = \begin{cases} 1 & \text{if } \delta \text{ is in the fluid} \\ 0 & \text{if } \delta \text{ is in the solid} \end{cases} \tag{3.6}$$

We cannot determine Eq. (3.6) and we cannot solve Eqs. (3.4) and (3.5). The method posed to overcome the dilemma is to volume average the equations of change.

The volume average of a point quantity $\underline{\psi}$ associated with the fluid is

$$\langle \underline{\psi} \rangle = \frac{1}{V} \int_{V_f} \underline{\psi} \, dV \tag{3.7}$$

As discussed in Chap. 2, the characteristic length (or size) of a porous medium when $\langle \underline{\psi} \rangle$ becomes smooth is ℓ. If δ is pore size, $\ell \gg \delta$. Whitaker (1970) requires $\langle\langle \underline{\psi} \rangle\rangle = \langle \underline{\psi} \rangle$, where

$$\langle\langle \underline{\psi} \rangle\rangle = \frac{1}{V} \int_V \langle \underline{\psi} \rangle \, dV \tag{3.8}$$

From a series expansion of

$$\langle\langle \underline{\psi} \rangle\rangle = \langle \underline{\psi} \rangle + 0\left[\langle \underline{\psi} \rangle \left(\frac{\ell}{L}\right)^2 \right] \tag{3.9}$$

$\ell \ll L$ for L associated with the size of the system.

In order to volume-average the equations of change we need gradients of averages and averages of gradients. Slattery (1969) describes a proceduce for the volume average of a gradient from the general transport theorem

$$\langle \underline{\nabla}\underline{\underline{\psi}} \rangle = \underline{\nabla}\langle \underline{\underline{\psi}} \rangle + \frac{1}{V} \int_{A_i} \underline{\underline{\psi}}\underline{n} \, dA \tag{3.10}$$

where $\underline{n}$ is the outward directed normal for V or V_f and A_i is the area of the solid-fluid interface. The divergence is described by

$$\langle (\underline{\nabla} \cdot \underline{\underline{\psi}}) \rangle = (\underline{\nabla} \cdot \langle \underline{\underline{\psi}} \rangle) + \frac{1}{V} \int_{A_i} (\underline{\underline{\psi}} \cdot n) \, dA \tag{3.11}$$

Consider a point $\underline{z}$ in a porous medium of closed surface S and volume V. $\underline{z}$ is either in the matrix or in a pore or on the boundary, as sketched in Fig. 3.1. Apply Eqs. (3.10) and (3.11) to the equation of continuity, Eq. (3.4), and the equation of motion, Eq. (3.5), for the region enclosed by S, sketched in Fig. 3.1. The mass balance for the fluid inside the closed surface S is

$$\int_{V_f} \left[\frac{\partial \rho}{\partial t} + (\underline{\nabla} \cdot \rho \underline{v}) \right] \, dV = 0 \tag{3.12}$$

Using Eqs. (3.10) and (3.11)

$$\frac{\partial \langle \rho \rangle}{\partial t} + (\underline{\nabla} \cdot \langle \rho \underline{v} \rangle) = 0 \tag{3.13}$$

Equation (3.13) simplifies for an incompressible fluid to

$$\underline{\nabla} \cdot \langle \underline{v} \rangle = 0 \tag{3.14}$$

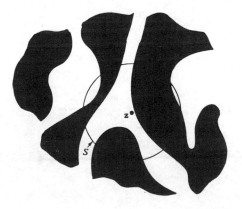

FIG. 3.1 The averaging surface S to be associated with every point z in the porous medium. (From Slattery, 1972.)

The equation of motion for the fluid inside the closed surface S is

$$\int_{V_f} \left[\frac{\partial \rho \underline{v}}{\partial t} + (\underline{\nabla} \cdot \rho \underline{v}\underline{v}) - (\underline{\nabla} \cdot \tau) - \rho \underline{F} \right] dV = 0 \qquad (3.15)$$

Applying Eqs. (3.10) and (3.11) again,

$$\frac{\partial (\rho \underline{v})}{\partial t} + (\underline{\nabla} \cdot \langle \rho \underline{v}\underline{v} \rangle) - (\underline{\nabla} \cdot \langle \underline{\underline{\tau}} \rangle) - \langle \rho \underline{F} \rangle - \frac{1}{V} \int_S (\underline{\underline{\tau}} \cdot n) \, dS = 0 \qquad (3.16)$$

For an incompressible fluid without any inertial effects present and with body forces represented by a potential function

$$\underline{F} = -\underline{\nabla} \hat{\phi} \qquad (3.17)$$

$$(\underline{\nabla} \cdot \{ \langle \underline{\underline{\tau}} \rangle - \rho \hat{\phi} \, \underline{\underline{I}} \}) + \frac{1}{V} \int_{S_w} (\{ \underline{\underline{\tau}} - \rho \hat{\phi} \, \underline{\underline{I}} \} \cdot \underline{n}) \, dS = 0 \qquad (3.18)$$

where S_w is the interface between the fluid and the solid.

Slattery (1972) shows that the second term of Eq. (3.18) is a function of the difference between the local average fluid velocity $\langle \underline{v} \rangle$ and the local average solid velocity $\langle \underline{u} \rangle$. Therefore, the force per unit volume an incompressible fluid exerts on an anisotropic (oriented) porous medium is in addition to the hydrostatic force and the local pressure. Let $\mathscr{P} = p - p_0 + \rho g$; then Eq. (3.18) can be written

$$\underline{\nabla} \mathscr{P} - \mu \nabla^2 \langle \underline{v} \rangle + \mu \underline{\underline{k}}^{-1} \langle \underline{v} \rangle = 0 \qquad (3.19)$$

The second term in Eq. (3.19) represents viscous drag at the boundaries and was recognized heuristically by Brinkman (1947). Howells (1974) and Hinch (1977) confirm the validity of Eq. (3.19) by considering slow flow in random arrays of fixed spheres and for suspensions, respectively. Normally, this viscous drag term is neglected and

$$\langle \underline{v} \rangle = -\frac{\underline{\underline{k}}}{\mu} \underline{\nabla} \mathscr{P} \qquad (3.20)$$

If $\langle \underline{v} \rangle$ is equal to $\underline{q}$, then Eq. (3.20) is Darcy's law in three dimensions for an anisotropic porous medium. The permeability tensor $\underline{\underline{k}}$ is symmetric.

Whitaker (1970) summarizes the assumptions and restrictions associated with volume averaging to obtain the equations of continuity and motion in porous media, Eqs. (3.13) and (3.20).

1. The averaged functions behave in such a manner that $\delta \ll \ell \ll L$.
2. The averaging volume is constant and of fixed orientation.
3. $\underline{v}$ is a unique function of $\langle \underline{v} \rangle$.
4. $\underline{\nabla} p = 0$ implies $\langle \underline{v} \rangle = 0$.
5. The anisotropic nature of a porous medium is represented by a symmetrical permeability tensor.

The following are the equations of change for single-fluid flow in a porous medium based on the preceding discussion. The equation of continuity is

$$\phi \, \frac{\partial \rho}{\partial t} + (\underline{\nabla} \cdot \rho \underline{q}) = 0 \tag{3.21}$$

where ϕ is the porosity of the medium, ρ is the density of the fluid in the pores. The equation of motion, Darcy's law, in three dimensions is

$$\underline{q} = -\frac{\underline{\underline{k}}}{\mu} \, \underline{\nabla} \left(p - \frac{d}{dz} \rho g z \right) \tag{3.22}$$

where g is the acceleration due to gravity. If we define a flow potential Φ such that

$$\Phi = p + \rho g z \tag{3.23}$$

and combine Eqs. (3.21) to (3.23), we obtain the equation of condition for single-fluid flow in a porous medium.

$$\phi \mu \, \frac{\partial \rho}{\partial t} = (\underline{\nabla} \cdot \rho \underline{\underline{k}} \nabla \Phi) \tag{3.24}$$

For flow of an incompressible fluid, ρ is constant and $\partial \rho / \partial t = 0$ so that

$$\underline{\nabla} \cdot \underline{\underline{k}} \, \underline{\nabla} \Phi = 0 \tag{3.25}$$

If the medium is homogeneous and isotropic, the permeability can be taken outside of the derivative

$$\nabla^2 \Phi = 0 \tag{3.26}$$

For compressible liquids

$$c_\ell = \frac{1}{\rho} \, \frac{\partial \rho}{\partial p} \tag{3.27}$$

If c_ℓ is constant, then

$$\rho \, \underline{\nabla} p = \frac{1}{c_\ell} \, \underline{\nabla} \rho \tag{3.28}$$

Assuming the effect of gravity is small compared to the pressure, Eq. (3.24) becomes

$$\phi \mu c_\ell \, \frac{\partial \rho}{\partial t} = \underline{\nabla} \cdot \underline{\underline{k}} \, \underline{\nabla} \rho \tag{3.29}$$

For slightly compressible liquids assume the following equation for density as a function of pressure.

$$\rho = \rho_0 e^{c_\ell (p-p_0)} = \rho_0 \left[1 + c_\ell (p - p_0) + \frac{1}{2!} c_\ell^2 (p - p_0)^2 + \cdots \right] \qquad (3.30)$$

Assume the higher order terms of Eq. (3.30) are negligible, and differentiate Eq. (3.30) to obtain

$$d\rho = \rho_0 c_\ell \, dp \qquad (3.31)$$

Substituting Eq. (3.31) into Eq. (3.29) gives

$$\phi \mu c_\ell \frac{\partial p}{\partial t} = \underline{\nabla} \cdot \underline{\underline{k}} \, \underline{\nabla} p \qquad (3.32)$$

which is the equation of condition for a slightly compressible liquid.

For flow of an ideal gas in a porous medium, substitute the ideal-gas equation

$$\rho = \frac{M}{RT} p \qquad (3.33)$$

into Eq. (3.24) and assume gravity is negligible, to obtain

$$\phi \mu \frac{\partial p}{\partial t} = \underline{\nabla} \cdot \underline{\underline{k}} p \, \underline{\nabla} p \qquad (3.34)$$

Upon rearrangement

$$\phi \mu \frac{\partial p}{\partial t} = \underline{\nabla} \cdot \underline{\underline{k}} \, \underline{\nabla} p^2 \qquad (3.35)$$

For the flow of real gas we use the appropriate equation of state in place of Eq. (3.33). Depending on the pressure and temperature ranges the resulting equation of condition may become very complex.

3.2 POROSITY, PERMEABILITY, TORTUOSITY, AND CONNECTIVITY OF NONIDEAL MEDIA

We will discuss four macroscopic properties of nonideal porous media which may be used to describe single-fluid flow. (Inversely, single-fluid flow may be used to infer these properties.) To calculate flow through a nonideal porous medium we consider the effects of heterogeneity, nonuniformity, and anisotropy on these macroscopic properties.

Porosity macroscopically characterizes the effective pore volume of the medium. The porosity is directly related to the size of the pores relative to the matrix. When porosity is substituted, we lose the details of the structure.

Permeability is the conductance of the medium defined with direct reference to Darcy's law. The permeability is related to the pore size distribution since the distribution of the sizes of entrances, exits, and lengths of the pore walls make up the major resistances to flow. The permeability is the single parameter that reflects the conductance of a given pore structure. The permeability and porosity are related since if the porosity is zero the permeability is zero. There may be a correlation between porosity and permeability, but they certainly do not regress; that is, permeability cannot be predicted from porosity alone since we need additional parameters which contain more information about pore structure. These additional parameters are the next two macroscopic properties.

Tortuosity is the relative average length of a flow path, the average length of the flow paths to the length of the medium. The tortuosity is a macroscopic measure of both the sinuousness of the flow path and the variation in pore size along the flow path. Like porosity, tortuosity correlates with permeability but cannot be used alone to predict permeability, except in some limiting cases.

Connectivity is the manner and number of pore connections. If all the pores were the same size, it is the average number of pores per junction (Fatt, 1956a). The connectivity is a macroscopic measure of the number of pores at a junction, and, further, the manner of connection, that is, the different sizes of the pores at the junction. Like porosity and tortuosity, connectivity correlates with permeability but cannot be used alone to predict permeability except in certain limiting cases.

There has been discussion in the literature concerning the need for tortuosity (Fatt, 1956a). Fatt uses connectivity to determine permeability from pore structure. The real problem conceptually lies in the simplifications which result in replacing the real porous medium with macroscopic parameters that are averages and relate to some idealized model of the medium. Tortuosity and connectivity represent different features of the pore structure and are useful to interpret macroscopic flow properties such as permeability, capillary pressure, and dispersion.

Kozeny (1927) represents a porous medium as an ensemble of channels of various cross sections of the same length and solves the Navier-Stokes equations for all channels passing a cross section normal to the flow to obtain

$$S^2 = \frac{c\phi^3}{k} \tag{3.36}$$

where the Kozeny constant c is a shape factor taking on different values depending on the shape of the capillary (c = 0.5 for a circular capillary). S is the specific surface of the channels. We can obtain a similar result by considering flow in circular capillaries of length L and radius r.

$$Q = \frac{n\pi r^4 \Delta p}{8\mu L} \tag{3.37}$$

The pore volume of a bundle of capillaries is $n\pi r^2 L$ and the bulk volume is AL; so

$$\phi = \frac{n\pi r^2 L}{AL} = \frac{n\pi r^2}{A} \tag{3.38}$$

Combining Eqs. (3.37) and (3.38) with Darcy's law

$$Q = \frac{kA\Delta p}{\mu L} \tag{3.39}$$

then

$$r^2 = \frac{8k}{\phi} \tag{3.40}$$

For other than circular capillaries Leverett (1941) included a shape factor

$$r^2 = \frac{ck}{\phi} \tag{3.41}$$

The surface area for cylindrical pores is

$$S_A = \frac{n2\pi rL}{n\pi r^2 L} = \frac{2}{r} \tag{3.42}$$

and

$$S_A = 2\sqrt{\frac{\phi}{8k}} \tag{3.43}$$

If we replace $2/\sqrt{8}$ with a shape parameter c and S_A with a specific surface

$$S = \phi S_A \tag{3.44}$$

we obtain the Kozeny equation.

The tortuosity τ was introduced as a modification to the Kozeny equation, to account for the fact that in a real medium the pores are not straight—the length of the most probable flow path is longer than the overall length of the porous medium and

$$S^2 = \frac{c\phi^3}{\tau k} \tag{3.45}$$

Many authors include a shape factor in the tortuosity. In any event there is an inferred relation between permeability, porosity, tortuosity, and pore structure.

The connectivity must be related to the permeability since the manner and number of intersections affects the flow resistance. Fatt (1956a, 1956b) uses connectivity to relate the structure of his model to permeability. Since the tortuosity of his model is one, tortuosity does not enter the problem. It would seem that the Kozeny relation must be modified so that tortuosity includes the geometric effect and a connectivity parameter than includes the effect of intersections (and thus shape) where a is connectivity and replaces the c in Eq. (3.45). As we shall see in Sec. 3.6 when considering the wave equation in a porous medium, the permeability is frequency dependent. This frequency dependence must be related to the connectivity of the medium.

In single-fluid flow situations we are concerned with determining properties in a real medium, that is, a medium having macroscopic heterogeneity, nonuniformity, and anisotropy. Although porosity can be a function of position, thus be heterogeneous, and can vary nonuniformly, as long as porosity is constant in time an average value is all that is required in the equation of condition, since

$$\phi \frac{\partial \rho}{\partial t} = -(\underline{\nabla} \cdot \rho \underline{q}) \tag{3.46}$$

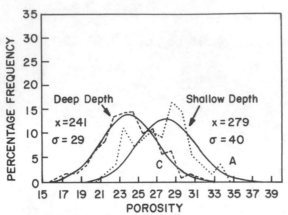

FIG. 3.2 Porosity distributions. (From J. Law, Statistical approach to the interstitial heterogeneity of sand reservoirs, Pet. Trans. AIME 155, SPEJ 202.
© 1944 SPE-AIME.)

Orientation of the pores does not cause a directional effect through porosity—
porosity is not a tensor—so anisotropy need not be considered.

    To find the average porosity of a homogeneous but nonuniform medium we
determine the correct mean of the distribution of porosity. It has been observed
that data on porosity of both natural and artificial media usually are normally distributed (Law, 1944; Bennion and Griffiths, 1966). Figure 3.2 shows the porosity
distribution of samples from a heterogeneous sandstone; there are two distinct
distributions with widely differing permeabilities. The two distributions are
nonuniform. Also, In Fig. 3.2 in both sets the distribution of the nonuniform porosity is approximately normal. The mean of the normal distribution is

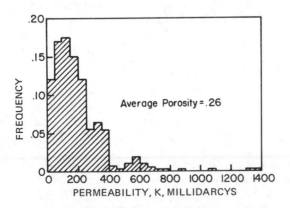

FIG. 3.3 Permeability distribution for a natural sandstone. (From J. Law, Statistical approach to the interstitial heterogeneity of sand reservoirs, Pet. Trans.
AIME 155, SPEJ 202. © SPE-AIME.)

$$\langle \phi \rangle = \frac{\Sigma_{i=1}^{n} \phi_i}{n} \tag{3.47}$$

the arithmetic average. The average porosity of a heterogeneous nonuniform media as in Fig. 3.2 is the volume-weighted average of the number average, Eq. (3.47)

$$\langle\langle \phi \rangle\rangle = \frac{\Sigma_{i=1}^{m} V_i \langle \phi_i \rangle}{\Sigma_{i=1}^{m} V_i} \tag{3.48}$$

For the case of Fig. 3.2, m = 2 there are two distinct distributions.

The average nonuniform permeability is a function of position. For a homogeneous but nonuniform media the average permeability is the correct mean (first moment) of the permeability distribution function. Permeability for a nonuniform media is usually skewed, as shown in Fig. 3.3. The correlation curves of porosity and permeability for "Bradford sand" are plotted on a normal probability scale and a log normal probability scale respectively in Fig. 3.4. Most data for nonuniform permeability show permeability to be distributed log normal (Csallany and Walton, 1963; Law, 1944; Seaber and Hollyday, 1966). The correct average for a homogeneous, nonuniform permeability assuming it is distributed as a log normal is the geometric mean.

$$\langle k \rangle = \Big(\prod_{i=1}^{n} k_i \Big)^{1/n} \tag{3.49}$$

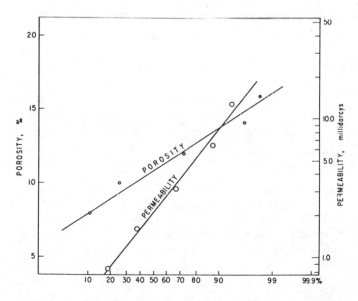

FIG. 3.4 Cumulative curves of porosity and permeability of "Bradford sand." Porosity is plotted on a normal probability scale, and permeability is plotted on a log-normal probability scale. (Data from Muskat, 1937.)

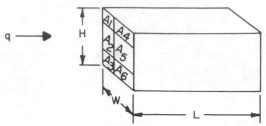

FIG. 3.5 Flow through parallel nonuniform elements of porous media.

For flow in heterogenous media, the average permeability will depend on the arrangement and geometry of the nonuniform elements, each of which will have a different average permeability. First consider flow through a set of nonuniform elements where the elements are parallel to the flow, as sketched in Fig. 3.5. Since flow is through parallel elements of different constant area, Darcy's law for each element, assuming overall length of each element is equal, is

$$Q_1 = -\frac{A_1 \langle k_1 \rangle \Delta p}{\mu L} \tag{3.50}$$

$$Q_2 = -\frac{A_2 \langle k_2 \rangle \Delta p}{\mu L} \tag{3.51}$$

. . . .

The flow rate through the entire system of elements is

$$Q = Q_1 + Q_2 + \cdots \tag{3.52}$$

Substituting Eqs. (3.50) and (3.51) into Eq. (3.52) and canceling terms,

$$A \langle\langle k \rangle\rangle_p = A_1 \langle k_1 \rangle + A_2 \langle k_2 \rangle + \cdots \tag{3.53}$$

or

$$\langle\langle k \rangle\rangle_p = \frac{A_1 \langle k_1 \rangle + A_2 \langle k_2 \rangle + \cdots}{A} \tag{3.54}$$

Thus the average permeability for this heterogeneous medium made up of parallel nonuniform elements is the area weighted average of the average permeability of each of the elements, which, if the permeability of each element is log normally distributed, are the geometric means. Many reservoirs and soils are composed of heterogeneities that are nonuniform layers so that only the thickness of the layers varies. So $\langle\langle k \rangle\rangle_p$ simplifies to

$$\langle\langle k \rangle\rangle_{ph} = \frac{h_i \langle k_i \rangle + h_2 \langle k_2 \rangle + \cdots}{h} \tag{3.55}$$

If all the layers have the same thickness,

$$\langle\langle k \rangle\rangle_{ph} = \frac{\sum_{i=1}^{h} \langle k_i \rangle}{n} \tag{3.56}$$

where n is the number of layers.

The other simple situation that we can consider is flow through a set of nonuniform elements that are in series, as sketched in Fig. 3.6. In this instance the volumetric flow rate through each element is equal

$$Q = -\frac{\langle k_1 \rangle A(p_1 - p_2)}{\mu L_1} \tag{3.57}$$

$$Q = -\frac{\langle k_2 \rangle A(p_2 - p_3)}{\mu L_2} \tag{3.58}$$

since

$$p_1 - p_n = (p_1 - p_2) + (p_2 - p_3) + \cdots + (p_{n-1} - p_n) \tag{3.59}$$

$$\frac{L}{\langle\langle k \rangle\rangle_s} = \frac{L_1}{\langle k_1 \rangle} + \frac{L_2}{\langle k_2 \rangle} + \cdots \tag{3.60}$$

and

$$\langle\langle k \rangle\rangle_s = \frac{L}{L_1 / \langle k_1 \rangle + L_2 / \langle k_2 \rangle + \cdots} \tag{3.61}$$

If all the elements are of equal length,

$$\langle\langle k \rangle\rangle_{sL} = \frac{n}{\sum_{i=1}^{n} \langle k_i \rangle} \tag{3.62}$$

where n is the number of elements. The correct average permeability for a hetero-

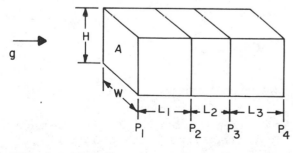

FIG. 3.6 Flow through serial nonuniform elements of porous media.

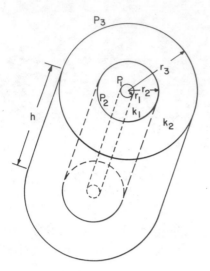

FIG. 3.7 Radial heterogeneous system.
(From Muskat, 1937.)

geneous series of nonuniform elements is the harmonic mean of the average permeability of each of the elements. For a radial system such as sketched in Fig. 3.7, the area is a function of radius and

$$\frac{Q}{A} = - \frac{\langle k \rangle}{\mu} \frac{dp}{dr} \tag{3.63}$$

Integrating this equation for one element

$$Q \int_{r_1}^{r_2} \frac{dr}{A} = - \frac{\langle k_1 \rangle}{\mu} \int_{p_1}^{p_2} dp \tag{3.64}$$

Assume h, the thickness of the element, is constant; then $A = 2\pi hr$ and for the two two elements of Fig. 3.7

$$Q = \frac{2\pi \langle k_1 \rangle h(p_1 - p_2)}{\mu \ln (r_1/r_2)} \tag{3.65}$$

$$Q = \frac{2\pi \langle k_2 \rangle h(p_2 - p_3)}{\mu \ln (r_2/r_3)} \tag{3.66}$$

Using the pressure relation of Eq. (3.59), we find

$$\langle\langle k \rangle\rangle_r = \frac{\ln (r_3/r_1)}{(1/\langle k_1 \rangle) \ln (r_2/r_1) + (1/\langle k_2 \rangle) \ln (r_3/r_2)} \tag{3.67}$$

and the correct average for heterogeneous elements that are radially distributed is the harmonic mean weighted by the logarithm of the ratio of the radii of the heterogeneous elements.

For systems of heterogeneous nonuniform elements that are not simple geometries, the computation of an average permeability for the entire system is difficult. We will discuss a few of the approximations that are used for realistic cases other-than using the averages above by assuming that the system is either layered or in series. For reservoirs where heterogeneity may be three-dimensional and flow is essentially two-dimensional, we usually assume layers at a given point and compute the average according to Eq. (3.55). To determine an overall average between points when the heterogeneity is restricted to layers, determine the geometric average of the layers and combine them into one average using Eq. (3.55). To determine an overall average where the heterogeneities are three-dimensional, compute the average at one point according to Eq. (3.55) and combine these averages as a weighted harmonic average (Eq. 3.61) (Johnson and Greenkorn, 1962). If we are interested in the details of the flow at an intermediate point in a heterogeneous porous medium, the average permeability will not provide correct pressure or rate values at that point. The average permeability gives a "black box" result which only yields the correct pressure and flow at the boundaries. If the data are available at intermediate points, we can include the effect of heterogeneity by solving the equation of condition, Eq. (3.24), with permeability as a function of position.

$$\phi \mu \frac{\partial \rho}{\partial t} = \underline{\nabla} \cdot \rho \left\langle k(x, \ y, \ z) \right\rangle \underline{\nabla} \Phi \qquad (3.68)$$

We need to know the average permeability (assuming it is nonuniform) as a function of position. Most often, the equation of condition is solved numerically. In some instances it is difficult to decide if the permeability is heterogeneous and/or nonuniform. Normally, with enough samples, one can identify the separate distributions in a heterogeneous system. In natural systems use the best values considering the use of these values.

The permeability in general is a tensor and we determine the elements of the tensor in anisotropic systems. Again, in natural systems we have the dilemma of deciding whether directional effects are ordered heterogeneities taken into account, as in Eq. (3.68), or whether permeability is really tensorial in that the directional effect is a point property. (Interestingly enough, in most situations either way will give a usable answer.) Permeability is a volume-averaged property for a finite but small volume of a medium. Anisotropy in natural or artificially packed media may result from particle (or grain) orientation, bedding of different sizes of particles, or layering of media of different permeability. It is difficult to say at what level we should treat a directional effect as anisotropy or as an oriented heterogeneity. Assuming the problem of scale is solved, consider the directional effects represented by the permeability tensor.

In principle, simply use the general form of Darcy's law, Eq. (3.3) or Eq. (3.20), where, neglecting gravity,

$$\underline{q} = -\frac{\underline{\underline{k}}}{\mu} \underline{\nabla} p \qquad (3.69)$$

The significance of the tensor $\underline{k}$ is that in an anisotropic (oriented) medium, the velocity $\underline{q}$ and the pressure gradient $\underline{\nabla} p$ are not in the same direction. Perhaps an analogy will help explain. Imagine you are flying from one point in the United States to another point, and, for the sake of argument, say from Indianapolis to Chicago. The wind is blowing out of the west. The direction from Indianapolis to Chicago is essentially north, and if you point your airplane directly at Chicago and start out, at the time you should reach Chicago you would find yourself much farther east, over Lake Michigan, because the wind has blown you in that direction; so you drifted. To take care of this problem, you must point your airplane not at Chicago but into the wind some—you actually point the airplane further to the west than you are going, and the combined effect of flying to the west and the wind blowing is eventually, if you do it correctly, to end in Chicago.

In an oriented porous medium, the resistance to flow is different depending on the direction. Thus, if there is a pressure gradient between two points and you mark a spot of fluid, unless the pressure gradient is parallel to oriented flow paths, you would find that the spot would not go from the original point to the point which you expected, but would drift. This is a result of anisotropy of the porous medium.

We have an additional dilemma in that in order to determine anisotropy experimentally finite chunks of porous media must be used and, unfortunately, there are two extremes of possible conditions plus the intermediate possibilities (Scheidegger, 1957; Marcus, 1962). For one extreme, permeability is measured in the direction of pressure gradient; the velocity component parallel to the pressure gradient is

$$q_n = \underline{n} \cdot \underline{q} \tag{3.70}$$

The directional permeability measured is

$$k_n = \underline{n} \cdot (\underline{\underline{k}} \cdot n) \tag{3.71}$$

and in two dimensions

$$k_n = k_{11} \cos^2 \alpha + k_{22} \cos^2 \beta \tag{3.72}$$

where k_{11} and k_{22} are the major and minor values of the permeability ellipse in a plot of $1/\sqrt{k_n}$ versus the angle of measurement. α and β are the angles of the principal axes of the tensor with the unit vector $\underline{n}$.

For the other extreme, permeability is measured in the direction of flow; the component of pressure parallel to the flow is

$$(\underline{\nabla}p)_n = n \cdot \underline{\nabla}p \tag{3.73}$$

The directional permeability measured is

$$k'_n = \frac{1}{(\underline{n} \cdot \underline{\underline{k}}^{-1}) \cdot \underline{n}} \tag{3.74}$$

and in two dimensions

$$\frac{1}{k'_n} = \frac{\cos^2 \alpha}{k_{11}} + \frac{\cos^2 \beta}{k_{22}} \tag{3.75}$$

thus the $\sqrt{k_n'}$ versus measured angle plots as an ellipse, where k_{11} and k_{22} are the major and minor values for the tensor.

Both Scheidegger (1957) and Marcus (1962) discuss the error involved when making one or the other measurement. This error is a maximum at 45° from the principal axis of the permeability tensor. Fortunately, the error is small unless the difference between the values of k_{11} and k_{22} is large.

Parsons (1964) makes this point, discussing the papers of Marcus (1962) and of Greenkorn et al. (1964) where measurements are made as in Fig. 3.8. The maximum variation of measured permeability as a function of anisotropy is shown in Fig. 3.9. The difference between the two extremes is not large if the ratio of k_{11}/k_{22} is small.

As a way of separating the effects of heterogeneity and anisotropy Greenkorn et al. (1964) assumed

$$\underline{\underline{k}} = k_m \underline{\underline{I}} + \underline{\underline{k}}' \tag{3.76}$$

where k_m is the minor principal axis of the permeability tensor (in two dimensions) and represents the overall (isotropic) permeability at a point. $\underline{\underline{k}}'$ is a tensor added to $k_m \underline{\underline{I}}$ to give the directional value at a point; it represents the directional effect. In two dimensions,

$$\underline{\underline{k}} = k_m \begin{bmatrix} 1 & 0 \\ 0 & 1 \end{bmatrix} + \begin{bmatrix} k_{11} - k_m & k_{12} \\ k_{21} & k_{22} - k_m \end{bmatrix} \tag{3.77}$$

Greenkorn et al. (1964) show for a sandstone system k_m correlates with grain size (heterogeneity) and $\underline{\underline{k}}'$ correlates with bedding, the directional effect.

The tortuosity is difficult to relate to the nonuniformity and the anisotropy of a medium. If we attempt to predict permeability from a pore structure model, then

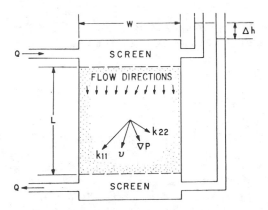

FIG. 3.8 Direct measurement of the apparent directional permeability. (From H. Marcus, Permeability of a sample of an anisotropic porous medium, J. Geophys. Res. 67(13), 5215. © 1962 The American Geophysical Union.)

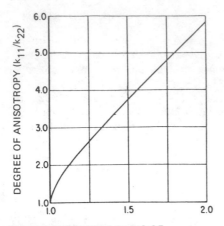

MAXIMUM POSSIBLE RATIO OF
MEASURED PERMEABILITIES $(k/k'$ at $a, \beta = 45^\circ)$

FIG. 3.9 Maximum variation of measured permeabilities as a function of anisotropy. (From R. W. Parsons, Discussion of directional permeability of heterogeneous anisotropic porous media, <u>Pet. Trans. AIME</u> <u>231</u>, SPEJ 364. © 1964
SPE–AIME.)

tortuosity is needed to correct the model for pore length. For a given model including nonuniformity and anisotropy; tortuosity is constant (2.25 for the model of
Haring and Greenkown, 1970; 1 for the model of Fatt, 1956a). Tortuosity enters
the problem significantly through fluid flow, that is, in diffusion in pores and in
dispersion when we are concerned with flow of miscible fluids. In mass transfer
by diffusion in pores, the tortuosity is used to relate diffusivity and effective diffusivity in the pores (Satterfield and Sherwood, 1963).

$$\mathcal{D}_{eff} = \frac{\mathcal{D}}{\tau} \tag{3.78}$$

With dispersion of two miscible liquids where diffusion effects are present (at slow
velocities) the effective diffusion or dispersion is described by tortuosity. According
to Whitaker (1967)

$$D_{eff} = D(\delta_{jk} + RB_{jk}) \tag{3.79}$$

where RB_{jk} is the tortuosity effect. We will discuss this topic further in Chap. 5.

During two–fluid immiscible flow as one fluid is displaced by another fluid,
the nonwetting fluid may become trapped in a pore. This trapping results in apparent
change in tortuosity to the displacing fluid since the trapped fluid changes the paths
of the wetting fluid. We will discuss this point further in Chap. 4.

It is interesting to speculate whether pure creeping flow is required everywhere through the pores to use permeability as the conductance parameter in single-
fluid flow. If the overall pressure drop is linear with velocity, and if there is form
drag (noncreeping flow) in addition to the wall resistance and entrance–exit effects,
then the length and shape of the flow path (tortuosity) and the manner and number of

intersections (connectivity) must affect permeability. Further, it is interesting to speculate whether permeability is reciprocal in a noncreeping flow, but linear regime since form drag will be different in different directions.

3.3 STEADY FLOW: COMPLEX POTENTIALS

Two-dimensional steady flow of an incompressible single fluid in heterogeneous, anisotropic porous media is often described using complex potentials. For steady single-fluid flow where we are mainly concerned with the pressure and velocity, we normally do not differentiate between uniform and nonuniform media other than determining the correct permeability distribution and its average. For steady flow of an incompressible single fluid, Eq. (3.25) applies and

$$\underline{\nabla} \cdot \underline{\underline{k}} \, \underline{\nabla}\Phi = 0 \tag{3.80}$$

where $\Phi = p + \rho gz$. To discuss steady flow in terms of complex potentials, we need some definitions used with the flow of ideal fluids. Solution of Eq. (3.80) in two dimensions models flow hypothetically as if the matrix were not present. We obtain results superficially imposed on the porous medium modeled.

A streamline is an imaginary line in the fluid which has direction of the local velocity at all points. In steady flow, fluid moves along streamlines. To iterate, a streamline for flow in porous media is hypothetical. If we put a dye in a fluid flowing in a porous medium, we would see, from a distance, a fuzzy streamline. However, if we investigate the actual fluid movement closely, we see that it is complex because of dispersion as it moves and mixes in the pores. Since a streamline has the direction of velocity at all points, there is no flow across the streamline, only along it.

The flowing fluid follows a path line. In steady flow, path lines and streamlines are identical. In unsteady flow, a streamline shifts as the direction of velocity changes and a fluid "particle" shifts from one streamline to another. In unsteady flow a path line is not a streamline.

A streak line is the instantaneous path exhibited by a tracer emitted at a point in the fluid.

A stream tube is a tube whose surfaces are made of all streamlines passing through a closed curve. There is no flow through the imaginary wall of the stream tube, since there can be no flow across streamlines. For steady flow, a stream tube is fixed in space.

The stream function is a quantity defined in such a way that it is constant along a streamline. The two-dimensional stream tube of Fig. 3.10 is constructed of two streamlines represented by the stream functions $\psi_1 = \text{constant}_1$ and $\psi_2 = \text{constant}_2$. Since the flow rate inside a stream tube is constant, we may integrate across any path to evaluate the mean flow rate. For path C_1 of Fig. 3.10

$$w = \int_{C_1} \rho \, (\underline{v} \cdot \underline{n}) \, ds = \int_{C_1} \rho v \cos \alpha \, ds \tag{3.81}$$

where α is the angle between the velocity vector and the outward directed normal. The value of the stream function is defined by

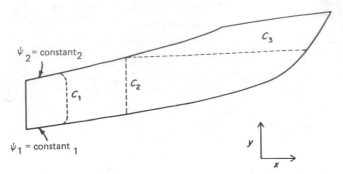

FIG. 3.10 Two-dimensional stream tube. (From Greenkorn and Kessler, 1972.)

$$w = (\Psi_1 - \Psi_2)\rho \tag{3.82}$$

for a stream tube

$$\int \rho \, d\Psi = \text{constant} \tag{3.83}$$

Equation (3.83) is true for any path joining two streamlines, such as C_1 or C_2 or C_3 of Fig. 3.10. For path C_2, $\delta s = \delta y$ and $v_x = v \cos \alpha_2$; for path C_3, $\delta s = \delta x$ and $v_y = v \cos \alpha_3$. Since $\cos \alpha_2 = -1$

$$\delta \Psi = -v_x |_x \delta y \tag{3.84}$$

and $\cos \alpha_3 = 1$

$$\delta \Psi = v_y |_y \delta x \tag{3.85}$$

In the limit

$$v_x = -\frac{\partial \Psi}{\partial y} \tag{3.86}$$

$$v_y = \frac{\partial \Psi}{\partial x} \tag{3.87}$$

A <u>potential</u> is the pressure, gravity, or their sum causing flow along a stream-line and is orthogonal to the streamline so that

$$v_x = -\frac{\partial \Phi}{\partial x} \tag{3.88}$$

$$v_y = -\frac{\partial \Phi}{\partial y} \tag{3.89}$$

The Cauchy-Riemann equations result from combining Eqs. (3.86) to (3.89).

$$\frac{\partial \Phi}{\partial x} = \frac{\partial \Psi}{\partial y} \tag{3.90}$$

$$\frac{\partial \Phi}{\partial y} = -\frac{\partial \Psi}{\partial x} \tag{3.91}$$

Equations (3.90) and (3.91) must be satisfied by the real and imaginary parts of an analytic function

$$w(z) = \Phi(x, y) + i\Psi(x, y) \tag{3.92}$$

where w and z are complex numbers. The real and imaginary parts of any analytic function satisfy the two-dimensional Laplace equation.

$$\frac{\partial^2 \Phi}{\partial x^2} + \frac{\partial^2 \Phi}{\partial y^2} = 0 \tag{3.93}$$

or

$$\frac{\partial^2 \Psi}{\partial x^2} + \frac{\partial^2 \Psi}{\partial y^2} = 0 \tag{3.94}$$

Both Φ and Ψ are complex potentials and they form a mutually orthogonal network.

$$\frac{\partial \Phi}{\partial x}\frac{\partial \Phi}{\partial y} + \frac{\partial \Psi}{\partial y}\frac{\partial \Psi}{\partial x} = 0 \tag{3.95}$$

Also

$$\frac{v_y}{v_x} = \frac{\partial \Phi/\partial y}{\partial \Phi/\partial x} = -\frac{\partial \Psi/\partial x}{\partial \Psi/\partial y} \tag{3.96}$$

Equation (3.96) shows that the direction of the fluid at any point coincides with the tangent at that point to the stream function Ψ.

For a homogeneous, isotropic medium, Eq. (3.80) is

$$\nabla^2 \Phi = 0 \tag{3.97}$$

and is equivalent to Eqs. (3.93) and (3.94). The significance of this equivalence is that an analytic function given by Eq. (3.92) represents the streamlines and equipotential lines for flow problems for which Eqs. (3.93) and (3.94) are true. Further, this allows conformal transformations to be used to transform flow in complex geometry into simple geometries.

Consider single-fluid steady flow in a homogeneous, isotropic porous medium in plane radial flow. We might use this to model flow into or out of a well drilled into a reservoir. The flow is sketched schematically in Fig. 3.11 as flow between concentric cylinders where r_w represents the well radius and r_e a drainage radius. Laplace's equation, Eq. (3.93), in polar coordinates is

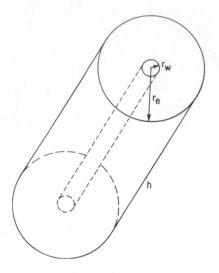

FIG. 3.11 Potential flow between two
concentric cylinders. (From Muskat,
1937.)

$$\frac{1}{r}\frac{\partial}{\partial r}\left(r\frac{\partial \Phi}{\partial r}\right) + \frac{1}{r}\frac{\partial^2 \Phi}{\partial \theta^2} = 0 \tag{3.98}$$

The problem is symmetrical and $\partial^2 \Phi / \partial \theta^2 = 0$; so

$$\frac{\partial}{\partial r}\left(r\frac{\partial \Phi}{\partial r}\right) = 0 \tag{3.99}$$

The stream function, Eq. (3.94), considering the symmetry, is

$$\frac{\partial^2 \Psi}{\partial \theta^2} = 0 \tag{3.100}$$

To solve this flow problem we look for an analytic function

$$w(re^{i\theta}) = \Phi(r, \theta) + i\Psi(r, \theta) \tag{3.101}$$

which satisfies the boundary conditions.

Since the flow is radial we know lines of constant Φ are concentric circles
about the origin and lines of constant Ψ are a family of radiating straight lines. An
analytic function which describes this situation is

$$w(z) = A \ln Z + B = A \ln r^{i\theta} + B \tag{3.102}$$

$$\Phi = A \ln r + b \tag{3.103}$$

$$\Psi = A\theta + b' \tag{3.104}$$

$$B = b + ib' \tag{3.105}$$

In using complex flow potentials, the problem is to determine an equation for w(z). There are books that contain solutions or different forms of w that give different flow lines and further transformations of these situations. The boundary conditions for C are

$$\Phi(r_w) = \Phi_w \tag{3.106}$$

$$\Phi(r_e) = \Phi_e \tag{3.107}$$

and therefore

$$\Phi = \frac{\Phi_e - \Phi_w}{\ln(r_e/r_w)} \ln \frac{r}{r_w} + \Phi_w \tag{3.108}$$

The boundary conditions for Ψ

$$\Psi(0) = 0 \tag{3.109}$$

$$\Psi(2\pi) = -\frac{\rho_w Q_w}{h} \tag{3.110}$$

and

$$\Psi = -\frac{\rho_w Q_w}{2\pi h} \theta \tag{3.111}$$

The value of A in Eq. (3.104) et seq. is determined from both Φ and Ψ, so equating A yields

$$-\rho_w Q_w = 2 \ln \frac{\Phi_e - \Phi_w}{\ln r_e/r_w} \tag{3.112}$$

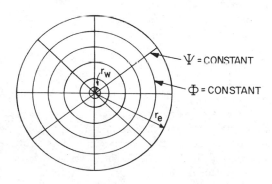

FIG. 3.12 Equipotentials and streamlines for potential flow between concentric cylinders. (From Churchill, 1960.)

The resulting flow field from plotting Eqs. (3.108) and (3.109) is sketched in Fig. 3.12.

We could also solve flow between concentric cylinders using a transformation. At each point where a function f(z) is analytic and df/dz = 0, the mapping w = f(z) is conformal. Consider the conformal transformation

$$w = \ln z \tag{3.113}$$

or in complex numbers

$$u + iv = \ln re^{i\theta} = \ln r + i\theta \tag{3.114}$$

then

$$u = \ln (x^2 + y^2)^{\frac{1}{2}} \tag{3.115}$$

and

$$v = \tan^{-1} \frac{y}{x} \tag{3.116}$$

are conformal.

Every harmonic function of x and y transforms into a harmonic function of u and v under the change of variables x + iy = f(u + iv), where f is an analytic function. Laplace's equation is

$$\frac{\partial^2 \Phi}{\partial x^2} + \frac{\partial^2 \Psi}{\partial y^2} = 0 \tag{3.117}$$

and may be transformed by z = f(w), where x = x(u, v), y = y(u, v) so that

$$\frac{\partial^2 \Phi}{\partial u^2} + \frac{\partial^2 \Phi}{\partial v^2} = 0 \tag{3.118}$$

Returning to the problem of plane radial flow in a homogeneous, isotropic porous medium—the equation to be solved is Eq. (3.117) with $r_w^2 < x^2 + y^2 < r_e^2$. The boundary conditions are

$$\Phi = \Phi_w \qquad \text{on } x^2 + y^2 = r_w^2 \tag{3.119}$$

$$\Phi = \Phi_e \qquad \text{on } x^2 + y^2 = r_e^2 \tag{3.120}$$

In the z = x + iy plane, the equipotential lines are concentric circles and the streamlines are a family of radiating straight lines. We can map this region using the logarithmic transformation of Eq. (3.113) and obtain the results shown in Fig. 3.13. The flow region in the w plane is a rectangle, and the form of Laplace's equation is given by Eq. (3.118). In the w plane the flow is linear and

$$\Phi = A + Bu \tag{3.121}$$

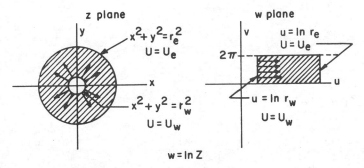

FIG. 3.13 Mapping of radial flow into rectangular flow. (From Churchill, 1960.)

The boundary conditions are

$$\Phi = \Phi_w \qquad \text{for } u = \ln r_w \tag{3.122}$$

$$\Phi = \Phi_e \qquad \text{for } u + \ln r_e \tag{3.123}$$

which gives

$$\Phi = \Phi_w + (\Phi_e - \Phi_w) \frac{u - \ln r_w}{\ln r_e - \ln r_w} \tag{3.124}$$

Using Eqs. (3.115)

$$\Phi = \Phi_w + \frac{\Phi_c - \Phi_w}{\ln (r_e/r_w)} \ln \frac{r}{r_w} \tag{3.125}$$

which is the same result as Eq. (3.108).

Although the procedure in this situation may seem trivial, in more complex flow situations the possibility of doing one or several mappings to transform a complicated flow geometry into a fairly simple one opens the possibilities for solving many kinds of flow problems. Additional techniques are useful for studying specific applications of single-fluid flow through a porous medium, such as Schwartz-Christoffel transformation (rectangular transformation), inclusion of sources and sinks, and method of images.

Greenkorn et al. (1964) used complex potentials to study the effect of the size and shape of a single heterogeneity in single-fluid flow in a Hele-Shaw model. A Hele-Shaw model is an analog to two-dimensional flow in porous media, and is discussed in more detail in Chap. 6. A Hele-Shaw model is constructed by placing two parallel flat plates, usually glass, close together and flowing fluid through the slit between the two plates. The equipotential lines and streamlines are the same for the model and a porous medium if

$$k = \frac{h^2}{12} \tag{3.126}$$

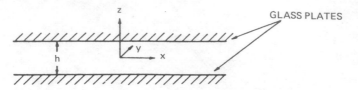

FIG. 3.14 Cross section through Hele-Shaw model. (From Greenkorn et al., 1964.)

where h is the space between the flat plates, as shown in Fig. 3.14. Experimentally, streamlines are produced by injecting a continuous filament of dye into the flowing fluid, as shown in Fig. 3.15. The dye lines for a homogeneous circumstance are shown in Fig. 3.15. Heterogeneity is introduced into the model by changing the spacing between the plates, by etching, or by creating a partial obstruction.

Greenkorn et al. (1964) determined streamlines for single heterogeneities of various shapes in a linear flow field. Some of their results are shown in Fig. 3.16. They show, as a result of their experiments, that heterogeneity is approximately defined for a mathematical model by knowing its size and shape. Further, they found the actual value of the transmissivity is not required if the transmissivity ratio between the heterogeneity and the flow field is less than 1/10 or greater than 10. To reach these conclusions, they determined the potential solution to the flow past a circular heterogeneity of infinite permeability (an infinite sink) or through a circular heterogeneity of zero permeability (an obstruction).

Consider two-dimensional flow around and through a single circular hetero-geneity of infinite or zero permeability in a uniform flow field of finite permeability. Define a complex plane

$$z = x + iy = re^{i\theta} \tag{3.127}$$

with origin at the center of a circle of radius r and real axis parallel to the direction of flow. Use index 1 for the interior of the circle and index 2 for the outside. The flow problem is solved if we find an analytic function

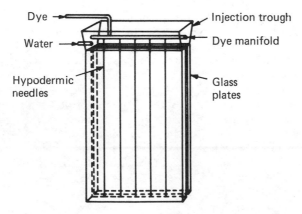

FIG. 3.15 Hele-Shaw model. (From R. A. Greenkorn et al., Flow in heterogeneous Hele-Shaw models, Pet. Trans. AIME 231, SPEJ 307. © 1964 SPE-AIME.)

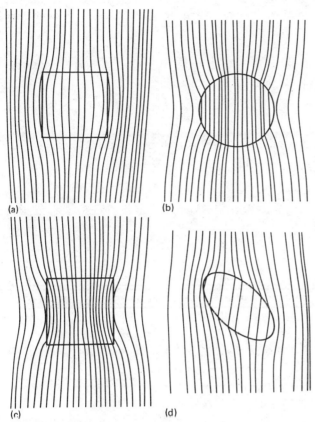

(a)   (b)

(c)   (d)

FIG. 3.16 (a) Streamlines in Hele-Shaw model with low-permeability square
(nominal transmissibility ratio = 0.216), (b) streamlines in Hele-Shaw model with
high-permeability circle (nominal transmissibility ratio = 4.63), (c) streamlines
in Hele-Shaw model with high-permeability square (nominal transmissibility ratio
= 4.63), (d) streamlines in Hele-Shaw model with low-permeability ellipse (nominal
transmissibility ratio = 0.216). (From R. A. Greenkorn et al., Flow in heterogeneous Hele-Shaw models, Pet. Trans. AIME 231, SPEJ 307. © 1964 SPE-AIME.)

$$w(z) = \Phi + i\Psi \tag{3.128}$$

for regions 1 and 2 that satisfies the boundary conditions for Φ and Ψ. These boundary conditions are as follows:

    1. For the absolute value of z approaching infinity the field must be uniform

$$\Phi_1 = -u_1 x \tag{3.129}$$

$$\Psi_1 = u_1 y \tag{3.130}$$

where u_1 is the velocity far from the circular heterogeneity.

2. At the boundary of the two regions, the absolute value of $|z| = a$, and the pressure and the normal component of the flux density (flow per unit length) are continuous

$$\frac{\Phi_1}{k_1} = \frac{\Phi_2}{k_2} \tag{3.131}$$

$$h_1 \frac{\partial \Phi_1}{\partial r} = h_2 \frac{\partial \Phi_2}{\partial r} \tag{3.132}$$

where both permeability and thickness may change at the boundary.

The equations

$$w_1(z) = u_1 \left(z + \beta \frac{a^2}{z} \right) \tag{3.133}$$

and

$$w_2(z) = u_2 z \tag{3.134}$$

satisfy these boundary conditions. Separating the real and imaginary parts of Eq. (3.133)

$$w_1(z) = \underbrace{u_1 x \left(1 + \beta \frac{a^2}{x^2 + y^2} \right)}_{\Phi} + \underbrace{i u_1 y \left(1 - \beta \frac{a^2}{x^2 + y^2} \right)}_{\Psi} \tag{3.135}$$

From Eq. (3.134)

$$\Psi_2(z) = u_2 r \cos \theta \tag{3.136}$$

Using the boundary conditions $r = a$, Eqs. (3.131) and (3.132) become

$$\frac{u_2}{k_2} = \frac{u_1}{k_1}(1 + \beta) \tag{3.137}$$

and

$$h_2 u_2 = h_1 u_1 (1 - \beta) \tag{3.138}$$

Solving for β and for u_2 gives

$$\beta = \frac{1 - R}{1 + R} \tag{3.139}$$

and

$$u_2 = u_1 \frac{h_1}{h_2} \frac{2R}{1 + R} \tag{3.140}$$

where R is the transmissivity ratio

$$R = \frac{h_2 k_2}{h_1 k_1} \tag{3.141}$$

Define $a' = \sqrt{|\beta|}\, a$, then for R < 1

$$\Phi_1(z) = u_1 x\left(1 + \frac{a'^2}{r^2}\right) \tag{3.142}$$

$$\Psi_1(z) = u_1 y\left(1 - \frac{a'^2}{r^2}\right) \tag{3.143}$$

For R < 1,

$$\Phi_1 = u_1 x\left(1 - \frac{a'^2}{r^2}\right) \tag{3.144}$$

$$\Psi_1 = u_1 y\left(1 + \frac{a'^2}{r^2}\right) \tag{3.145}$$

β has a limiting value of +1 for an impermeable circle and -1 for a circle of infinite permeability. In these cases, $a' \to a$; in all other cases, $a' < a$. Equations (3.142) to (3.145) were used to calculate the streamlines and potential lines for circular heterogeneities. The results for a circle of infinite permeability are shown in Fig. 3.17 compared to the experimental results for a high permeability circle (R greater than 10).

The numerical results shown in Fig. 3.17 were found by solving two-dimensional form of Eq. (3.25)

$$\frac{\partial}{\partial x}\left(\frac{kh}{\mu}\frac{\partial p}{\partial x}\right) + \frac{\partial}{\partial y}\left(\frac{kh}{\mu}\frac{\partial p}{\partial y}\right) = 0 \tag{3.146}$$

using an alternating direction iteration numerical procedure and the boundary conditions $p = p_0$ at $x = 0$ and $p = p_1$ at $x = L$, where L is length of the model. A grid of 40×32 elements was used. Streamlines were obtained from the pressure (potential), results by

$$\frac{\partial \Psi}{\partial x} = -\frac{kh}{\mu}\frac{\partial p}{\partial y} \tag{3.147}$$

The problem of including anisotropy in calculations of steady flow of a single fluid using complex potentials is handled using a coordinate transformation. Consider the equation of condition for steady flow of a single fluid in an anisotropic medium in two dimensions, Eq. (3.80)

$$k_{11}\frac{\partial^2 \Phi}{\partial x^2} + k_{22}\frac{\partial^2 \Phi}{\partial y^2} = 0 \tag{3.148}$$

Equation (3.148) is not Laplace's equation. We can solve the equation numerically or we can apply complex potentials by introducing the transformations

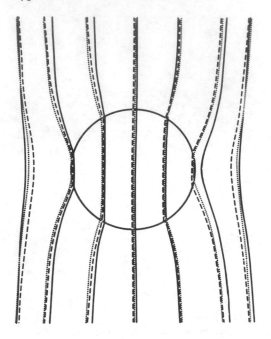

FIG. 3.17 Comparison of observed and computed flow lines for high-permeability circle. Solid line, experimental; dotted line, calculated; dashed line, analytical. (From R. A. Greenkorn et al., Flow in heterogeneous Hele-Shaw models, Pet. Trans. AIME 231, SPEJ 307. © 1964 SPE-AIME.)

$$\tilde{x} = \frac{x}{\sqrt{k_{11}}} \tag{3.149}$$

$$\tilde{y} = \frac{y}{\sqrt{k_{22}}} \tag{3.150}$$

Equation (3.148) becomes

$$\frac{\partial^2 \Phi}{\partial \tilde{x}^2} + \frac{\partial^2 \Phi}{\partial \tilde{y}^2} = 0 \tag{3.151}$$

Equation (3.151) is Laplace's equation in the transformed coordinates.

In the transformed system, equipotential lines and streamlines are orthogonal. In the original system, equipotential lines and streamlines are not orthogonal. The angle between the equipotential lines and the streamlines in the original system is given by

$$\cos \theta = \frac{k_{11}\left(\dfrac{\partial \Phi}{\partial x}\right)^2 + k_{22}\left(\dfrac{\partial \Phi}{\partial y}\right)^2}{|\tilde{q}| \, |\underline{i}\, \partial \Phi/\partial x + \underline{j}\, \partial \Phi/\partial y|} \tag{3.152}$$

and

$$k_n = \left(\frac{\cos^2 \theta_{11}}{k_{11}} + \frac{\cos^2 \theta_{22}}{k_{22}} \right)^{-1} \tag{3.153}$$

The average isotropic permeability is

$$\langle k \rangle = \sqrt{k_{11}k_{22}} \tag{3.154}$$

3.4 STEADY COMPRESSIBLE FLOW

The discussion of steady-state flow in Secs. 3.2 and 3.3 assumed the flowing fluid is incompressible. We are often concerned with compressible fluids when studying transient flow, that is, where time is a variable. However, for certain combinations of compressibility and system size for liquids and for gases it is possible to have steady-state or pseudo-steady-state compressible flow. Such systems of compressible fluid flow are either in permanent steady state or in a succession of steady states. Time enters the problem in the succession of steady states as a parameter; time is not explicit so Laplace's equation applies at each successive time. It is assumed that boundary conditions vary as a function of time. The pressure distribution changes immediately to a new steady-state distribution. This assumption means the velocity of propagation of a pressure disturbance is infinite. Muskat (1937) shows that this assumption is reasonable for most situations, even though the system is large. Compared to the normal flow velocities, the actual velocity that a disturbance propagates is relatively very large. Also assume change in mass flux, change in boundary conditions is small compared to the steady-state flux. This means there is no attenuation of a disturbance, a reasonable assumption for most systems. The equation of condition for flow of a compressible fluid, Eq. (3.29) is

$$\phi \mu c_\ell \frac{\partial \rho}{\partial t} = \underline{\nabla} \cdot \underline{\underline{k}} \, \underline{\nabla} \rho \tag{3.155}$$

If the density is not time dependent,

$$\underline{\nabla} \cdot \underline{\underline{k}} \, \underline{\nabla} \rho = 0 \tag{3.156}$$

If the porous medium is isotropic,

$$\nabla^2 \rho = 0 \tag{3.157}$$

Thus the steady-state results discussed in Secs. 3.2 and 3.3 can be reinterpreted by replacing the potential with density. In order to use the results from Secs. 3.2 and 3.3 we need to know density as a function of position. For a linear system with densities ρ_1 and ρ_2 at the boundaries,

$$\rho = \rho_2 + \frac{1}{L}(\rho_1 - \rho_2)x \tag{3.158}$$

where L is the length of the system. Assume an equation of state, such as Eq. (3.30),

$$\rho = \rho_0 e^{c_\ell p}$$

(3.159)

Substituting Eq. (3.159) into Eq. (3.158) gives the pressure distribution of the system

$$p = \frac{1}{c_\ell} \ln\left[\frac{1}{L}(e^{c_\ell p_1} - e^{c_\ell p_2})x + e^{c_\ell p_2}\right]$$

(3.160)

The flow rate per unit area is

$$Q = -\frac{k}{\mu c_\ell L}(p_1 - p_2)$$

(3.161)

For a radial system as discussed in Section 3.3 and sketched in Fig. 3.12 with the densities at the boundaries of ρ_w and ρ_e the analog of Eq. (3.125) is

$$\rho = \rho_w + \frac{\rho_e - \rho_w}{\ln(r_e/r_w)}\ln\frac{r}{r_w}$$

(3.162)

The pressure distribution for the radial system using Eq. (3.159) is

$$p = \frac{1}{c_\ell}\ln\ (e^{c_\ell p_e} - e^{c_\ell p_w})\frac{\ln(r/r_w)}{\ln(r_e/r_w)} + e^{c_\ell p_w}$$

(3.163)

the flow rate for the system of thickness h is

$$Q = \frac{2\pi kh}{c_\ell \mu}\frac{p_e - p_w}{\ln(r_e/r_w)}$$

(3.164)

The equation of condition for flow of an ideal gas is

$$\phi\mu\frac{\partial p}{\partial t} = \underline{\nabla}\cdot\underline{\underline{k}}\ \underline{\nabla}p^2$$

(3.165)

For steady state in an isotropic medium

$$\nabla^2 p^2 = 0$$

(3.166)

From the ideal gas law, ρ is proportional to p; therefore, in terms of density

$$\nabla^2 \rho^2 = 0$$

(3.167)

The pressure distribution for a linear system is

$$p^2 = \frac{1}{L}(p_1^2 - p_2^2)x + p_2^2$$

(3.168)

The flow rate per unit area is

$$Q = -\frac{k\rho}{2\mu L}(p_2^2 - p_1^2) \qquad (3.169)$$

The pressure distribution for the ideal system of Fig. 3.12 is

$$p^2 = p_w^2 + \frac{p_e^2 - p_w^2}{\ln(r_e/r_w)} \ln \frac{r}{r_w} \qquad (3.170)$$

The flow rate for a system of constant thickness h is

$$Q = \frac{\pi k h \rho}{\mu} \frac{p_e^2 - p_w^2}{\ln(r_e/r_w)} \qquad (3.171)$$

To determine the pressure distribution and flow rate for real gases an appropriate equation of state is used in place of the ideal-gas equation.

For systems that are heterogeneous and anisotropic $\underline{\underline{k}} = \underline{\underline{k}}(x, y, z)$ in Eq. (3.156) the approach for a purely anisotropic system is to use a transformation to change Eq. (3.156) to Laplace's equation. For heterogeneous systems the solution for real situations is normally numerical.

3.5 TRANSIENT FLOW

Transient flow is described by including time as an independent variable in the equations used to model flow. The equation of condition for a compressible fluid is given by

$$\phi\mu \frac{\partial \rho}{\partial t} = \nabla \cdot \underline{\underline{k}} \nabla \rho \qquad (3.172)$$

The variation in pressure and flux in a compressible system includes the time variation of density. For slightly compressible systems Eq. (3.32) is used with pressure the dependent variable

$$\phi\mu c_\ell \frac{\partial p}{\partial t} = \nabla \cdot \underline{\underline{k}} \nabla p \qquad (3.173)$$

Generally, Eq. (3.24) or Eq. (3.172) correctly describe transient flow, but for most situations of practical interest we use Eq. (3.173) and its solutions to model transient flow. For transient flow of highly compressible fluids (gases), we consider variation of density and pressure through appropriate equations of state. It may also be necessary to include the variation of viscosity.

Equations (3.172) and (3.173) and their simplified forms and boundary conditions that follow are variations of the diffusion equation. This equation and its solutions have been used extensively in the study of the transport of momentum, heat, and mass. Solutions for the equations for physical situations analogous to those discussed here may be found in books and the scientific literature. (See, for example, Crank, 1956; Carslaw and Jaeger, 1959; Bird et al., 1960.)

For a linear system of semi-infinite length, Eq. (3.173) becomes

$$\frac{\partial p}{\partial t} = \frac{k}{\phi \mu c_\ell} \frac{\partial^2 p}{\partial x^2} \qquad (3.174)$$

If we let $\alpha = \dfrac{k}{\phi \mu c_\ell}$, then

$$\frac{\partial p}{\partial t} = \alpha \frac{\partial^2 p}{\partial x^2} \qquad (3.175)$$

Equation (3.175) is analogous to diffusion of mass in one dimension, conduction of heat in one dimension, and transport of momentum in one dimension.

Consider a porous medium occupying the semi-infinite space from $x = 0$ to $x = \infty$. The system is initially at pressure p_0. At time $t = 0$ the pressure is raised to p_1 on the face of the porous medium at $x = 0$, and maintained at p_1 for all times greater than zero. What is the time-dependent pressure $p(x, t)$? The problem just posed is a classic problem in transport phenomena. By analogy with the solution to the transport of momentum, heat, and mass, we write Eq. (3.175) in terms of the dimensionless pressure $p^* = (p - p_0)/(p_1 - p_0)$, then the initial conditions and the boundary conditions are

$$p^*(x, 0) = 0 \qquad \text{all } x \qquad\qquad (3.176)$$
$$p^*(0, t) = 1 \qquad t > 0 \qquad\qquad (3.177)$$
$$p^*(\infty, t) = 0 \qquad t > 0 \qquad\qquad (3.178)$$

Analogously, we let $\eta = x/\sqrt{4\alpha t}$ and rewrite Eq. (3.175) in terms of p^* and η

$$\frac{d^2 p^*}{d\eta^2} + 2\eta \frac{dp^*}{d\eta} = 0 \qquad (3.179)$$

with initial and boundary conditions

$$p^*(0) = 1 \qquad\qquad (3.180)$$
$$p^*(\eta) = \infty \qquad\qquad (3.181)$$

The solution of Eq. (3.179) with boundary conditions (3.180) and (3.181) is

$$p^* = 1 - \frac{2}{\sqrt{\eta}} \int_0^\eta e^{-\eta^2} \, d\eta \qquad (3.182)$$

or

$$\frac{p - p_0}{p_1 - p_0} = 1 - \operatorname{erf} \frac{x}{\sqrt{4\alpha t}} \qquad (3.183)$$

For an isotropic radial system of thickness h, the equation of condition, Eq. (3.175), is

$$\frac{1}{\alpha}\frac{\partial p}{\partial t} = \frac{\partial^2 p}{\partial r^2} + \frac{1}{r}\frac{\partial p}{\partial r} \qquad (3.184)$$

Solutions to this equation are numerous, especially for reservoir situations. Earlougher (1977) writes a generalized solution to Eq. (3.184) in terms of the initial pressure p_i, for a reservoir of porous rock with a single well with source q as

$$p_i - p(t, r) = 141.2\frac{qB\mu}{kh}[p_D(t_D, r_D, C_D, \text{geometry}, \ldots) + s] \qquad (3.185)$$

where B is the formation volume factor, p_D is the dimensionless pressure solution to Eq. (3.184) for appropriate boundary conditions, C_D is a wellbore storage constant, s is the skin effect. r_D is a dimensionless radius defined by

$$r_D = \frac{r}{r_w} \qquad (3.186)$$

where r_w is the well radius. t_D is the dimensionless time based on wellbore radius

$$t_D = \frac{0.0002637\,kt}{\phi\mu c_t r_w^2} \qquad (3.187)$$

sometimes t_{DA}, a dimensionless time based on total drainage area, is used.

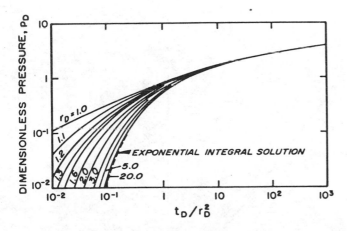

FIG. 3.18 Dimensionless pressure function at various dimensionless distances from a well located in an infinite system. (From T. D. Mueller and P. A. Witherspoon, Pressure interference effects within reservoirs and aquifers, Pet. Trans. AIME 234, SPEJ 471. © 1965 SPE–AIME.)

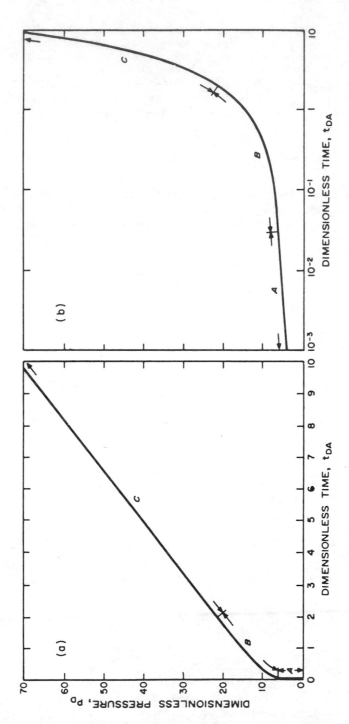

FIG. 3.19 Transient flow regimes: (a) infinite acting; (b) transition; (c) pseudo–steady state. 4:1 rectangle with the well at x/L = 0.75, y/W = 0.5. (From Erlougher and Ramey, 1973.)

$$t_{DA} = \frac{0.0002637\,kt}{\phi\mu c_t A} = t_D \frac{r_w^2}{A} \tag{3.188}$$

In Eqs. (3.187) and (3.188), c_t is total compressibility (including the small effect of rock compressibility). Figure 3.18 is a plot of p_D versus t_D/r_D^2 for a single source in an infite radial system. Figure 3.19 is a plot of p_D versus t_{DA} for a 4:1 rectangle with well at $1/L = 0.75$ and $y/W = 0.5$.

Examine the solution of Eq. (3.184) for the case of an infinite radial system. Assume at a point the pressure increases or declines (flow into or flow out of) over a line of length h at a point in the medium. The initial condition and boundary conditions are similar to those for the linear system of semi-infinite length

$$p(r,\ 0) = p_0 \qquad \text{all } r \tag{3.189}$$

$$q(0,\ t) = \text{constant} \qquad t > 0 \tag{3.190}$$

$$p(\infty,\ t) = p_0 \qquad t > 0 \tag{3.191}$$

The solution to Eq. (3.184) with these initial and boundary conditions is

$$p - p_0 = \frac{q\mu}{4\pi kh}\left[-\text{Ei}\left(-\frac{r^2}{4\alpha t}\right)\right] \tag{3.192}$$

where the Ei function or exponential integral is defined

$$-\text{Ei}\,(-x) = \int_x^\infty \frac{e^{-\tau}}{\tau}\,d\tau \tag{3.193}$$

and is tabulated for values of x. The curve labeled exponential integral in Fig. 3.18 is this solution. This solution, Eq. (3.192), is satisfactory for use with sources (−q) and sinks (+q) of finite radius as long as

$$\frac{r^2}{4\alpha t} > 100 \tag{3.194}$$

so this solution is good for large systems or for short times.

The exponential integral solution is sometimes approximated with a logarithmic expression so that

$$p - p_0 = \frac{q\mu}{4\pi kh}\left(\ln\frac{4\alpha t}{r^2} + 0.80907\right) \tag{3.195}$$

The error in this approximation is about 2 percent. Since Eq. (3.184) is a linear differential equation, the effect of all point (line) sources and sinks may be superimposed to obtain solutions for multiple sources or sinks in a porous medium, or for successive changes in q (or p) at a given source or sink. Since n sources or sinks with q_i occurs at time $> t_i$,

$$p - p_0 = \frac{\mu}{4\pi kh}\sum_{i=1}^{n}(q_i - q_{i-1})\left[-\text{Ei}\left(-\frac{r^2}{4\alpha(t - t_i)}\right)\right] \tag{3.196}$$

As mentioned, Eq. (3.196) can be used with a single source (or sink) where q changes as a function of time.

If at time t_S a source is turned off, the fluid in the infinite porous medium will redistribute itself so as to establish a uniform pressure. The pressure as a function or r and t is represented by superposition of the source of strength q and the existing sink starting at t_S:

$$p - p_0 = -\frac{q\mu}{4\pi kh}\left[-Ei\left(-\frac{r^2}{4\alpha t}\right) + Ei\left(-\frac{r^2}{4\alpha(t - t_S)}\right)\right] \tag{3.197}$$

If we use a logarithmic approximation,

$$p - p_0 = -\frac{q\mu}{4\pi kh}\ln\frac{t}{t - t_S} \tag{3.198}$$

Define the period of time the source is turned off as $\Delta t = t - t_S$; then

$$p - p_0 = -\frac{q\mu}{4\pi kh}\ln\frac{t_S + \Delta t}{\Delta t} \tag{3.199}$$

is valid for large Δt. A plot of p versus $\ln[(t_S + \Delta t)/\Delta t]$ will yield a straight line of slope $-q\mu/4\pi kh$. If we know q, μ, and h, the value of k for the porous medium can be determined.

If the flow rate or pressure of a line source is varied as a function of time, the response at some radius r can be determined. Johnson et al. (1966) generated a series of pressure pulses at one well in a reservoir and observed the response in an adjacent well to determine the properties of the region between the wells. Figure 3.20 is a schematic of the resultant pulses and responses showing the time

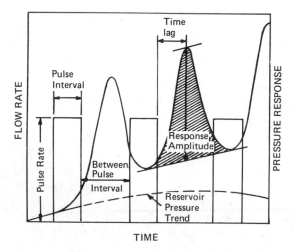

FIG. 3.20 Pulse-test terminology. (From C. R. Johnson, R. A. Greenkorn, and E. G. Woods, Pulse-testing: A new method for describing reservoir flow properties between wells, Pet. Trans. AIME 237, SPEJ 1599. © 1966 SPE-AIME.)

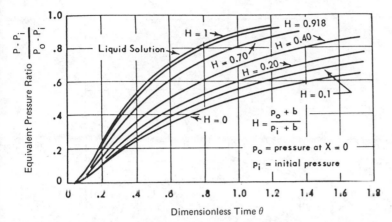

FIG. 3.21 Flow of an ideal gas in a linear system showing the pressure variation with time at the closed end of a tube when the pressure at the other end is suddenly reduced. (From Aronofsky, 1954.)

lag and response amplitude. The dimensionless time lag $t_{DL} = t_L/\Delta t$ can be shown to be approximately

$$t_{DL} = \frac{r^2}{11.36\alpha\,\Delta t} \qquad\qquad (3.200)$$

The pulse response amplitude is

$$\Delta p_m = \frac{70.6qB}{T}(\bar{p}_m - \bar{p}_{tm}) \qquad\qquad (3.201)$$

where $T = kh/\mu$. $\bar{p}_m$ is from the solution of Eq. (3.196) at $\Delta t + t_L$; $\bar{p}_m$ is the pressure response observed at this time. Knowing response amplitude and time lag, we can determine α from Eq. (3.196).

The equation of condition for flow of an ideal gas is given by Eq. (3.35). Rewriting Eq. (3.35)

$$\frac{\phi\mu}{2p}\frac{\partial p^2}{\partial t} = \underline{\nabla}\cdot\underline{\underline{k}}\,\nabla p^2 \qquad\qquad (3.202)$$

where the Klinkenberg effect has not been included explicitly.

Aronofsky (1954) solved Eq. (3.202) for a linear isotropic system numerically. His results are shown in Fig. 3.21. Jenkins and Aronofsky (1953) solved Eq. (3.202) numerically for radial flow with a single constant rate source in an infinite radial system. Their results are shown in Fig. 3.22.

The effects of anisotropy and heterogeneity can be included. For the infinite two-dimensional radial case, anisotropy is included by writing

$$p - p_0 = -\frac{q\mu}{4\pi h\sqrt{k_{11}k_{22}}}\left\{-\mathrm{Ei}\left[-\frac{\phi\mu c_t}{4t}\left(\frac{(x-x')^2}{k_{11}} + \frac{(y-y')^2}{k_{22}}\right)\right]\right\} \qquad (3.203)$$

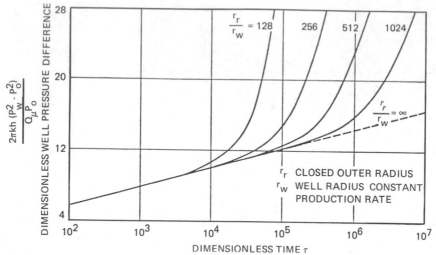

FIG. 3.22 Well-pressure change with time for various reservoir radii. (From Jenkins and Aronofsky, 1953.)

where k_{11} and k_{22} are the principal values of the permeability tensor when the axes coincide with the coordinates. x and y locate the observation point in the medium. (x', y') is the location of the source (or sink). The effect of heterogeneity is included in the solution of Eq. (3.173) by keeping $\underline{\underline{k}}$ inside the derivative. In this case we need to know $\underline{\underline{k}}$ as a function of position. Solutions for heterogeneous media are usually numerical.

3.6 INERTIAL EFFECTS: WAVE PROPAGATION

We assumed inertial effects are negligible in arriving at the various steady- and unsteady-state models in Sec. 3.1. However, inertia is included in the volume averaged equation Eq. (3.16). There are flow circumstances where inertial effects may be present, such as high velocity gas flow or in transient flow. It may also be possible to use wave propagation to obtain estimates of the permeability of a medium (Pascal, 1969).

Recall Eq. (3.16) and obtain the equations of change including inertial effects. Assume gravity is negligible (or contained in the potential term) and ignore viscous drag on the impermeable boundaries of the medium. The volume-averaged equation of motion for an isotropic medium

$$\underline{\nabla} p + K^{-1} \underline{\nabla} \langle \underline{v} \rangle + \rho \lambda \langle \underline{v} \rangle \langle \underline{v} \rangle = 0 \tag{3.204}$$

where K^{-1} is the conductance parameter and, for steady flow, λ is an inertial parameter. The inertial parameter λ accounts for small vibrations of the matrix and for the fact that the pressure gradient and pore velocity are not in the same direction (or in phase). Assume the equation of state is

$$(\rho - \rho_0) = \rho_0 c_\ell (p - p_0) \tag{3.205}$$

where the slight matrix compressibility is accounted for by assuming

$$c_f = a c_\ell \tag{3.206}$$

where a > 1. The equation of continuity for the medium is

$$\phi \frac{\partial \rho}{\partial t} + (\underline{\nabla} \cdot \rho \underline{q}) = 0 \tag{3.207}$$

Combining Eqs. (3.204), (3.205), and (3.207),

$$\frac{\rho_0 \lambda}{\phi} \frac{\partial^2 p}{\partial t^2} + \frac{1}{K} \frac{\partial p}{\partial t} = \frac{1}{\phi c_f} \nabla^2 p \tag{3.208}$$

Consider one dimension: for a plane wave,

$$p = p_0 e^{(i\omega t - \Gamma x)} \tag{3.209}$$

Substitute Eq. (3.209) into Eq. (3.208), and solve for the propagation constant

$$\Gamma^2 = \rho_0 c_f \omega \left(-\lambda \omega + i \frac{\phi}{\rho_0 k} \right) \tag{3.210}$$

Geertsma (1974) defines a "Reynolds" number, $\hat{Re}$ for flow in porous media in terms of the inertial parameter

$$\hat{Re} = \frac{\lambda \rho q}{K^{-1}} \tag{3.211}$$

This Reynolds number represents the ratio of inertial to viscous forces and may be used to describe the upper limit of Darcy's law. Figure 3.23 shows the correlation of friction factor for all porous media versus Re. The straight-line portion of the curve is represented by

$$\frac{k \Delta p}{L \mu q} = 1 + \hat{Re} \tag{3.212}$$

or

$$\frac{k \Delta p / L \mu q}{\hat{Re}} = \frac{1}{\hat{Re}} + 1 = -H \tag{3.213}$$

For an ideal gas we use MK $\Delta p^2 / 2LRT\mu$ on the left-hand side of Eq. (3.212). We need to know λ to use the correlation of Fig. 3.23. Smith and Greenkorn (1972) imagined the porous medium replaced by a bundle of capillaries to model wave propagation. The equation of motion for a circular capillary is

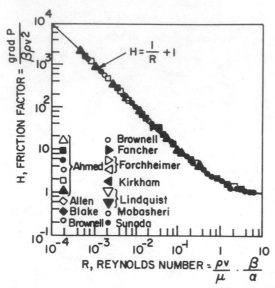

FIG. 3.23 Graphical correlation of friction factor versus Reynolds number. (From N. Ahmed and D. K. Sunada, Nonlinear flow in porous media, Proc. ASCE J. Hydraulic Div. 95, HY 6, 1847. © 1969 American Society of Civil Engineers.)

$$\rho_0 \frac{\partial \bar{v}}{\partial t} = -\frac{\tau_w}{R} - \frac{\partial p}{\partial x} \tag{3.214}$$

where $\bar{v}$ is an area average velocity. If we assume

$$\bar{v} = \frac{\lambda}{\phi} \langle v \rangle \tag{3.215}$$

Eq. (3.214) becomes

$$\frac{\rho_0 \lambda}{\phi} \frac{\partial \langle v \rangle}{\partial t} = -\frac{2\tau_w}{R} - \frac{\partial p}{\partial x} \tag{3.216}$$

Combining Eqs. (3.208) and (3.216),

$$K = \frac{\phi R}{2\lambda} \frac{\bar{v}}{\tau_w} \tag{3.217}$$

For steady state,

$$\frac{\bar{v}}{\tau_w} = \frac{K}{4} \tag{3.218}$$

and

$$K_0 = \frac{\phi R^2}{8\mu\lambda} \tag{3.219}$$

If we assume all variables are sinusoidal functions of time, then for the dynamic situation (Biot, 1956)

$$\frac{\bar{v}}{\tau_w} = \frac{R}{4\mu} S\{i\delta_k\} \tag{3.220}$$

where

$$S\{i\delta_k\} = \frac{4 + i(8T\{ik\}/\delta_k)}{\delta_k T\{ik\}} \tag{3.221}$$

$$T\{i\delta_k\} = \frac{ber'\delta_k + i\ bei'\delta_k}{ber\ \delta_k + i\ bei\ \delta_k} \tag{3.222}$$

where prime means differentiation. The dimensionless frequency parameter is defined as

$$\delta_k = R\left(\frac{\omega\rho_0}{\mu}\right)^{\frac{1}{2}} \tag{3.223}$$

Biot (1956) shows that irregular pore shape can be represented by defining an effective frequency parameter δ_{ek} by

$$\delta_{ek} = \frac{\delta_k}{n} \tag{3.224}$$

where n is a shape factor. Using Eq. (3.219) as a definition of K, in general

$$K = K_0 S\{i\ \delta_{ek}\} \tag{3.225}$$

Zwikker and Kosten (1949) derived an expression for compressibility as a function of frequency similar to the above

$$c_f = \frac{c_I}{\gamma} + i\ \frac{2(1 - \gamma)\ T\{i\ \delta_t\}}{\gamma\ \delta_t} \tag{3.226}$$

where c_I is isothermal compressibility and γ is the heat capacity ratio. The frequency parameter is defined as

$$\delta_t = R\left(\frac{\omega r}{\nu_t}\right)^{\frac{1}{2}} \tag{3.227}$$

where

$$\nu_t = \frac{\hat{k}}{\rho_0 C_V} \tag{3.228}$$

$\hat{k}$ is thermal conductivity, and C_V is the heat capacity at constant volume. An effective frequency parameter is defined as

$$\delta_{et} = \frac{\delta_t}{n} \tag{3.229}$$

Since

$$\delta_{et} = \frac{1}{n}\left(\frac{8\mu_0 \lambda r k_0 \omega}{\phi \nu_t}\right)^{\frac{1}{2}} \tag{3.230}$$

then

$$\delta_{et} = Pr^{\frac{1}{2}} \delta_{et} \tag{3.231}$$

The Prandtl number is defined as

$$P_r = \frac{C_p \mu}{\hat{k}} \tag{3.232}$$

Writing c_f in terms of δ_{et},

$$c_f = \frac{c_I}{\gamma} + i \frac{2(1-\gamma) T\{i\delta_{et}\}}{\gamma \delta_{et}} \tag{3.233}$$

Smith and Greenkorn (1972) determined expressions for the attenuation and phase factors α and β in the expression:

$$\Gamma(iw) = \alpha + i\beta \tag{3.234}$$

Since

$$\Gamma^2 = \Gamma^2(k, c_f) = \Gamma^2(\delta_{ek}, \delta_{et}) \tag{3.235}$$

then

$$\Gamma(iw) = \alpha(\delta_{ek}, \delta_{et}) + i\beta(\delta_{ek}, \delta_{et}) \tag{3.236}$$

Generally, for low frequencies and/or low permeability, $\delta_{ek} \ll 1$, $\delta_{et} \ll 1$, and

$$K \sim K_0 \tag{3.237}$$

$$c_f \sim c_I \tag{3.238}$$

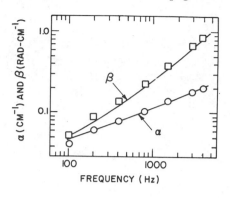

FIG. 3.24 Data compared to theory with $\ell = 1.05$ and $n = 0.5$: $\phi = 0.97$, $K_0 = 0.116$ cm^3 sec g^{-1}. (From Smith and Greenkorn, 1972; Nakamura, 1959.)

For high frequencies and/or high permeabilities, $\delta_{ek} \gg 1$, $\delta_{et} \gg 1$ and

$$K \sim \frac{1}{\delta_{ek}} \tag{3.239}$$

$$c_f \sim \frac{1}{\delta_{et}} \tag{3.240}$$

$$\alpha \sim n \tag{3.241}$$

$$\beta \sim \lambda n \tag{3.242}$$

Figure 3.24 shows a comparison of calculations for the theory of Smith and Greenkorn (1972) with the data of Nakamura (1959).

Smith et al. (1974a) presented experimental measurements and compared them with the above theory. Figure 3.25 shows the comparison for nitrogen in

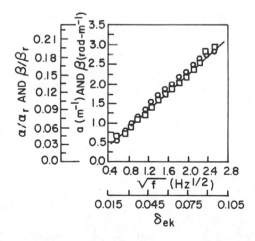

FIG. 3.25 Measured and predicted propagation characteristics of sample 1: o, experimental α results; □, experimental β results; –, theoretical α and β results. (From Smith et al., 1974a.)

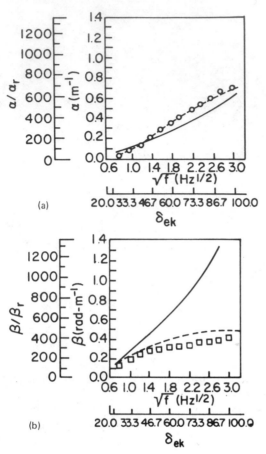

FIG. 3.26 (a) Measured and (b) predicted phase characteristics of sample 4. Experimental results: –, present theory; ---, first-order response theory with $\tau_1 = 0.125$ sec. (From Smith et al., 1974a.)

Ottawa sand. Figure 3.26 shows the comparison for water in Ottawa sand. A first-order response model for α and β where t_1 was determined from the data:

$$\alpha = \frac{\ln\left[(\omega t_1)^2 + 1\right]}{2L} \tag{3.243}$$

$$\beta = \tan^{-1}\frac{\omega t_1}{L} \tag{3.244}$$

fits the data quite well.

Equation (3.29) is used to describe small-amplitude transient flow in rigid media (Fatt, 1959; Johnson et al., 1966; Javendel and Witherspoon, 1968). In these studies as well as with conventional buildup and fall-off analysis an inertial effect may be present. Foster et al. (1967) applied a constant parameter model to study

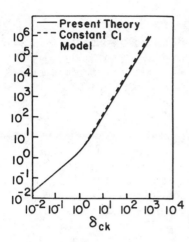

FIG. 3.27 Effects of a dynamic conductance and compressibility on the phase factor. (From Smith et al., 1974b.)

propagation velocity of pressure pulses. They obtained a very large range of possible values for the inertial parameter λ. The large range is questionable; their theory does not account for the effect of high-frequency attenuation.

Smith et al. (1974b) extended the theory of Smith and Greenkown (1972) to transient pressure response. The range of values of λ in their theory is between 1 and 10, corresponding to experiment. Figure 3.27 shows the effect of dynamic conductance and compressibility on the attenuation and phase factor. Figure 3.28 shows the effect of dynamic conductance and compressibility on response to a step increase. The results of this transient analysis show inertial effects exist for short distances and high permeabilities. One must be careful in interpreting pressure transient data for short times or high permeabilities to see if it is necessary to include inertial effects in interpreting such data.

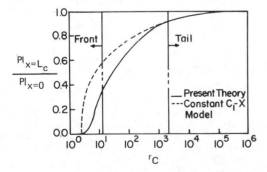

FIG. 3.28 Effects of a dynamic conductance and compressibility on a step response. (From Smith et al., 1974b.)

3.7 NON-NEWTONIAN FLOW

Rheologically complex flow behavior in porous media has received increasing attention since the 1960s because of the use of polymer solutions in enhanced oil recovery processes and for drilling fluids. Newton's law of viscosity in one dimension is

$$\tau_{yz} = -\mu \frac{dv_z}{dy} \tag{3.245}$$

Equation (3.245) relates the shear stress in the z direction on a fluid of constant surface y, to the velocity gradient. Newton's law of viscosity, Eq. (3.245), defines the momentum transport coefficient viscosity, μ. Fluids which behave according to Eq. (3.245) are called <u>Newtonian fluids</u>. A large number of fluids, such as polymer solutions, polymer melts, suspensions, slurries, emulsions, and pastes, do not follow Newton's law of viscosity. If we plot shear stress versus the negative velocity gradient for a Newtonian fluid, the result will be a straight line with slope μ and as curve a on Fig. 3.29. For fluids which do not follow Newton's law, a plot of shear stress versus the negative velocity gradient will plot in a variety of ways, including curves b, c, and d in Fig. 3.29.

The three types of non-Newtonian fluids represented by curves b, c, and d on Fig. 3.29 are classified as follows.

Curve c of Fig. 3.29 represents a Bingham plastic. Such a fluid does not flow until the applied stress reaches a yield stress τ_0. After reaching the yield stress, the plot of shear stress versus the negative velocity gradient is linear. A Bingham plastic is described by

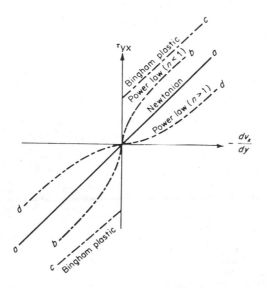

FIG. 3.29   Newtonian and non-Newtonian behavior. (From Bird et al., 1960.)

$$\tau_{yz} = \tau_0 - \mu_p \frac{dv_z}{dy}$$

(3.246)

μ_p is called the plastic viscosity. Toothpaste is a typical Bingham plastic.

Curve b of Fig. 3.29 represents a pseudoplastic fluid. A pseudoplastic fluid flows in such a way that the apparent viscosity decreases as the velocity gradient increases. A power law model is used to describe pseudoplastic behavior

$$\tau_{yz} = -m \left| \frac{dv_z}{dy} \right|^{n-1} \frac{dv_z}{dy}$$

(3.247)

where n < 1. When n = 1, m is identical to μ, and Eq. (3.247) is identical to Eq. (3.245). Polymer solutions are pseudoplastic.

A dilatant fluid is represented by curve d on Fig. 3.29. A dilatant fluid flows in such a way that the apparent viscosity increases as velocity gradient increases. The power law model, Eq. (3.247), describes a dilatant fluid using n > 1. Suspensions, such as drilling muds, are dilatant fluids.

Fluids that have a stress recovery (recoil) when shear is relaxed are called viscoelastic. Such fluids have both elastic and viscous behavior. A simple model of viscoelastic behavior includes an elastic constant:

$$\tau_{yz} + \lambda \frac{\tau_{yz}}{dt} = -m \left| \frac{dv_z}{dy} \right|^{n-1} \frac{dv_z}{dy}$$

(3.248)

Rubber cement is viscoelastic.

There are many other types of complex fluid behavior including time-dependent behavior. Thixotropic fluids have a decrease in apparent viscosity with constant stress; rheopectic fluids have an increase in apparent viscosity with constant stress.

Savings (1970) in a review of non-Newtonian flow in porous media identifies three major approaches to describing such flow: (1) A generalized method which adapts Darcy's law to non-Newtonian fluids without using a specific rheological model; (2) coupling of a capillary or hydraulic radius model of a porous medium with a function relationship between shear stress and shear rate to describe flow; (3) using a simple fluid concept in applying dimensional analysis for scaling non-Newtonian flow.

McKinley et al. (1966) developed a generalized model by direct analogy with results obtained for flow of a non-Newtonian fluid through a uniform capillary. This generalized model is valid over the entire range of shear stress behavior. The shear stress for steady flow of fluid in a uniform capillary at radius r is

$$\sigma\{r\} = -\frac{r}{2}\frac{dp}{dr}$$

(3.249)

from conservation of momentum

$$\sigma\{r\} = \psi_1 \frac{dv}{dr}$$

(3.250)

where ψ_1 is an unknown function of $(dv/dr)^2$. Combining Eqs. (3.249) and (3.250)

$$\psi_1\left[\left(\frac{dv}{dr}\right)^2\right]\frac{dv}{dr} = -\frac{r}{2}\frac{\Delta P}{L} \qquad (3.251)$$

If ψ_1 is assumed constant and equal to viscosity μ, then

$$Q = \frac{\pi R^4}{8\mu}\frac{\Delta p}{L} \qquad (3.252)$$

which is the Hagen-Poisuelle law. For an arbitrary ψ_1

$$Q = \frac{\pi R^4}{8}\frac{\Delta p}{L}\int_0^1 \frac{4(r/R)}{\psi_1}\,d\left(\frac{r}{R}\right) = \frac{\pi R^4}{8\mu\{\sigma_R\}}\frac{\Delta p}{L} \qquad (3.253)$$

$\mu\{\sigma_R\}$ is an apparent viscosity at the shear stress of the capillary wall.

$$\sigma_R = \frac{R\Delta p}{2L} \qquad (3.254)$$

Define a dimensionless viscosity ratio,

$$F(\sigma_R) = \frac{\mu_0}{\mu\{\sigma_R\}} \qquad (3.255)$$

where μ_0 is apparent viscosity at some reference stress σ_0. Writing Eq. (3.253) including F, Eq. (3.255)

$$Q = F(\sigma_R)\frac{\pi R^4}{8\mu_0}\frac{\Delta p}{L} \qquad (3.256)$$

Dividing Eq. (3.256) by the area of the capillary,

$$\frac{Q}{A} = F\left(\frac{R\Delta p}{2L}\right)\frac{R^2}{8\mu_0}\frac{\Delta p}{L} \qquad (3.257)$$

Use the Dupuit-Forchheimer relation with $R = (k/\phi)^{\frac{1}{2}}$ for the porous medium

$$q = F\left(\frac{R_{eq}\,\Delta p}{2\tau L}\right)\frac{k}{\mu_0}\frac{\Delta p}{L} \qquad (3.258)$$

and

$$\frac{R_{eq}}{2\tau} = \alpha_0\left(\frac{k}{\phi}\right)^{\frac{1}{2}} \qquad (3.259)$$

where τ is tortuosity and α_0 is a constant for a particular type of porous medium; α_0 reflects the pore size distribution. Defining a shear stress for the porous medium σ_k,

$$\sigma_k = \alpha_0 \left(\frac{k}{\phi}\right)^{\frac{1}{2}} \frac{\Delta p}{L} \qquad (3.260)$$

then the modified form of Darcy's law for a non-Newtonian fluid is

$$q = F(\sigma_k) \frac{k}{\mu_0} \frac{\Delta p}{L} \qquad (3.261)$$

In terms of the apparent viscosity

$$q = \frac{k}{\mu(\sigma_k)} \frac{\Delta p}{L} \qquad (3.262)$$

According to the model, the functions F in Eq. (3.256) and in Eq. (3.261) are the same function. Thus if a Newtonian fluid is used to determine the permeability of a porous medium, then Eq. (3.256) and (3.261) represent two independent expressions for determining μ versus σ for non-Newtonian fluid; they differ only in the scale of the shear stress σ. By proper choice of α_0, the two rheograms of μ versus σ should superimpose.

Figure 3.30 shows the experimental results of McKinley et al. (1966) using a 1.5% solution of Dextran (a soluble polysaccharide) in brine solution which flowed through both a capillary rheometer and a linear sandstone core. The data from the two experiments are superimposed using α_0. Figure 3.31 shows the results of using this value of α_0 and the 1.5% Dextran-brine solution to compute the permeability of several sandstone cores (of the same type) but with varying permeability (nonuniform).

The Blake-Kozeny equation has been coupled with rheologic models to describe non-Newtonian fluid in porous media. The Blake-Kozeny model may be written

$$f_p = \frac{150}{Re_p} \qquad (3.263)$$

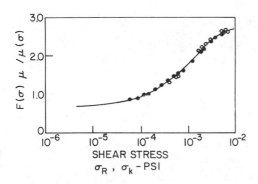

FIG. 3.30 Superposition of capillary and core rheograms. $\mu = 9.60$ cP; solid line, capillary data: $\mu = R\,\Delta p/2L$; solid circle, core data: $\sigma = a\,\sqrt{h/t}\;\Delta p/L$, $k = 0.20$ darcys, $a = 0.00240 \pm 0.00076$; open circle, additional core data for $= H.1$ cP. (From McKinley et al., 1966.)

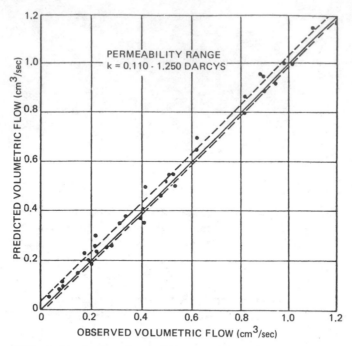

FIG. 3.31 Comparison of theory with routine measurements of the flow rate of 1.5 wt % Dextran through sandstone cores of various permeabilities. (From McKinley et al., 1966.)

In terms of the equivalent particle diameter

$$v_\infty = \frac{\Delta p}{L} \frac{D_p^2 \phi^3}{150\mu(1 - \phi)^2}$$

(3.264)

The permeability from the Blake-Kozeny model is

$$k = \frac{D_p^2 \phi^3}{150(1 - \phi)^2}$$

(3.265)

Christopher and Middleman (1965) combined a power law model of a non-Newtonian fluid with the Blake-Kozeny model for the medium. The equation for flow of a power law fluid in a capillary is

$$\langle v \rangle = \frac{n}{3n + 1} D^{1+1/n} \left(\frac{\Delta p}{2mL}\right)^{1/n}$$

(3.266)

Including the hydraulic radius and using a tortuosity of 25/12 for a porous medium, the velocity of approach is

$$\langle v_\infty \rangle = \frac{n\phi}{3n+1} \left[\frac{D_p \phi}{3(1-\phi)} \right]^{1+1/n} \left(\frac{6\,\Delta p}{25\text{mL}} \right)^{1/n} \tag{3.267}$$

Christopher and Middleman (1965) defined a non-Newtonian bed factor

$$H = \frac{m}{12} \left(9 + \frac{3}{n} \right)^n (150 k \phi)^{(1-n)/2} \tag{3.268}$$

to represent a modified Darcy's law for flow of a power law fluid in a porous medium

$$v_\infty = \left(\frac{k}{H} \frac{\Delta p}{L} \right)^{1/n} \tag{3.269}$$

Figure 3.32 shows a test of Eq. (3.269).

Sadowski and Bird (1965) used an Ellis model to modify the Blake-Kozeny equation. The Ellis model is

$$\frac{1}{\eta_{\text{eff}}} = \frac{1}{\eta_0} \left[1 + \left(\frac{\tau}{\tau_{\frac{1}{2}}} \right)^{\alpha-1} \right] \tag{3.270}$$

where η_0 is the lower limiting viscosity and $\tau_{\frac{1}{2}}$ is the shear stress at which the non-Newtonian viscosity has dropped to $(1/2)\eta_0$. The Ellis model is a superposition

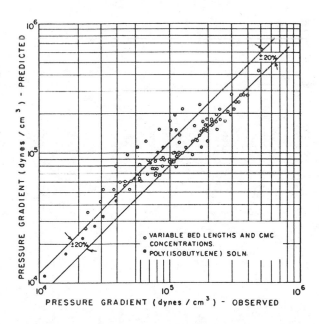

FIG. 3.32 Test of Christopher-Middleman capillary model for power law flow through porous media. [Reprinted with permission from R. H. Christopher and S. Middleman, Power law flow through a packed tube, Ind. Eng. Chem. Fund. 4, 422 (1965). Copyright by the American Chemical Society.]

of Newton's law of viscosity and the power law model. Including $\tau_{\frac{1}{2}}$ as a parameter allows the Ellis model to account for viscoelastic effects. The equation for flow of an Ellis fluid in a capillary is

$$\langle v \rangle = \frac{R\tau_R}{4\eta_0} \left[1 + \frac{4}{\alpha + 3} \left(\frac{\tau_R}{\tau_{\frac{1}{2}}} \right) \right]^{\alpha - 1} \tag{3.271}$$

Introducing the hydraulic radius

$$v_\infty = \frac{D_p \phi^2 \tau_{R_H}}{12\eta_0(1 - \phi)} \left[1 + \frac{4}{\alpha + 3} \left(\frac{\tau_{R_H}}{\tau_{\frac{1}{2}}} \right) \right]^{\alpha - 1} \tag{3.272}$$

where

$$\tau_{R_H} = \frac{R_H \Delta p}{L} \tag{3.273}$$

Defining the effective non–Newtonian viscosity as

$$\frac{1}{\eta_{eff}} = \frac{1}{\eta_0} \left[1 + \frac{4}{\alpha + 3} \left(\frac{\tau_{R_H}}{\tau_{\frac{1}{2}}} \right) \right]^{\alpha - 1} \tag{3.274}$$

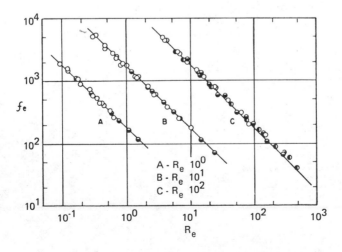

FIG. 3.33 Reynolds number versus friction factor for Ellis fluid flow through bead packs. Fluid systems containing different concentrations of (A) polyethylene glycol, (B) polyvinyl alcohol, and (C) hydroxyethylcellulose. (From Sadowski and Bird, 1965.)

The Darcy law equivalent for an Ellis fluid is

$$v_\infty = \frac{k}{\eta_{eff}} \frac{\Delta p}{L} \tag{3.275}$$

Figure 3.33 shows the test of Eq. (3.275) in terms of friction factor defined as

$$f_e = \frac{k\phi^2 \Delta p}{\rho v_\infty^2 L} \tag{3.276}$$

Gaitonde and Middleman (1967) extended the earlier work of Christopher and Middleman (1965) to elastic fluids to see if viscoelastic effects were significant during flow in a porous medium. In their experimental work they did not locate any significant effects. Marshall and Metzner (1967) show the Deborah number, the ratio of duration of fluid memory θ_f to duration of the deformation θ_p,

$$De = \frac{\theta_f}{\theta_p} \tag{3.277}$$

may be used to describe transition from viscous to viscoelastic flow. The Deborah number for a porous medium is

$$De_p = \frac{\theta_f v_\infty}{\phi D_p} \tag{3.278}$$

Sadowski and Bird (1965) use the Ellis number, which is

$$El_p = \frac{\eta_0}{\tau_{\frac{1}{2}}} \frac{v_\infty}{D_p} \tag{3.279}$$

to describe viscoelastic effects. Marshall and Metzner (1967) found that for $De_p < 0.05$ the power law model describes non-Newtonian flow for a fluid that may be viscoelastic. However if De_p is greater than 0.1, the power law model shows significant deviations. The data of Gaitonde and Middleman (1967) were for $De_p < 0.1$. Savins (1970) used the generalized method of McKinley et al. (1966) to encompass all the results discussed to this point.

Slattery (1967) obtained the equation of flow in a porous medium by volume averaging and obtained a generalized Darcy's law.

$$\underline{\nabla}(\overline{\mathscr{P}} - p_0) + K\underline{v} = 0 \tag{3.280}$$

The resistance factor K should be a function of characteristic length of the system L, the magnitude of the characteristic velocity $|v|$, and the material constants for a Noll simple fluid, where μ_0 is the characteristic viscosity and t_0 is the characteristic time.

$$K = f(L, |v|, \mu_0, t_0)$$ (3.281)

Applying dimensional analysis to Eq. (3.281):

$$k* = f_1\left(\frac{|v| t_0}{L}\right)$$ (3.282)

where

$$k* = \frac{KL}{\mu_0}$$ (3.283)

Alternatively, if we assume K is a function of the magnitude of $\underline{\nabla}(\mathscr{P} - p$

$$K = g(L, |\underline{\nabla}(\overline{\mathscr{P} - p_0})| \mu_0, t_0)$$ (3.284)

and

$$k* = g_1\left(\frac{|\underline{\nabla}(\mathscr{P} - p_0)| Lt}{\mu_0}\right)$$ (3.285)

We can correlate data for similar geometric situations for flow of a non-New-tonian fluid in a porous medium using Eq.s. (3.284) or (3.285). The data should correlate as

$$L^2 K = h_1\left(\frac{|v|}{L}\right)$$ (3.286)

or

$$L^2 K = h_2(L | \underline{\nabla}(\overline{\mathscr{P} - p_0})|)$$ (3.287)

Assuming a capillary model for the porous medium, $L^2 = k/\phi$ and

$$L^2 K = \frac{\Delta p k}{v_\infty L}$$ (3.288)

so

$$\frac{|\underline{v}|}{L} = \frac{v_\infty}{\sqrt{k\phi}}$$ (3.289)

$$|\underline{\nabla}(\overline{\mathscr{P} - p_0})| = \frac{\Delta p}{L}\left(\frac{k}{\phi}\right)^{\frac{1}{2}}$$ (3.290)

Figure 3.34 shows the test of this approach using the data of Dauben (1966).

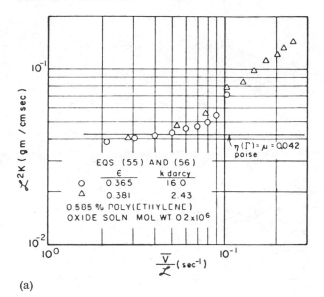

(a)

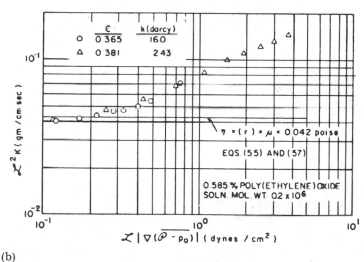

(b)

FIG. 3.34 (a) Slattery-type correlation of non-Newtonian flow through bead packs using data of Dauben (1966); and (b) Slattery-type correlation of Darcy viscosity of nonnewtonian flow through bead packs using data of Dauben (1966). (From J. G. Savins, in Flow through Porous Media, Richard J. Nunge (ed.), American Chemical Society, Washington, D.C., 1970.)

    All three methods of modeling non-Newtonian flow discussed above agree for specific cases.

    The power law equations for flow of a non-Newtonian fluid in steady-state linear or radial flow are summarized in the following: For linear flow,

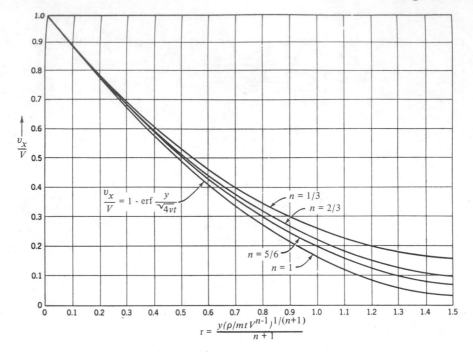

FIG. 3.35 Velocity distribution (in dimensionless form) for flow in the neighborhood of a wall suddenly set in motion. The curves are drawn for the power model with parameters m and n. For n = 1 (Newtonian flow), the variable r becomes $y/\sqrt{4vt}$. (From Bird et al., 1960.)

$$\log \frac{\Delta p}{q/A} = \log \frac{mL}{k} = n \log \frac{q}{A} \tag{3.291}$$

For radial flow ($-1 < n < 0$, r_w/r_e small),

$$\log \frac{\Delta p}{q} = \log \frac{-m\phi^{n/2}}{2^{n+1} nk(\pi h)^{n/2+1}} + \frac{n}{2} \log \frac{q^2}{Q} \tag{3.292}$$

The transient flow of a non-Newtonian fluid for a power law fluid for the linear and radial cases are briefly summarized below. For the linear case, the analogous equation for flow near a wall suddenly set in motion is discussed by Bird et al. (1960). Figure 3.35 shows the velocity distribution in dimensionless form for flow in the neighborhood of a wall suddenly set in motion with curves for a power law fluid. If we replace the ordinate in Fig. 3.35 with $p - p_0/p_1 - p_0$ and the abscissa with $x/\sqrt{4\alpha t}$ for a porous medium, then these curves represent the pressure for a power law fluid for the transient linear case.

Van Poolen and Jargon (1969) calculated the transient case for radial flow by solving Eq. (3.32) numerically with an apparent viscosity in each of the cells using a power law.

$$\mu_{app} = m\left(\frac{q}{A}\right)^n_{avg} \tag{3.293}$$

Figure 3.36 shows p_D versus t_D for a 5, 10, and 100 percent pore volume of a non-Newtonian fluid represented as a power law fluid. These curves do not reach the usual straight line relations encountered with non-Newtonian fluids.

Ikoku and Ramey (1978) use the power law model in the transient flow equations directly. For single-fluid, radial flow, the non-Newtonian analog of Darcy's law is

$$q^n = -\frac{k}{\mu_{eff}}\frac{\partial p}{\partial r} \tag{3.294}$$

where

$$\mu_{eff} = \frac{H}{12}\left(9 + \frac{3}{n}\right)^n (150\ k\phi)^{1-n/2} \tag{3.295}$$

$$H = \frac{\mu_{app}}{\gamma^{n-1}} \tag{3.296}$$

and γ is the shear rate. The equation of condition is

$$Gr^{1-n}\frac{\partial p}{\partial t} = \frac{\partial^2 p}{\partial r^2} + \frac{n}{r}\frac{\partial p}{\partial r} \tag{3.297}$$

where

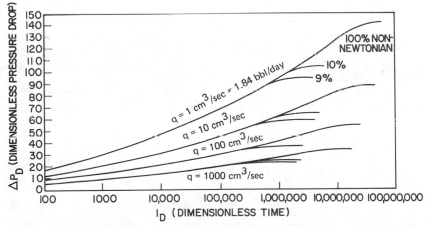

FIG. 3.36 Drawdown curves at various rates and 5, 10, and 100% PV of non-Newtonian fluid located at the well bore. (From H. K. van Poolen and J. R. Jargon, Steady-state and unsteady-state flow of non-Newtonian fluids through porous media, Pet. Trans. AIME 246, SPEJ 80. © 1969 SPE-AIME.)

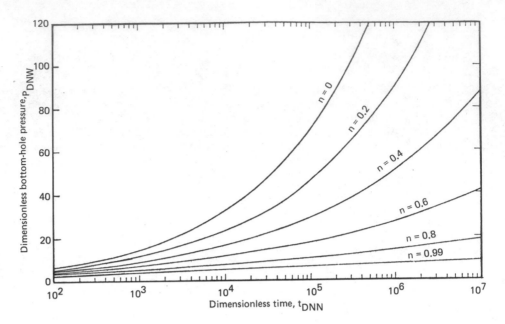

FIG. 3.37 Dimensionless bottom-hole pressure versus dimensionless time for pseudoplastic non-Newtonian fluids in an infinitely large system. (From C. U. Okoku and H. J. Ramey, Jr., Transient flow of non-Newtonian power-law fluids in porous media, SPE Meeting, SPE 7139. © 1978 SPE-AIME.)

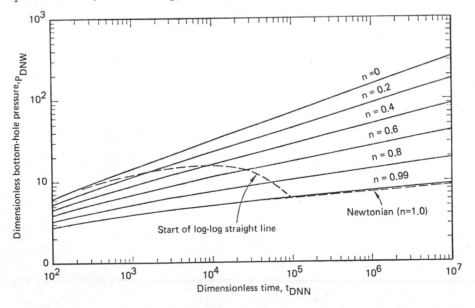

FIG. 3.38 Dimensionless bottom-hole pressure versus dimensionless time for pseudoplastic fluids in an initely large system. (From C. U. Okoku and H. J. Ramey, Jr., Transient flow of non-Newtonian power-law fluids in porous media, SPE Meeting, SPE 7139. © 1978 SPE-AIME.)

$$G = \frac{n\phi c_t \mu_{eff}}{k} \left(\frac{2\pi h}{q} \right)^{1-n} \tag{3.298}$$

The initial and boundary conditions are

$$p(r_1, 0) = p_i \qquad r > 0 \tag{3.299}$$

$$\left(-\frac{\partial p}{\partial r} \right)_{r_w} \left(\frac{q}{2\pi k r_w} \right)^n \frac{\mu_{eff}}{R} \qquad t > 0 \tag{3.300}$$

$$p(\infty, t) = p_i \qquad t > 0 \tag{3.301}$$

Figures 3.37 and 3.38 are solutions of Eq. (3.297) by Laplace transform and numerical inversion. These results are similar to those of van Poolen and Jargon (1969), since they are solving the same problem but using a different method.

REFERENCES

Ahmed, N., and D. K. Sunada (1969): Nonlinear Flow in Porous Media, Proc. ASCE J. Hydraulic Div. 95, HY6, 1847.

Aronofsky, J. S. (1954): Effect of Gas Slip on Unsteady Flow of Gas through Porous Media, J. Appl. Phys. 25, 48.

Bennion, D. W., and J. C. Griffiths (1966): Stochastic Model for Predicting Variations in Reservoir Rock Properties, Pet. Trans. AIME 237, 9.

Biot, M. A. (1956): Theory of the Propagation of Elastic Waves in a Fluid-Saturated Porous Solid II Higher Frequency Range, J. Acoust. Soc. Amer. 28, 179.

Bird, R. B., W. E. Stewart, and E. N. Lightfoot (1960): Transport Phenomena, Wiley, New York, p. 127.

Brinkman, H. C. (1947): A Calculation of the Viscous Force Exerted by a Flowing Fluid on a Dense Swarm of Particles, Appl. Sci. Res. A1, 27.

Crank, J. (1956): The Mathematics of Diffusion, Oxford University Press, Oxford.

Christopher, R. H., and S. Middleman (1965): Power Law Flow through a Packed Tube, Ind. Eng. Chem. Fund. 4, 422.

Csallany, S., and W. C. Walton (1963): Yields of Shallow Dolomite Wells in Northern Illinois, Ill. State Water Surv. Dept. Invest. 26.

Dauben, D. L. (1966): Flow of Polymer Solutions through Porous Media, Ph.D. Thesis, Univ. of Oklahoma, Norman.

Earlougher, R. C., Jr. and H. J. Ramey, Jr. (1973): Interference Analysis in Bounded Systems, J. Can. Pet. Tech., Oct.-Dec., 33.

Earlougher, R. C., Jr. (1977): Advances in Well Test Analysis, Monograph Vol. 5, Soc. Pet. Eng. AIME, New York.

Fatt, I. (1956a): The Network Model of Porous Media: I. Capillary Pressure Characteristics, Pet. Trans. AIME 207, 144.

Fatt, I. (1956b): The Network Model of Porous Media: II. Dynamic Properties of a Single Size Tube Network, Pet. Trans. AIME 207, 160.

Fatt, I. (1959): A Demonstration of the Effect of "Dead-End" Volume on Pressure Transients in Porous Media, Pet. Trans. AIME 216, 449.

Ferrandon, J. (1948): Les Lois de l'Écoulement die Filtration, Genie civl. 125(2), 24.

Foster, W. R., J. M. McMillen, and A. S. Odeh (1967): The Equations of Motion of Fluids in Porous Media: I. Propagation Velocity of Pressure Pulses, Pet. Trans. AIME 240, SPEJ 333.

Gaitonde, N. Y., and S. Middleman (1967): Flow of Viscoelastic Fluids through Porous Media, Ind. Eng. Chem. Fund. 6, 145.

Geertsma, J. (1974): Estimating the Coefficient of Inertial Resistance in Fluid Flow through Porous Media, Pet. Trans. AIME 257, SPEJ 445.

Greenkorn, R. A., C. R. Johnson, and L. K. Shallenberger (1964): Directional Permeability of Heterogeneous Anisotropic Porous Media, Pet. Trans. AIME 231, SPEJ 124.

Greenkorn, R. A., R. E. Haring, H. O. Jahns, and L. K. Shallenberger (1964): Flow in Heterogeneous Hele-Shaw Models, Pet. Trans. AIME 231, SPEJ 307.

Haring, R. E., and R. A. Greenkorn (1970): A Statistical Model of a Porous Medium with Nonuniform Pores, A.I.Ch.E. J. 16(3), 471.

Hinch, E. J. (1977): An Averaged-Equation Approach to Particle Interactions in a Fluid Suspension, J. Fluid Mech. 83, 695.

Howells, I. D. (1974): Drag Due to the Motion of a Newtonian Fluid through a Sparse Random Array of Small Fixed Rigid Objects, J. Fluid Mech. 64, 449.

Ikoku, C. U., and H. J. Ramey, Jr. (1978): Transient Flow of Non-Newtonian Power-Law Fluids in Porous Media, San Francisco Sec. Meet. SPE, April, SPE 7139.

Javendel, I., and P. A. Witherspoon (1968): Application of the Finite Element Method to Transient Flow in Porous Media, Pet. Trans. AIME 243, SPEJ 241.

Jenkins, R., and J. S. Aronofsky (1953): Unsteady Radial Flow of Gas through Porous Media, Proc. 1st U.S. Natl. Cong. Appl. Mech. Chicago, ASME Applied Mech. Div. Paper 52-A26.

Johnson, C. R., and R. A. Greenkorn (1962): Comparison of Core Analysis and Drawdown-Test Results from a Water-Bearing Upper Pennsylvanian Sandstone of Central Oklahoma, Bull. Int. Assoc. Sci. Hydrol. VII, 46.

Johnson, C. R., R. A. Greenkorn, and E. G. Woods (1966): Pulse Testing: A New Method for Describing Reservoir Flow Properties between Wells, Pet. Trans. AIME 237, SPEJ 1599.

Kozeny, J. (1927): Ober kapillare Leitung das Wassers im Boden, S. Ber. Wiener Akad. Abt. IIa 136, 271.

Law, J. (1944): A Statistical Approach to the Interstitial Heterogeneity of Sand Reservoirs, Pet. Trans. AIME 155, SPEJ 202.

Leverett, M. C. (1941): Capillary Behavior in Porous Solids, Pet. Trans. AIME 142, SPEJ 161.

Marcus, H. (1962): The Permeability of a Sample of an Anisotropic Porous Medium, J. Geophys. Res. 67(13), 5215.

Marshall, R. J., and A. B. Metzner (1967): Flow of Viscoelastic Fluids through Porous Media, Ind. Eng. Chem. Fund. 6, 393.

McKinley, R. M., H. O. Jahns, W. W. Harris, and R. A. Greenkown (1966): Non-Newtonian Flow in Porous Media, A.I.Ch.E. J. 12, 17.

Mueller, T. D., and P. A. Witherspoon (1965): Pressure Interference Effects within Reservoirs and Aquifers, Pet. Trans. AIME 234, SPEJ 471.

Muskat, M. (1937): The Flow of Homogeneous Fluids through Porous Media, Chap. X, McGraw-Hill, New York.

Nakamura, A. (1959): On the Mechanism of Sound Absorption by Porous Materials, Mem. Inst. Sci. Ind. Res. Osaka Univ. 16 1.

Parsons, R. W. (1964): Discussion of Directional Permeability of Heterogeneous Anisotropic Porous Media, by Greenkorn, Johnson, and Shallenberger, Pet. Trans. AIME 231, SPEJ 364.

Pascal, H. (1969): Sur Quelques Methodes De Determination "In Situ" De La Perméabilitié Du Milieu Poreux, Revue De L'Institut Francais Du Petrole 24(3), 275.

Sadowski, T. J., and R. B. Bird (1965): Non-Newtonian Flow through Porous Media, I. Theory, Trans. Soc. Rheol. 9, 243.

Savins, J. G. (1970): Non-Newtonian Flow through Porous Media, in Flow through Porous Media, Amer. Chem. Soc., Washington, D.C., Chapter 5, p. 71.

Satterfield, C. N., and T. K. Sherwood (1963): The Role of Diffusion in Catalysis, Addison-Wesley, Reading, Mass.

Scheidegger, A. E. (1954): Directional Permeability of Porous Media to Homogeneous Fluids, Geofis. Pura. Appl. 28, 75.

Scheidegger, A. E. (1957): The Physics of Flow through Porous Media, Macmillan, New York.

Seaber, P. R., and E. F. Hollyday (1966): Statistical Analysis of Regional Aquifers, Ann. Meeting Geol. Soc. of Am., San Francisco.

Slattery, J. C. (1967): Flow of Viscoelastic Fluids through Porous Media, A.I.Ch.E. J. 13, 1066.

Slattery, J. C. (1969): Single-Phase Flow through Porous Media, A.I.Ch.E. J. 15(6), 866.

Slattery, J. C. (1972): Momentum, Energy and Mass Transfer in Continua, McGraw-Hill, New York.

Smith, P. G., and R. A. Greenkorn (1972): Theory of Acoustical Wave Propagation in Porous Media, J. Acoust. Soc. Amer. 52, 247.

Smith, P. G., R. A. Greenkorn, and R. G. Barile (1974a): Infrasonic Response Characteristics of Gas and Liquid-Filled Porous Media, J. Acoust. Soc. Amer. 56, 781.

Smith, P. G., R. A. Greenkorn, and R. G. Barile (1974b): Theory of the Transient Pressure Response of Fluid-Filled Porous Media, J. Acoust. Soc. Amer. 56, 789.

van Poolen, H. K., and J. R. Jargon (1969): Steady-State and Unsteady-State Flow of Non-Newtonian Fluids through Porous Media, Pet. Trans. AIME 246, SPEJ 80.

Whitaker, S. (1967): Diffusion and Dispersion in Porous Media, A.I.Ch.E. J. 13, 420

Whitaker, S. (1970): Advances in Theory of Fluid Motion in Porous Media, in Flow through Porous Media, Amer. Chem. Soc., Washington, D.C., Chapter 2, p. 31.

Zwikker, C., and C. W. Kosten (1949): Sound Absorbing Materials, Elsevier, New York, p. 23.

SUGGESTED READING

Bear, J., D. Zaslavsky, and S. Irmay (eds.), Physical Principles of Water Percolation and Seepage, UNESCO, Paris, 1968.

Bear, J., Dynamics of Fluids in Porous Media, Elsevier, New York, 1972.

Biot, M. A., Generalized Theory of Acoustic Propagation in Porous Dissipative Media, J. Acous. Soc. Amer. 34, 1254 (1962).

Calhoun, J. C., Fundamentals of Reservoir Engineering, University of Oklahoma Press, Norman, 1953.

Carslaw, H. S., and J. C. Jaeger, Heat Conduction in Solids, 2d ed., Oxford University Press, Oxford, 1959.

Churchill, R. V., Complex Variables and Applications, McGraw-Hill, New York, 1960.

Davis, S. N., Porosity, Permeability of Natural Materials, in Flow through Porous Media, R. S. M. DeWiest (ed.), Academic, London, 1969, Chapter 2.

Dullien, F. A. L., Porous Media Fluid Transport and Pore Structure, Academic, New York, 1979.

Gassman, F., Elastic Waves through a Packing of Spheres, Geophysics 16, 673 (1951).

Greenkorn, R. A., and C. R. Johnson, Variation of a Natural Sandstone Reservoir Element: An Objective Analysis of Core Measurement, Paper No. 1577-6, Annual Fall Meeting SPE of AIME, Denver, 1960.

Greenkorn, R. A., Matrix Properties of Porous Media, Proc. Second Symp. IAHR-ISSS Fundamentals of Transport Phenomena in Porous Media, Guelph, 1972.

Greenkorn, R. A., and D. P. Kessler, Transfer Operations, McGraw-Hill, New York, 1972.

Ivanov, V. V., Yu A. Medvedev, and B. M. Stepanev, Sound Propagation in a Wet Porous Medium, Sov. Phys.—Acoustics 14(1), 51 (1968).

Kidder, R. E., Unsteady Flow of Gas through a Semi-Infinite Porous Medium, Paper 57-APM-13, ASME Appl. Mech. Div., June, 1956.

Milne-Thomson, L. M., Theoretical Hydrodynamics, Macmillan, New York, 1955.

Payatakes, A. C., Chi Tien, and R. Turian, A New Model for Granular Porous Media, Part I: Model Formulation; Part II: Numerical Solution of Steady State Incompressible Newtonian Flow through Periodically Constricted Tubes, A.I.Ch.E. J. 19, 58, 67 (1973).

Raghavan, R., and S. S. Marsden, Jr., Theoretical Aspects of Emulsification in Porous Media, Pet. Trans. AIME 251, SPEJ 153.

Slider, H. C., Practical Petroleum Reservoir Engineering Methods, The Petroleum Publishing Co., Tulsa, Okla., 1976.

Whitaker, S., The Equations of Motion in Porous Media, Chem. Eng. Sci. 21, 291 (1966).

4

Multifluid Immiscible Flow

The equations of change for multifluid immiscible flow are similar to those for single-fluid flow. However, the equations are written for each fluid and connected using capillary pressure and saturation of the various fluids. This brings additional parameters to the models which result from interfacial tension in immiscible systems. Additional parameters wettability, capillary pressure, relative tortuosity, and relative permeability are introduced.

Immiscible flow is modeled ignoring capillary pressure as in the Buckley-Leverett displacement model or by assuming that a distinct boundary exists and moves as in the Muskat model. When more than one fluid is flowing, unstable flow (fingering) may result due to a variety of regimes depending on whether one fluid is more dense than the other (gravity fingering) or whether one of the fluids is less viscous than the other (viscous fingering). The criteria for unstable flow result from considering complex interaction of gravity, viscosity, interfacial tension, and properties of the media.

Transient flow is normally modeled much in the same manner as with single fluids by writing an average expression. Phase changes are possible during multifluid flow. These flow situations are dominated by the thermodynamics of the system.

4.1 EQUATIONS OF CHANGE

For single-fluid flow in a porous medium, Darcy's law is

$$\underline{q} = -\frac{\underline{\underline{k}}}{\mu} \underline{\nabla} p \qquad (4.1)$$

Gravity may be included in Eq. (4.1) by replacing p with a potential including gravity, as follows:

$$\Phi = p + \rho g z \qquad (4.2)$$

The continuity equation for single–fluid flow is

$$\phi \frac{\partial \rho}{\partial t} = -(\underline{\nabla} \cdot \rho \underline{q}) \qquad (4.3)$$

Density and pressure are related by an appropriate equation of state

$$\rho = \rho(p) \qquad (4.4)$$

Muskat et al. (1937) assumed Darcy's law is valid for each flowing fluid, when the fluids are a gas, subscript 1, and a liquid, subscript 2.

$$\underline{q}_1 = - \frac{\underline{\underline{k}}k_{r1}}{\mu_1} \nabla p_1 \qquad (4.5)$$

and

$$\underline{q}_2 = - \frac{\underline{\underline{k}}k_{r2}}{\mu_2} \underline{\nabla} p_2 \qquad (4.6)$$

The product $\underline{\underline{k}}k_{r1}$ is the effective permeability of the medium to fluid 1 with fluid 2 present $\underline{\underline{k}}_1$. The product $\underline{\underline{k}}k_{r2}$ is the effective permeability of the medium to fluid 2 with fluid 1 present, $\underline{\underline{k}}_2$. Effective permeabilities depend on both pore structure and the fluids present in the porous medium. The values of k_{r1} and k_{r2} are called the relative permeabilities and are defined by

$$k_{r1} = \frac{\underline{\underline{k}}_1}{\underline{\underline{k}}} \qquad (4.7)$$

$$k_{r2} = \frac{\underline{\underline{k}}_2}{\underline{\underline{k}}} \qquad (4.8)$$

The relative permeabilities depend on saturation (and wettability and saturation history). The continuity equations for each fluid are

$$\phi \frac{\partial \rho_1 S_1}{\partial t} = -(\underline{\nabla} \cdot \rho_1 \underline{q}_1) \qquad (4.9)$$

and

$$\phi \frac{\partial \rho_2 S_2}{\partial t} = -(\underline{\nabla} \cdot \rho_2 \underline{q}_2) \qquad (4.10)$$

where S_1 is fractional saturation of fluid 1, and S_2 is fractional saturation of fluid 2. These saturations must add up to 1 so that

$$S_1 + S_2 = 1 \qquad (4.11)$$

The equations of state for each fluid are

$$\rho_1 = \rho_1(p_1) \tag{4.12}$$

$$\rho_2 = \rho_2(p_2) \tag{4.13}$$

For immiscible fluids, the interfacial tension between the two fluids causes a pressure discontinuity at the boundary separating the two fluids. This pressure is the capillary pressure and is related to saturation so that

$$p_c(S_1) = p_2 - p_1 \tag{4.14}$$

The extension of Darcy's law to the flow of two immiscible fluids in a porous medium is heuristic. Slattery (1968, 1970) introduced the concept of local averaging to each fluid in multifluid immiscible flow in a porous medium. Consider a particular point in a porous medium with volume V contained by a closed surface S. Let V_i be the volume containing fluid i. The volume and shape of V_i may change from point to point in the medium. The volume average of any quantity is defined by

$$\langle B \rangle = \frac{1}{V} \int_{V_i} B \, dV \tag{4.15}$$

Applying Eq. (4.15) to obtain the local average of the equation of continuity yields

$$\frac{\partial \langle \rho \rangle^i}{\partial t} + (\underline{\nabla} \cdot \langle \rho \underline{v} \rangle^i) = 0 \tag{4.16}$$

The local average of the equation of motion is

$$\frac{\partial \langle \rho \underline{v} \rangle^i}{\partial t} + (\underline{\nabla} \cdot \langle \rho \underline{vv} \rangle^i = (\underline{\nabla} \cdot \langle \underline{\underline{\tau}} \rangle^i) + \langle \rho \underline{F} \rangle^i + \frac{1}{V} \int_{S_{mi}+S_{fi}} (\underline{\underline{\tau}} \cdot \underline{n}) \, dS \tag{4.17}$$

where S_i is the closed boundary surface of V_i and is the sum of S_{ei}, S_{mi}, and S_{fi}. S_{ei} is that part of the surface S_i that coincides with S; S_{mi} is that part of S_i that coincides with the boundaries with other fluids; S_{fi} is that part of S_i that coincides with pore walls.

Consider the movement of two incompressible fluids, 1 and 2. Equation (4.16) is written as

$$\phi \frac{\partial S_i}{\partial t} + (\underline{\nabla} \cdot \langle \underline{v} \rangle^i) = 0 \tag{4.18}$$

If we neglect inertial effects and represent internal forces by the gradient of a potential function, so that

$$\underline{F} = -\underline{\nabla} \hat{\phi} \tag{4.19}$$

with pressure defined for each fluid as

$$\mathscr{P} = p + \rho \hat{\phi} \tag{4.20}$$

then

$$\underline{\nabla} \langle \mathscr{P} - p_o \rangle^i - (\underline{\nabla} \cdot \langle \underline{\underline{\tau}} + p\underline{\underline{I}} \rangle^i) - \frac{1}{V} \int_{S_{mi} + S_{fi}} (\{\underline{\underline{\tau}} + (p_o - \rho_i \hat{\phi}\underline{\underline{I}}\} \cdot \underline{n}) \, dS = 0 \tag{4.21}$$

Slattery (1970) shows that the last term on the right of Eq. (4.21), call the term g_i, is

$$\underline{g}_i = R_i(\langle \underline{v} \rangle^i - \langle \underline{u} \rangle) \tag{4.22}$$

where R_i is a resistance coefficient and $\langle u \rangle$ is the local average solid velocity. The stress tensor in the second term of Eq. (4.21) can be replaced by

$$\langle \underline{\underline{\tau}} + p\underline{\underline{I}} \rangle^i = \mu_i [\underline{\nabla} \langle \underline{v} \rangle^i + (\underline{\nabla}\langle \underline{v} \rangle^i)^\dagger] \tag{4.23}$$

Now, if we assume that $\langle \underline{u} \rangle^i$ is approximately zero,

$$\underline{\nabla} \langle \mathscr{P} - p_o \rangle^i - \mu_i [(\underline{\nabla} \cdot \underline{\nabla} \langle \underline{v} \rangle^i) + \underline{\nabla}(\underline{\nabla} \cdot \langle \underline{v} \rangle^i)] + R_i \langle \underline{v} \rangle^i = 0 \tag{4.24}$$

The second term is small compared to the other terms and can be dropped so that

$$\nabla \langle \mathscr{P} - p_o \rangle^i + R_i \langle \underline{v} \rangle^i = 0 \tag{4.25}$$

If we assume $\langle \underline{v} \rangle^i$ is equivalent to $\underline{q}^i$ and capillary pressure as in Eq. (4.14), Eqs. (4.5) and (4.6) are given by

$$\underline{\nabla} \langle \mathscr{P} - p_o \rangle^i + L_i \underline{q}_i = 0 \tag{4.26}$$

Thus a relative permeability concept results from volume averaging the equations of continuity and motion for multifluid immiscible flow. The difference in Eqs. (4.25) and (4.26) is that L_i is equivalent to R_i. The velocity of the solid, $\langle \underline{u} \rangle^i$, must be zero.

   In summary, for multifluid immiscible flow of incompressible fluids, in an isotropic porous medium with no gravity

$$\underline{q}_i = -\frac{k_i(S_i)}{\mu_i} \nabla p_i \tag{4.27}$$

and

$$\phi \frac{\partial S_i}{\partial t} = -(\underline{\nabla} \cdot \underline{q}^i) \tag{4.28}$$

where

$$\sum_{i=1}^{n} S_i = 1 \tag{4.29}$$

and where n and w stand for nonwetting and wetting fluids, respectively,

$$p_c(S_w) = p_n - p_w \tag{4.30}$$

4.2 WETTABILITY, CAPILLARY PRESSURE, RELATIVE TORTUOSITY, AND RELATIVE PERMEABILITY OF NONIDEAL MEDIA

Morrow (1970) discusses the thermodynamics of surfaces between immiscible fluids and between fluids and solids with reference to the superficial surface free energy, surface free energy, and surface (interfacial) energy. The surface between two immiscible fluids is not a surface of infinitesimal thickness but a nonhomogeneous zone of finite thickness. This zone is modeled by a superficial mathematical surface (a line in two dimensions). In the model, this surface divides the volume containing the two immiscible fluids into two regions. It is assumed the bulk properties of the two immiscible fluids are maintained up to the dividing surface. The Gibbs expression for the change in Helmholtz free energy for a system of k components is

$$dF = -S \, dT + \sum_{i=1}^{k} \mu_i \, dn_i - p_1 \, dv_1 - p_2 \, dv_2 + \sigma \, dA \tag{4.31}$$

where the intensive quantities on the right-hand side of Eq. (4.31), T, μ_i, p, and σ, are the temperature, chemical potential, pressure, and interfacial (surface) tensions, respectively. S is entropy, n_i is the number of moles of component i, V is volume, and A is interfacial area. For isothermal conditions the change in superficial surface free energy is

$$dF_S = \sum_{i=1}^{k} \mu_i \, d(n_i)_S + \sigma \, dA \tag{4.32}$$

where the surface excess of unit area of component i is Γ_i; then

$$(n_i)_S = \Gamma_i A \tag{4.33}$$

Differentiating Eq. (4.33),

$$d(n_i)_S = \Gamma_i \, dA + A \, d\Gamma_i \tag{4.34}$$

and

$$dF_S = \sum_{i=1}^{k} \mu(\Gamma_i \, dA + A \, d\Gamma_i) + \sigma \, dA \tag{4.35}$$

Differentiating Eq. (4.35) with respect to area for constant T, V, and n, yields

$$\left(\frac{\partial F_S}{\partial A}\right)_{T,V,n} = \sum_{i=1}^{k} \mu_i \left[\Gamma_i + A\left(\frac{\partial \Gamma_i}{\partial A}\right)_{T,V,n}\right] + \sigma \tag{4.36}$$

When the properties of the surface per unit area are independent of the total area, the second term in brackets following the summation sign in Eq. (4.36) is zero, and upon integration Eq. (4.36) becomes

$$\frac{F_S}{A} = \sum_{i=1}^{k} \mu_i \Gamma_i + \sigma \tag{4.37}$$

Equation (4.37) defines the superficial surface free energy per unit area. For a single component system composed of a pure liquid and its vapor the dividing surface in the model can be located such that $\Gamma - 0$ and the superficial surface free energy per unit area F_S/A and the surface tension are equal.

The interfacial tension is numerically equal to the reversible work of extension per unit area, from Eq. (4.31)

$$\sigma = \left(\frac{\partial F}{\partial A}\right)_{T,V,n} \tag{4.38}$$

If the interfacial tension is constant, then

$$\sigma \, \Delta A = (\Delta F)_{T,V,n} \tag{4.39}$$

which defines the surface free energy. In systems where the interfacial tension depends on the surface area, such as systems containing surface active agents

$$(\Delta F)_{T,V,n} = \int_{A_1}^{A_2} \sigma(A) \, dA \tag{4.40}$$

Consider two immiscible fluids, oil and water, contained in the interstices of a porous medium. Further, concentrate on an individual idealized pore containing the two immiscible fluids and a hypothetical interface as in Fig. 4.1 (Melrose and Brandner, 1974). At equilibrium, two hydrostatic equations define the state of the interface, Laplace's equation and Young's equation. Laplace's equation can be written as

$$p_o - p_w = \sigma_{ow} R_{ow} \tag{4.41}$$

where p_o is the pressure in the oil, p_w is the pressure in the water, σ_{ow} is the

118 Multifluid Immiscible Flow

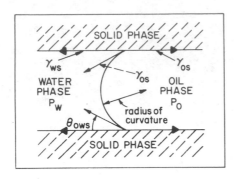

FIG. 4.1 Hydrostatic equilibrium: two
fluid phases on contact with a solid phase.
(From Melrose and Brandner, 1974.)

interfacial tension between the two fluids, and R_{ow} represents the curvature of the
interface. Young's equation is

$$\sigma_{os} = \sigma_{ws} + \sigma_{ow} \cos \theta_{ows} \tag{4.42}$$

where the subscript s stands for the solid, σ_{os} is the interfacial tension between the
oil and the solid, σ_{ws} is the interfacial tension between the water and the solid, and
θ_{ows} is the angle of contact measured through the water that the fluid-fluid interface
makes with the solid surface.

The contact angle θ_{ows} in Eq. (4.42) is used to define the wettability of the
solid surface. In Fig. 4.1 the solid surface is water wet if $\theta_{ows} < 40°$; the solid
surface is oil wet if $\theta_{ows} > 140°$ for $40° < \theta_{ows} < 140°$, the solid surface is said to
have neutral wettability. The contact angle, θ_{ows} (and therefore the wettability)
may show hysteresis; that is, the angle may be different depending on whether the
fluid-fluid interface is moving to the left or to the right. Wettability of a given surface may also depend on surface viscosity, which influences the speed at which a
given wetting fluid covers the surface. It is possible for solid surfaces to have mixed
wettability, that is, for different areas of the solid surface to have different wetta-
bilities (different contact angle or speed of wetting).

For porous media of intermediate and low permeabilities (glass beads and
sand), the pores are small enough that we can assume pressure is constant in each
of the fluids, the interfacial tension between the two fluids does not depend on the
radius of curvature, and the curvature of the fluid-fluid interface is constant. The
expressions for the curvature of the interface R_{ow} are still complex with these
assumptions except for simple geometries (here o and w represent nonwetting and
wetting). The expression for curvature R_{ow} will contain both first and second derivatives. Laplace's equation may be written in terms of the two principal radii of
curvature r_1 and r_2 and Eq. (4.41) defines the capillary pressure p_c at a point as

$$p_c = \sigma\left(\frac{1}{r_1} + \frac{1}{r_2}\right) \tag{4.43}$$

where we dropped the subscripts ow on the interfacial tension.

The sketches in Fig. 4.2 represent three simple geometries for which the
terms in Eq. (4.43) are replaced with a dimension defined by the geometry. For the
two flat plates close together separated by a distance h = $r_1 \cos \theta$, r_2 approaches
infinity, and

$$p_c = \frac{\sigma \cos \theta}{h} \qquad (4.44)$$

where we dropped the subscripts o, w, and s on θ. For a capillary tube, $2/r$ is the harmonic average of r_1 and r_2 so that

$$p_c = \frac{2\sigma \cos \theta}{r} \qquad (4.45)$$

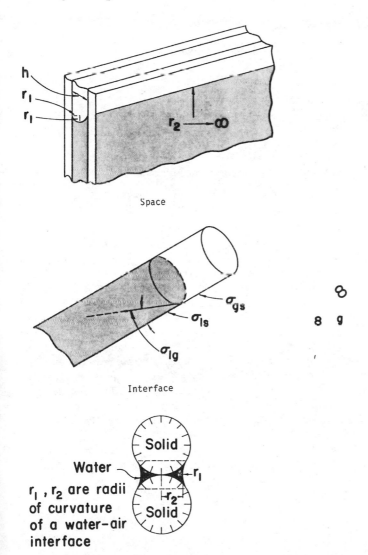

FIG. 4.2 Equilibrium at a curved interface between two immiscible fluids. (a) Interface across space between parallel flat plates. (b) Liquid–air interface across a capillary tube. (c) r_1, r_2 are radii of curvature of a water–air interface. (From Corey, 1977.)

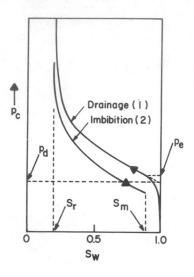

FIG. 4.3 Capillary pressure as a function of saturation. (From Corey, 1977.)

For a pendular ring in the point-to-point packing of uniform spheres, $r_1 = r_2 = r$, the radius of the sphere and

$$p_c = \frac{2\sigma}{r} \qquad\qquad (4.46)$$

The capillary pressure p_c in Eq. (4.43) is defined at a point on the interior of an interface of constant curvature without reference to the contact angle θ. The contact angle is assumed constant and defines the interfacial curvature along with an average distance as in Eqs. (4.44) to (4.46).

Consider a porous medium saturated with a wetting fluid and assume the contact angle θ is zero. For displacement of the wetting fluid by a nonwetting fluid, the capillary pressure increases as the saturation of the wetting fluid decreases. A plot of capillary pressure p_c versus wetting fluid saturation S_w is sketched in Fig. 4.3. The curve just described is called the drainage or drying curve—the upper curve in Fig. 4.3. The reverse process—where the medium starts saturated with a nonwetting fluid and the wetting fluid imbibes displacing the nonwetting fluid—is the lower curve of Fig. 4.3. This curve is called the imbibition or wetting curve. For imbibition, the capillary pressure decreases as the wetting saturation increases.

The hysteresis in the drainage and wetting curves of Fig. 4.3 is due to the instability of certain fluid–fluid interface configurations and to hysteresis in the

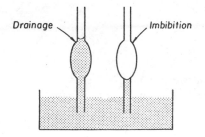

FIG. 4.4 Capillary hysteresis. (From
Corey, 1977.)

contact angle. The bistability of the interface in a nonuniform capillary can be used to explain this hysteresis. Figure 4.4 is a sketch of two nonuniform capillary tubes with one end of each tube immersed in a wetting fluid. The tube on the left is first filled with the wetting fluid and allowed to drain, until it reaches the stable interface configuration shown. The tube on the right is initially empty, and it is allowed to imbibe the wetting fluid, until it reaches the stable interface configuration shown. The fluid in the tube on the left is higher than that of the tube on the right; the left tube has a higher capillary pressure p_c than the tube on the right. In an actual porous medium which contains a complex network of nonuniform pores, many stable interface configurations in the pores are possible. Since there are many stable configurations if one stops imbibition and begins draining in a medium at arbitrary points, a series of scanning loops will be obtained as shown in Fig. 4.5. The scanning curves result from intermediate stable configurations in the pores. Notice that if we measure the drainage curve a second time after imbibition, the loop starts where imbibition saturation stopped.

Fluid-fluid interfaces in a pore may be stable or unstable in certain configurations depending on the geometry of the pores, the saturation, and the pressure drop across each pore. Figure 4.6 shows, schematically, stable and unstable interface conditions in an arbitrary pore. The limiting number of stable configurations

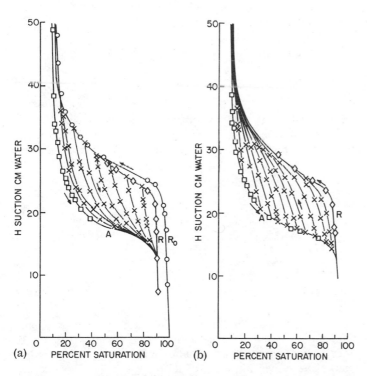

FIG. 4.5 (a) Imbibition scanning curves originating from the secondary desaturation curve R; and (b) desaturation scanning curves originating from the pendular imbibition curve A. (From N. R. Morrow and C. C. Harris, Capillary Equilibrium in Porous Materials, Pet. Trans. AIME 234, SPEJ 15. © 1965 SPE-AIME.)

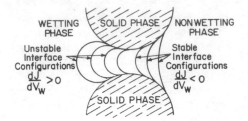

gives rise to the hysteresis described earlier. If the interface takes on an unstable
shape (or corresponding position in a pore), there will be sudden local flow in which
one fluid will displace the other. This stuttering flow which restores equilibrium
was first reported by Haines (1930) and the jumps in flow are often referred to as
Haines jumps. In Fig. 4.4 the two interfaces in the two capillaries are stable. However, if in the left tube pressure was applied to push the interface into the bulge,
the interface would be unstable and the bulge would immediately drain and come to
equilibrium, as in the right tube. The stability conditions for the interface of Fig.
4.6, assuming the nonwetting fluid is oil and the wetting fluid is water, are (Melrose,
1970)

$$\frac{dR_{ow}}{dV_w} < 0 \qquad \theta < 40° \tag{4.47}$$

$$\frac{dR_{ow}}{dV_w} > 0 \qquad \theta > 140° \tag{4.48}$$

The maximum and minimum curvature as determined from Eqs. (4.47) and (4.48)
gives rise to the drainage and imbibition curvatures R_{dr} and R_{imb}, respectively.
The effect of varying contact angle on these curvatures is given by Melrose (1965)
as

$$R_{dr} = \frac{2H_{dr}Z_{dr}(\theta) \cos \theta}{r} \tag{4.49}$$

and

$$R_{imb} = \frac{2H_{imb}Z_{imb}(\theta) \cos \theta}{r} \tag{4.50}$$

H_{dr} and H_{imb} in Eqs. (4.49) and (4.50) are the limiting curvatures for $\theta = 0$ normalized by $2/r$, where r is the characteristic distance for the pore model assumed. For
a capillary r is the capillary radius, for packing of uniform spheres r is the radius
of the spheres. Experiments show values $H_{dr} = 2.70$ and $H_{imb} = 1.75$. For drainage,
$Z_{dr} > 1$ if $\theta < 0$ and for imbibition, $Z_{imb} < 1$ if $\theta > 0$. At $\theta = 0$, $Z_{dr} = Z_{imb} = 1$.
    The drainage and wetting capillary pressure curves sketched in Fig. 4.3 have
certain limiting values of capillary pressure and saturation. The drainage capillary
pressure curve approaches a minimum wetting fluid saturation greater than zero.

This minimum saturation, the residual saturation of the wetting fluid S_r appears to be reached asymptotically as the capillary pressure increases. An explanation for the residual saturation is that at this saturation the wetting fluid exists only as unconnected pendular rings in the pore spaces and the wetting fluid does not flow. Another explanation for the residual saturation is that the wetting fluid remains in very small pores. In oil-water systems, the residual saturation is referred to as connate water—in air-water systems as the minimum water content. It is generally observed that the broader the pore size distribution, the higher the value of S_r. Any change in wettability of a system changes the residual saturation.

Another characteristic of the drainage curve in Fig. 4.3 is that the capillary pressure drops off from a relatively constant value decreasing rapidly to zero very close to $S_w = 1$. When a certain threshold value is reached, pressure decreases rapidly. This threshold or entry pressure value is ambiguous. In oil-water systems the entry or threshold pressure is called the displacement pressure, p_d. It is defined as the first p_c at which desaturation of the drainage curve occurs. p_d is determined by extrapolating the drainage curve in a straight line fashion to $S_w = 1$, as shown in Fig. 4.3. The saturation at which p_d occurs is called S_d. In air-water systems the entry or threshold pressure p_e is defined at the inflection point of the p_c curve; it is called the <u>bubble pressure</u>, the air pressure required to force air through an initially water-saturated sample.

The wetting curve has a critical saturation S_m. For imbibition, the nonwetting fluid becomes trapped in a pore or cluster of pores and the wetting fluid bypasses this trapped nonwetting fluid. Prediction of S_m has not been accomplished; it would be a most useful number for oil-water systems. Values may be as high as 30 to 40 percent of the nonwetting fluid. Since the nonwetting fluid for $S_w > S_m$ is not interconnected, it does not have a unique capillary pressure. It is generally observed that the broader the pore size distribution, the larger S_m.

Brooks and Corey (1964, 1966) empirically define the effective saturation

$$S_e = \frac{S_w - S_r}{1 - S_r} \tag{4.51}$$

(since the pore space containing the wetting fluid at S_r contributes relatively little to the flow) to the inverse of the dimensionless capillary pressure made dimensionless by the displacement pressure p_d so that

$$S_e = \left(\frac{p_d}{p_c}\right)^\lambda \tag{4.52}$$

A plot of $\ln S_e$ versus $\ln p_c$ as in Fig. 4.7 yields straight-line behavior and provides a precise way of determining p_d. λ depends on the pore size distribution. An effective or drainable porosity can be defined as

$$\phi_e = (1 - S_r)\phi \tag{4.53}$$

Leverett (1941) suggested plotting a reduced capillary pressure function versus saturation as a means of correlating capillary pressure data for different media. The J function defined by Leverett is usually written

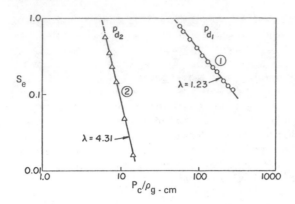

FIG. 4.7 Effective saturation as a function of capillary pressure. (From R. H. Brooks and A. T. Corey, Hydraulic Properties of Porous Media, Hydraulic Paper Number 3, Colorado State University. © 1964 American Society of Civil Engineers.)

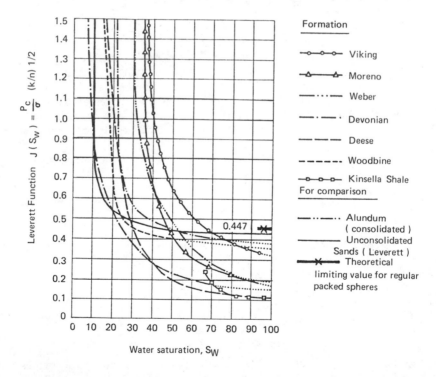

FIG. 4.8 Leverett function for various formations. (From W. Rose and W. A. Bruce, Evaluation of Capillary Character in Petroleum Reservoir Rock, Pet. Trans. AIME 186, SPEJ 127. © 1949 SPE-AIME.)

$$J = \frac{p_c}{\sigma f(\theta)} \left(\frac{k}{\phi}\right)^{\frac{1}{2}} \tag{4.54}$$

Data from many porous media plot within a narrow band using Eq. (4.54) to define J. Rose and Bruce (1949) plotted capillary pressure for measurements on samples from different formations in terms of the J function as in Fig. 4.8.

Based on observations that the capillary pressure curve and its parameters depend on the pore structure, J should be directly related to the parameters of the pore size distribution. If we use the results from the statistical model of a nonuniform porous medium discussed in Sec. 2.2 for p_c^* and k/ϕ with Eq. (4.54), we find

$$J = \frac{p_c^* \{S_{w, \alpha, \beta}\}}{\sigma f(\theta)} \left(\frac{R^2}{24} \frac{(\alpha + 2)(\alpha + 1)}{(\alpha + \beta + 3)(\alpha + \beta + 2)}\right)^{\frac{1}{2}} \tag{4.55}$$

For a given fluid system assume $\sigma f(\theta)$ is constant and R or $\langle r \rangle^2$ are constant. The effect of nonuniformity on the behavior of J can be determined from the model. Figure 4.9 is a plot of the distribution of pore sizes for the statistical model using the beta function. In each of the five cases, the distribution function has average pore size $\langle r^* \rangle = 0.4$ but the spread or variance of the pore size distribution is different. Figure 4.10 shows the values of J calculated from Eq. (4.55) and the parameters alpha and beta for the five distributions of Fig. 4.9. Figure 4.10 shows quantitatively for the model that for a given value of J the wetting fluid saturation increases as the pore size distribution narrows.

If the wetting fluid saturation of a porous medium decreases, the fluid "particles" of the wetting fluid must take an increasingly longer path to move between two points (since the nonwetting fluid is in the way). Burdine (1953) suggested

$$\tau_w(S_e) = \frac{\tau_1}{S_c^2} \tag{4.56}$$

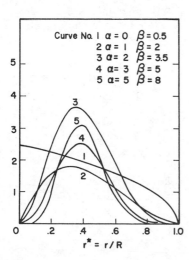

FIG. 4.9 Sample beta functions with mean of 0.4. (From Smith, 1976.)

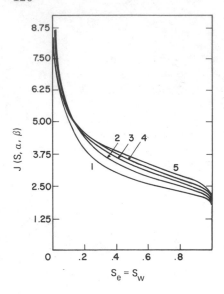

FIG. 4.10 Capillary pressure behavior
of sample beta function. (From Smith,
1976.)

where τ_w is the relative tortuosity of the wetting fluid. τ_1 is the tortuosity of the
wetting fluid at $S_e = 1$. For the nonwetting phase

$$\tau_n(S_e) = \frac{\tau_1}{(1 - S_e)^2} \qquad\qquad (4.57)$$

Equations (4.56) and (4.57) are not valid for anisotropic media.

We need effective permeabilities to use the extended form of Darcy's law for
multifluid flow such as water and oil, Eqs. (4.5) and (4.6). The effective permeabilities can be found if we know the intrinsic permeability of a medium and the relative
permeabilities of the wetting (water) and nonwetting (oil) fluids as functions of saturation. Relative permeability may be based on intrinsic permeability or dry air
permeability, or $S_w = 1$ and S_r. We will use the latter in the discussion that follows.
The relative permeability is a function of saturation and also depends on pore size

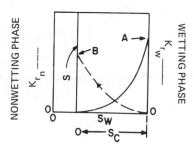

FIG. 4.11 Relative permeability versus
saturation. (From Standing, 1975.)

distribution, wettability, and saturation history. Figure 4.11 is a sketch of effective wetting and nonwetting fluid relative permeability curves determined from drainage measurements. The relative permeability of the wetting fluid is usually defined as the ratio of the effective permeability at a given wetting fluid saturation to the effective permeability of the wetting fluid at a wetting fluid saturation $S_w = 1$ (point A of Fig. 4.11). The relative permeability of the nonwetting fluid is similarly defined as the ratio of the effective permeability of the nonwetting fluid at a given saturation to the effective permeability at residual wetting fluid saturation (point B of Fig. 4.11). If we plot relative permeability as a function of effective saturation, S_e as defined by Eq. (4.51) for media with a different pore size distribution index λ, introduced in Eq. (4.52), we will obtain different relative permeability curves. Figure 4.12 is a sketch of wetting and nonwetting fluid relative permeabilities as a function of S_e for $\lambda = 2$ (a wide pore size distribution) and $\lambda = 4$ (a medium width pore size distribution).

As with capillary pressure, relative permeability values will show hysteresis depending on whether they were determined under drainage or imbibition conditions. Figure 4.13 shows a sketch of relative permeability curves for drainage and imbibition separately, and these are overlaid where the data were obtained from sequential measurements, that is, draining and imbibition. The value S_d on the nonwetting drainage relative permeability curve is the saturation at the displacement pressure (see Fig. 4.3). The dashed line in each sketch in Fig. 4.13 is the residual saturation of the wetting phase, S_r. The value S_c on the nonwetting imbibition relative permeability curve is the critical saturation (see Fig. 4.3). The third sketch of Fig. 4.13 combines the two sets of curves and shows hysteresis in the nonwetting relative permeability. For the wetting imbibition relative permeability the value only reaches the point where it intersects the saturation associated with the nonwetting curve since the trapped oil at this saturation allows no additional wetting fluid flow.

Burdine (1953) proposed the following relations for relative permeability based on drainage capillary pressure written in terms of effective saturation defined by Eq. (4.51):

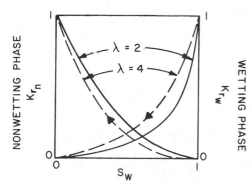

FIG. 4.12 Effect of pore size distribution. (From Standing, 1975.)

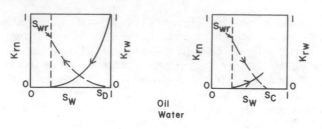

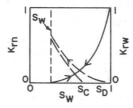

FIG. 4.13 Hysteresis effects. (From Standing, 1975.)

$$k_{rw} = S_e^2 \frac{\int_0^{S_e} p_c^{-2} dS_e}{\int_0^1 p_c^{-2} dS_e} \tag{4.58}$$

and

$$k_{rn} = (1 - S_e)^2 \frac{\int_{S_e}^1 p_c^{-2} dS_e}{\int_0^1 p_c^{-2} dS_e} \tag{4.59}$$

Purcell (1949) used a capillary model to relate relative permeability and capillary pressure. Corey (1977) writes equations similar to Purcell in terms of the average of the square of the hydraulic radius of the pores as

$$\langle R_w^2 \rangle = \frac{\sigma^2 \cos^2 \theta}{S} \int_0^S \frac{dS}{p_c^2} \tag{4.60}$$

and

$$\langle R_n^2 \rangle = \frac{\sigma^2 \cos^2 \theta}{S} \int_0^S \frac{dS}{p_c^2} \tag{4.61}$$

The Kozeny-Carman equation for permeability for single-fluid flow (see Sec. 3.2) is

$$k = \frac{\phi^3}{a\tau^2}$$

(4.62)

where for sand the tortuosity τ is about 2 and the connectivity factor a is about 2.5. Combining the generalized form of Darcy's law Eqs. (4.5) and (4.6) with Eqs. (4.60) and (4.61) leads to a generalized Kozeny-Carman expression for effective permeabilities

$$k_w = \left(\frac{\sigma^2 \cos^2 \theta \, \phi e}{a\tau_1}\right) S_e^2 \int_0^{S_e} \frac{dS_e}{p_c^2}$$

(4.63)

and

$$k_n = \left(\frac{\sigma^2 \cos^2 \theta \, \phi e}{a\tau_1}\right)(1 - S_e)^2 \int_{S_e}^1 \frac{dS_e}{p_c^2}$$

(4.64)

which are equivalent to Eqs. (4.58) and (4.59).

Brooks and Corey (1964) used Eq. (4.52) to integrate Eqs. (4.63) and (4.64), and obtained

$$k_w = k S_e^{(2+\lambda)/\lambda}$$

(4.65)

$$k_n = k(1 - S_e^2)(1 - S_e^{(2+\lambda)/\lambda})$$

(4.66)

From Eqs. (4.58) and (4.59), or Eqs. (4.65) and (4.66),

$$k_{rw} = S_e^{(2+3\lambda)/\lambda}$$

(4.67)

$$k_{rn} = (1 - S_e^2)(1 - S_c^{(2+\lambda)/\lambda})$$

(4.68)

These expressions containing λ and S_e are semiempirical since λ must be determined experimentally and S_e is normalized based on S_r. From statistical considerations using the model of Sec. 2.2, one would expect at least two parameters are required to relate structure and capillary (α and β, which describe the pore radius distribution).

For the nonwetting relative permeability saturation, S_d is not 1, and further, Corey and Rathjens (1956) have shown the value depends on anisotropy. Where there are layers of different material (different permeabilities) perpendicular to flow, S_d is very small. Where there are layers of different material (different permeability) parallel to flow, S_d approaches 1. Average values for S_d are of the order of 0.85.

Standing (1975) proposes using the following expressions in place of Eq. (4.68) to correct for S_d not equal to 1. Let

$$\hat{S}_e = \frac{S_w - S_r}{S_d - S_r} \tag{4.69}$$

and

$$k_{rn} = (1 - \hat{S}_e)^2 (1 - S_e^{(2+\lambda)/\lambda}) \tag{4.70}$$

Figure 4.14 shows relative permeability for an air–water–sand system as a function of capillary pressure. The discrepancy on the drainage cycle in the region $p_c < p_e$ may be a result of the difference in the drainage process, which is by nature discontinuous, and the steady-state measurements (Corey and Brooks, 1972). Since an air permeability does not exist for $p_c < p_e$, the erratic behavior in this region is due to air breaking through. Once the air is continuous, the erratic behavior stops. Brooks and Corey (1966) show relative permeability of air as a function of p_c compared to

$$k_{rn} = \left[1 - \left(\frac{p_d}{p_c} \right)^\lambda \right]^2 \left[1 - \left(\frac{p_d}{p_c} \right)^{2+\lambda} \right] \qquad p_c \geq p_e \tag{4.71}$$

in Fig. 4.15. For the wetting phase they suggest

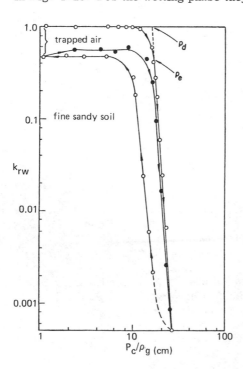

FIG. 4.14 Wetting phase relative permeability as a function of capillary pressure on the drainage and wetting cycle. (From Brooks and Corey, 1964.)

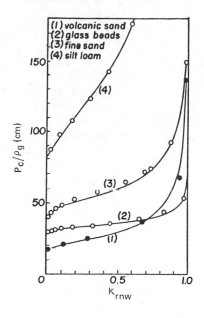

(1) volcanic sand
(2) glass beads
(3) fine sand
(4) silt loam

FIG. 4.15 Relative permeability of air as a function of p_c compared with theoretical function. (From R. H. Brooks and A. T. Corey, Properties of Porous Media Affecting Fluid Flow, ASCE Proc. J. Irrigation Drainage Div. 92, IR 2, 61. © 1966 American Society of Civil Engineers.)

$$k_{rw} = \left(\frac{p_d}{p_c} \right)^{2+3\lambda} \qquad p_c \geq p_d \qquad (4.72)$$

There are practical flow situations in oil reservoirs where three immiscible fluids, water, oil, and gas may be present. In such situations we need three fluid relative permeabilities, usually called three-phase relative permeabilities, to use the modified form of Darcy's law. We can extend the treatment of Burdine (1953) which was used above for two fluids. Assume a porous medium contains water, oil, and gas (use w for water, o for oil, and g for gas). Assume that the medium is water wet, and the effective saturation of each fluid is defined by the following equations

$$S_w^* = \frac{S_w - S_r}{1 - S_r} \qquad (4.73)$$

$$S_o^* = \frac{S_o}{1 - S_r} \qquad (4.74)$$

$$S_g^* = \frac{S_g}{1 - S_r} \qquad (4.75)$$

For the total liquid saturation

$$S_L^* = \frac{S_L - S_r}{1 - S_r} = S_w^* + S_o^* \qquad (4.76)$$

and in addition

$$S_w + S_o + S_g = 1 \tag{4.77}$$

We assume in defining Eqs. (4.73) to (4.75) that there is not any residual oil or residual gas. The definition of an effective saturation will depend on whether the relative permeability is on an imbibition or drainage cycle, because the amount of oil trapped (residual oil) will depend on the cycle (see Fig. 4.13). For different situations it may be more convenient to define effective saturation including residual (trapped) oil as well as residual (connate) water.

The equations analogous to Eqs. (4.58) and (4.59) in terms of S_L^* are

$$k_{rw} = (S_w^*)^2 \frac{\displaystyle\int_0^{S_w^*} p_c^{-2}\, dS_L^*}{\displaystyle\int_0^1 (p_c^*)^{-1}\, dS_2^*} \tag{4.78}$$

$$k_{ro} = (S_o^*)^2 \frac{\displaystyle\int_{S_w^*}^{S_L^*} p_c^{-2}\, dS_2^*}{\displaystyle\int_0^1 (p_c^*)^{-1}\, dS_2^*} \tag{4.79}$$

$$k_{rg} = (S_g^*)^2 \frac{\displaystyle\int_{S_i^*}^1 p_c^{-2}\, dS_2^*}{\displaystyle\int_0^1 (p_c^*)^{-1}\, dS_2^*} \tag{4.80}$$

A sketch of capillary pressure versus total liquid saturation is shown in Fig. 4.16. Equations (4.78) to (4.80) can be integrated in terms of λ to obtain equations analogous to Eqs. (4.67) and (4.68). Figure 4.17 shows k_{ro} and k_{rg} for $S_L = .2$, .4, and for $\lambda = 2$.

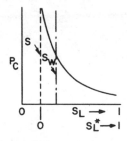

FIG. 4.16 Capillary pressure versus S_L^*: three phase. (From Standing, 1975.)

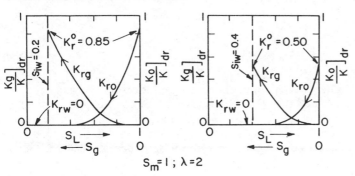

FIG. 4.17 Three phase drainage relative permeability curves. (From Standing, 1975.)

        Schneider and Owens (1970) measured three-phase water-oil-gas relative
permeabilities in sandstones and carbonates. They summarized general relative
permeability characteristics based on their measurements and those of other investigators (Corey et al., 1956; Land, 1968; Narr and Wygal, 1961; Saraf and Fatt,
1967; Sarem, 1966). The relative permeabilities in the two materials, sandstone
and carbonate, did not show different behavior. Wettability has the most influence.
The effects of heterogeneity, nonuniformity, and anisotropy were not considered.
The water relative permeability is a function of its own saturation. Therefore,
water relative permeabilities as determined in two-phase systems can be used in
three-phase systems that have the same wettability characteristics. Oil-phase relative permeabilities are related to their own saturations and saturation histories.
The relative permeability of a gas in an oil-wet system is not sensitive to residual
oil saturation. The relative permeability of gas in a water-wet system is decreased
by the presence of residual oil. The relative permeability of oil in an imbibition
process does not depend on the flowing gas phase when gas saturation is increasing.
The relative permeability of oil in an imbibing process increases when gas saturation decreases. In systems with no strong wetting preference, three-phase relative
permeabilities cannot be predicted from two-phase relative permeability.
        Stone (1970) presents a model to estimate three-phase relative permeabilities
for systems with a strong wetting preference, that is, based on two-phase data. In
the following assume a strongly water-wet system containing water, oil, and gas.
Normalized saturations are defined including both residual water and residual oil.
The flowing fluids are oil and gas with a connate water phase. The dimensionless
saturations are defined

$$S_o^* = \frac{S_o - S_{or}}{1 - S_w - S_{or}} \qquad S_o \geq S_{or} \tag{4.81}$$

$$S_w^* = \frac{S_w - S_{wr}}{1 - S_{wr} - S_{or}} \qquad S_w \geq S_{wr} \tag{4.82}$$

$$S_g = \frac{S_g}{1 - S_{wr} - S_{or}} \tag{4.83}$$

and

$$S^*_g + S^*_w + S^*_o = 1 \qquad (4.84)$$

When $S^*_o = 1$, $k_{ro} = 1$, but decreasing S^*_o by increasing either water or gas causes a decrease in k_{ro} which is greater than the decrease in S^*_o. A coefficient β_w is defined to correct for this disproportionate decrease in k_{ro} due to water. β_w is assumed to be a function of S^*_w. Similarly, for changes due to gas, β_w is defined as a function of S^*_g. Assuming impedance of oil due to gas or due to water to be independent, then

$$k_{ro} = S^*_o \beta_w \beta_g \qquad (4.85)$$

The values of β_w are based on values of k_{row} data for two-phase flow of oil and water

$$\beta_w = \frac{k_{row}}{1 - S^*_w} \qquad (4.86)$$

The values of β_g are based on values of k_{rog} for two-phase flow of oil and gas

$$\beta_g = \frac{k_{rog}}{1 - S^*_g} \qquad (4.87)$$

This development is restricted to $S_w \geq S_{wr}$. Figure 4.18 shows the two-phase data of Corey et al. (1956) for Berea cores. Figure 4.19 shows the comparison of the

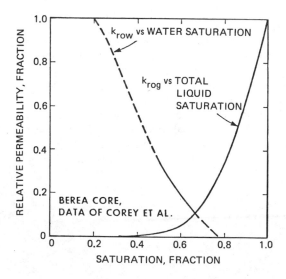

FIG. 4.18 Two-phase relative permeability data used in tests of probability model. (From H. L. Stone, Probability Model for Estimating Three-Phase Relative Permeability, <u>Pet. Trans. AIME</u> <u>249</u>, SPEJ 214. © 1970 SPE-AIME.)

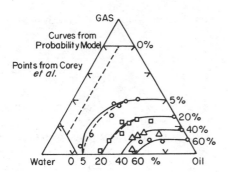

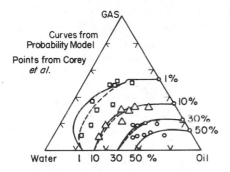

FIG. 4.19 Oil-phase relative isoperms
(Berea core). (From H. L. Stone, Probability Model for Estimating Three-Phase
Relative Permeability, Pet. Trans. AIME
249, SPEJ 214. © 1970 SPE-AIME.)

calculations using the two-phase data and the probability model with the data of
Corey et al. (1956).

For the movement of water in dry soil, a diffusion-type equation is sometimes
used since p_c is a single-valued function of wetting fluid saturations (Richards,
1931). The basic assumptions are (Handy, 1960) as follows: Both water and air are
continuous behind the imbibing front; pressure gradient in the gas is negligible both
ahead of and behind the imbibing water front; the capillary pressure gradient over
any increment of length provides the driving force for overcoming the viscous
forces in that same increment of length. Combining Darcy's law, the equation of
continuity, and the capillary pressure equation yields

$$\psi \frac{\partial S_w}{\partial t} = -\frac{\partial}{\partial x}\left[\left(\frac{k_w}{\mu_w}\frac{\partial p_c}{\partial S_w}\right)\frac{\partial S_w}{\partial x}\right]$$

(4.88)

We assume the term in parentheses is a saturation-dependent diffusion-type term.

$$\hat{D} = \frac{k_w}{\mu_w}\frac{\partial p_c}{\partial S_w}$$

(4.89)

and

$$\phi \frac{\partial S_w}{\partial t} = -\frac{\partial}{\partial x}\left(\hat{D}\frac{\partial S_w}{\partial x}\right)$$

(4.90)

For horizontal flow (no gravity) one can use the boundary value results from heat and mass transfer to calculate $S_W = S_W(t, x)$.

4.3 DISPLACEMENT IN IMMISCIBLE FLOW

In general, analytical solutions to the equations of change for immiscible flow, Eqs. (4.5) to (4.14), or for incompressible flow, Eqs. (4.27) and (4.28), are not possible. The nonlinearity of the equations results from the relative permeability saturation relationships. Most solutions of the immiscible flow equations are numerical. A description of numerical solutions and numerical methods used to solve the relative permeability equations and used to simulate reservoirs is given by Crichlow (1977). There is a short discussion of numerical solution techniques in Chap. 6.

The initial efforts used to solve the equations of change provide both useful limiting solutions as well as a description of the mechanism of displacement in immiscible flow. Buckley and Leverett (1941) simplified the problem by assuming one-dimensional incompressible flow and neglected the capillary pressure across the fluid-fluid interface. The capillary effects implied in the relative permeability are included. For relatively high flow rates or for flow over long distances the assumptions are reasonable. Consider linear displacement of oil either by gas or water, as sketched in Fig. 4.20. The equations for the water displacing oil are

$$q_w = -\frac{k_w A}{\mu_w}\left(\frac{\partial p_w}{\partial x} + \rho_w \sin \alpha\right) \tag{4.91}$$

$$q_o = -\frac{k_o A}{\mu_o}\left(\frac{\partial p_o}{\partial x} + \rho_o g \sin \alpha\right) \tag{4.92}$$

We ignore capillary pressure, the difference between the pressure in the two fluids. Since we assume the fluids incompressible,

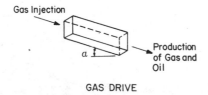

Gas Injection

Production of Gas and Oil

a

GAS DRIVE

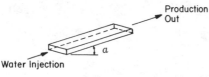

Production Out

Water Injection

a

WATER DRIVE

FIG. 4.20 Linear displacement. (From Muskat, 1937.)

$$\phi A \, \frac{\partial S_w}{\partial t} = -\frac{\partial q_w}{\partial x} \tag{4.93}$$

$$\phi A \, \frac{\partial S_o}{\partial t} = -\frac{\partial q_o}{\partial x} \tag{4.94}$$

Since

$$S_w + S_o = 1 \tag{4.95}$$

we combine Eqs. (4.93), (4.94), and (4.95)

$$\frac{\partial}{\partial x}(q_w + q_o) = 0 \tag{4.96}$$

Define the fraction of water flowing as

$$f_w = \frac{q_w}{q} \tag{4.97}$$

Then the fraction of oil flowing must be

$$f_o = \frac{q_o}{q} = 1 - f_w \tag{4.98}$$

Substituting (4.97) and (4.98) into Eq. (4.93) and (4.94), respectively,

$$\frac{\partial S_w}{\partial t} = -\frac{q}{\phi A} \, \frac{\partial f_w}{\partial x} \tag{4.99}$$

and

$$\frac{\partial S_o}{\partial t} = -\frac{q}{\phi A} \, \frac{\partial f_o}{\partial x} \tag{4.100}$$

Combining Eqs. (4.91) and (4.92), assuming no interfacial tension and no density difference and the density differences in the fluids small or $\alpha = 0$,

$$f_w = \frac{1}{1 + (k_o \mu_w / k_w \mu_o)} \tag{4.101}$$

f_w is a function of saturation since k_o and k_w contain relative permeability; the viscosity ratio is a parameter. Applying the chain rule in one dimension

$$\frac{\partial f_w}{\partial x} = \frac{df_w}{dS_w} \, \frac{\partial S_w}{\partial x} \tag{4.102}$$

Substituting Eq. (4.102) into Eq. (4.99)

$$\frac{\partial S_w}{\partial t} = -\left(\frac{q}{\phi A}\frac{df_w}{dS_w}\right)\frac{\partial S_w}{\partial x} \tag{4.103}$$

This equation is nonlinear since $\partial S_w/\partial x$ is a function of S_w. Since

$$S_w = S_w(x,\,t) \tag{4.104}$$

$$\frac{dS_w}{dt} = \frac{\partial S_w}{\partial x}\frac{dx}{dt} + \frac{\partial S_w}{\partial t} \tag{4.105}$$

Choose $x = x(t)$ to coincide with a fixed S_w; then

$$\frac{dS_w}{dt} = 0 \tag{4.106}$$

and

$$\left(\frac{dx}{dt}\right)_{S_w} = -\frac{dS_w/dt}{\partial S_w/\partial x} \tag{4.107}$$

Combining Eqs. (4.103) and (4.107), eliminating dS_w/dt, we obtain the Buckley-Leverett equation

$$\left(\frac{dx}{dt}\right)_{S_w} = \frac{q}{\phi A}\frac{df_w}{dS_w} \tag{4.108}$$

If the permeability ratio, k_o/k_w (or k_{ro}/k_{rw}) is known as a function of S_w, then we can determine df_w/dS_w as a function of S_w. Since S_w is known at $t = $ zero, Eq. (4.108) can be integrated to determine saturation at any $t > 0$.

$$x_{S_w}(t) - x_{S_w}(0) = \frac{Q(t) - Q(0)}{\phi A}\frac{df_w}{dS_w} \tag{4.109}$$

$$Q(t) = \int_0^{x(t)} \phi[S_w(t) - S_w(0)]\,A\,dx \tag{4.110}$$

Unfortunately, df_w/dS_w is a multiple-valued function of S_w in that more than one value may occur for a given value of S_w. This problem is overcome by postulating a discontinuity in S_w at the front and requiring the material balance to be satisfied. Using the relative permeabilities of Fig. 4.21, a curve of f_w versus S_w for a given μ_w/μ_o is constructed on Fig. 4.22. Considering an initial saturation distribution of $S_c < S_w < 1$ existing at $x = 0$ and $t = 0$ [$S_c \equiv S_w(0)$] and applying Eq. (4.109),

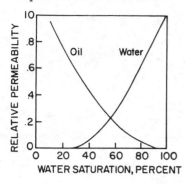

FIG. 4.21 Effect of saturation on relative permeabilities to water and oil in unconsolidated sands. (From S. E. Buckley and M. C. Leverett, Mechanism of Fluid Displacement in Sands, Pet. Trans. AIME 146, SPEJ 107. © 1941 SPE-AIME.)

we obtain a saturation distribution as in Fig. 4.23 for t > 0. The multiple values of saturation are eliminated by the following procedure. At t = 0 $S_w = S_c$, the wetting fluid injected at x = 0 ($f_w = 1$ at x = 0). A volumetric balance for the wetting fluid is given by Eq. (4.110). For $x > x_c$, $S_w = S_c$; this introduces a discontinuity of saturation at $x = x_c$. Integrate Eq. (4.110) by parts with $f_w = 1$ at $S_w = 1 - S_n(0)$

$$Q = \phi A (S'_w - S_c) x_c - Q[f_w(S_w) - 1] \tag{4.111}$$

where S'_w is from Fig. 4.22 and is the upper saturation at the discontinuity. Using Eq. (4.109) again gives

$$\phi A x_c - Q \frac{df_w}{dS_w} S'_w \tag{4.112}$$

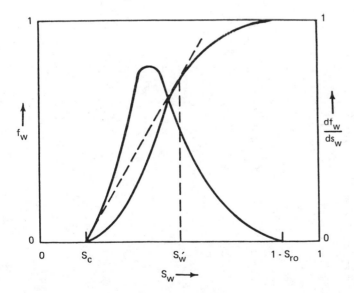

FIG. 4.22 Fraction of wetting fluid in flowing fluid as a function of wetting-fluid saturation; also the first-derivative curve showing the determination of saturation at the displacement front. (From Buckley and Leverett, 1941.)

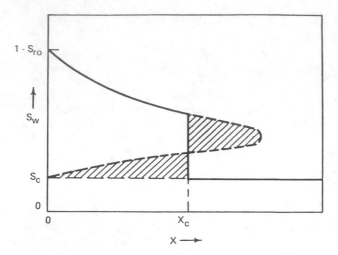

FIG. 4.23 Saturation distribution during linear immiscible displacement as computed from the Buckley-Leverett equation showing the discontinuity in saturation as required by a material balance. (From Buckley and Leverett, 1941.)

and

$$\frac{df_w(S'_s)}{dS_w} = \frac{f_w(S'_w)}{S'_w - S_c}$$

(4.113)

The saturation will remain single valued if all saturations below S'_w are removed. Figure 4.24 shows the results from Buckley and Leverett (1941) for the relative permeabilities of a typical oil-bearing sand displaced with water. Similar equations can be derived for a nonwetting fluid displacing oil (the gas drive in Fig. 4.20).

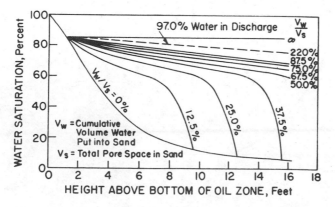

FIG. 4.24 Saturation history of oil-bearing sand under water flood. (From S. E. Buckley and M. C. Leverett, Mechanism of Fluid Displacement in Sands, Pet. Trans. AIME 146, SPEJ 107. 1941 SPE-AIME.)

Welge (1952) noted in the continuity equation that the change in fraction of the wetting fluid flowing to the saturation at the outflow face of system of length L is

$$(Q_w)_{in} = \frac{1}{(df_w/dS_w)_{out}} \tag{4.114}$$

For the outflow end

$$f_o = \frac{q_o}{q} = \frac{dQ_o}{(dQ_w)_{in}} \tag{4.115}$$

and

$$\frac{k_o}{k_w} = \frac{(\mu_o/\mu_w) \, dQ_o/(dQ_w)_{in}}{1 - dQ_o/(dQ_w)_{in}} \tag{4.116}$$

The ratio of the permeabilities can be estimated if $dQ_o/(dQ_w)_{in}$ is known. From a material balance on the wetting fluid

$$\phi AL(1 - S_c) - Q_o = \phi A \int_0^1 S_o \, dx \tag{4.117}$$

Integrating by parts and noting $dS_w = - dS_o$

$$\phi A x_{S_w} = (Q_w)_{in} \frac{df_w}{dS_w} \tag{4.118}$$

Then

$$S_o(L) = \frac{1}{\phi AL} \left[\phi AL(1 - S_c) - Q_o + (Q_w)_{in} \frac{dQ_o}{(dQ_w)_{in}} \right] \tag{4.119}$$

If one knows the relative permeabilities, the saturation history for the displacement can be determined. Or one can perform a displacement and determine the relative (or effective) permeability ratio.

Johnson et al. (1959) extended the Welge integration of the Buckley-Leverett equation to determine relative permeabilities of each fluid using a high rate two-fluid immiscible displacement on laboratory samples of a porous medium. If we consider a particular instant in the displacement experiment, the pressure drop across a system of length L is

$$\Delta p = - \int_0^L \frac{dp}{dx} \, dx \tag{4.120}$$

Substituting for dp/dx in terms of f_n

$$f_o = q \frac{kk_{ro}}{\mu_o} \frac{dp}{dx} \qquad (4.121)$$

gives

$$\Delta p = \frac{q\mu_o}{k} \int_0^L \frac{f_o}{k_{ro}} \, dx \qquad (4.122)$$

At a given instant in the displacement

$$\frac{x}{L} = \frac{df_w/dS_w}{(df_w/dS_w)_{out}} \qquad (4.123)$$

Combining Eqs. (4.122) and (4.123)

$$\int_0^{(df_o/dS_w)_{out}} \frac{f_o}{k_{ro}} \, d\left(\frac{df_o}{dS_w}\right) = \left(\frac{df_w}{dS_w}\right)_{out} \frac{q_s/\Delta p_s}{q/\Delta p} \qquad (4.124)$$

where q_s and Δp_s are the values when the displacement is started. Differentiating Eqs. (4.124) defining the right-hand side of Eq. (4.124) as $(df_w/dS_w)_{out}/I_r$

$$\frac{f_o}{k_{ro}} = \frac{d\,[(df_w/dS_w)_{out}/I_r]}{d\,(df_w/dS_w)_{out}} \qquad (4.125)$$

Since $(df_w/dS_w)_{out}$ is the reciprocal of the cumulative volume injected V_i, we can write Eq. (4.125) as

$$\frac{f_o}{k_{ro}} = \frac{d(1/V_i I_r)}{d(1/V_i)} \qquad (4.126)$$

Thus for any value of cumulative injection we can determine the fraction on oil f_o and from Eq. (4.126) determine k_{ro}. Knowing k_{ro}, we can obtain k_{rw}.

Douglas et al. (1958) formulated the displacement equation including the effects of capillary pressure. Since the total flow is

$$q = q_o + q_w \qquad (4.127)$$

Substituting Darcy's law for q_w and q_o,

$$\frac{\partial p_w}{\partial x} = -\frac{q}{A(k_w/\mu_w + k_o/\mu_o)} - \frac{dp_c/dS_o}{1 + (k_w\mu_o/k_o\mu_w)}\frac{\partial S_o}{\partial x} + \frac{(k_w\rho_w/\mu_w + k_o\rho_o)g\sin\alpha}{k_w/\mu_w + k_o/\mu_o}$$

(4.128)

and

$$\frac{\partial p_c}{\partial x} = \frac{dp_c}{dS_o}\frac{\partial S_o}{\partial x}$$

(4.129)

From Darcy's law for $\alpha = 0$

$$q_w = \frac{q}{1 + k_o\mu_w/k_w\mu_o} + \frac{A(k_w/\mu_w)\,dp_c/dS_o}{1 + k_w\mu_o/k_o\mu_w}\frac{\partial S_o}{\partial x}$$

(4.130)

Substituting into the continuity equation for the wetting field

$$\frac{\partial q_w}{\partial x} = \phi A\frac{dS_w}{dt}$$

(4.131)

This coupled set of differential equations must be solved numerically.

Capillary pressure is the only driving force for displacement by imbibition. The total flow of fluid in the medium is constant, so for any cross section

$$q = q_w + q_o = 0$$

(4.132)

For a cylinder of material of length L and cross section A,

$$q_w = -\frac{k_w A}{\mu_w}\frac{\partial p_w}{\partial x}$$

(4.133)

$$q_o = -\frac{k_o A}{\mu n}\frac{\partial p_o}{\partial x}$$

(4.134)

$$p_c = p_o - p_w$$

(4.135)

For the x direction positive into the face over which imbibition is taking place,

$$\frac{\partial p_w}{\partial x} = -\frac{k_o/\mu_o}{k_n/\mu_n + k_w/\mu_w}\frac{dp_c}{dS_o}\frac{\partial S_o}{\partial x}$$

(4.136)

Combining with Eq. (4.133),

$$q_w = \frac{k_w k_o A}{k_w\mu_o + k_o\mu_w}\frac{dp_c}{dS_o}\frac{\partial S_o}{\partial x}$$

(4.137)

The continuity equation is (where $q_w = -q_o$)

$$\frac{\partial q_o}{\partial x} = \phi A \frac{\partial S_o}{\partial t} \tag{4.138}$$

We again have a coupled set of differential equations that must be solved numerically.

Special problems involved in the field of soils science with the drainage and imbibition of water in air-water fluid systems are special applications of the preceding equations. These equations are discussed by Corey (1977) and will be considered again in Chap. 9.

4.4 MOVING BOUNDARIES: THE MUSKAT MODEL

Displacement of one immiscible fluid by another in a porous medium, under certain assumptions, can be represented as a moving boundary value problem (Muskat, 1937). The assumptions are the medium is initially saturated with one fluid and then this fluid is displaced by another; the residual amount of either fluid (for example, in an oil reservoir, connate water or residual oil) is ignored; the permeability-viscosity (mobility) ratios are constant; and there is a front between the two fluids, a region of abrupt change in the saturation of the two fluids. We model the front as a mathematical surface separating the two fluids. There is no mixing (dispersion) at the front. Assume only one fluid is moving ahead of the front and only the other fluid is moving behind the front.

Suppose a fluid, 1, displaces another fluid, 2, from a porous medium with straight parallel boundaries as sketched in Fig. 4.25. The plane of the two-dimensional motion is tilted an angle β from the vertical, and the bulk motion differs in angle α from the horizontal. Neglecting inertia Darcy's law describing the motion of each fluid on either side of the front is

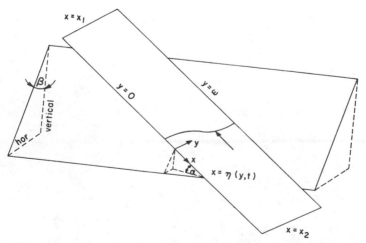

FIG. 4.25 Plane of the two-dimensional motion. (From Varnon, 1971.)

$$q = -\frac{\underline{\underline{k}}}{\mu} \underline{\nabla} \Phi \tag{4.139}$$

where

$$\Phi = p + \rho g z \tag{4.140}$$

The distance z in Fig. 4.25 is given by

$$z = y \cos \alpha \cos \beta - x \sin \alpha \tag{4.141}$$

The continuity equation for each fluid, assuming the fluids are incompressible, is

$$\underline{\nabla} \cdot \underline{q} = 0 \tag{4.142}$$

Combining Eqs. (4.139) and (4.142) yields Laplace's equation for each fluid

$$\nabla^2 \Phi_1 = 0 \tag{4.143}$$
$$\nabla^2 \Phi_2 = 0 \tag{4.144}$$

Equations (4.143) and (4.144) along with the appropriate boundary conditions describe the motion of each fluid. For impermeable boundaries the velocity normal to the boundary is zero; so

$$\frac{\partial \Phi_1}{\partial y} = 0 \qquad y = 0, w \tag{4.145}$$

$$\frac{\partial \Phi_2}{\partial y} = 0 \qquad y = 0, w \tag{4.146}$$

The potential functions are normally specified at the inlet and outlet so that

$$\Phi_1 = f_1(y, t) \qquad x = x_1 \tag{4.147}$$
$$\Phi_2 = f_2(y, t) \qquad x = x_2 \tag{4.148}$$

At the front, the normal velocity must be continuous to satisfy the mass balance, assuming an isotropic medium,

$$\frac{k_1}{\mu_1} \frac{\partial \Phi_2}{\partial n} = \frac{k_2}{\mu_2} \frac{\partial \Phi_2}{\partial n} \tag{4.149}$$

where n is normal to the front.

In addition there must be some relationship between the potentials on either side of the front

$$\Phi_1 = \Phi_1(\Phi_2) \tag{4.150}$$

The conditions expressed by Eqs. (4.149) and (4.150) refer to a floating or moving boundary. The position of the moving boundary is determined in the following manner. Let the front be represented by

$$x = \eta(y, t) \tag{4.151}$$

Apply the chain rule to Eq. (4.151) and differentiate with respect to time to obtain

$$\frac{dx}{dt} = \frac{\partial \eta}{\partial y} \frac{dy}{dt} + \frac{\partial \eta}{\partial t} \tag{4.152}$$

Since dx/dt and dy/dt are the velocity components in the x and y direction, Eq. (4.152) may be rewritten as

$$\frac{\partial \eta}{\partial t} = \frac{k}{\mu} \left(\frac{\partial \Phi}{\partial x} - \frac{\partial \eta}{\partial y} \frac{\partial \Phi}{\partial y} \right) \tag{4.153}$$

Equation (4.153) applies to each fluid. The frontal boundary position is given by an initial value problem; an initial condition of (y, 0) must be specified.

The problem may be restated as follows. Find the potential functions Φ_1 and Φ_2 which satisfy the conditions given by Eqs. (4.145 to (4.148) on the external boundaries and satisfy conditions given by Eqs. (4.149) and (4.150) on the mutual boundary $x = \eta$ given by Eq. (4.153). The problem is unsteady-state because the shape of the boundary $x = \eta$ is time dependent. If the shape of the boundary at $x = \eta$ is a constant, the time dependence is a translation. Solutions for this model require simultaneous generation of Φ_1, Φ_2, and η. Therefore Φ_1, Φ_2, and η are dependent variables. The frontal advance equation, Eq. (4.153), and the condition, Eq. (4.149), are nonlinear. The frontal boundary presents two problems in obtaining a solution. The geometry of the boundary may be irregular, the boundary conditions of Φ_1, and Φ_2 are not known at the outset; they are given implicitly by Eqs. (4.149) and (4.150).

The model is generally stated in three dimensions as follows. Determine the potential function Φ_1 between a surface S_w and $\eta(x, y, z, t) = 0$, and determine the potential function Φ_2, between $\eta(x, y, z, t) = 0$, and a surface S_e so that the following boundary conditions are satisfied:

$$\Phi_1 = \Phi_w \qquad \text{on } S_w \tag{4.154}$$

$$\Phi_2 = \Phi_e \qquad \text{on } S_e \tag{4.155}$$

On the front, $\eta(x, y, z, t) = 0$

$$\Phi_1 = \Phi_2 \tag{4.156}$$

$$\frac{k_1}{\mu_1} \frac{\partial \Phi_1}{\partial \eta} = \frac{k_2}{\mu_2} \frac{d\Phi_2}{\partial n} \tag{4.157}$$

with a moving boundary given by

$$\frac{\partial \eta}{\partial t} = \left(\frac{k}{\mu} \right)_{1,2} \nabla \Phi_{1,2} \nabla \eta \tag{4.158}$$

The simplest case of applying the preceding model is in one-dimensional flow with $\partial \Phi / \partial y = 0$. In this case the frontal advance equation, Eqs. (4.158) and (4.157)

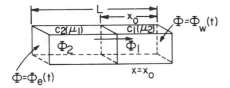

FIG. 4.26 A linear encroachment system. (From Muskat, 1937.)

are linear; the front has a specified constant shape. Figure 4.26 is a sketch of a linear system. The length of the system is L. Gravity and capillary effects are assumed negligible. From Darcy's law and the continuity equation, Eqs. (4.143) and (4.144) become for fluids 1 and 2

$$\frac{\partial^2 p_1}{\partial x^2} = 0 \qquad 0 \le x = x_f \tag{4.159}$$

$$\frac{\partial^2 p_2}{\partial x^2} = 0 \qquad x_f \le x \le L \tag{4.160}$$

At the front, $x = x_f$,

$$p_1 = p_2 \tag{4.161}$$

$$\frac{k_1}{\mu_1}\frac{\partial p_1}{\partial x} = \frac{k_2}{\mu_2}\frac{\partial p_2}{\partial x} \tag{4.162}$$

where k_1 is the permeability to fluid 1 at residual saturation of fluid 2, k_2 is the permeability to fluid 2 at residual saturation of fluid 1. The boundary conditions are

$$p_1 = p_e \qquad x = 0 \tag{4.163}$$

$$p_2 = p_w \qquad x = L \tag{4.164}$$

Integrating Eqs. (4.159) and (4.160) gives

$$p_1 = Ax + B \tag{4.165}$$
$$p_2 = Cx + D \tag{4.166}$$

Define the mobility ratio as

$$m = \frac{k_1 \mu_2}{k_2 \mu_1} \tag{4.167}$$

Applying the boundary conditions defining

$$\Delta p = p_e - p_w \tag{4.168}$$

yields

$$A = -\frac{\Delta p}{mL + (1 - m)x_f} \qquad (4.169)$$

$$B = p_e \qquad (4.170)$$

$$C = -\frac{m\,\Delta p}{mL + (1 - m)x_f} \qquad (4.171)$$

$$D = -\frac{(1 - m)x_f\,\Delta p}{mL + (1 - m)x_f} + p_e \qquad (4.172)$$

If S_{r2} is the residual saturation of fluid 2 behind the front and S_{r1} is the residual saturation of fluid 1 ahead of the front, then from a material balance the velocity of the front in terms of these residual saturations is

$$v_f = \frac{q_1}{\phi(1 - S_{r2} - S_{r1})} \qquad (4.173)$$

and

$$v_f = \frac{q_2}{\phi(1 - S_{r2} - S_{r1})} \qquad (4.174)$$

On the moving boundary the normal velocities on either side of the front are equal. From the above,

$$q_1 = -\frac{\underline{\underline{k}}_1}{\mu_1}\frac{\partial p_1}{\partial x} = -\frac{\underline{\underline{k}}_1}{\mu_1}A \qquad (4.175)$$

From Eq. (4.173),

$$\frac{dx_f}{dt} = \frac{\underline{\underline{k}}_1\,\Delta p}{\mu_1\phi(1 - S_{r2} - S_{r1})}\frac{1}{mL + (1 - m)x_f} \qquad (4.176)$$

Integrating Eq. (4.176) with the initial condition, $x_f = 0$ at $t = 0$,

$$t = \frac{\mu_1\phi(1 - S_{r2} - S_{r1})}{\underline{\underline{k}}_1\,\Delta p}[mLx_f + \frac{1}{2}(1 - m)x_f^2] \qquad (4.177)$$

Figure 4.27 is a plot of this result. For $m > 1$ the front accelerates; for $m < 1$, the front decelerates. The pressure distribution is independent of the front in the model.

Muskat (1937) solves the problem of radial flow, as sketched in Fig. 4.28. The solution for this problem is

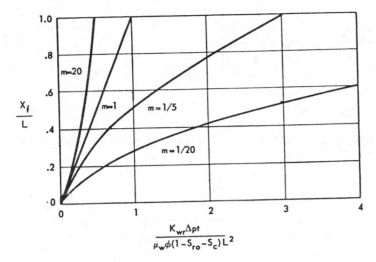

FIG. 4.27 Fraction of length traversed versus time for linear frontal displacement showing the effect of mobility ratio. (From Muskat, 1937.)

$$\frac{r_o^2}{r_e^2}\left[\ln\frac{r_e^2}{r_w^2} - \left(1 - \frac{1}{m}\right) + \left(1 - \frac{1}{m}\right)\ln\frac{r_o^2}{r_e^2}\right] - 4\frac{k_{=1}}{\mu_1}t\frac{\Phi_e - \Phi_w}{r_e^2} + \ln\frac{r_e^2}{r_w^2} - \left(1 - \frac{1}{m}\right)$$

$$(4.178)$$

Figure 4.29 shows the progress of the interface for $r_e = 2000r_w$, where

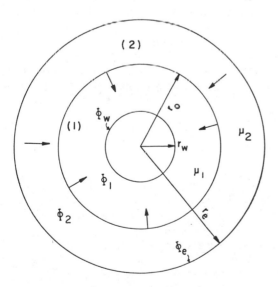

FIG. 4.28 Geometrical layout in Muskat's approximate solution of a radial displacement problem. (From Muskat, 1937.)

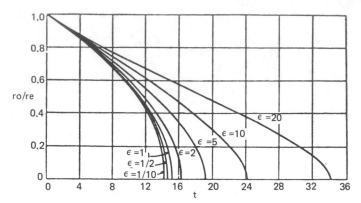

FIG. 4.29 The progress of the interface in a radial-encroachment system as a function of the time. r_o = radius of interface at time t (arbitrary units). r_e = external boundary radius. ϵ = (viscosity of encroaching liquid)/(viscosity of liquid displaced). r_w = well radius = $r_e/2000$. (From Muskat, 1937.)

$$4\frac{k_{=1}}{\mu_1}(\Phi_e - \Phi_w)\frac{1}{r_e^2} = 1 \tag{4.179}$$

Muskat (1937) and others have used this technique for estimating the frontal advance under various operating conditions.

Bear (1972) discusses perturbation methods to linearize the equations for certain approximate solutions and for cases of a phreatic surface, p = 0. The velocity potential and its derivatives on the free surface can be related to the position of the free surface y = h(x, t) (Brahma and Harr, 1962; Dvinoff, 1970). Frontal advance equations are then written in terms of h(x, t). The problem is reduced from a moving boundary value problem in two dependent variables to a fixed boundary value problem in one dependent variable; the problem is still nonlinear. The moving boundary formulation is also used in studying filtering of solid particles.

4.5 UNSTABLE FLOW

The discussion of multifluid flow to this point has ignored instabilities; we have tacitly assumed a front where a steep saturation gradient exists when discussing the relative permeability and Muskat models for multifluid flow. When two or more fluids are flowing simultaneously, flow is characterized by interface changes between the fluids, as a result of changes in relative forces (Varnon and Greenkorn, 1969). For simplicity, consider immiscible two-fluid flow such as flow of a gas and a liquid in a porous medium. The two fluids have different viscosities and densities. At low displacement velocities the gravity forces will dominate the flow. As the velocity increases, viscous forces affect the flow and eventually there is a balance between gravity and viscous forces. As velocity increases further, viscous forces dominate the flow.

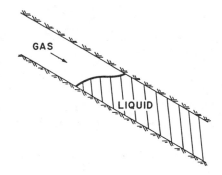

FIG. 4.30 Gravity-induced stable flow.
(From J. E. Varnon and R. A. Greenkorn, Unstable Two Fluid Flow in a Porous
Medium, Soc. Pet. Eng. J. 9, 293.
© 1969 SPE-AIME.)

In the plane parallel to gravity, four flow regimes result as velocity increases.
The first regime, a gravity-induced stable flow regime, is sketched in Fig. 4.30.
The general flow pattern represented in Fig. 4.30 persists with increasing velocity
until the interface becomes parallel with the bulk flow—then an unstable gravity
finger forms. The length of this gravity finger grows, and the fluid behind the nose
of the finger is practically immobile. This unstable gravity-dominated flow regime
is sketched in Fig. 4.31. As the velocity of the displacing fluid continues to increase,
the gravity finger thickens until it spans the medium, creating a stable interface
and all of the in-place fluid is mobile again. This stable regime sketched in Fig.
4.32 results from a balance between gravity and viscous forces. This regime will
not occur in the absence of gravity forces (in the plane perpendicular to gravity or
if the two fluids have the same density). As the velocity of the displacing fluid continues to increase, the viscous forces begin to dominate and the interface breaks
into viscous fingers. Figure 4.33 is a sketch of the viscous-force-induced unstable
flow regime. In general, displacement is three-dimensional and interface motion
must be characterized by considering both gravity and viscous forces. In a given
situation it is possible to have a combination of flow regimes. In the plane perpendicular to gravity there may be viscous fingers, while in the plane parallel to
gravity there may be gravity, stable or viscous dominated flow.

On a microscopic level, every real porous medium is nonuniform, and mixing
of the fluid particles will occur. This mixing or dispersion is essentially a microscopic phenomenon. Under close inspection in two-fluid flow, there is no single
sharp front but a transition zone in which both fluids may be moving. Mixing is
accentuated if the two fluids have different mobilities. The macroscopic front is not

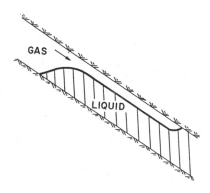

FIG. 4.31 Gravity-dominated unstable
flow. (From J. E. Varnon and R. A.
Greenkorn, Unstable Two Fluid Flow in
a Porous Medium, Soc. Pet. Eng. J. 9,
293. © 1969 SPE-AIME.)

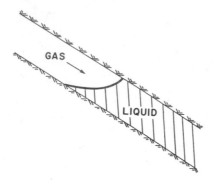

FIG. 4.32 Gravity-viscous balanced
stable flow. (From J. E. Varnon and
R. A. Greenkorn, Unstable Two Fluid
Flow in a Porous Medium, Soc. Pet.
Eng. J. 9, 293. © 1969 SPE-AIME.)

distinct, but is "blurred" by the microscopic mixing. This mixed zone increases in
size as the displacement progresses.

The unstable flow discussed previously for adverse situations is called fingering, and is a macroscopic phenomenon. Fingering will take place in two-fluid flow
under unstable conditions even if there is no porous structure. Dispersion, on the
other hand, results from microscopic irregularities in the pore structure. If there
were no pore structure, the only microscopic mixing of the two fluids would be due
to molecular diffusion and changes in the containing boundaries. Fingers grow, that
is, there is macroscopic mixing with propagation only if conditions exist for unstable
flow—gravity or adverse mobility ratios. Dispersion and unstable fingering may
occur simultaneously, but they are different mechanisms of mixing the fluids. When
we refer to unstable flow or flow instability or fingering, we are referring to a
macroscopic phenomenon related to the growth of a disturbance in the overall shape
of the interface of the saturation transition zone. When we refer to dispersion, we
are referring to a predominantly microscopic phenomenon which thickens the transition zone. Large-scale nonuniformities or heterogeneities may also affect dispersion
since this large-scale difference also means a change in the microscopic nature of
the pores.

It is at the point of including dispersion that models for immiscible and
miscible displacements diverge. In the miscible displacement model the dispersion
transport is introduced directly as a velocity superimposed on the Darcy velocity.
At each point there is a single fluid, a single Darcy velocity, a single pressure,
plus a dispersion velocity for each component in the fluid. For immiscible displace-

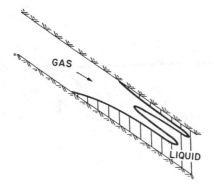

FIG. 4.33 Viscous-dominated unstable
flow. (From J. E. Varnon and R. A.
Greenkorn, Unstable Two Fluid Flow in a
Porous Medium, Soc. Pet. Eng. J. 9,
293. © 1969 SPE-AIME.)

ments, Darcy's law is assumed to describe the velocity of each fluid. In the Muskat model we assume a sharp interface with essentially a single fluid or phase flowing on either side of the interface with no mixing at the interface. In the relative permeability model, the relative permeability of each fluid depends on the fluid saturation. The saturation of each fluid is assumed to be continuous and to have a value for each fluid for every point in the medium. The displacement is treated as the simultaneous flow of two fluids. The relative permeabilities take into account any mixing since the saturation dependence is determined experimentally. The mixing is included in Darcy's law as the difference in viscosity and relative permeabilities of the two fluids. Mixing takes place only in the direction of flow; if the velocity has no component in a given direction, neither does the mixing. The widening of the transition zone is due to viscous-capillary interaction. Since there is no lateral mixing, the relative permeability model may not be applicable for gravity and viscous fingers in the plane parallel to gravity since it does not allow for lateral mixing.

Viscous fingering in unstable flow has been studied experimentally in two dimensions in the plane perpendicular to gravity, no gravity effect, in both packed and Hele-Shaw models. The Hele-Shaw model is a two-dimensional fluid analog of flow in porous media (see Chapter 6). The experimental results have been modeled using the Muskat model for the linear region of finger growth, infant fingers or fully developed fingers. Chuoke et al. (1959) observed macroscopic instabilities during the displacement of oil by water-glycerine in a Hele-Shaw model and of oil by water in glass powder packs. Figures 4.34 and 4.35 show the interfaces in these models. They determined the critical rate at which the instability occurs and the wavelength of maximum instability—that is, the number of infant fingers—were functions of permeability, viscosity, and effective interfacial tensions. Chuoke et al. (1959) posed the problem using the Muskat model as follows. When a finger first forms because of its small amplitude both $\partial \Phi / \partial y$ and $\partial \eta / \partial y$ are small and their product is neglected in Eq. (4.153). The frontal advance equation is nonlinear

$$\frac{\partial \eta}{\partial t} = \frac{k}{\mu} \frac{\partial \Phi}{\partial x} \qquad (4.180)$$

for fluids 1 and 2, and on the interface

$$\frac{\partial \Phi_1}{\partial x} = \frac{\partial \Phi_2}{\partial x} \qquad (4.181)$$

The solution of Chuoke et al. (1959) results from a perturbation analysis. The following solution parallels that in the previous section where the problem is solved as an initial value problem. Consider an infinitely long system at a uniform input velocity U (for a porous medium, U is equivalent to q_1, V is equivalent to q_2):

$$\Phi_1 = Ux \qquad x = -\infty \qquad (4.182)$$

$$\Phi_2 = Ux \qquad x = +\infty \qquad (4.183)$$

For the pressure discontinuity relationship, Eq. (4.150),

$$p_1 - p_2 = -\sigma^* \frac{\partial^2 \eta}{\partial y^2} + p_c \qquad (4.184)$$

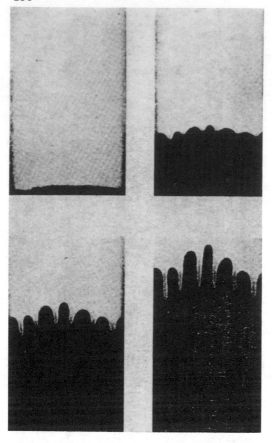

(a)

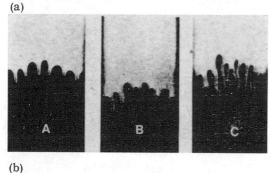

(b)

FIG. 4.34 (a) Instability of a water–oil interface moving with uniform normal speed of 0.41 cm sec^{-1} as compared to critical speed of 0.23 cm sec^{-1} in a tilted channel formed by parallel plates. Angle of tilt, 45°; $\mu_1 = 0.552$ P; and $\mu_2 = 1.30$ P. (b) The effect of rate on finger spacing as observed in tilted channel formed by parallel plates. Fluids and angle of tilt are same as in (A) U = 0.41 cm sec^{-1}, (B) U = 0.87 cm sec^{-1}, and (C) U = 1.66 cm sec^{-1}. (From R. L. Chuoke, P. van Muers, and C. van der Poel, The Instability of Slow, Immiscible, Viscous Liquid–Liquid Displacements in Permeable Media, Pet. Trans. AIME 216, SPEJ 188. © 1959 SPE-AIME.)

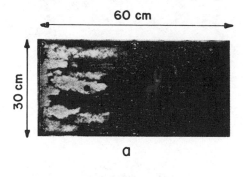

60 cm

30 cm

a

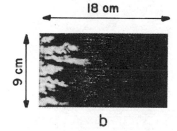

18 om

9 cm

b

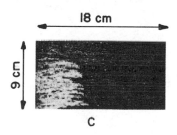

18 cm

9 cm

c

FIG. 4.35 Three examples showing the effect of oil viscosity and bulk interfacial tension on finger spacing. Displacing water appears white. (a) $\mu_O = 9.45$ cP, k = 96 D, $\mu_O = 1$ cP, U = 0.00932 cm^{-1}, and $\sigma = 42$ dyne cm^{-1}; (b) $\mu_O = 66$ cP, k = 65 D, $\mu_W = 0.936$ cP, U = 0.0106 cm sec^{-1}, and $\sigma = 48$ dyne cm^{-1}; (c) $\mu^O = 202$ cP, k = 65 D, $\mu^W = c$ cP, U = 0.00525 cm sec^{-1}, and $\sigma = 3.5$ dyne cm^{-1}. (From R. L. Chuoke, P. van Muers, and C. van der Poel, The Instability of Slow, Immiscible, Viscous Liquid-Liquid Displacements in Permeable Media, Pet. Trans. AIME 216, SPEJ 188. © 1959 SPE-AIME.)

where σ^* is the effective interfacial tension (σ^* is equal to σ in a Hele-Shaw model). As the initial condition use a wavelike corrugation of wavelength

$$\Omega = \frac{2\pi}{\gamma} \tag{4.185}$$

described by

$$\eta(y, 0) = Ee^{i\gamma y} \tag{4.186}$$

If

$$e^{\gamma\eta} = e^{-\gamma\eta} = 1 \tag{4.187}$$

the solutions of Eqs. (4.180), (4.182), and (4.183) are

$$\eta = Ut + Ee^{nt+i\gamma y} \tag{4.188}$$

$$\Phi_1 = Ux + \frac{nE}{\gamma} e^{\gamma x + nt + i\gamma y} \tag{4.189}$$

$$\Phi_2 = Ux - \frac{nE}{\gamma} e^{-\gamma x + nt + i\gamma y} \tag{4.190}$$

where

$$n = \left| \frac{(\mu_2/k_2 - \mu_1/k_1)U - (\rho_2 - \rho_1) g \sin \alpha}{\mu_2/k_2 + \mu_1/k_1} \right| \gamma - \left| \frac{\sigma^*}{\mu_2/k_2 + \mu_1/k_1} \right| \gamma^3 \tag{4.191}$$

For the distortion to grow, n must be positive. Since γ is positive, the condition for $n > 0$ is

$$\left(\frac{\mu_2}{k_2} - \frac{\mu_1}{k_1}\right)U - (\rho_2 - \rho_1)g \sin \alpha - \sigma^*\left(\frac{2\pi}{\Omega}\right)^2 > 0 \tag{4.192}$$

The distortions will grow if the velocity and wavelength exceed critical values given by

$$U_c = \frac{(\rho_2 - \rho_1)g \sin \alpha}{\mu_2/k_2 - \mu_1/k_1} \tag{4.193}$$

and

$$\Omega_c = 2\pi \left[\frac{\sigma^*}{(\mu_2/k_2 - \mu_1/k_1)(U - U_c)}\right]^{\frac{1}{2}} \tag{4.194}$$

The index of stability n given by Eq. (4.191), possesses an absolute maximum as a function of γ. Maximizing n with respect to γ gives

$$\Omega_m = \sqrt{3}\Omega_c \tag{4.195}$$

It is reasonable to assume Ω_m should be the observed singular wavelength since it corresponds at early times to the fastest growing finger.

Gupta et al. (1973) and Gupta and Greenkorn (1974) showed that the theory of Chuoke et al. (1959) correctly predicted the number of infant fingers for Hele-Shaw models and a porous medium model of various widths. Gupta et al. (1973, 1974) showed that the effect of local heterogeneities overrode the predicted wavelength. Further, in every case one finger rapidly dominated, growing at the expense of the others to a single parallel-sided steady-state finger. When the finger is mature, i.e., when it reaches a steady-state parallel-sided finger, the velocity of the nose is uniform. If the input velocity is constant, the finger increases linearly with time and does not change in shape. Figure 4.36 is a sketch of a fully developed steady-state finger. Since the velocity is uniform (for a porous medium $q_1 = U$, $q_2 = V$),

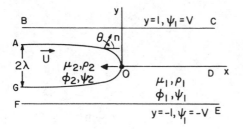

FIG. 4.36 Finger moving into a channel.
(From Saffman and Taylor, 1958.)

$$\frac{\partial \Phi_1}{\partial x} = U \tag{4.196}$$

$$\frac{\partial \Phi_2}{\partial y} = 0 \tag{4.197}$$

The frontal advance equation, Eq. (4.153), becomes

$$\frac{\partial \eta}{\partial t} = U \tag{4.198}$$

Since mass is conserved, $\psi_1 = \psi_2$ for $x = \eta$, for all y

$$\psi_1 = \psi_2 = Uy \qquad x = \eta \tag{4.199}$$

Saffman and Taylor (1958) solved Laplace's equation

$$\nabla^2 \Phi = 0 \tag{4.200}$$

for this case. The solution is difficult because of the interface shape. The boundary conditions are

$$\frac{\partial \Phi_2}{\partial y} = 0 \qquad y = 0, \ 1 \tag{4.201}$$

$$\frac{\partial \Phi_2}{\partial x} = U^*x \qquad x = -\infty \tag{4.202}$$

$$\frac{\partial \Phi_2}{\partial x} = Vx \qquad x = +\infty \tag{4.203}$$

where U^* is the velocity adjacent to a fully developed steady-state finger. If the pressure on either side of the finger is assumed equal,

$$\mu_1 \Phi_1 = \mu_2 \Phi_2 \tag{4.204}$$

or

$$\mu_1 Ux = \mu_2 \Phi_2 \tag{4.205}$$

Combining Eqs. (4.204) and (4.205)

$$\Phi_2 = \frac{\mu_1}{\mu_2} Ux \tag{4.206}$$

On the finger, $\eta(y)$

$$\Phi_2 = U^*x \tag{4.207}$$

Saffman and Taylor (1958) solved this problem by letting y be the dependent variable and writing Laplace's equation as follows:

$$\frac{\partial^2 y}{\partial \Phi^2} + \frac{\partial^2 y}{\partial \psi^2} = 0 \tag{4.208}$$

using the boundary conditions

$$y = 0 \qquad \Psi = 0 \tag{4.209}$$

$$y = 1 \qquad \Psi = V \tag{4.210}$$

$$y = \frac{\Psi}{V} \qquad \Phi > Vx \ (x \to \infty) \tag{4.211}$$

$$y = \frac{\Psi}{U^*} \qquad \Phi > U^*x \ (x \to -\infty) \tag{4.212}$$

$$y = \frac{\Psi}{U} \qquad \Phi = U^*x \tag{4.213}$$

Defining

$$\tilde{\Phi} = \Phi - U^*x \tag{4.214}$$

$$\tilde{\Psi} = \Psi - U^*y \tag{4.215}$$

the domain becomes the semi-infinite strip bounded by $\Phi = 0$, ∞, and $\tilde{\Psi} = 0$, $V - U^*$ as in Fig. 4.37. The transformed problem is solved to yield

$$x = \frac{1 - \lambda}{\pi} \ln \frac{1}{2} \left(1 + \cos \frac{\pi y}{\lambda} \right) \tag{4.216}$$

where

$$\lambda = \frac{V - U^*}{U - U^*} \tag{4.217}$$

Since λ is not specified, the solution is nonunique. Saffman and Taylor (1958) observed $\lambda = 1/2$ in their experiments.

Greenkorn et al. (1967) solved the steady-state problem in the plane parallel to gravity including the effect of gravity. Figure 4.38 is a sketch of the finger for this problem. They used the same procedures as Saffman and Taylor (1958) in that they transformed the problem into the $\tilde{\Phi}$ and $\tilde{\Psi}$ plane. The boundary conditions for the transformed problem are shown in Fig. 4.39. The velocity potential for fluid 1 is

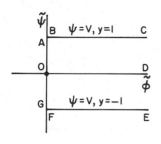

FIG. 4.37 The potential plane for motion in a channel. (From Saffman and Taylor, 1958.)

$$\tilde{\Phi}_1 - \frac{(\rho_1 - \rho_2)gk}{\mu_1 U} \tilde{\Psi}_1 = G\tilde{\Psi}_1 \tag{4.218}$$

where we assume $\mu_1 \Phi_1$ is much greater than $\mu_2 \Phi_2$. The solution of Eq. (4.218) is

$$\tilde{\Phi}_1 = \frac{Vh}{\lambda} \sum_{i=1}^{\infty} A_n \exp\left(\frac{-n\pi\tilde{\Phi}_1}{Vh}\right) \sin\frac{n\pi\tilde{\Psi}_1}{Vh} \tag{4.219}$$

where the A_n's were determined by collocation. In terms of the coordinates of the interface

$$x = \frac{Gy}{\lambda} - \sum_{i=1}^{\infty} A_n \exp\left(-\frac{n\pi Gy}{\lambda h}\right) \cos\frac{n\pi y}{\lambda h} \tag{4.220}$$

Greenkorn et al. (1967) were able to match their Hele-Shaw experiments by adjusting λ except in the region where the top of the finger intersects the upper boundary. The reason the finger would not match here is that interfacial forces between the boundary and the fluid are not included in the mathematical model. The finger width in this plane depends strongly on velocity. Local heterogeneities in this plane have virtually no effect on these gravity-dominated viscous fingers.

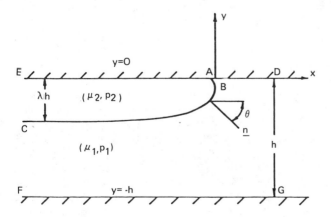

FIG. 4.38 Interface between two immiscible fluids. (From Greenkorn, Matar, and Smith, 1967.)

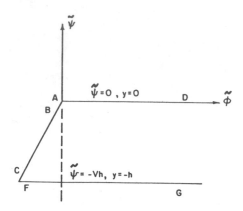

FIG. 4.39 Interface in potential plane.
(From Greenkorn et al., 1967.)

The nonuniqueness of the steady-state formulation for growth of unstable fingers has received attention. Outmans (1963) assumed the problem should be posed with more than one finger. Saffman and Taylor (1958) and Scheidegger (1960) suggested conservation of energy or production of entropy must be included in the model formulation. Varnon and Greenkorn (1970) determined the general form of the multifinger problem and concluded that λ is nonunique.

There has been discussion of the nonlinear region—the growth region from infant to steady-state fingers. Outmans (1962) used a first-order theory with symmetrical fronts to consider this region. Jacquard and Seguier (1962) developed an analytical solution to the Muskat model for sinusoidal initial conditions and a pressure condition, $p_1 = p_2$. This solution considers the growth of a small disturbance to a parallel-sided finger. The solution maintains the symmetry of the initial condition and gives $\lambda = 1/2$ for a fully developed finger.

Gupta et al. (1973) present experimental data from both Hele-Shaw and porous media models that show the tail and the nose of a viscous finger grow at constant velocity and further this constant velocity behavior begins almost immediately when the finger is formed. Figure 4.40 shows the growth of a viscous finger in a plane perpendicular to gravity. Figure 4.41 shows the growth of a finger in the same system for the planes perpendicular to and parallel to gravity. These figures illustrate the effect of gravity on the viscous finger.

Rachford (1964) solved the relative permeability model numerically to calculate a first-order stability analysis for immiscible displacement with adverse mobility ratios. The awkward dependence of relative permeability and capillary pressure on saturation makes it difficult to develop cleancut stability criteria. He compared predicted stability behavior against finite difference solutions for some completely specified displacement problems. Perturbations for the numerical solution were introduced by letting permeability vary. His calculated results never predicted gross instability. The computed stability behavior was insensitive to changes in viscosity ratio and velocity. Rachford (1964) concludes that scaling in models is correct and further that in water-wet oil systems the macroscopic fingers are not as severe as in Hele-Shaw models. There is nothing inconsistent with the Hele-Shaw scale model results, or the relative permeability model calculations. For immiscible displacements without gravity, that is, in the plane perpendicular to flow, macroscopic fingers are replaced by microscopic mixing in the direction of flow. The mixing is inherently included in the relative permeability.

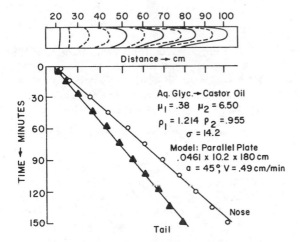

FIG. 4.40 Experiment suggesting linear development rate. (From Varnon, 1971.)

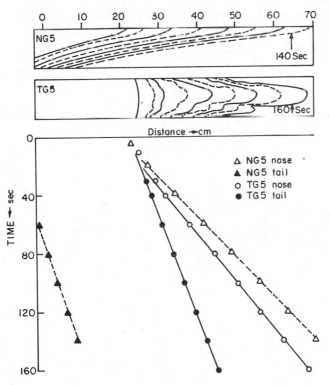

FIG. 4.41 Illustration of linear development rate. (From Varnon, 1971.)

4.6 TRANSIENT FLOW

The problem of predicting pressure and saturation distributions as a function of time and position (and the inverse of the problem, determining fluid-media properties) is usually solved using the single-fluid equations. There is some research into using the nonlinear multifluid equations to model pressure transients, but the results of this approach have not been experimentally verified. We will discuss using the nonlinear multifluid equations to model pressure transients briefly at the end of this section.

The equations describing the immiscible multifluid flow problem are generally Eqs. (4.5) to (4.14) or in their summarized form, Eqs. (4.27) and (4.28). In addition, if there is phase change we would use the equations including formation volume factors (see Sec. 4.7). The usual approach is to simplify the expressions and then make them look like the single fluid equations by using an average expression for the relative permeability and compressibility. Perrine (1956) states heuristically that the single-fluid model applies after substituting effective total properties of the multifluid system in terms of equivalent fluid properties. He assumed the effective total mobility and production rate are the sums of the mobilities and production rates for the separate fluids. For the three fluids oil, gas, and water in a porous medium,

$$\left(\frac{k}{\mu}\right)_T = \left(\frac{k}{\mu}\right)_o + \left(\frac{k}{\mu}\right)_g + \left(\frac{k}{\mu}\right)_w \qquad (4.221)$$

Perrine (1956) suggests using the method of Miller et al. (1950) to determine the mobilities from the slope of transient pressure data from a well in a radial system and from appropriate formation volume factors. From the measured pressure transient data

$$\left(\frac{k}{\mu}\right)_T = \frac{162.6}{mh} (Bq)_T \qquad (4.222)$$

where m is the slope of the buildup curve, B is formation volume factor, q refers to production rate, and h is the thickness of the radial system. All pressure-dependent quantities are determined at estimated average pressures of the system. The mobilities of each fluid are

$$\left(\frac{k}{\mu}\right)_o = \frac{162.6 B_o q_o}{mh} \qquad (4.223)$$

$$\left(\frac{k}{\mu}\right)_g = \frac{162.6 B_g}{mh} (q_g - q_o R_s) \qquad (4.224)$$

$$\left(\frac{k}{\mu}\right)_w = \frac{162.6 B_w q_w}{mh} \qquad (4.225)$$

where R_s is the solution gas/oil ratio. The total compressibility is defined as the total decrease in reservoir volume per unit volume per psi pressure increase.

$$c_T = c_{of} + c_{gf} + c_{wf} \qquad (4.226)$$

Compressibilities at the average system pressure are

$$c_{of} = \frac{S_o}{S_o + S_g + S_w}\left(\frac{B_g}{B_o}\frac{dR_s}{dp} - \frac{1}{B_o}\frac{dB_o}{dp}\right) \qquad (4.227)$$

$$c_{gf} = \frac{S_g}{S_o + S_g + S_w}\frac{1}{p} \qquad (4.228)$$

$$c_{wf} = \frac{S_w}{S_o + S_g + S_w}c_w \qquad (4.229)$$

Martin (1959) showed the heuristic approach of Perrine (1956) was a valid approximation for multifluid transient flow if the pressure and saturation gradients are small. Assume the effects of gravity and compressibility of the matrix may be neglected; then the simultaneous flow of oil, gas, and water are given by Eqs. (4.27) and (4.28)

$$\phi\frac{\partial}{\partial t}\left(\frac{R_s S_o}{B_o} + \frac{R_{sw} S_w}{B_w} + \frac{S_g}{B_g}\right) = \underline{\nabla}\cdot\left(\frac{R_o k_o}{B_o \mu_o} + \frac{R_{sw} k_w}{B_w \mu_w} + \frac{k_g}{B_g \mu_g}\right)\underline{\nabla}p \qquad (4.230)$$

$$\phi\frac{\partial(S_o/B_o)}{\partial t} = \underline{\nabla}\cdot\frac{k_o}{B_o \mu_o}\underline{\nabla}p \qquad (4.231)$$

$$\phi\frac{\partial(S_w/B_w)}{\partial t} = \underline{\nabla}\cdot\frac{k_w}{B_w \mu_w}\underline{\nabla}p \qquad (4.232)$$

In those cases where the pressure gradient and saturation gradients are small, the scalar products of the gradients $\underline{\nabla}p\cdot\underline{\nabla}p$, $\underline{\nabla}p\cdot\underline{\nabla}S$, and $\underline{\nabla}p\cdot\underline{\nabla}S_w$ are small compared to $\underline{\nabla}p$, $\underline{\nabla}S_o$, and $\underline{\nabla}S_w$. Neglecting the above scalar products in the expansions of Eqs. (4.230) to (4.232) yields

$$\frac{\phi}{R_s k_o/B_o\mu_o + R_{sw} k_w/B_w\mu_w + k_g/B_g\mu_g}\frac{\partial}{\partial t}\left(\frac{R_s S_o}{B_o} + \frac{R_{sw} S_w}{B_w} + \frac{S_g}{B_g}\right) = \nabla^2 p \qquad (4.233)$$

$$\phi\frac{B_o\mu_o}{k_o}\frac{\partial(S_o/B_o)}{\partial t} = \nabla^2 p \qquad (4.234)$$

$$\phi\frac{B_w\mu_w}{k_w}\frac{\partial(S_w/B_w)}{\partial t} = \nabla^2 p \qquad (4.235)$$

Eliminating $\nabla^2 p$ from Eqs. (4.233) through (4.235) leads to the following ordinary differential equations:

$$\frac{dS_o}{dp} = \frac{S_o}{B_o} \frac{\partial B_o}{\partial p} + \frac{(k/\mu)_o}{(k/\mu)_T} c_T \tag{4.236}$$

and

$$\frac{dS_w}{dp} = \frac{S_w}{B_w} \frac{\partial B_w}{\partial p} + \frac{(k/\mu)_w}{(k/\mu)_T} c_T \tag{4.237}$$

Equations (4.236) and (4.237) apply for short time periods. The initial pressure and saturations correspond to those values at the start of the time period. Combining Eqs. (4.236) and (4.237) with any one of the equations (4.233), (4.234), or (4.235) yields

$$\frac{\phi c_T}{(k/\mu)_T} \frac{\partial p}{\partial t} = \nabla^2 p \tag{4.238}$$

Equation (4.238) is the usual form of the pressure transient equation for single-fluid flow with $(k/\mu)_T$ and c_T given by Eqs. (4.221) and (4.226), respectively. Earlougher et al. (1967) used the described method to analyze pressure buildup behavior in two-well gas-oil systems and compared the results favorably to computer simulations for the same problem. Pascal and co-workers (Pascal and Dranchuk, 1972; Pascal, 1972; Pascal, 1973; Pascal and Spulber, 1977) have examined transient multifluid flow for several fluid combinations and geometries by obtaining approximate solutions to the nonlinear differential equations. They have chosen to solve the equation by perturbation solutions. This approach needs to be verified by experiment with numerical simulations of the problem.

4.7 FLOW WITH PHASE CHANGE

Multifluid flow in media with phase change is possible in hydrocarbon systems under elevated pressures and temperatures. Before discussing this type of flow we will review briefly some of the thermodynamics associated with the problem. The reader is referred to discussions of high pressure equilibria in thermodynamics texts, for example, Chao and Greenkorn (1975), or to discussions of phase behavior associated with oil reservoir problems in Standing (1952) and Craft and Hawkins (1959).

Experimentally observed pressure-volume-temperature behavior of a pure fluid is like that shown in Fig. 4.42. The coordinates of Fig. 4.42 are in terms of dimensionless pressure, volume, and temperature based on the critical properties p_c, V_c, and T_c.

$$p_r = \frac{p}{p_c} \tag{4.239}$$

$$V_r = \frac{V}{V_c} \tag{4.240}$$

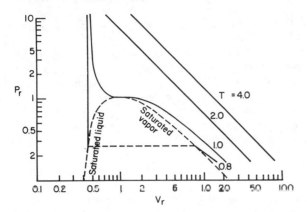

FIG. 4.42 Isotherms of simple fluids. (From Chao and Greenkorn, 1975.)

$$T_r = \frac{T}{T_c} \tag{4.241}$$

The area contained inside the dashed lines on Fig. 4.42 represents the condition where two-phase liquid and vapor exist in equilibrium. Figure 4.43 is a plot of the phase boundary, p_r versus T_r, and isochors (constant volume) for methane. The temperature and pressure of the liquid and vapor are the same on the phase boundary—the liquid and vapor are in equilibrium. A single component system can exist in three phases: solid, liquid, and vapor. The state of a single-component system is described by an equation of state, functionally

$$p_r = p_r(V_r, T_r) \tag{4.242}$$

Figure 4.42 is a graphical representation of Eq. (4.242). Pressure p_r and temperature T_r are not independent when the liquid and vapor are in equilibrium. This is

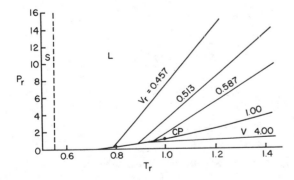

FIG. 4.43 Phase boundaries and isochores of methane. (From Vennix, 1966.)

shown by the phase boundary on Fig. 4.42 and the single line from the triple point (TP) to the critical point (CP) on Fig. 4.43.

The Gibbs phase rule is a systematic and general expression for counting the number of independent intensive (do not depend on mass) variables that can be specified for heterogeneous systems at equilibrium. The number of independent variables in a system that are free to be specified is called the <u>degrees of freedom</u> F. The degrees of freedom are equal to the number of variables in the problem minus the number of equations available to replate the variables; in terms of the number of components and the number of phases in the system,

$$F = 2 - P + C \qquad\qquad (4.243)$$

For a pure substance, $C = 1$. When a pure substance exists in one phase, $P = 1$. Thus, in the single-fluid area of Fig. 4.43, we are free to specify p_r and T_r to find the corresponding V_r. When a pure substance exists in two phases, say, liquid and vapor, $P = 2$. Thus, on the phase boundary of Fig. 4.43, we are free to specify p_r or T_r but not both. Specifying p_r specifies T_r. When a pure substance exists in three phases, $P = 3$ and $F = 0$. In this instance there is only one point and we have no freedom to specify a variable on Fig. 4.43.

The experimentally observed phase behavior for mixtures is much more complex than for pure fluids. The composition of the mixture creates an additional set of variables i equal to the number of components in the mixture

$$p = p(V, T, z_i) \qquad\qquad (4.244)$$

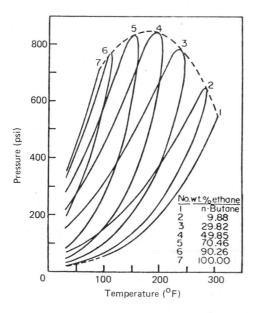

FIG. 4.44 Pressure-temperature diagrams at constant composition of mixtures of ethane and n-butane. [Reprinted with permission from W. B. Kay, Liquid-Vapor Equilibrium Relations in Binary Systems. The Ethane-Butane System, <u>Ind. Eng. Chem.</u> <u>32</u>, 353 (1940). Copyright by the American Chemical Society.]

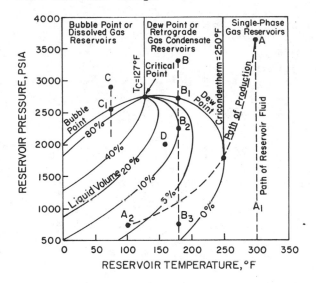

FIG. 4.45 Pressure-temperature phase diagram of a reservoir fluid. (From B. C.
Craft and M. F. Hawkins, Applied Petroleum Engineering. © 1959, pp. 64, 66.
Reprinted by permission of Prentice-Hall, Inc., Englewood Cliffs, N.J.)

Figure 4.44 shows the pressure-temperature diagram for the phase boundaries at
constant composition for mixtures of ethane and n-butane. Figure 4.45 is meant to
represent the phase envelope for a given composition of a typical gas or oil reservoir fluid, which is assumed for the following discussion to be a mixture of hydrocarbons. The area enclosed by the bubble-point and the dew-point lines are the
pressure-temperature-compositions in which both a liquid and a gas phase exist.
The curves within this two-phase region are for the quality of the hydrocarbon
liquid.

      Consider a pressure-temperature condition A, above the critical point to the
right of the cricondentherm, the maximum two-phase temperature. As the pressure
declines at constant temperature (in the reservoir) to point A_1, the composition is
constant, and the fluid does not change phase—it remains a gas. Consider the curved
path AA_2 which represents the decline in pressure and temperature of the produced
fluid outside the reservoir. This fluid remains at constant composition, but it may
fall below the cricondentherm (it does for the system represented in Fig. 4.45) and
enter the two-phase region. When the pressure and temperature on path AA_2 reach
the dew point, liquid will begin to condense.

      Consider a different situation represented by point B on Fig. 4.45. Point B is
above the critical point but below, to the left, of the cricondentherm. The fluid at
B is a gas. Letting the pressure decline along the line BB_1, the dew point is reached
at B_1 and liquid begins to condense in the reservoir. This condensed liquid is
immobile in the reservoir and the gas produced will have a lower liquid content than
the reservoir fluid. As the dew point is reached, the composition of the produced
fluid and the reservoir fluid change and the phase envelope shifts to the right as
pressure is lowered. For the sake of this discussion, assume this shift in the phase
envelope is so slight that we can continue to use the sketch in Fig. 4.45. At this

point we begin the process called <u>retrograde condensation</u> because normally as pressure is lowered one would expect vaporization rather than condensation to take place. This is due to the relative location of the critical point and the cricondentherm. After point B_2 the liquid begins to revaporize and continues on to B_3 or after.

Consider another situation represented by the fluid at the pressure and temperature of point C. Point C is above the critical pressure but below the critical temperature, and it is a liquid. As the pressure declines to the bubble point C_1, gas begins to evolve and this evolution continues as we continue to lower the pressure. As the gas evolves less liquid will flow in the porous medium.

Point D on Fig. 4.45 represents a two-phase situation containing a liquid with a gas cap. The composition of the liquid and of the gas will not be the same and each will have a phase diagram. The liquid will be at its bubble point and the gas may be at its dew point either below or above the cricondentherm (retrograde or nonretrograde) as shown in Fig. 4.46.

We use the single-fluid or multifluid flow equations for those appropriate situations on Fig. 4.45. However, for the situation represented by the dissolved gas CC_1, we must consider multifluid flow with phase change.

Before considering multifluid flow with phase change, we need to define some additional terms. Vapor-liquid equilibria at elevated temperatures and pressures is usually expressed in terms of the vaporization equilibrium ratio

$$K_i = \frac{y_i}{x_i} \qquad\qquad (4.245)$$

where y_i is the mole fraction of species i in the vapor, and x_i is the mole fraction of species i in the liquid. The K_i values of components in the mixture are a quantitative index of their volatilities. At any point in a pressure reduction such as indicated in Fig. 4.45 on line CC_1 below the bubble point the liquid and vapor separate. If the pressure and temperature are reduced to atmospheric, some vapor evolves. If the vapor is continuously removed from contact with the remaining liquid as the

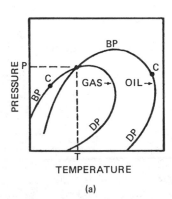
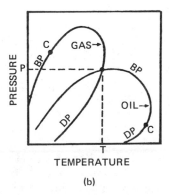

(a)   (b)

FIG. 4.46 Phase diagrams of a cap gas and oil zone fluid showing (a) retrograde cap gas, and (b) nonretrograde cap gas. (From B. C. Craft and M. F. Hawkins, <u>Applied Petroleum Reservoir Engineering</u>. © 1959, pp. 64, 66. Reprinted by permission of Prentice-Hall, Inc., Englewood Cliffs, N.J.)

pressure is lowered, the process is called <u>differential vaporization</u>. In differential vaporization the overall composition of the system changes as the process proceeds. If the evolved vapor is not removed, the process of evolution of vapor when pressure and temperature of the system is lowered to atmospheric is called <u>flash vaporization</u>.

Standing (1952) discusses multifluid flow in a system with dissolved gas in terms of the phase behavior and relative permeability. The rates of each of the phases of gas, oil, and water (assume water is wetting the medium and is at or near its residual value) are proportional to the pressure gradient and mobility of each fluid or phase.

$$q_g \sim \frac{k_g}{\mu_g}\left(\frac{dp}{dx}\right)_g \qquad (4.246)$$

$$q_o \sim \frac{k_o}{\mu_o}\left(\frac{dp}{dx}\right)_o \qquad (4.247)$$

$$q_w \sim \frac{k_w}{\mu_w}\left(\frac{dp}{dx}\right)_w \qquad (4.248)$$

Since the water is not moving, if we divide Eq. (4.246) by Eq. (4.247) and assume pressure gradient the same

$$\frac{q_g}{q_w} \sim \frac{k_g}{k_o}\frac{\mu_o}{\mu_g} \qquad (4.249)$$

Assume the phase behavior of the system is represented by the data of Fig. 4.47. Standing (1952) considers two fluid cases: one at 2800 psia and 70°F and the other at 2800 psia and 190°F. The original fluid is a single phase; the reservoir permeability is 50 mdarcy. As the pressure declines to the bubble point pressure (2690 psia and 70°F) gas will begin to evolve within the pores. Since the lowest pressure

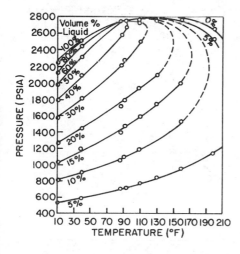

FIG. 4.47 Phase behavior of a natural gas-natural gasoline mixture. [Reprinted with permission from D. L. Katz and F. U. Kurata, Retrograde Condensation. <u>Ind. Eng. Chem.</u> <u>32</u>, 817 (1940). Copyright by the American Chemical Society.]

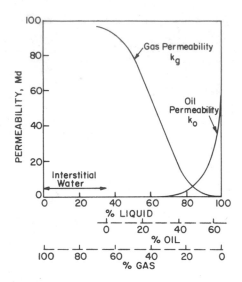

FIG. 4.48 Multiphase permeability curves for a consolidated sandstone. (From Standing, 1977.)

is at the well, gas will appear first in the media next to the well. Initially the evolved gas will remain in the pores. The relative permeability curve representing this system (Fig. 4.48) shows the gas relative permeability is zero until a gas saturation of approximately 5 percent. There is a decrease in relative permeability of the oil so the produced oil will decrease slightly at the well. Once the critical gas saturation is reached (5%) the ratio k_g/k_o increases rapidly, and in conjunction with the viscosity ratio—see Eq. (4.247)—causes a high ratio of gas to liquid flow. When the gas saturation reaches 30 to 35 percent, the soil relative permeability is so low that oil no longer flows.

We can describe multifluid flow with phase change—the system in the two-phase region of Fig. 4.45 along the line CC_1—by writing the partial differential equations for the concentration of the components as a function of position and time, by writing partial differential equations in terms of pressure and overall saturation of the liquid and gas (the relative permeability model), or by a material balance which averages pressure and gas/oil ratios as a function of recovery. Consider a volume element of a medium containing gas, oil, and water where water is present as residual or connate water. The gas and oil are at temperatures below the bubble point and the critical point in the two-phase region. A differential mass balance for the gas and liquid

$$\phi \frac{\partial}{\partial t} (w_{i\ell}\rho_\ell S_\ell + w_{ig}\rho_g S_g) = - [\underline{\nabla} \cdot (w_{i\ell}\rho_\ell q_\ell + w_{ig}\rho_g q_g)] \qquad (4.250)$$

where the subscript i refering to components, ℓ for liquid, g for gas, all at the pressure and temperature at that point in the porous medium. Since water is not moving, it is not considered part of the hydrocarbon liquid,

$$S_\ell + S_g = 1 \qquad (4.251)$$

Darcy's law for each phase

$$q_\ell = -\frac{k_\ell}{\mu_\ell} \nabla (p_\ell + \rho_\ell gz) \qquad (4.252)$$

$$q_g = -\frac{k_g}{\mu_g} \nabla (p_g + \rho_g g_z) \qquad (4.253)$$

The pressures p_ℓ and p_g are related by the capillary pressure

$$p_c = p_g - p_\ell \qquad (4.254)$$

The viscosities and densities of the gases and liquids are functions of pressure, temperature, and composition of each phase

$$\rho_\ell = \rho_\ell(p_\ell, \ T, \ x_i) \qquad (4.255)$$

$$\rho_g = \rho_g(p_g, \ T, \ y_i) \qquad (4.256)$$

$$\mu_\ell - \mu_\ell(p_\ell, \ T, \ x_i) \qquad (4.257)$$

$$\mu_g = \mu_g(p_g, \ T, \ y_i) \qquad (4.258)$$

Mass fractions and mole fractions are related by

$$w_{i\ell} = \frac{x_i M_{i\ell}}{\sum_{j=1}^{n} x_j M_{j\ell}} \qquad (4.259)$$

$$w_{ig} = \frac{y_i M_{ig}}{\sum_{j=1}^{n} y_j M_{jg}} \qquad (4.260)$$

Define a mean molecular weight for the liquid and the gas as

$$M_\ell = \sum_{j=1}^{n} x_j M_{j\ell} \qquad (4.261)$$

$$M_g = \sum_{j=1}^{n} y_j M_{jg} \qquad (4.262)$$

Combine Eqs. (4.251) to (4.253) and Eqs. (4.259) to (4.262) to obtain the following equation of condition in terms of mole fraction of the components in the liquid and gas.

$$\phi \frac{\partial}{\partial t} \left(x_i \frac{\rho_\ell}{M_\ell} S_\ell + y_i \frac{\rho_g}{M_g} S_g \right) = -\left[\nabla \cdot \left(x_i \frac{\rho_\ell}{M_\ell} q_\ell + y_i \frac{\rho_g}{M_g} q_g \right) \right] \qquad (4.263)$$

We need vapor equilibrium ratios to find x_i and y_i. Define V as the moles of vapor mixture and L as the moles of liquid mixture, then for one mole of the overall system

$$z_i = x_i L + y_i V \tag{4.264}$$

where z_i is the mole fraction of the overall mixture. Substituting Eq. (4.245) for y_i in Eq. (4.264) and solving for x_i

$$x_i = \frac{z_i}{L + VK_i} \tag{4.265}$$

Since the x_i must add up to 1 and $L + V = 1$

$$\sum_{i=1}^{n} \frac{z_i}{L + (1 - L)K_i} \tag{4.266}$$

Equation (4.266) describes the flash equilibrium. Similarly, if we substitute Eq. (4.245) in terms of y_i into Eq. (4.264)

$$y_i = \frac{z_i}{L/K_i + V} \tag{4.267}$$

and

$$\sum_{i=1}^{n} \frac{z_i}{(1 - V)/K_i + V} = 1 \tag{4.268}$$

Neglecting capillary pressure, Eq. (4.263) becomes:

$$\phi \frac{\partial}{\partial t}\left(\frac{x_i \rho_\ell S_\ell}{M_\ell} + \frac{y_i \rho_\ell S_g}{M_g}\right) = \nabla \cdot \left(\frac{x_i \rho_\ell k_\ell}{M_\ell \mu_\ell} = \frac{y_i \rho_g k_g}{M_g \mu_g}\right) \nabla p \tag{4.269}$$

Taylor (1966) solved Eq. (4.269) in one dimension to find the moles of C_{7+} and CH_4 in a porous medium as a function of distance. He found the gas liberation process depends on the nature of the relative permeability.

The equations of change for flow with phase change may be written with saturation of the gas and liquid as a function of time and position. Call the residual liquid in the porous medium at atmospheric conditions oil, where the vaporization process, since we are removing gas continually, is a differential vaporization. At any time, an element of the porous medium will contain a volume of liquid which at atmospheric conditions will yield a certain volume of liquid and volume of gas. The ratio of gas volume to oil volume at standard conditions (atmospheric) conditions is called solubility S:

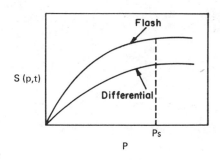

FIG. 4.49 Gas solubility versus pressure
for a typical petroleum hydrocarbon
system. (From Collins, 1961.)

$$S(p, T) = \frac{V_g \text{ (std. cond.)}}{V_o \text{ (std. cond.)}}$$ (4.270)

The solubility as a function of pressure is different depending on whether the evolution of gas is a flash or differential process as sketched in Fig. 4.49. The element
of the porous medium also contains a volume of gas which is some different volume
at atmospheric conditions. For a gas saturation S_g and a liquid saturation S_ℓ in a
unit volume of the medium, there is a mass of oil given by $\phi \rho_{os} S_\ell / B_o$, where ρ_{os}
is oil density and B_o is the formation volume factor defined as

$$B = \frac{V(p, t)}{V_s}$$ (4.271)

where V is volume (of gas or oil) at the pressure and temperature of the system,
and V_s is a volume (of gas or oil) at standard conditions. The formation volume
factor is different depending on whether the evolution of gas is a flash or differential
process as sketched in Fig. 4.50. In this same unit volume there is a mass of free
gas $\phi \rho_g S_g$ and a mass of gas dissolved in the oil, $\phi S \rho_{gs} S_\ell / B_o$. The mass of gas per
unit volume of reservoir is

$$m_g = \phi \rho_g S_g + \frac{\phi S \rho_{gs} S_\ell}{B_o}$$ (4.272)

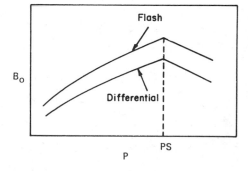

FIG. 4.50 Oil formation volume factor
versus pressure for a typical petroleum
hydrocarbon system. (From Collins, 1961.)

The mass of oil per unit volume of reservoir is

$$m_o = \frac{\phi \rho_{os} S_\ell}{B_o} \tag{4.273}$$

From a mass balance and Darcy's law, the equations of condition for the liquid and gas are (note: oil is in the liquid phase, gas is in both liquid and gas phases) neglecting gravity and capillary pressure

$$\phi \frac{\partial}{\partial t} \left(\frac{\rho_{os} S_\ell}{B_o} \right) = \underline{\nabla} \cdot \frac{\rho_{os} k_\ell}{B_o \mu_1} \underline{\nabla} p \tag{4.274}$$

and

$$\phi \frac{\partial}{\partial t} \left(\rho_g S_g + \frac{S \rho_{gs} S_\ell}{B_o} \right) = \underline{\nabla} \cdot \left(\frac{\rho_s k_g}{\mu_g} + \frac{S \rho_{gs} k_g}{B_o \mu_\ell} \right) \underline{\nabla} p \tag{4.275}$$

Equations (4.274) and (4.275) are approximate since overall composition is the same everywhere and the same functions apply for the parameters at the pressure and temperature of the system. The initial conditions are

$$p(x, y, z, 0) = \hat{p}(x, y, z) \tag{4.276}$$

$$S_\ell(x, y, z, 0) = \hat{S}_\ell(x, y, z) \tag{4.277}$$

where $\hat{p}$ and $\hat{S}_\ell$ are the initial distributions of pressure and liquid saturation. The boundary conditions are

$$j_o = - \left(\underline{n} \ \frac{\rho_{os} k_\ell}{B_o \mu_\ell} \underline{\nabla} p \right) \tag{4.278}$$

$$j_g = - \left[\underline{n} \ \left(\frac{\rho_g k_g}{\mu_g} + \frac{S \rho_{gs} k_\ell}{B_o \mu_\ell} \right) \underline{\nabla} p \right] \tag{4.279}$$

where j is the mass flux—in this case, normal to the boundaries of the system. Equation (4.274) is often written with $S_o \equiv S_\ell$, $k_o \equiv k_\ell$, $\mu_o \equiv \mu_\ell$, and removing ρ_{os} from inside the derivative

$$\phi \frac{\partial}{\partial t} \left(\frac{S_o}{B_o} \right) = \underline{\nabla} \ \frac{k_o}{B_o \mu_o} \underline{\nabla} p \tag{4.280}$$

Equation (4.275) is often written with $S_o \equiv S_\ell$, $k_o \equiv k_\ell$, $\mu_o \equiv \mu_\ell$ and in terms of the gas/oil ratio defined by

$$R_s = \frac{\rho_g k_g / \mu_g + S\rho_{os} k_o / B_o \mu_o}{k_o / B_o \mu_o} \qquad (4.281)$$

so that

$$\phi \frac{\partial}{\partial t}\left(\frac{R_s S_o}{B_o} + \frac{S_g}{B_g}\right) = \nabla \cdot \left(\frac{R_s k_o}{B_o \mu_o} + \frac{k_g}{B_g \mu_g}\right)\nabla p \qquad (4.282)$$

B_g replaces ρ_{gs}/ρ_g by Eq. (4.271).

Another means of representing flow with phase change is by approximating the flow with the material balance equations. Collins (1961) obtains these equations by formally integrating Eqs. (4.274) and (4.275) averaging over the reservoir volumes and areas. Integrating (4.280) and (4.282) to obtain averages or writing the equations based on material balance and average values gives

$$Q_o = -\frac{\phi V}{B_o}\frac{dS_o}{dt} - \left(\phi V S_o \frac{d(1/B_o)}{dp}\right)\frac{dp}{dt} \qquad (4.283)$$

and

$$R_s Q_o = -\phi V\left(\frac{S}{B_o} - \frac{1}{B_g}\right)\frac{dS_o}{dt} - \phi V\left[\frac{d(1/B_g)}{dp} + \frac{d}{dp}\left(\frac{S}{B_o} - \frac{L}{B_g}\right)S_o\right]\frac{dp}{dt} \qquad (4.284)$$

If Q_o is known as a function of t, p, and S_o, given at t = 0, Eqs. (4.283) and (4.284) can be solved numerically.

West et al. (1954) solved Eq. (4.280) and (4.282) for one dimension in linear and radial geometries with a single input and output and compared the results to the material balance solutions Eqs. (4.283) and (4.284). Figures 4.51 and 4.52 show the comparison of the results where the curves labeled approximate theory are the material balance results.

Stone and Garder (1961) solved Eqs. (4.280) and (4.282) including gravity and more than one source or sink. They found that gravity is an important factor.

Carlson and Land (1976) solved Eqs. (4.280) and (4.282) in one dimension including gravity and the effects of compressibility and solution. They compared their results to those calculated using a Buckley-Leverett (1941) solution and a material balance solution by Kern (1952). They showed that for small pressure drops across the system when input and output are near the initial bubble point pressure, the results of the numerical solution, the Buckley-Leverett solution, and the Kern solution agree. For high-pressure gradients the various results are shown in Fig. 4.53. A modified Kern solution agrees fairly closely with the gas breakthrough and the gas moves as in a linear series calculation. They conclude from their numerical comparisons that compressibility and solution effects are significant.

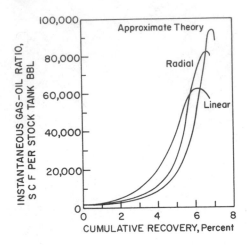

FIG. 4.51 Calculated gas-oil ratio histories of hypothetical solution gas-drive radial and linear flow systems. Calculated gas-oil ratio history from approximate solution gas-drive theory. (From W. J. West, W. W. Garvin, and J. W. Sheldon, Solution of the Equations of Unsteady State Two-Phase Flow in Oil Reservoirs, Pet. Trans. AIME 201, SPEJ 217. © 1954 SPE-AIME.)

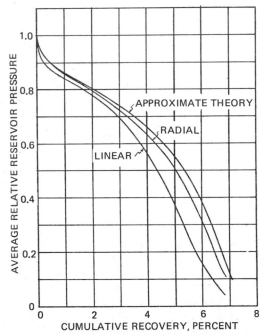

FIG. 4.52 Calculated average pressure histories of hypothetical solution gas-drive radial and linear flow systems. Calculated average pressure history from approximate solution gas-drive theory. (From W. J. West, W. W. Garvin, and J. W. Sheldon, Solution of the Equations of Unsteady State Two-Phase Flow in Oil Reservoirs, Pet. Trans. AIME 201, SPEJ 217. © 1954 SPE-AIME.)

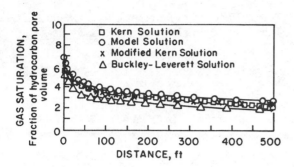

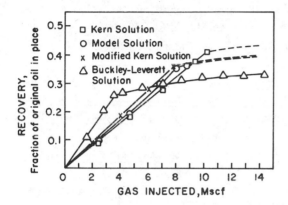

FIG. 4.53 Fractional oil recoveries for an injection pressure of 3800 psia. (From F. M. Carlson and C. S. Land, Effects of Compressibility and Solution Gas on the Gas Drive Mechanism for Linear and Radial Systems, Pet. Trans. AIME 261, SPEJ 1079. © 1976 SPE-AIME.)

REFERENCES

Bear, J. (1972): Dynamics of Fluids in Porous Media, Elsevier, New York.

Brahma, S. P., and M. E. Harr (1962): Transient Development of the Free Surface in a Homogeneous Earth Dam, Geotechnique, December.

Brooks, R. H., and A. T. Corey (1964): Hydraulic Properties of Porous Media, Hydraulic Paper Number 3, Colorado State University, Ft. Collins, Colo.

Brooks, R. H., and A. T. Corey (1966): Properties of Porous Media Affecting Fluid Flow, ASCE Proc. J. Irrig. Drainage Div., 92 IR2, 61.

Buckley, S. E., and M. C. Leverett (1941): Mechanism of Fluid Displacement in Sands, Pet. Trans. AIME 146, 107.

Burdine, N. T. (1953): Relative Permeability Calculations from Pore Size Distribution Data, Pet. Trans. AIME 198, 71.

Carlson, F. M., and C. S. Land (1976): Effects of Compressibility and Solution Gas on the Gas Drive Mechanism for Linear and Radial Systems, Pet. Trans. AIME 261, 1079.

Chao, K. C., and R. A. Greenkorn (1975): Thermodynamics of Fluids—An Introduction to Equilibrium Theory, Marcel Dekker, New York.

Chuoke, R. L., P. van Muers, and C. van der Poel (1959): The Instability of Slow, Immiscible, Viscous Liquid-Liquid Displacements in Permeable Media, Pet. Trans. AIME 216, 188.

Collins, R. E. (1961): Flow of Fluids through Porous Materials, Reinhold, New York.

Corey, A. T., C. H. Rathjenes, J. H. Henderson, and M. R. J. Wyllie (1956): Three-Phase Relative Permeability, Pet. Trans. AIME 207, 349.

Corey, A. T., and R. H. Brooks (1972): Drainage Characteristics of Soils, SSSA Proc. 39(2).

Corey, A. T. (1977): Mechanics of Heterogeneous Fluids in Porous Media, Water Resources Publications, Fort Collins, Colo.

Craft, B. C., and M. F. Hawkins (1959): Applied Petroleum Reservoir Engineering, Prentice-Hall, Englewood Cliffs, N.J.

Crichlow, H. B. (1977): Modern Reservoir Engineering—A Simulation Approach, Prentice-Hall, Englewood Cliffs, N.J.

Douglas, J., Jr., P. M. Blair, and R. J. Wagner (1958): Calculation of Linear Waterflood Behavior Including the Effects of Capillary Pressure, Pet. Trans. AIME 213, 96.

Dvinoff, A. H. (1970): Responses of the Phreatic Surface in Earthen Dams to Headwater Fluctuation, Ph.D. Thesis, Purdue University, West Lafayette, Ind.

Earlougher, R. C., Jr., F. G. Miller, and T. D. Mueller (1967): Pressure Buildup Behavior in a Two-Well Gas-Oil System, Pet. Trans. AIME 240, SPEJ 195.

Greenkorn, R. A., J. E. Matar, and R. C. Smith (1967): Two-Phase Flow in Hele-Shaw Models, A.I.Ch.E. J. 13, 273.

Gupta, S. D., J. E. Varnon, and R. A. Greenkorn (1973): Viscous Finger Wavelength Degeneration in Hele-Shaw Models, Water Resour. Res. 9(4), 1039.

Gupta, S. P., and R. A. Greenkorn (1974): An Experimental Study of Immiscible Displacement with Unfavorable Mobility Ratio in Porous Media, Water Resour. Res. 10(2), 371.

Haines, W. B. (1930): Studies in the Physical Properties of Soil V. The Hysteresis Effect in Capillary Properties and the Modes of Moisture Distribution Associated Therewith, J. Agr. Sci. 20, 97.

Handy, L. L. (1960): Determination of Effective Capillary Pressure for Porous Media from Imbition Data, Pet. Trans. AIME 219, 73.

Jacquard, P., and P. Seguier (1962): Movement de deux Fluides en contans Dans un Milieu Poreux, J. de Mecanique 1, 367.

Johnson, E. F., D. P. Bossler, and V. O. Nauman (1959): Calculation of Relative Permeability from Displacement Experiments, Pet. Trans. AIME 216, 370.

Katz, D. L., and F. U. Kurata (1940): Retrograde Condensation, Ind. Eng. Chem. 32, 817.

Kay, W. B. (1940): Liquid-Vapor Equilibrium Relations in Binary Systems: The Ethane-n-Butane System, Ind. Eng. Chem. 32, 353.

Kern, L. R. (1952): Displacement Mechanism in Multi-well Systems, Pet. Trans. AIME 195, 39.

Land, C. S. (1968): Calculation of Imbibition Relative Permeability for Two- and Three-Phase Flow from Rock Properties, Pet. Trans. AIME 243, SPEJ 149.

Leverett, M. C. (1941): Capillary Behavior in Porous Solids, Pet. Trans. AIME 142, 152.

Martin, J. C. (1959): Simplified Equations of Flow in Gas Drive Reservoirs and the Theoretical Foundation of Multiphase Buildup Analyses, Pet. Trans. AIME 216, 309.

Melrose, J. C. (1965): Wettability as Related to Capillary Action in Porous Media, Pet. Trans. AIME 234, SPEJ 259.

Melrose, J. C. (1970): Interfacial Phenomena as Related to Oil Recovery Mechanisms, Can. J. Chem. Eng. 48, 638.

Melrose, J. C., and C. F. Brandner (1974): Role of Capillary Forces in Determining Displacement Efficiency for Oil Recovery by Waterflooding. J. Can. Pet. Tech. Oct.-Dec. 1.

Miller, C. C., A. B. Dyes, and C. A. Hutchinson, Jr. (1950): The Estimation of Permeability and Reservoir Pressure from Bottom Hole Pressure Build-up Characteristics, Pet. Trans. AIME 189, 91.

Morrow, N. R., and C. C. Harris (1965): Capillary Equilibrium in Porous Materials, Pet. Trans. AIME 234, SPEJ 15.

Morrow, N. R. (1970): Physics and Thermodynamics of Capillary Action in Porous Media, in Flow through Porous Media, Amer. Chem. Soc., Washington, D.C., Chapter 6, p. 103.

Muskat, M. (1934): Two Fluid Systems in Porous Media. The Encroachment of Water into an Oil Sand. Physics 5, 255.

Muskat, M. (1937): Flow of Homogeneous Fluids, McGraw-Hill, New York.

Muskat, M., R. D. Wyckoff, H. G. Botset, and M. W. Meres (1937): Flow of Gas-Liquid Mixtures through Sands, Pet. Trans. AIME 123, 69.

Naar, J., and R. J. Wygal (1961): Three-Phase Imbibition Relative Permeability, Pet. Trans. AIME 222, SPEJ 254.

Outmans, H. D. (1962): Nonlinear Theory for Frontal Stability and Viscous Fingering in Porous Media, Pet. Trans. AIME 225, 165.

Outmans, J. D. (1963): On Unique Solutions for Steady-State Fingering in a Porous Medium, J. Geophys. Res. 68, 5735.

Pascal, H. (1972): Some Volterra Type Nonlinear Integral Equations in Multiphase Flow through Porous Media, Rev. Roum. Math. Puresset. Appl. 17(10), 1681.

Pascal, H., and P. M. Dranchuk (1972): Simultaneous Nonsteady Flow of Gas and Oil through Porous Media, J. Méc. Paris 11(2), 251.

Pascal, H. (1973): Nonsteady Multiphase Flow through Porous Medium, Rev. Roum. Sci. Techn. Méc. Appl. 18(2), 329.

Pascal, H., and F. Spulber (1977): On Some Problems of Nonsteady Multiphase Flow through Porous Media, Rev. Roum. Sci. Techn. Méc. Appl. 18(2), 329.

Perrine, R. L. (1956): Analysis of Pressure-Buildup Curves, Drill Prod. Proc. API, 482.

Purcell, W. R. (1949): Capillary Pressures—Their Measurement Using Mercury and the Calculation of Permeability Therefrom, Pet. Trans. AIME 186, 139.

Rachford, H. H., Jr. (1964): Instability in Water Flooding Oil from Water-Wet Porous Media Containing Connate Water, Pet. Trans. AIME 231, SPEJ 133.

Richards, L. A. (1931): Capillary Conduction of Liquids through Porous Mediums, Physics 1, 231.

Rose, W., and W. A. Bruce (1949): Evaluation of Capillary Pressure Characteristics in Petroleum Reservoir Rock, Pet. Trans. AIME 186, 127.

Saffman, P. G., and G. Taylor (1958): The Penetration of a Fluid into a Porous Medium or a Hele-Shaw Cell Containing a More Viscous Liquid, Proc. Roy. Soc. A245, 312.

Saraf, D. N., and I. Fatt (1967): Three-Phase Relative Permeability Measurement Using a Nuclear Magnetic Resonance Technique for Estimating Fluid Saturation, Pet. Trans. AIME 240, SPEJ 235.

Sarem, A. M. (1966): Three-Phase Relative Permeability Measurements by Unsteady-State Method, Pet. Trans. AIME 237, SPEJ 199.

Scheidegger, A. E. (1960): On the Stability of Displacement Fronts in Porous Media: A Discussion of the Muskat-Aranofsky Model, Can. J. Phys. 38, 153.

Schneider, F. N., and W. W. Owens (1970): Sandstone and Carbonate Two- and Three-Phase Relative Permeability Characteristics, Pet. Trans. AIME 249, SPEJ 75.

Slattery, J. C. (1968): Multiphase Viscoelastic Flow through Porous Media, A.I.Ch.E. J. 14, 50.

Slattery, J. C. (1970): Two-Phase Flow through Porous Media, A.I.Ch.E.J. 16, 345.

Smith, M. F. (1976): An Experimental Investigation of Capillary Pressure Behavior as a Function of Pore-Size Distribution, M.S. Thesis, Purdue University, West Lafayette, Ind.

Standing, M. B. (1977): Volumetric and Phase Behavior of Oil Field Hydrocarbon Systems, Society of Petroleum Engineers.

Standing, M. B. (1975): Notes on Relative Permeability Relationships, Class Notes NTH Trondheim, Pet. Eng. Stanford.

Stone, H. L., and A. O. Garder, Jr. (1961): Analysis of Gas-Cap or Dissolved-Gas Drive Reservoirs, Pet. Trans. AIME 222, SPEJ 92.

Stone, H. L. (1970): Probability Model for Estimating Three-Phase Relative Permeability, Pet. Trans. AIME 249, 214.

Taylor, J. G. V. (1966): Distribution of Hydrocarbon Fluids and Their Compositions in Volatile Oil Reservoirs During Depletion, Ph.D. Thesis, Stanford University, Stanford, Calif.

Varnon, J. E., and R. A. Greenkorn (1969): Unstable Two-Fluid Flow in a Porous Medium, Soc. Pet. Eng. J. 9, 293.

Varnon, J. E., and R. A. Greenkorn (1970): On Uniqueness of Finger Theory, Water Resour. Res. 6, 1411.

Varnon, J. E. (1971): Porous Media Displacements with Unfavorable Viscosity Ratios, Ph.D. Thesis, Purdue University, West Lafayette, Ind.

Vennix, A. J. (1966): Low Temperature Volumetric Properties and the Development of an Equation of State of Methane, Ph.D. Thesis, Rice University, Houston.

Welge, H. J. (1952): A Simplified Method for Computing Oil Recovery by Gas or Water Drive, Pet. Trans. AIME 195, 91.

West, W. J., W. W. Garvin, and J. W. Sheldon (1954): Solution of the Equations of Unsteady State Two-Phase Flow in Oil Reservoirs, Pet. Trans. AIME 201, 217.

SUGGESTED READING

Corey, A. T. The Interrelation between Gas and Oil Relative Permeabilities, Prod. Monthly 19(1), 38 (1954).

Collins, R. E. Flow of Fluids through Porous Materials, Van Nostrand, New York, 1961.

Gray, W. G., and P. C. Y. Lee, On the Theorems for Local Volume Averaging of Multiphase Systems, Int. J. Multiphase Flow 3, 333 (1976).

Greenkorn, R. A., Matrix Properties of Porous Media, Proc. Second IAHR-ISSS Symp. Fundamentals of Transport Phenomena in Porous Media, Geulph, 1972.

Richardson, J. G., Flow through Porous Media, in Handbook of Fluid Dynamics, V. L. Streeter (ed.), McGraw-Hill, New York, 1961, Section 16.

Scheidegger, A. E., Growth of Instabilities on Displacement Fronts in Porous Media, Phys. Fluids 3, 94 (1960).

Scheidegger, A. E., The Physics of Flow through Porous Media, 3d ed., University of Toronto Press, Toronto, 1974.

Wooding, R. A., Growth of Fingers at an Unstable Diffusing Interface in a Porous Media or Hele-Shaw Cell, J. Fluid Mech. 39(3), 477.

Wyllie, M. R. J., and G. H. F. Gardner, I The Generalized Kozeny-Carman Equation. II A Novel Approach to Problems of Fluid Flow, World Oil, March 121, April 210 (1958).

5

Multifluid Miscible Flow

The equations of change for multifluid miscible flow are the same as with single-fluid flow; however, if the two fluids begin flowing separately a mixed region separates them. Much of the modeling for miscible flow is concerned with modeling the mixed region and the parameter associated with the mixing, the dispersion coefficient. Dispersivity results from the complex nature of the flow paths in the media, although diffusion and turbulence affect mixing.

Steady flow displacements are modeled with average equations or with the equation describing the mixed region, the dispersion equation. Unstable flow results from imbalances of forces due to gravity and viscosity, but the balancing force of interfacial tension is not present. Transient flow in miscible systems uses the single-fluid equations, although significant fluid property (or matrix property) changes can be taken into account.

The additional problem of adsorption enters the miscible flow problem since as the fluids mix, the possibility of a given material in solution adsorbing confounds the mixing problem. When the movement of chemicals (surface-active agents, fertilizers, salt, etc.) is considered, the problem of adsorption is a major factor in calculating the mixed region.

5.1 EQUATIONS OF CHANGE

In principle the overall equations of motion and continuity for multifluid miscible flow are the same as for single-fluid flow. The equation of condition resulting from the combination of Darcy's law and the overall continuity equation is the same. Appropriate single-fluid viscosities and densities are used if the fluids are commingled. In multifluid miscible flow we are sometimes interested in the regions between the fluids where the relative concentrations of the fluids are different, i.e., in the mixed region. Outside of the mixed region, the single-fluid equations apply directly.

The first problem in the description of multifluid miscible flow is to describe the mixing or dispersion of the fluids when one fluid displaces another during flow through a porous medium. The second problem is related to how to cope with the fact that even though the fluids are miscible, they may have different densities and viscosities, and until they are completely commingled and the resulting "single fluid"

of an average density and average viscosity is flowing), there is possibility of either
stable or unstable flow.

Consider two miscible fluids flowing in a porous medium, each with different
densities but the same viscosity, with the more dense fluid initially "on top." In
this instance the denser fluid will not only disperse but may cause gravity instability
(gravity fingering) depending on the balance between gravity and viscous forces. If
the less dense fluid is on top, flow will be stable with transverse mixing.

Consider two miscible fluids flowing in a porous medium, each with a different
viscosity but the same density. If the less viscous fluid is displacing the more viscous fluid, there will be viscous instability (viscous fingering). If the more viscous
fluid is displacing the less viscous fluid, stable flow with mixing will result. If the
two fluids have different viscosities and different densities, then all combinations
can occur. In unstable flow, normally the macroscopic mixing due to fingering is
not included as part of the mixing description.

To describe the mixing or dispersion of multifluid miscible flow, assume two
fluids of equal viscosity and equal density, where one of the fluids is displacing the
other fluid from a porous medium. Assume initially that the flow is in one dimension.
Darcy's law for the flow of each fluid, and their sum, is

$$q = -\frac{k}{\mu}\frac{dp}{dx} \qquad (5.1)$$

Note: Miscible flow permeability is only a function of the medium—a relative permeability is not used—interfacial tension is zero. The equation of continuity for the
total flow is

$$\phi\frac{\partial\rho}{\partial t} + \frac{\partial\rho q}{\partial x} = 0 \qquad (5.2)$$

Combining Eqs. (5.1) and (5.2) with an equation of state, the equation of condition is

$$\phi c_f \frac{\partial p}{\partial t} = \frac{k}{\mu}\frac{\partial^2 p}{\partial x^2} \qquad (5.3)$$

For an incompressible fluid

$$\frac{\partial^2 p}{\partial x^2} = 0 \qquad (5.4)$$

The description of the total flow of the two fluids is in terms of the overall
pressure drop. Define the relative amount or concentration of each fluid as a function of time and space in order to track the fluids separately. Conceptually, imagine
for t < 0 an element of a porous medium contains a fluid and at t = 0 we start a displacement by another fluid with the same properties. Experimentally, the same
fluid is used but add a tracer, i.e., a tag that does not change density, viscosity,
or interfacial tensions. For the moment, assume the tagged fluid is injected over
the entire inlet face of the medium. As the displacement progresses the two fluids
will mix due to pore structure (mechanical dispersion) and due to the diffusion and
hydrodynamic dispersion. Diffusion is a molecular mechanism. Hydrodynamic dispersion is mixing resulting from the velocity profile in the pores; fluid at the walls

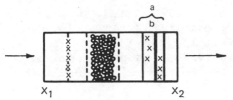

FIG. 5.1 Dispersion of a marked fluid.
(From Greenkorn and Kessler, 1970.)

moves slower than fluid in the center of the pores. Mechanical dispersion is the mixing resulting from the fluid particles moving through the tortuous paths of the medium. For the present discussion it suffices to describe dispersion as the macroscopic mixing caused by uneven cocurrent flow in the medium.

The behavior of the traced fluid is sketched in Fig. 5.1. At $t < 0$, $c = 0$, where c is the concentration of the tracer of the produced (in-place) fluid. At $t \geq 0$ a traced fluid of concentration c_0 is injected, a step increase in concentration of tracer. This abrupt switch in concentration is indicated as the line of X's in the left of Fig. 5.1. As the traced fluid moves from left to right, this line of X's smears, i.e., the X's spread, due to velocity variation in the pores and actual differences in distance due to path selection, so at some time after $t = 0$ at a distance on the right the X's are contained in a region identified as a in Fig. 5.1. This spreading due to flow is dispersion. If the line of X's had not dispersed, the X's would still be on a line at b in Fig. 5.1—translation of the X's at the average velocity for an ideal medium (homogeneous, uniform, and isotropic).

Figure 5.2 is a record of the experiment where c/c_0 of the produced fluid is plotted versus V/V_0, the pore volume. If there is no mixing of any sort, the plot of c/c_0 versus V/V_0 is a step change from $c/c_0 = 0$ to $c/c_0 = 1$ at $V/V_0 = 1$, showing the line of X's translated through the element. If the only mixing that occurs is due to diffusion, the line of X's spreads slightly and the plot of c/c_0 versus V/V_0 (Fig. 5.2) shows a slight breakthrough of tracer prior to $V/V_0 = 1$ and a small tail as c/c_0 approaches 1 after $V/V_0 = 1$. The real situation represented by the S-shaped curve on Fig. 5.2 shows significant early breakthrough and a long tail after $V/V_0 = 1$. This corresponds to the width of the region containing the X's in Fig. 5.1.

The problem is to describe the growth of the mixed region, i.e., to find concentration as a function of time and position as the two miscible fluids flow through the medium. Outside of the mixed zone (on either side) the single-fluid equations describe the motion.

The problem is more complicated, even in one-dimension with fluids of equal properties, since the mixing takes place both longitudinally (in the direction of flow) and transversely (perpendicular to the flow). Imagine at $t = 0$ we inject a "dot" of traced fluid of concentration c_0 rather than over the entire face. This situation is

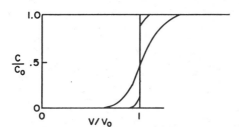

FIG. 5.2 Breakthrough curve. (From Greenkorn and Kessler, 1970.)

FIG. 5.3 Longitudinal and transverse dispersion. (From R. A. Greenkorn and D. P. Kessler, in Flow Through Porous Media, p. 149, R. J. Nunge (ed.). © American Chemical Society, Washington, D.C., 1970.)

sketched in Fig. 5.3. As the dot moves from left to right it will spread in the direction of flow and perpendicular to the flow. At the right the dot has transformed into an ellipse with concentration varying from c to c_0 across it.

There are several possibilities for obtaining the partial differential equations describing the concentration behavior of the mixed zone as a function of time and position, to model the phenomenon shown in Figs. 5.1 to 5.3. Kramers and Albreda (1953) apply the mixing cell model for studying the frequency response of continuous flow in a packed tube to obtain

$$\frac{\partial c}{\partial t} + v_x \frac{\partial c}{\partial x} = D \frac{\partial^2 c}{\partial x^2} \tag{5.5}$$

where D is the dispersion coefficient and v_x is the average interstitial velocity q_x/ϕ. Scheidegger (1954) assumes that the motion of a tracer through a porous medium at each point is uncorrelated. He applies the central limit theorem to show the concentration of marked particles is represented by an error function which is the solution to the differential equation

$$\frac{\partial c}{\partial t} + v_x \frac{\partial c}{\partial x} = \frac{\partial}{\partial x}\left(D \frac{\partial c}{\partial x}\right) \tag{5.6}$$

Nikolaevskii (1959) obtained a model for dispersion by analog to diffusion in homogeneous isotropic turbulence. He applies the central limit theorem to marked particles of fluid moving with the mean velocity; so at an arbitrary instant in time there exists a three-dimensional probability density of finding particles at some point (x, y, z). The concentration of marked particles is

$$c(x, y, z, t) = \frac{(2\pi)^{-3/2}}{\sqrt{\sigma_x^2 \sigma_y^2 \sigma_z^2}} \exp\left[-\frac{1}{2}\left(\frac{x^2}{\sigma_x^2} + \frac{y^2}{\sigma_y^2} + \frac{z^2}{\sigma_z^2}\right)\right] \tag{5.7}$$

Eq. (5.7) is a solution of the diffusion equation of the form

$$\frac{\partial c}{\partial t} = \frac{\partial}{\partial x} D_x \frac{\partial c}{\partial x} + \frac{\partial}{\partial y} D_y \frac{\partial c}{\partial y} + \frac{\partial}{\partial z} D_z \frac{\partial c}{\partial z} \tag{5.8}$$

where the dispersion D_x is

$$D_x = \frac{\sigma_x^2}{2t} \tag{5.9}$$

and similarly for D_y and D_z.

Since the coefficient of dispersion is different along each of the axes $\underline{\underline{D}}$ must be a second-order tensor. Assuming $\underline{\underline{D}}$ is invariant to rotation about the direction of the mean velocity and mirror reflections relative to planes including the mean velocity vector or perpendicular to this vector, the dispersion coefficient has the following form

$$D_{ij} = A\tilde{v}_i\tilde{v}_j + B\delta_{ij} \tag{5.10}$$

where $\tilde{v}_i$ and $\tilde{v}_j$ are the components of the perturbation of the average velocities. Now superimpose the average motion of the fluid, Darcy's law, and Eq. (5.8) in fixed coordinates

$$\frac{\partial c}{\partial t} + v_i\frac{\partial c}{\partial x_i} = \frac{\partial}{\partial x_i}\left(D_{ij}\frac{\partial c}{\partial x_j}\right) \tag{5.11}$$

All of the preceding approaches are statistical in that they obtain an equation for the mass balance by postulating a solution and then relating the solution to the partial differential equation. The partial differential equation whose solution will describe the breakthrough curve of Fig. 5.2 can be used to determine D experimentally. The procedure is to curve-fit the appropriate solution of Eq. (5.11) by adjusting the coefficient D.

Rifai et al. (1956) followed similar reasoning to that of Scheidegger (1954) to arrive at the expression for the breakthrough curve of Fig. (5.2) from the normal probability density function.

$$\frac{c}{c_0} = \frac{1}{2}\,\text{erfc}\,\frac{x - v_x t}{2\sqrt{Dt}} \tag{5.12}$$

Equation (5.12) is a solution of Eq. (5.11) in one dimension. However, if we use the initial and boundary conditions indicated in Fig. 5.2,

$$\frac{c}{c_0}(x,\,0) = 0 \qquad x \geq 0 \tag{5.13}$$

$$\frac{c}{c_0}(0,\,t) = 1 \qquad t \geq 0 \tag{5.14}$$

$$\frac{c}{c_0}(\infty,\,t) = 0 \qquad t \geq 0 \tag{5.15}$$

the solution of Eq. (5.11) in one dimension with these conditions is

$$\frac{c}{c_0} = \frac{1}{2}\left(\text{erfc}\,\frac{x - v_x t}{2\sqrt{Dt}} + e^{v_x x/D}\,\text{erfc}\,\frac{x + v_x t}{2\sqrt{Dt}}\right) \tag{5.16}$$

Ogata and Banks (1961) analyzed experimental results in linear systems using both Eqs. (5.16) and (5.12). Considering experimental errors, they determined if $D/v_x x < 0.002$ the maximum error in approximating Eq. (5.16) with Eq. (5.12) is

3% or less. Equation (5.12) is used to determine dispersion coefficients from measured breakthrough data. For an ideal medium D may be determined from the slope of the breakthrough curve at $V/V_0 = 1$ since

$$D = \frac{1}{4\pi} \frac{Lv_x}{\left(\frac{\partial c/c_0}{\partial V/V_0}\right)^2_{V/V_0=1}} \tag{5.17}$$

For nonideal media, Eq. (5.16) is fit to the breakthrough data.

The true mixing or dispersion is the variance or spread of the dot in Fig. 5.3 in the longitudinal and transverse directions represented by

$$K_L = \frac{\overline{(X - v_x t)^2}}{2t} \tag{5.18}$$

$$K_T = \frac{\overline{Y}^2}{2t} \tag{5.19}$$

The more appropriate measurement is to measure the spread in the x and y directions, say, for a salt tracer using electrodes buried at several distances in the x and y direction. Pakula and Greenkorn (1971) used this measurement to compare dispersion measurements to those predicted by a statistical model. Using Eq. (5.17) or curve-fitting Eq. (5.16) assumes Eq. (5.11) and the appropriate boundary conditions satisfactorily model the dispersion.

If diffusion is not considered part of the mixing process, D is sometimes called the coefficient of mechanical dispersion. The dispersion coefficient will always be the hydrodynamic dispersion including molecular diffusion and mechanical dispersion, unless it is specifically stated otherwise.

Whitaker (1967) determined that dispersion should be modeled by an equation of the form of Eq. (5.11) by volume-averaging the equations of motion and continuity and assuming diffusion is explained by

$$\frac{\partial c}{\partial t} + \frac{\partial (v_i c)}{\partial x_i} = \frac{\partial}{\partial x_i}\left(\mathscr{D}\frac{\partial c}{\partial x_j}\right) \tag{5.20}$$

where $\mathscr{D}$ is the molecular diffusion and c is the concentration of the diffusing species. The volume-averaged equations are assumed valid everywhere; however, c is zero in the solid. The volume average of Eq. (5.20) is

$$\frac{\partial \langle c \rangle}{\partial t} + \frac{\partial \langle v_i c \rangle}{\partial x_i} = \frac{\partial}{\partial x_j}\left[\mathscr{D}\left(\frac{\partial \langle c \rangle}{\partial x_j} + \frac{A}{V}\tau_j\right)\right] \tag{5.21}$$

V is the volume over which the averaging takes place, A is the area of this volume, and τ_j is the tortuosity vector given by

$$\tau_j = \int_A [c]n_j\, dA \tag{5.22}$$

where [c] represents the jump in going from the fluid to the solid ([c] = αc, where α = 1 in the fluid and 0 in the solid). Consider point velocity and concentration to be the sum of a volume-averaged quantity and a fluctuation

$$v_i = \langle v_i \rangle + \tilde{v}_i \tag{5.23}$$

$$c = \langle c \rangle + \tilde{c} \tag{5.24}$$

where $\langle \tilde{v}_i \rangle = 0$ and $\langle \tilde{c} \rangle = 0$. Writing Eq. (5.21) defines a dispersion vector $\langle \tilde{v}_i \tilde{c} \rangle$, then

$$\frac{\partial \langle c \rangle}{\partial t} + \langle v_i \rangle \left(\frac{\partial \langle c \rangle}{\partial x_i} \right) + \frac{\partial \langle \tilde{v}_i \tilde{c} \rangle}{\partial x_i} = \frac{\partial}{\partial x_j} \mathscr{D} \left(\frac{\partial \langle c \rangle}{\partial x_j} + \frac{\Lambda}{V} \tau_j \right) \tag{5.25}$$

Expanding $\langle \tilde{v}_i \tilde{c} \rangle$ in a Taylor series about $\langle v_i \rangle = 0$ and $\partial \langle c \rangle / \partial x_j = 0$ a dispersion tensor results which is the sum of several terms including diffusion, so that

$$D_{jk} = \mathscr{D} \left(\delta_{jk} + \frac{A}{V} B_{jk} \right) + A_{jik} \langle v_i \rangle + A_{jilk} \langle v_k \rangle \langle v_i \rangle \tag{5.26}$$

As the bulk motion stops $\langle v_i \rangle = 0$ and

$$\mathscr{D}_{eff} = \mathscr{D} \left(\delta_{jk} + \frac{A}{V} B_{jk} \right)$$

where the terms in parentheses represent the effect of tortuosity, including both the sinuousness of the paths and the expansion and contraction of the pores. Using Eq. (5.26), Eq. (5.25) becomes

$$\frac{\partial \langle c \rangle}{\partial t} + \langle v_i \rangle \frac{\partial \langle c \rangle}{\partial x_i} = \frac{\partial}{\partial x_i} \left(D_{ij} \frac{\partial \langle c \rangle}{\partial x_j} \right) \tag{5.28}$$

which compares to Eq. (5.11) if c and v_i are replaced by their volume averages $\langle c \rangle$ and $\langle v_i \rangle$. The form of the dispersion tensor in Eq. (5.26) includes molecular diffusion and an implied dependence on velocity. Even though we reached the form of Eq. (5.28) by volume averaging, the diffusion equation was used as the starting point.

Based on statistical arguments and the volume-averaging technique, it is reasonable to accept Eq. (5.11) or Eq. (5.28) as a model for the mixed zone as a function of time and position.* The form of the dispersion tensor and its dependence on the matrix structure and velocity will be discussed in detail in the following section. As was mentioned in Chapter 4, this is a mixing zone model; the relative permeability model is also a mixing zone model. With the dispersion model, mixing results from fluctuations of the individual fluid velocities superimposed on the average velocity of the composite. With the relative permeability model, each fluid has its own velocity, and mixing is built into the relative permeability-saturation relationship.

---

*This assumes the system is large enough for complete mixing to occur; i.e., mixing is described by Fick's law.

5.2 DISPERSION: DISPERSIVITY OF NONIDEAL MEDIA

One might imagine a variety of mechanisms that cause macroscopic mixing. The
following nine mechanisms account for most of the mixing:

1. Molecular diffusion: If time scales are sufficiently long, dispersion results
    from molecular diffusion.
2. Mixing due to obstructions: The fact that the flow channels in a porous
    medium are tortuous means that fluid elements starting a given distance
    from each other and proceeding at the same velocity will not remain the
    same distance apart, as shown in Fig. 5.4.
3. Presence of autocorrelation in flow paths: Dispersion can result from the
    fact that all pores in the porous medium are not accessible to a fluid ele-
    ment after it has entered a particular flow path. In other words, the con-
    nectivity of the medium is not complete (Fig. 5.5).
4. Recirculation caused by local regions of reduced pressure: Dispersion
    can be caused by a recirculation arising from flow restrictions. The con-
    version of pressure energy into kinetic energy gives a local region of low
    pressure, and if this region is accessible to fluid which has passed through
    the region previously, a recirculation is set up much as in a venturi-
    manometer combination which contains no manometer fluid (Fig. 5.6).
5. Macroscopic or megascopic dispersion: Caused by nonidealities which
    change gross streamlines.
6. Hydrodynamic dispersion: Macroscopic dispersion is produced in a capil-
    lary even in the absence of molecular diffusion because of the velocity
    profile produced by the adhering of the fluid to the wall. This causes fluid
    particles at different radial positions to move relative to one another so a
    series of mixing-cup samples at the end of the capillary exhibits dispersion.
7. Eddies: If the flow within the individual flow channels of the porous medium
    becomes turbulent, mixing results from eddy migration.
8. Dead-end pores: Dead-end pore volumes cause mixing in unsteady flow
    (concentration profiles varying) because, as a solute-rich front passes the
    pore, diffusion into the pore occurs by molecular diffusion. After the front
    passes, this solute will diffuse back out, thus dispersing, for example, a
    step concentration input to the system. This pore volume also causes con-
    fusion in experimental interpretation because it is measured as porosity,
    but this porosity does not contribute to the available flow cross section.

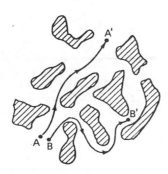

FIG. 5.4 Mixing caused by obstructions.
(From Greenkorn and Kessler, 1970.)

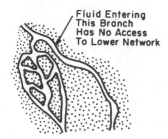

Fluid Entering
This Branch
Has No Access
To Lower Network

FIG. 5.5 Incomplete connectivity of medium.
(From Greenkorn and Kessler, 1970.)

9. Adsorption: Adsorption is an unsteady-state phenomenon. Just as with dead-end pores, a concentration front will deposit or remove material and therefore tends to flatten concentration profiles in the interstitial fluid.

Generally we include the first seven mechanisms in the dispersion coefficient. Dead-end pores usually are assumed to cause early breakthrough and increase the length of the tail of the breakthrough curve. The early breakthrough results since the effective flow porosity is less than that used to calculate the velocity, therefore, the concentration front moves faster than predicted when the porosity is not corrected for dead-end pores. The tail-off results because even though dead-end pores do not contribute to flow, there will be diffusion of the displacing fluid into and out of pores, thus dragging out the breakthrough curve. Coats and Smith (1964) considered the possible effect of dead-end pores using the three-parameter capacitance model of of Deans (1963) to account for the mass transfer into and out of the dead-end pores. Their results showed that the solutions of Eq. (5.11) for three different sets of boundary conditions gave the same dispersion coefficient as the capacitance model. In some cases the three-parameter capacitance model gave a better fit to the asymmetrical tail. All four solutions seemed to fit the breakthrough data about equally well. The effects of adsorption are usually modeled by adding a rate-dependent term to Eq. (5.11). Dispersion with adsorption is discussed in Section 5.6.

Pfannkuch (1962) plotted the existing data for longitudinal dispersion as in Fig. 5.7 plotting

$$\frac{D_L}{\mathscr{D}} = f\left(\frac{vd}{\mathscr{D}}\right) \tag{5.29}$$

where d is a characteristic dimension perpendicular to flow usually referred to as the effective particle diameter, $\mathscr{D}$ is the diffusion coefficient of the tracer, and the

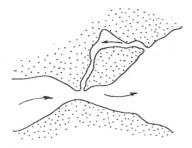

FIG. 5.6 Recirculation caused by local regions of reduced pressure. (From Greenkorn and Kessler, 1970.)

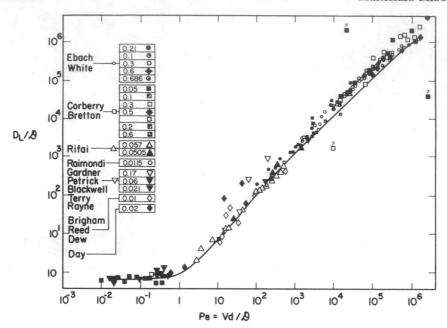

FIG. 5.7 Relationship between molecular diffusion and convective dispersion.
(From Pfannkuch, 1962.)

dimensionless group vd/$\mathscr{D}$ is the Peclet number Pe. If the dispersion coefficient
replaces the diffusion coefficient in the dimensionless group so that it is vd/D, the
group is called the <u>Bodenstein number</u> Bo.

    The following are possible explanations of the mechanisms or combination of
mechanisms of mixing as the Peclet number increases as a result of increase in
the velocity of the displacing fluid. For 0 < Pe < 0.01 mixing is by molecular diffusion. Up to Pe = 0.01 the velocity is so slow that diffusion dominates. In this region
dispersion does not depend on velocity. The ratio $D_L/\mathscr{D}$ is less than one because of
the increased path length for diffusion—the effect of tortuosity. The effective diffusion in a porous medium is

$$\mathscr{D}_{eff} = \frac{\mathscr{D}}{\tau} \tag{5.30}$$

For a homogeneous, uniform, isotropic porous medium, fluids move on the average
at about 45° to the average direction of flow; therefore the distance traveled is the
$\sqrt{2}$ times the net distance. Tortuosity in this case is equal to $\sqrt{2}$. (Carman, 1939)
and Klinkenberg (1951) discuss the analogy between diffusion and electrical conductivity in porous media so that in this region

$$\frac{D_L}{\mathscr{D}} = \frac{1}{F\phi} \tag{5.31}$$

where F is the medium electrical resistivity factor. Over the region $0.01 < Pe < 4$ mechanisms 2 to 5 begin to cause mixing and as velocity increases the mixing gradually overrides diffusion in the direction of flow. Flow is still slow enough, however, that diffusion transverse to flow helps maintain plug flow, and the dispersion is directly proportional to velocity.

Perkins and Johnston (1963) use the mixing-cell approximation of Aris and Amundson (1957) to model this region. The model imagines the medium is a series of pore spaces connected by small openings with complete mixing in each chamber. Therefore the dispersion of a series of mixing cells shows the dispersion is directly proportioned to velocity and further that the dispersivity of the medium is 0.5, so that for $0 < Pe < 50$

$$\frac{D_L}{\mathcal{D}} = \frac{1}{F\phi} + 0.5\,Pe \qquad Pe < 50 \tag{5.32}$$

As the velocity increases to $Pe > 50$ the dependence of velocity increases with an average for the data of Fig. 5.7 of about 1.2 so D_L is proportional to $v^{1.2}$. This approximate dependence is maintained for several decades of Pe so that over the region $50 < Pe < 10^4$ several mixing mechanisms are involved. At the lower end of the region transverse diffusion is still a factor, but this influence lessens as velocity increases. Over most of this region mechanical mechanisms come into play. Several possible factors create the velocity dependence greater than 1. First, as velocity increases and the effect of transverse diffusion decreases the flow in the individual pores no longer is approximated by plug flow. A velocity profile develops, and if fully developed flow existed such as in capillaries, then as Taylor (1953) shows, dispersion in a capillary depends on velocity squared (v^2). Mixing due to obstructions as discussed by Brigham et al. (1961) would create a first-power dependence of velocity. Mixing due to recirculation would increase with velocity also creating a first-power dependence. The presence of autocorrelation—physically, the connectivity of the medium—affects the mixing significantly in this region. Torelli and Scheidegger (1972) constructed a random maze model (a statistical model where randomness is applied to the medium) and inferred a velocity dependence on dispersion. In their model connectivity dominates; the velocity dependence of dispersion is $D_L \sim v^{1.2}$. In summary in the range of $50 < Pe < 10^4$ as velocity increases, the effect of transverse diffusion lessens and profiles develop in the pores, tending to create a velocity dependence $D_L \sim v^2$. Also, as velocity increases, mechanical dispersion increases, tending to create a velocity dependence $D_L \sim v$. The autocorrelation or connectivity creates an intermediate velocity dependence. The combination of mechanisms results in an intermediate velocity dependence approximating $v^{1.2}$ but the power of v varies depending on the nature of the medium.

For the approximate range $10^4 < Pe < 10^6$ mechanical dispersion dominates, and near the upper end of the range eddies start to occur. The velocity dependence again approaches $D_L \sim v$. Gradually the flow becomes turbulent and the velocity dependence of dispersion is dominated by turbulence. The data at higher velocities are scarce so values of Pe for this range are at best approximate.

Perkins and Johnston (1963) plotted the existing data (Blackwell, 1962; Grane and Gardner, 1961) for transverse dispersion as in Fig. 5.8 plotting

$$\frac{D_T}{\mathcal{D}} = f\left(\frac{vd}{\mathcal{D}}\right) \tag{5.33}$$

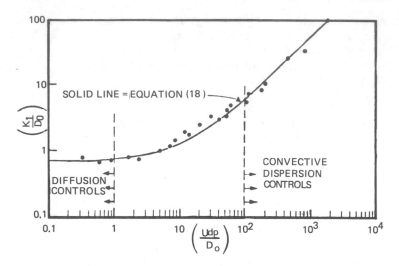

FIG. 5.8 Transverse dispersion coefficients for unconsolidated, random packs of
uniform-size sand or beads. (From T. R. Perkins and D. C. Johnston, A Review
of Diffusion and Dispersion in Porous Media, Pet. Trans. AIME 228, SPEJ 70.
© 1963 SPE-AIME.)

They use an equation similar to Eq. (5.32) to fit the data

$$\frac{D_T}{\mathcal{D}} = \frac{1}{F\phi} + 0.055\,\text{Pe} \qquad \text{Pe} < 100 \tag{5.34}$$

The preceding discussion of mechanism applies in general to transverse dispersion.
For transverse dispersion, the various ranges of Pe are higher since diffusion is
additive to the transverse dispersion. Figure 5.9 summarizes Eqs. (5.31) and
(5.33) graphically. Perkins and Johnston (1963) summarized the regimes by plotting
the Bodenstein number Bo versus the Reynolds number for both longitudinal and
transverse dispersion

$$\frac{vd}{D} = f\left(\frac{v\,d\rho}{\mu}\right) \tag{5.35}$$

Figure 5.10 is this plot for a water system. Figure 5.11 is this plot for an air
system.

      The continuum approach, using the volume averaged equations of change,
ignores the microscopic nature of the mixing process. To better understand the
relationship between pore structure, dispersion, and the effects of nonideal media,
statistical models are used. In the extreme, it is conceivable to solve the equations
for flow in individual pores knowing the geometry of the pore structure. Practically,
determining the motion of individual "fluid particles" is mathematically intractable.
Another approach to understanding relationships between pore structure and dispersion is to construct statistical models of the pore structure which in their average
properties show a relationship. The model of Torelli and Scheidegger (1972) is an

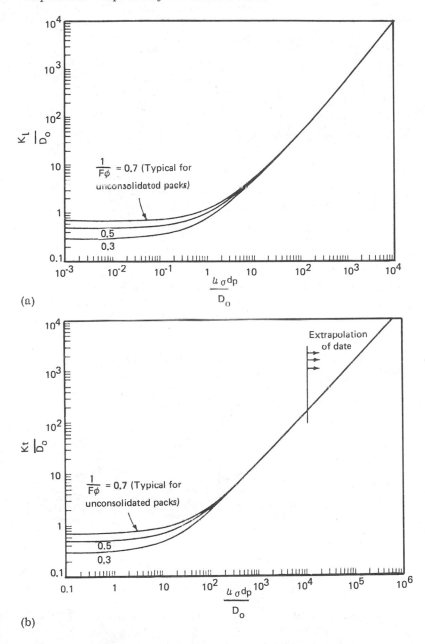

(a)

(b)

FIG. 5.9 Longitudinal and transverse dispersion coefficients. L = longitudinal dispersion coefficient, $cm^2 sec^{-1}$; K_t = transverse dispersion coefficient, $cm^2 sec^{-1}$; D_0 = diffusion coefficient, $cm^2 sec^{-1}$; U = average interstitial velocity, $cm sec^{-1}$; d_p = average particle diameter, cm; σ = inhomogeneity factor; F = formation electrical resistivity factor; ϕ = porosity. (From T. R. Perkins and D. C. Johnston, A Review of Diffusion and Dispersion in Porous Media, Pet. Trans. AIME 228, SPEJ 70. © 1963 SPE-AIME.)

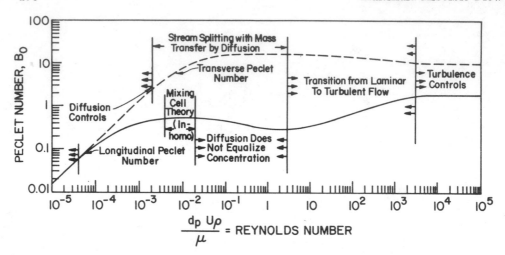

FIG. 5.10 Bodenstein numbers for an aqueous system. Assumptions: $d_p = 0.2$ cm (10 mesh), $\rho = 1$ g cm^{-3}, $\mu = 1$ cP, $D_0 = 1 \times 10^{-5}$ cm^2 sec^{-1}, $1/F\phi = 0.7$, $\sigma = 3.5$. (From T. R. Perkins and D. C. Johnston, A Review of Diffusion and Dispersion in Porous Media, Pet. Trans. AIME 228, SPEJ 70. © 1963 SPE-AIME.)

approach where randomness is applied to the media. In general this approach cannot be related directly to media structure and pore size distributions. The usual approach is to apply the randomness to the fluid and construct a network model. Here again there are two approaches: we can consider flow around objects or flow

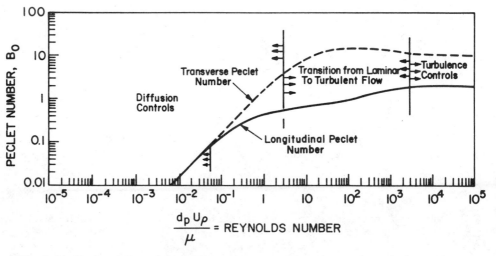

FIG. 5.11 Bodenstein numbers for a gaseous system (air). Assumptions: $d_p = 0.2$ cm (10 mesh), $\rho = 1$ g cm^3, $\mu = 1$ cP, $D_0 = 1 \times 10^{-5}$ cm^2 sec^{-1}, $1/F\phi = 0.7$, $\sigma = 3.5$. (From T. R. Perkins and D. C. Johnston, A Review of Diffusion and Dispersion in Porous Media, Pet. Trans. AIME 228, SPEJ 70. © 1963 SPE-AIME.)

in channels. The latter approach is the one most used. There are several ways of constructing a statistical model using this latter approach, such as networks of capillaries or networks of regular polyhedrons.

Much of the discussion concerning dispersion originated in reference to capillary network models for uniform isotropic porous media (de Josselin de Jong, 1958; Saffman, 1959; Scheidegger, 1959) and for nonuniform and anisotropic porous media (Greenkorn and Kessler, 1969; Haring and Greenkorn, 1970; Guin et al., 1972). We use the model of Haring and Greenkorn (1970) introduced in Section 2.2 for nonuniform media to discuss dispersion. Even though the model, a capillary network, may not have a real-world homolog, it does represent a porous medium structure; so the form of the equations and the form of the dispersion that apply to it must apply to porous media. The model is not meant to be used for practical calculations; rather, it is meant to help us understand the relationship between pore structure and dispersion and to rationalize forms of the dispersion tensor.

Assume a marked particle takes a random walk through the statistical model introduced in Section 2.2, selecting pores proportional to the volumetric flow rate. The probability of pore selection is

$$dP = \frac{v \pi r^2}{M} dE \tag{5.36}$$

where M normalizes the probability density dP from 0 to 1 and dE is the probability of a pore in a given range for length, radius, and orientation, Eq. (2.32). The dispersion is the variance of the average position of the marked particle. For dispersion in the direction of flow (in this case the z direction)

$$K_L = \frac{\overline{(Z - \tilde{V}T)^2}}{2T} \tag{5.37}$$

where $\tilde{V}$ is the velocity of the particle following the most probable path. For an ideal medium

$$\tilde{V} = \langle v_{DF} \rangle = \frac{q}{\phi} \tag{5.38}$$

In nonuniform and/or anisotropic media the average velocity, v_x in Eq. (5.5) (the motion of line b in Fig. 5.1), is a function of the medium and is not equal to q/ϕ. Further, the pressure drop and velocity are not necessarily parallel. The dispersion perpendicular to the flow is

$$K_T = \frac{\overline{X}^2}{2T} = \frac{\overline{Y}^2}{2T} \tag{5.39}$$

The value of Z in the longitudinal dispersion, Eq. (5.37) and values of X and Y in the transverse dispersion, Eq. (5.39) are the distances traveled after a random walk of n steps, where

$$Z = \sum_{i=1}^{n} z_i \tag{5.40}$$

$$X = \sum_{i=1}^{n} x_i \tag{5.41}$$

$$Y = \sum_{i=1}^{n} y_i \tag{5.42}$$

and

$$T = \sum_{i=1}^{n} t_i \tag{5.43}$$

The average position of the particle at the end of n steps of the random walk is

$$\bar{Z}_n = nL \int_p \ell^* \cos \alpha \, dP \tag{5.44}$$

and

$$\bar{X}_n = \bar{Y}_n = 0 \tag{5.45}$$

The average time to take the n steps is

$$\bar{T}_n = nL \int_p \frac{\ell^*}{v_z} \, dP \tag{5.46}$$

We need the variances in position and time and the covariance of time and longitudinal motion to calculate K_L and K_T. They are

$$\overline{(Z_n - \bar{Z}_n)^2} = n \int_p (z - \bar{z})^2 \, dP = nL^2 \sigma_Z^2 \tag{5.47}$$

$$\overline{(X_n - \bar{X}_n)^2} = n \int_p x^2 \, dP = nL^2 \sigma_x^2 \tag{5.48}$$

$$\overline{(Y_n - \bar{Y}_n)^2} = n \int_p y^2 \, dP = nL^2 \sigma_y^2 \tag{5.49}$$

$$\overline{(T_n - \bar{T}_n)^2} = n \int_p (t - \bar{t})^2 \, dP = \frac{nL^2}{V^2} \sigma_T^2 \tag{5.50}$$

$$\overline{(Z_n - \bar{Z}_n)(T_n - \bar{T}_n)} = n \int_p (z - \bar{z})(t - \bar{t}) \, dP = \frac{nL^2}{V} \sigma_{ZT} \tag{5.51}$$

The σ's in the preceding expressions are all complex functions of the properties of the pore size distribution. Combining the required variances allows us to determine K_L and K_T from Eqs. (5.37) and (5.39). For the model

$$K_L = \frac{1}{12} \frac{(a+2)(a+b+2)}{(a+1)(a+b+3)J^2} \langle \ell \rangle \, V \ln \left(\frac{27}{2} \frac{(a+b+2)^2}{(a+1)^2 J^3} \frac{VT}{\langle \ell \rangle} \right) \qquad (5.52)$$

and

$$K_T = \frac{3}{16} \frac{(a+2)(a+b+3)}{(a+1)(a+b+2)J} \langle \ell \rangle V \qquad (5.53)$$

where

$$J = \frac{(\alpha+1)(\alpha+2)(\alpha+\beta+4)(\alpha+\beta+5)}{(\alpha+3)(\alpha+4)(\alpha+\beta+2)(\alpha+\beta+3)} \qquad (5.54)$$

The ratio of longitudinal to transverse dispersion is

$$\frac{K_L}{K_T} = \frac{4}{9J} \ln \left(\frac{27}{2} \frac{(a+b+2)^2}{(a+1)^2 J^3} \frac{VT}{\langle \ell \rangle} \right) \qquad (5.55)$$

The effect of nonuniformity—variation of pore size distribution parameters—is shown in Table 5.1 for the distribution functions of Fig. 5.12. The data of Grane and Gardner (1961) for unconsolidated glass beads and consolidated Berea sandstone are compared to that calculated by Eqs. (5.52) and (5.53) in Figs. 5.13 and 5.14.

The equation for dispersion from the statistical model does not include diffusion. According to Biggar and Nielsen (1962) mixing due to coupling of molecular diffusion and dispersion is important in field systems.

Pakula and Greenkorn (1971) measured the properties of the pore size distribution of unconsolidated glass beads and other structure-related properties to determine a and b and α and β for a physical model of the beads. They measured values of K_L and K_T by imbedding electrodes in the model and actually measuring the variance of longitudinal and transverse dispersion during flow. The results of their experiments and the calculations based on the values of a, b, α, and β and Eqs. (5.52) and (5.53) are shown in Table 5.2. The agreement in Figs. 5.13 and 5.14 and Table 5.2 seems adequate to assume the statistical model is credible for relating pore structure and dispersion. We note that probability of path selection is based on volumetric flow rate which means the model inherently follows a velocity relation for dispersion.

Moranville et al. (1973) used specific distribution functions in the statistical model of Guin et al. (1972) to estimate the general effect of nonuniformity on dispersion. They used the three functions summarized in Table 5.3 (a spike, a uniform, and an exponential distribution) to calculate longitudinal dispersion. Figure 5.15 is a plot of these calculated longitudinal dispersion coefficients plotted as D_L/ν versus $L\nu/\nu$. The dashed line is from the correlation of dispersion coefficients for unconsolidated sand of Harleman et al. (1963).

The effects of nonuniformity on longitudinal dispersion have been investigated experimentally by Raimondi et al. (1959), Brigham et al. (1961), and Nieman (1969).

TABLE 5.1 Calculated Values of Dispersion (cm^2 sec^{-1}) for Various Values of the Parameters of the Beta Distribution

a	b	α	β	$K_L \times 10^3$	$K_T \times 10^4$	K_L/K_T
2	2	0	0	1.77	4.85	3.66
2	2	$-\frac{1}{2}$	$-\frac{1}{2}$	7.28	7.87	9.37
2	2	$-\frac{1}{2}$	1	33.2	16.8	19.8
2	2	2	4	4.82	6.54	7.40
2	2	2	2	3.78	5.87	6.42
2	2	4	2	2.43	4.89	4.96
2	2	1	1	4.54	6.32	7.18
2	4	2	4	3.96	5.15	7.70
4	2	2	4	4.30	5.98	7.20
2	4	1	1	4.78	4.96	9.65
4	2	1	1	7.95	11.6	6.85
2	4	4	2	2.02	4.00	5.05
4	2	4	2	4.26	8.95	4.76
1	1	4	2	1.98	4.28	4.68
1	1	2	4	3.54	5.75	6.16

Source: Haring and Greenkorn (1970).

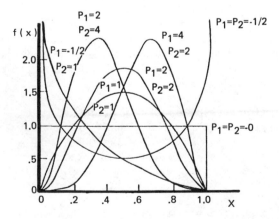

FIG. 5.12 Beta distribution. (From Haring and Greenkorn, 1970.)

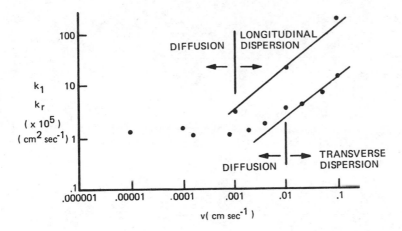

FIG. 5.13 Comparison of measured and calculated dispersion for glass beads. (From Haring and Greenkorn, 1970.)

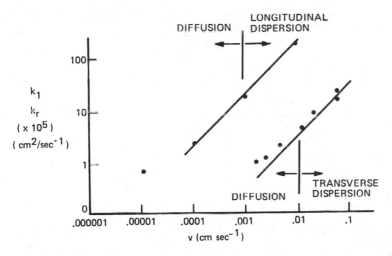

FIG. 5.14 Comparison of measured and calculated dispersion for Berea sandstone. (From Haring and Greenkorn, 1970.)

TABLE 5.2 Experimental and Calculated Dispersion Coefficients

Velocity (cm sec^{-1})	K_L (cm^2 sec^{-1}) $\times 10^3$		K_T (cm^2 sec^{-1}) $\times 10^4$	
	Meas.	Calc.	Meas.	Calc.
0.0273	8.62	(8.62)[a]	6.08	(6.08)[a]
0.0156	6.02	5.55	3.53	3.48
0.0078	1.84	2.78	2.05	1.74

[a]These values are the same because the model parameters a and b were determined at this velocity.
Source: Pakula and Greenkorn (1971).

TABLE 5.3 Distributions Used

Name	Equation	Moments
Spike	$P_A(A) = \delta(A - b) \quad A \geq 0$	$\langle A^i \rangle = b^i$
Uniform	$P_A(A) = \begin{cases} 1/b & 0 \leq A \leq b \\ 0 & \text{otherwise} \end{cases}$	$\langle A^i \rangle = \dfrac{b^i}{1+i}$
Exponent	$P_A(A) = \dfrac{1}{b} \exp\left(-\dfrac{A}{b}\right) \quad A \geq 0$	$\langle A \rangle = b$ $\langle A^2 \rangle = 2b$

Source: Moranville et al. (1973).

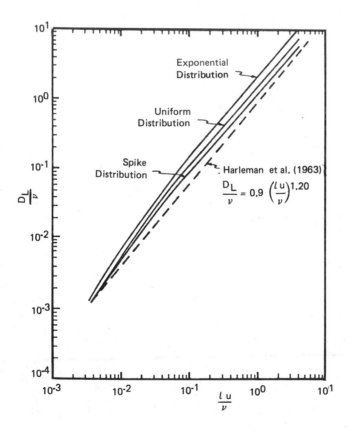

FIG. 5.15 Comparison between the results of the statistical model and the correlation of Harleman et al. (From Moranville et al., 1973.)

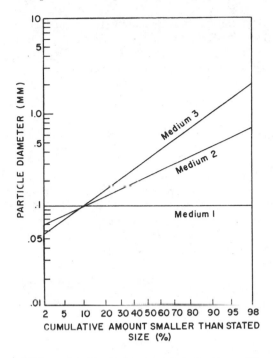

FIG. 5.16 Particle size distribution. (From Raimondi et al., 1959.)

Raimondi et al. (1959) constructed packed porous media where the permeability of the media is constant, i.e., the mean pore size is the same, but with different pore size distributions as shown in Fig. 5.16. Their results are represented by

$$D_T = 0.5\sigma \, dv \qquad\qquad (5.56)$$

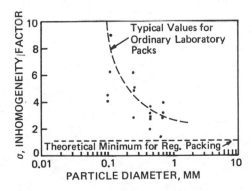

FIG. 5.17 Inhomogeneity factor for random packs of spheres. (From T. R. Perkins and D. C. Johnston, A Review of Diffusion and Dispersion in Porous Media, Pet. Trans. AIME 228, SPEJ 70. © 1963 SPE-AIME.)

where σ is a measure of the nonuniformity. Perkins and Johnston (1963) used the data of Raimondi et al. (1959) and Brigham et al. (1961) to plot σ versus particle diameter for random packs of spheres as in Fig. 5.17. They use σ to modify their empirical correlation of dispersion coefficients for nonuniform media so that

$$D_L = \frac{1}{F\phi} + 0.5\sigma\, Pe \qquad Pe < 50$$

$$D_T = \frac{1}{F\phi} + 0.055\sigma\, Pe \qquad Pe < 100$$

Nieman (1969) constructed six glass bead packs where three of the models had the same mean diameter d[10] = 0.036 mn and a different spread of sizes. The other three had the same mean diameter d[10] = 0.170 mn and a different spread of sizes (but the same spread for each set) as in Fig. 5.18. Figure 5.19 shows that the dispersion of the models is a function of the velocity, permeability, and the slope of the particle size (the pore distribution variance). Nieman shows that combining the models by screwing them together yields the same dispersion regardless of the order of combining the models. The dispersion in nonuniform media is reciprocal.

There seems to be general agreement among investigators that the appropriate equation for mixing is

$$\frac{\partial c}{\partial t} + v_i \frac{\partial c}{\partial x_i} = \frac{\partial}{\partial x_i}\, D_{ij}\, \frac{\partial c}{\partial x_j} \tag{5.59}$$

The coefficient of dispersion is a symmetric second-order tensor of the form

$$D_{ij} = a_{ijkl}\, \frac{v_k v_l}{|v|} \tag{5.60}$$

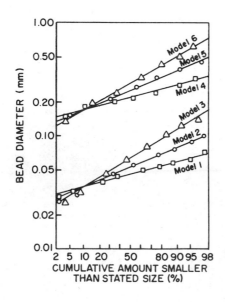

FIG. 5.18 Size distributions of beads used in packing the six homogeneous models. (From Niemann, 1969.)

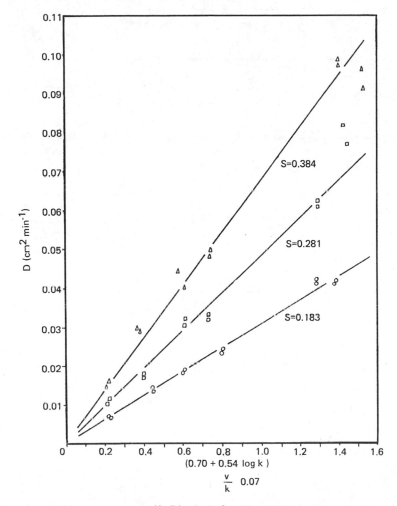

FIG. 5.19 D versus $v^{(0.70+0.54\log K)}/k^{0.07}$. (From Niemann, 1969.)

The coefficient of dispersivity is a fourth order tensor and represents the structure of the media. Two of the elements of the dispersivity a_{ijkl} represent the nonuniformity; the remaining ones represent the orientation of the pore structure. If the velocities are the correct ones, relative to the most probable path, then dispersivity should not be a function of velocity. Guin et al. (1972) show that for the equation of change represented by Eq. (5.59) the a_{ijkl} depend on velocity.

Greenkorn and Kessler (1970) summarize the effect of nonuniformity by tabulating the ratio of the longitudinal to transverse dispersion as represented by various models as in Table 5.4. In general, the most effort for real media seems to be in the continuum theories. There is relatively less on the statistical models reported, and meager amounts of data are interpreted in nonideal media.

If one accepts the form of Eq. (5.59), the effect of heterogeneity can be incorporated by defining D_{ij} as a function of position. Any useful predicted results will be made using numerical techniques for solving Eq. (5.59).

TABLE 5.4   Ratio of Longitudinal to Transverse Dispersion Coefficients:
Effect of Nonuniformity Experimental Data $(3 \leq D_L/D_T \leq f)$

Statistical model

   Haring and Greenkorn (1970)

$$\frac{D_L}{D_T} = \frac{4}{95} \ln\left[\frac{27}{2} \frac{(a + b + 2)^2}{(a + 1)J^2} \frac{vT}{\langle \ell \rangle}\right]$$

Continuum models

   Nikolaevskii (1959)

$$\frac{D_L}{D_T} = \frac{\lambda_1}{\lambda_2}$$

   Bear (1961), Scheidegger (1961), and de Josselin de Jong and Bossem (1961)

$$\frac{D_L}{D_T} = \frac{A_I}{A_{II}}$$

   Bachmat and Bear (1964)

$$\frac{D_L}{D_T} = \frac{\lambda + 2\mu}{\lambda}$$

   Poreh (1964)

$$\frac{D_L}{D_T} = \frac{\alpha_1 + (\alpha_2 + \alpha_3)\, \ell^2 v_1^2 / D_0^2}{\alpha_1 + \alpha_2\, \ell^2 v_1^2 / D_0^2}$$

   Whitaker (1967)

$$\frac{D_L}{D_T} = 3$$

     The effect of nonuniformity is incorporated in the form of the dispersion tensor
and is summarized in terms of the ratio of the longitudinal to transverse dispersion
coefficients for one-dimensional flow in Table 5.4. These models and the data can
be made to agree, except in the case of the results of Whitaker (1967) which seem
to show that for the assumptions made to get to a comparable result the medium is
uniform. (Otherwise, the form of the ratio is a complex function of several tensors.)
The results of Poreh (1965) include three elements of the dispersivity tensor and a
dependence on velocity such that in the limit the dispersion matches hydrodynamic
dispersion in a single capillary. [See Taylor (1953) and Aris (1956).]
     Including anisotropy of the media in the description of dispersion is complex
since the mixing due to orientation of the pores is stored in the fourth-order

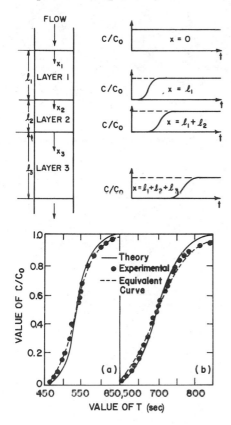

FIG. 5.20 Longitudinal dispersion behavior in flow perpendicular to layers. [From V. Y. Shamir and D. R. F. Harleman, Dispersion in Layered Porous Media, Proc. ASCE J. Hyd. Div., HY5, 237 (1967). Copyright by the American Society of Civil Engineers.]

dispersivity tensor. Scheidegger (1961) introduces several symmetry relations to show that the dispersivity tensor may be reduced to 36 independent elements to describe the effects of nonuniformity and anisotropy. In this discussion he shows that two elements remain for isotropic media. Shamir and Harleman (1967) modeled, physically and mathematically, layered porous media where one model had layers of sand of different permeability, perpendicular to flow, and other had two layers parallel to flow. In their investigation they determined the dispersion in each system experimentally; they fit solutions for one dimension of Eq. (5.59) and calculated an overall dispersion coefficient from the breakthrough curve. Ignoring the changing layers (that is, the system is a black box), they used convolution integrals to solve Eq. (5.59) for each layer and calculated the dispersion in each layer as the fluid flowed. Figure 5.20 shows a schematic of the system for flow perpendicular to layers and the three results: experimental, black box, convolution. Figure 5.21 shows a schematic of the system for flow parallel to layers and the three results: experimental, black box, convolution. They show within a close approximation that all three methods agree and suggest that layering is a possible way of approximating an anisotropic medium, on a macroscopic scale.

Moranville et al. (1977a) used the model of Guin et al. (1972) to obtain forms of the dispersion tensor for transversely isotropic porous media. Moranville et al. (1977a) apply Eq. (5.52) to several different cases including the measurements of

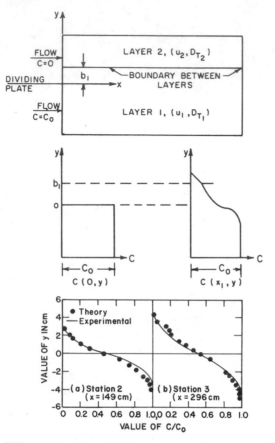

FIG. 5.21 Laterial dispersion in layered media. [From V. Y. Shamir and D. R. F. Harleman, Dispersion in Layered Porous Media, Proc. ASCE J. Hyd. Div., HY5, 237 (1967). Copyright by the American Society of Civil Engineers.]

Goad (1970) for cylinders approximating various angles of layered systems. In his systems, Goad (1970) found dispersion is greater for flow perpendicular to the axis of symmetry than parallel to it.

A transversely isotropic porous medium has an axis of symmetry. Many naturally occurring anisotropic porous media can be approximated as transversely anisotropic. For transversely anisotropic media the dispersivity tensor has six elements; the dispersion tensor can be characterized by six scalar invariants.

$$D_{ij} = a_0 \delta_{ij} v + a_1 \lambda_i \lambda_j v + a_2 \delta_{ij} \frac{v^2_{(\lambda)}}{v} + \alpha_3 \lambda_i \lambda_j \frac{v^2_{(\lambda)}}{v} + a_4 \frac{v_i v_j}{v} + a_5 (v_i \lambda_j + v_j \lambda_i) \frac{v_{(\lambda)}}{v}$$

$$(5.61)$$

Moranville et al. (1977b) constructed models of transversely isotropic media by packing alternate layers of glass bead and sintered glass at various angles β between the direction of symmetry $\underline{\lambda}$ and the direction of flow $\underline{n}$ as in Fig. 5.22. The layers were packed at angles of $\beta = 0°$, 15°, 30°, 45°, 60°, 75°, 90°; $\beta = 0°$ represents flow parallel to the layers; $\beta = 90°$ represents flow perpendicular to the layers. For this type of layered system the directional dispersion is given by

$$\frac{D_{(n)}}{v} = \frac{A_0 + A_I \cos^2 \beta + A_{II} \cos^4 \beta}{\cos^2 \beta + r^2 \sin^2 \beta} \qquad (5.62)$$

The ratio of the major to minor axis of the permeability tensor is

$$r = \frac{k_{22}}{k_{11}} \qquad (5.63)$$

The A's represent contractions of the a's in Eq. (5.61)

$$A_0 = r^2(a_0 + a_4) \qquad (5.64)$$

$$A_I = (a_0 + a_2) + 2r(a_4 + a_5) + r^2(a_1 - a_0 - 2a_4) \qquad (5.65)$$

$$A_{II} = (a_1 + a_3 + a_4 + 2a_5) - 2r(a_4 + a_5) + r^2(a_4 - a_1) \qquad (5.66)$$

The directional permeability of the models is

$$k_{(n)} = k_{22} + (k_{11} - k_{22}) \cos^2 \beta \qquad (5.67)$$

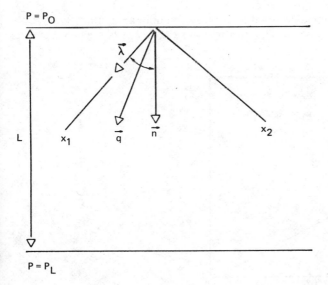

FIG. 5.22 The velocity vector in a transversely isotropic slab. (From Moranville et al., 1977a.)

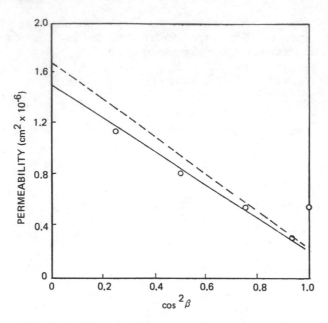

FIG. 5.23 Permeability data. (From Moranville et al., 1977b.)

and the experimental results for permeability of the model are shown in Fig. 5.23. The dashed line results from calculating by layers, using the permeability of each layer. The circles are the experimental points. (The deviation of the point at $\cos^2 \beta = 1.0$ is due to experimental error.)

The results of the experiments for dispersion are summarized in Fig. 5.24, where D_n/v is plotted versus $\cos^2 \beta$. The minimum in Fig. 5.24 is expected because

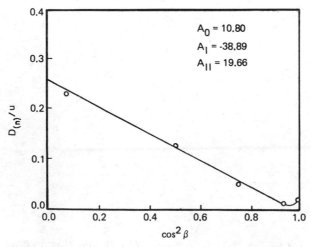

FIG. 5.24 Comparison between Eq. (5.53) and experimental results. (From Moranville et al., 1977b.)

of a minimum in v. It must be emphasized that these results are for transversely isotropic slabs and represent a macroscopic mechanism. Although it seems reasonable to expect similar behavior for orientation of the pores, which is the model used for the experiment, no experiments exist for dispersion with microscopic orientation effects. The effect of velocity on dispersivity seems to be a second-order effect at this level; however, for the microscopic mechanisms it may be more pronounced.

It was inferred in some of the earlier discussion that the effect of heterogeneity on dispersion must be handled by making the dispersion coefficient position dependent. At one level the experiments discussed above could be considered as oriented heterogeneities and handled this way.

Currently there exist several dilemmas concerning dispersion and the interpretation of laboratory measurements for use on a field scale. Field dispersivity may be orders of magnitude larger than for the same material in the laboratory. Biggar and Nielsen (1962) have emphasized the importance of considering molecular diffusion on a field scale. The possible dependence of the dispersivity on velocity seems to be indicated by theoretical studies. This means that the velocity with which we imagine the coordinate system to be translated when we interpret laboratory measurements may be incorrect. It is not clear that if we scale the ratio of average microscopic pore diameter to permeability that this assures scaling of microscopic and macroscopic lengths. In other words, can the macroscopic mechanism, e.g., change in gross streamlines, be included in the dispersion coefficient? The macroscopic mechanisms are dependent on heterogeneity and nonuniformity—mixing may be a nongradient phenomenon (non-Fickian) dependent on system size. Current research into the effects of spatial variability and stochastic representations of the apparent nongradient behavior in certain size systems and understanding of true scaling at the mixing may help understand these dilemmas. The question of velocity dependence, whether due to connectivity or the limiting cases of plug and parabolic flow, is not clearly answered. Finally, can we really treat anisotropy at the macroscopic and microscopic levels using the same constitutive relationships?

5.3 STEADY FLOW DISPLACEMENTS

The approach to modeling multifluid miscible flow depends on the velocity of the displacement relative to the size of the system, the properties of the fluid, and the degree of heterogeneity of the system. Brigham et al. (1961) suggest representing longitudinal dispersion over the range of Fig. 5.9 by an equation similar to Eq. (5.32)

$$\frac{D_L}{\mathscr{D}} = \frac{1}{F\phi} + aPe^{1.2} \tag{5.68}$$

where a is the geometric dispersivity of the medium. If the length of the transition zone between the displaced and the displacing fluid is defined as $x_{90} - x_{10}$, where x_{90} is the distance traveled to the point of 90 percent concentration of the displacing fluid and x_{10} is the distance to the point of 10 percent concentration of the displacing fluid, then based on a normal distribution

$$D_L = \frac{1}{t}\left(\frac{x_{90} - x_{10}}{3.625}\right) \tag{5.69}$$

The time in Eq. (5.69) is

$$t = \frac{LV}{vV_p} \tag{5.70}$$

where L is the length of the porous medium, V is volume of the fluid displaced in time t, v is the interstitial velocity, and V_p is the pore volume of the medium. Combining Eqs. (5.68) through (5.70),

$$\left(\frac{x_{90} - x_{10}}{L}\right)^2 = 13.10\left[\frac{V^*}{Pe_L}\left(\frac{1}{F\phi} + aPe^{1.2}\right)\right] \tag{5.71}$$

where

$$V^* = \frac{V}{V_p} \tag{5.72}$$

and Pe_L is the Peclet number based on the length of the system

$$Pe_L = \frac{Lv}{\mathscr{D}} \tag{5.73}$$

Equation (5.71) has a minimum. Brigham et al. (1961) calculated the minimum for a close random pack of glass beads 83 cm long with $1/F\phi = 0.7$ cm and a = 0.49 cm to be at $v = 3.3 \times 10^{-3}$ cm sec^{-1}. The measured velocity versus zone length for the glass bead pack compared with the calculated minimum is shown in Fig. 5.25. The mixed zone for this system is a maximum in the diffusion-controlled region and about 6 percent of the length of the model at 10^{-4} cm sec^{-1}. At the minimum the mixed zone is 2.5 percent of the length of the model—dispersion dominates here.

Blackwell et al. (1959) calculated the length of the mixed zone for typical oil field recovery rates for $D_L/\mathscr{D} = 0.67$. These results are shown in Table 5.5.

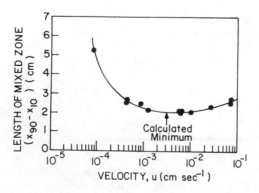

FIG. 5.25 Effect of velocity on zone length, bead pack no. 113-1, viscosity ratio 0.175. (From W. E. Brigham, P. W. Reed, and J. N. Dew, Experiments on Mixing During Miscible Displacement in Porous Media, Pet. Trans. AIME 222, SPEJ 1. © 1961 SPE-AIME.)

TABLE 5.5 Typical Mixing Zone Lengths for Reservoirs

Rate	Length of mixing zone (ft)[a]		
(ft/day)	L^b = 100 ft	L^b = 1000 ft	L^b = 10,000 ft[c]
0.1	4.0	12.7	40
0.5	1.8	5.7	18

[a] The value of D was assumed to be 5×10^{-8} ft^2 sec^{-1}.
[b] L is the length of the reservoir.
Source: R. S. Blackwell, W. M. Terry, and J. R. Rayne, Factors Influencing the Efficiency of Miscible Displacement, Pet. Trans. AIME 216, SPEJ 1. © 1959 SPE-AIME
[c] Current field data seem to imply these values may be several orders of magnitude too low (i.e., dispersion is larger).

Results for laboratory models are shown in Table 5.6. We see from Fig. 5.25 and Tables 5.5 and 5.6 and Eq. (5.71) that the mixed zone varies with velocity; the mixed zone increases as the square root of the distance traveled; diffusion dominates at oil field rates; the relative length of the mixed zone is small at rates typical of oil reservoirs; and the length of the mixed zone reaches a minimum, then increases as velocity increases.

The fluid properties density and viscosity can cause unstable flow depending on their relative values. If the flow is unstable, the effect on displacement overrides the mixing. If the flow is stable, the dispersion model Eq. (5.11) may be satisfactory for calculating mixing. Experiments by Brigham et al. (1961) indicate that dispersion increases proportional to the change in favorable viscosity ratio. As the viscosity ration becomes unfavorable—as low as 1.002—fingering starts, and the dispersion model does not satisfactorily describe mixing.

The effect of density difference is to cause unstable flow; however, the combination of density difference and viscosity difference with rate may stabilize flow.

TABLE 5.6 Typical Mixing Zone Lengths for Laboratory Models

Rate	Length of mixing zone (ft)[a]					
(ft day^{-1})	For 20-30 mesh sand			For 120-270 mesh sand		
laboratory	L^b = 1 ft	L^b = 10 ft	L^b = 100 ft	L^b = 1 ft	L^b = 10 ft	L^b = 100 ft
1	0.33	1.0	3.3	0.14	0.46	1.4
10	0.37	1.1	3.7	0.10	0.31	1.0
50	0.40	1.2	4.0	0.10	0.32	1.0
100	0.41	1.3	4.1	0.11	0.34	1.1

[a] The value of D was assumed to be 2×10^{-8} ft^2 sec^{-1}.
[b] L is the length of the model.
Source: R. S. Blackwell, W. M. Terry, and J. R. Rayne, Factors Influencing the Efficiency at Miscible Displacement, Pet. Trans. AIME 216, SPEJ 1. © 1959 SPE-AIME.

Hill (1952) and Craig et al. (1957) show that a critical velocity exists below which gravity segregation prevents fingering. This velocity is given by

$$\left(\frac{q}{A}\right)_{crit} = \frac{k \, \Delta\rho}{\Delta\mu} \sin \alpha \qquad (5.74)$$

The effects of heterogeneity dominate flow in miscible systems—adverse fluid conditions accentuate heterogeneous effects.

Assume stable flow exists for the remainder of the discussion in this section. For the moment assume a homogeneous medium. Under these assumptions, mixing, caused by diffusion and dispersion, influences the displacement depending on the velocity of the fluids and the size of the system. Consider a situation where velocity is so low that mixing is mainly by diffusion. Displacement in this situation is mod eled by mass transport due to average velocity plus the diffusion. The molar flux relative to a stationary coordinate system is

$$\underline{N} = -\frac{\mathscr{D}}{\tau} \underline{\nabla}c + \langle \underline{v} \rangle c \qquad (5.75)$$

Unsteady state diffusion is described by

$$\phi \frac{\partial c}{\partial t} + (\langle \underline{v} \rangle \cdot \underline{\nabla})c = \frac{\mathscr{D}}{\tau} \nabla^2 c \qquad (5.76)$$

Assume the fluids are incompressible and

$$\rho = \rho(c) \qquad (5.77)$$

Combining the continuity equation and Darcy's law

$$\phi \frac{\partial \rho}{\partial c} \frac{\partial c}{\partial t} = (\underline{\nabla} \cdot \frac{k\rho}{\mu} \{ \underline{\nabla}p + \rho gz \}) \qquad (5.78)$$

Also

$$\mu = \mu(c) \qquad (5.79)$$

Equations (5.75), (5.77), and (5.78) describe the displacement process. If the relative length of the mixed zone is small, Eq. (5.71), then the displacement can be approximated as a moving boundary problem and the Muskat model discussed in Section 4.4 is used to calculate the displacement.

For velocities where dispersion enters or dominates, the Muskat model is used if the relative size of the mixed zone is small. For situations where the relative size of the mixed zone is large, or where it is important to know the concentration gradient between fluids, the dispersion model Eq. (5.11) is used to calculate the displacement. Note that in radial systems the velocity varies; near sources and sinks it may be necessary to consider mixing even though the system is large. Equation (5.11) for linear one-dimensional flow is

$$\frac{\partial c}{\partial t} + v_x \frac{\partial c}{\partial x} = D \frac{\partial^2 c}{\partial x^2} \tag{5.80}$$

with initial and boundary conditions

$$\frac{c}{c_0}(x, 0) = 0 \qquad x \geq 0 \tag{5.81}$$

$$\frac{c}{c_0}(0, t) = 1 \qquad t \geq 0 \tag{5.82}$$

$$\frac{c}{c_0}(\infty, t) = 0 \qquad t \geq 0 \tag{5.83}$$

The solution to Eq. (5.80) for a step increase $c = c_0$ for these conditions is

$$\frac{c}{c_0} = \frac{1}{2}\left[\text{erfc}\, \frac{x - v_x t}{2\sqrt{Dt}} + \exp\left(\frac{v_x x}{D}\right) \text{erfc}\, \frac{x + v_x t}{2\sqrt{Dt}} \right] \tag{5.84}$$

Figure 5.26 is a plot of Eq. (5.84) from Ogata and Banks (1961) for various conditions expressed in terms of

$$\eta = \frac{D}{v_x x} = \frac{1}{\text{Bo}} \tag{5.85}$$

and

$$\xi = \frac{v_x t}{x} \tag{5.86}$$

Gelhar and Collins (1971) present a general approach to obtain solutions of the one-dimensional dispersion equation using similarity transformations. They assume that transverse dispersion is small compared to longitudinal dispersion and that $D_L \sim v$. (Their method allows including $D_L \sim v^m$.) They write the dispersion equation with $\mathcal{D}_{\text{eff}}$ separated from D_L as

$$\frac{\partial c}{\partial t} = \frac{\partial}{\partial x_i}\left(D_{ij} \frac{\partial c}{\partial x_j} - v_j c \right) + \mathcal{D}_{\text{eff}} \frac{\partial^2 c}{\partial x_i^2} \tag{5.87}$$

Assuming

$$D_L = \alpha v \tag{5.88}$$

Eq. (5.87) is rewritten in terms of $ds^{(i)}$ an element of arc length along the curvilinear coordinate line $y^{(i)}$.

$$ds^{(i)} = h_i dy^{(i)} \tag{5.89}$$

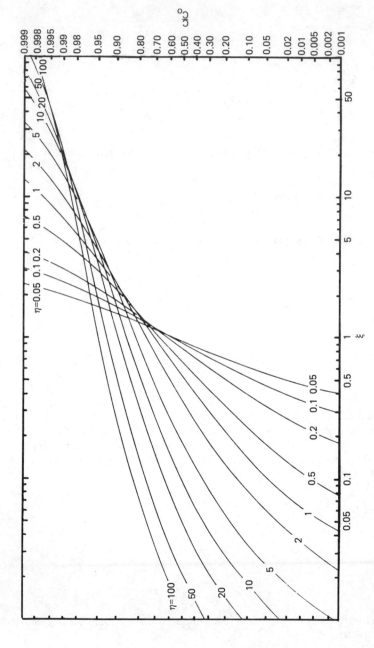

FIG. 5.26 Equation (5.84). (From Ogata and Banks, 1961.)

where h_i is a scale factor. A curvilinear coordinate system is chosen so that $v_1 = v$, $v_2 = v_3 = 0$, and Eq. (5.87) becomes

$$\frac{\partial c}{\partial t} + v \frac{\partial c}{\partial s^{(1)}} = \frac{\alpha}{h_2 h_3} \frac{\partial}{\partial s^{(1)}} \left(h_2 h_3 v \frac{\partial c}{\partial s^{(1)}} \right) + \frac{\mathscr{D}_{eff}}{h_2 h_3} \frac{\partial}{\partial s^{(1)}} \left(h_2 h_3 \frac{\partial c}{\partial s^{(1)}} \right) \tag{5.90}$$

For an incompressible fluid

$$\underline{\nabla} \cdot \underline{v} = \frac{1}{h_2 h_3} \frac{\partial}{\partial s^{(1)}} (h_2 h_3 x) = 0 \tag{5.91}$$

Substituting Eq. (5.91) into Eq. (5.90) yields

$$\frac{\partial c}{\partial t} + v \frac{\partial c}{\partial s} = \alpha v \frac{\partial^2 c}{\partial s^2} + \mathscr{D}_{eff} \left(\frac{\partial^2 c}{\partial s^2} - \frac{1}{v} \frac{\partial v}{\partial s} \frac{\partial c}{\partial s} \right) \tag{5.92}$$

Using the similarity transformation

$$\eta = x - t \tag{5.93}$$

where at $t = x$, $s = s'$,

$$x = \int_{s_t = 0}^{s_t} \frac{ds}{v(s)} \tag{5.94}$$

and

$$\omega = \int_{s'_t = 0}^{s'_t} \frac{v(s') + \mathscr{D}_{eff}/\alpha}{v^3(s')} \tag{5.95}$$

Then

$$\frac{\partial c}{\partial \omega} = \alpha \frac{\partial^2 c}{\partial \eta^2} \tag{5.96}$$

We can use $D \sim v^m$ by defining

$$\tilde{\omega} = \int_{s'_t = 0}^{s'_t} \frac{v^m(s') + \mathscr{D}_{eff}/\tau}{v^3(s')} \tag{5.97}$$

Consider uniform flow in one dimension

$$v(s) = \frac{A}{s^k} \tag{5.98}$$

TABLE 5.7 Solutions for Uniform, Radial, and Spherical Flows

Flow configuration	A	k	Frontal position $s'(t)$	Solution variable θ [in Eq. (5.32)]
Uniform	U	0	At	$(s - s')(4\alpha s')$
Radial	$q/(2\pi m)$	1	$(2At)^{1/2}$	$(s^2 - s'^2)/(16\alpha s'^3/3)^{1/2}$
Spherical	$Q/(4\pi m)$	2	$(3At)^{1/3}$	$(s^3 - s'^3)/(36\alpha s'^5/5)^{1/2}$

Source: M. Gelhar and M. A. Collins, General Analysis of Longitudinal Dispersion in Nonuniform Flow, Water Resour. Res. 7, 6, 1511 (1971). Copyrighted by the American Geophysical Union.

where A and k are constants. Evaluating Eqs. (5.94) and (5.95 for the velocity field of Eq. (5.96) with $s_{t=0} = 0$

$$x = \frac{s^{k+1}}{A(k+1)} \tag{5.99}$$

$$t = \frac{s'^{k+1}}{A(k+1)} \tag{5.100}$$

$$\omega = \frac{s'^{2k+1}}{A^2(2k+1)} \tag{5.101}$$

For a step input of concentration, solutions of Eq. (5.96) are of the form

$$\frac{c}{c_0} = \frac{1}{2}\,\text{erfc}\,\theta \tag{5.102}$$

where θ is given for three cases in Table 5.7. Figure 5.27 is a plot of c/c_0 versus ξ, where

$$\xi = \frac{s - s'}{(4\alpha s')^{\frac{1}{2}}} \tag{5.103}$$

    Other analytical solutions or analytical approximations are reported for one and two dimensions by Burch and Street (1967), Dagan (1971), Eldor and Dagan (1972), and Shamir and Harlemen (1967).
    Numerical solutions to the dispersion model have been studied extensively. Peaceman and Rachford (1962) and Lantz (1971) point out that for certain situations, especially long systems, the error in numerical analysis may create a "numerical dispersion" larger than the mixing modeled by the dispersion. Shamir and Harleman (1967), Garder et al. (1964), Nalluswami (1971), and Gupta and Greenkorn (1973) present numerical solutions of the dispersion equation using techniques that minimize numerical dispersion. (See Chapter 6, Appendix.)

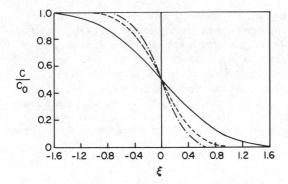

FIG. 5.27 Comparison of concentration distributions for uniform (solid line), radial (dashed line), and spherical (interrupted line with dot). [From L. W. Gelhar and M. A. Collins, General Analysis of Longitudinal Dispersion in Nonuniform Flow, Water Resour. Res. 7 (6), 1511 (1971). Copyrighted by the American Geophysical Union.]

     The effect of heterogeneity and viscosity can be included in the Muskat model by calculating the streamlines taking into account heterogeneity. The frontal movement in a stream tube is calculated using viscosity of the two fluids as a function of position. These effects can be included in Eq. (5.11) by including the dispersion tensor inside the derivative and making it a function of position. Viscosity is also a function of position and/or concentration. In practical situations one seldom has the information necessary to perform these calculations—even if a satisfactory numerical scheme exists.

     The effects of heterogeneity and viscosity ratio can be estimated from scaled model studies. The effect of heterogeneity in miscible displacement even at favorable

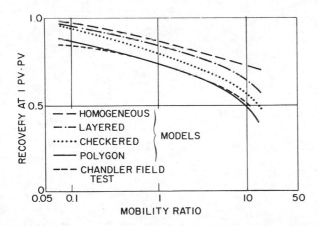

FIG. 5.28 Effect of mobility ratio on recovery for miscible floods. (From R. A. Greenkorn, C. R. Johnson, and R. E. Haring, Miscible Displacement in a Controlled Natural System, Pet. Trans. AIME 234, SPEJ 1329. © 1965 SPE-AIME.)

viscosity ratios is larger than the dispersion. Figure 5.28 shows a summary of model and field data for miscible displacements at favorable and unfavorable viscosity ratios for systems of increasing heterogeneity: a homogeneous model, a layered model, a randomly heterogeneous model, and a field-type heterogeneous model (modeled by polygons) determined by Greenkorn et al. (1965). Greenkorn et al. (1965) conclude that miscible displacement can be modeled if the degree and location of permeability variation are preserved in the model and the model is operated as the prototype is operated. We will discuss in more detail the results of unstable displacements, M > 1, in the next section.

5.4 UNSTABLE FLOW

As Fig. 5.28 shows, the effects of adverse viscosity ratio are significant. If unstable flow occurs, fingers form and grow, changing the nature of the displacement. The effect of dispersion is minor unless the concentration gradient at the front between the two fluids is important. As with immiscible flow, it is difficult to calculate unstable flow; much of the mathematics is applied to understand incipient fingering. In the linear region where infant fingers are starting, the Muskat model is used, since the sharp front approximation is certainly reasonable at the start of the displacement. (See Section 4.5.) In the nonlinear region, and for fully developed fingers, it is possible to solve the dispersion model numerically to approximate fingering. However, most information for this region results from scaled model studies.

In the miscible displacement model—the dispersion model—mixing is introduced directly as a component velocity superimposed on the Darcy velocity. At each point we imagine a single fluid, a single velocity, a single pressure, plus a dispersion velocity for each component of the fluid. The component velocity is determined by the sum of the dispersion velocities and the Darcy flow.

$$\underline{v}^i = -\underline{\underline{D}} \, \underline{\nabla} c^i + c^i \underline{v} \qquad i = 1, \, 2, \, \ldots \tag{5.104}$$

where

$$\underline{v}^i = -\frac{k}{\phi \mu^i(c^i)} \, \underline{\nabla} \, (p + \rho(c^i) \, gz) \qquad i = 1, \, 2, \, \ldots \tag{5.105}$$

The component mass balance on the displacing fluid is

$$\frac{\partial c^i}{\partial t} = \underline{\nabla} \cdot (\underline{\underline{D}} \, \underline{\nabla} c^i - c^i v) \qquad i = 1, \, 2, \, \ldots \tag{5.106}$$

From the overall mass balance

$$\underline{\nabla} \cdot \underline{v} = 0 \tag{5.107}$$

and since

$$\underline{\nabla} \cdot c^i v = c^i (\underline{\nabla} \cdot \underline{v}) + (\underline{\nabla} c^i \cdot \underline{v}) = \underline{\nabla} c^i \cdot \underline{v} \qquad i = 1, \, 2, \, \ldots \tag{5.108}$$

Eq. (5.106) becomes

$$\frac{\partial c^i}{\partial t} = (\underline{\nabla} \cdot \underline{\underline{D}} \; \underline{\nabla} c^i) - \left\{ \frac{k}{\phi\mu(c^i)} \; \underline{\nabla} [p + \rho(c^i) gz] \cdot \underline{\nabla} c^i \right\} \tag{5.109}$$

Write Eq. (5.109) for two fluids, two dimensions, with flow in the x direction using a longitudinal and a transverse dispersion coefficient. Assume the dispersion coefficients are constant at values corresponding to the average velocity. Diffusion is assumed to be included in the dispersion, then

$$\frac{\partial c}{\partial t} = D_L \frac{\partial^2 c}{\partial x^2} + D_T \frac{\partial^2 c}{\partial y^2} + \left\{ \frac{k}{\phi\mu(c)} \; \underline{\nabla} [p + \rho(c) gz] \cdot \underline{\nabla} c \right\} \tag{5.110}$$

An additional equation is needed since there are two dependent variables c and p. Using the overall mass balance

$$\underline{\nabla} \cdot \underline{v} = -\left\{ \underline{\nabla} \cdot \frac{k}{\phi\mu(c)} \; \underline{\nabla} [p + \rho(c) gz] \right\} = 0 \tag{5.111}$$

With appropriate initial and boundary conditions Eqs. (5.110) and (5.111) describe the miscible displacement process and in principle unstable flow in such a process.

If the concentration gradient is not important, the equations degenerate to the Muskat model, and one calculates using a sharp front. A perturbation analysis can be performed about one-dimensional stable flow to find stability criteria. For steady state in long, thin systems dispersion may be important; otherwise the mixing due to dispersion is not of major importance.

Let us describe some of the results of miscible unstable displacements. For the linear region the sharp front theory discussed in Section 4.5 can be used to model miscible fingers. Benham and Olson (1963) and Kyle and Perrine (1965) show that fingering occurs macroscopically regardless of the microscopic nature of the media. Figure 5.29 shows fingering in displacement of two miscible fluids from inception to developed fingers. Hawthorne (1960) shows that if gravity is involved a single finger persists. Any initial condition leads to fingering; there is no steady-state configuration. At zero velocity the steady state is a horizontal transition zone. As velocity increases the zone tilts more in the direction of the input-output boundaries. At critical velocity it parallels these boundaries. At higher velocities the angle exceeds the tilt of the system.

Blackwell et al. (1959) shows an entire unstable displacement. Early in the displacement many small fingers form. Eventually one finger dominates and moves ahead of the others (Fig. 5.30). As this dominant finger continues to move, it begins to mushroom at the nose; the smaller fingers behind it begin to disappear or join the base of the dominant finger. Lateral symmetry begins to disappear. The displacement continues with one dominant finger. The smaller the viscosity ratio, the less likely a dominant finger will occur. Hall and Geffen (1957) show that the transition zone length increases rapidly at first, then becomes essentially constant in unstable miscible flow.

In summary, for miscible fingers: The transition zone length containing the fingers varies linearly with velocity or time (Benham and Olsen, 1963; Kyle and Perrine, 1965; Perkins et al., 1965). This is the growth rate of a sharp-front

Run No. 26 M_1 =10.0-1 M_2= 5.1-1 V=1 Ft Hr^{-1}

Run No. 26 M_1 =10.0-1 M_2= 5.1-1 V=1 Ft Hr^{-1}

Run No. 26 M_1 =10.0-1 M_2= 5.1-1 V=1 Ft Hr^{-1}

Run No. 26 M_1 =10.0-1 M_2= 5.1-1 V=1 Ft Hr^{-1}

FIG. 5.29 Fingering for a miscible slug in an open model. (From A. L. Benham and R. W. Olson, A Model Study of Viscous Fingering, Pet. Trans. AIME 228, SPEJ 138. © 1963 SPE–AIME.)

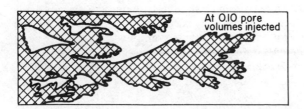

FIG. 5.30 Displacement front for mobility ratio of 383. (From R. S. Blackwell, W. M. Terry, and J. R. Rayne, Factors Influencing the Efficiency at Miscible Displacement, Pet. Trans. AIME 216, SPEJ 1. © 1959 SPE–AIME.)

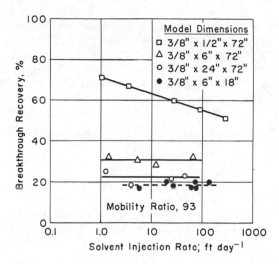

FIG. 5.31 Effect of injection rate on recovery at breakthrough. (From R. S. Blackwell, W. M. Terry, and J. R. Rayne, Factors Influencing the Efficiency at Miscible Displacement, Pet. Trans. AIME 216, SPEJ 1. © 1959 SPE-AIME.)

"steady-state" finger. Recovery is independent of velocity except in long, narrow models (Fig. 5.31) (Blackwell et al., 1959; Handy, 1959; Slobod and Thomas, 1963). For adverse miscible displacements in homogeneous media without gravity, the viscosity ratio is the major effect influencing the transition zone (Fig. 5.32). Even at small adverse viscosity ratios and/or at slower velocities transverse dispersion does not distort the fingers. Only if the system is long and narrow will transverse dispersion affect the fingering—the system becomes rate dependent— see Fig. 5.31. A graded viscosity zone may increase the influence of dispersion.

Consider the effect of gravity in more detail. A number of regimes exist: Domination by tangential gravity—flow is near stable—the transition zone has an

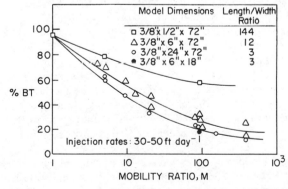

FIG. 5.32 Effect of mobility ratio on recovery at breakthrough. (From R. S. Blackwell, W. M. Terry, and J. R. Rayne, Factors Influencing the Efficiency at Miscible Displacement, Pet. Trans. AIME 216, SPEJ 1. © 1959 SPE-AIME.)

almost straight line shape. Domination by normal gravity—a single parallel-sided finger appears at the top or bottom depending on density and viscosity. Domination by viscous forces—unstable flow as with no gravity. At very high velocities secondary fingers grow again and a piston like displacement develops behind the nose of the primary finger (van der Poel, 1962). For the regime where tangential gravity is important, the viscous forces try to align the transition zone parallel to flow. If the velocity is less than critical, a steady-state exists, and the zone is tilted from the horizontal more parallel to the input-output boundary. As velocity increases and slightly exeeceeds critical velocity, the zone continues to tilt toward alignment with the boundaries. For high velocities or near horizontal flow, the regime dominated by normal gravity exists. Craig et al. (1957) show very narrow fingers and very low recoveries for this regime. The displacement depends strongly on viscosity, density, and velocity. Pozzi and Blackwell (1963) show that in this regime as rate increases normal gravity maintains single fingers. Varnon and Greenkorn (1969) show that gravity-dominated behavior is similar for both immiscible and miscible flow.

The perturbation analyses for determining instability in miscible flow cannot be written in compact mathematical form as with immiscible flow, because of the concentration dependence of viscosity and density in the mixed region. Schowalter (1965) gives the best estimate of the gravity stabilizing effect; he does not hold density constant in Darcy's law. Heller (1966) approximates the concentration variation with a ramp function and analyzes the growth of perturbations on the center profile. He shows one effect of transverse dispersion is to create a critical wavelength below which the perturbation is stable. Perrine (1961) treats the transition zone that would occur if flow were stable. The stabilizing effect of transverse dispersion decreases as the wavelength of the disturbance increases. There is no effect on stability from longitudinal dispersion. He explicitly determines the maximum permissible concentration gradient compatible with stable flow. The more influential gravity and dispersion, the greater the permissible gradients. The more influential the viscous forces the smaller the permissible gradients. Perrine and Gay (1966) tried to use higher order perturbation analyses to consider large fingers, but they had to restrict themselves to second-order truncations limiting the results to small disturbances.

The mathematical and semiempirical calculations for unstable flow are summarized below. Peaceman and Rachford (1962) solved Eqs. (5.110) and (5.111) numerically. They treated the disturbances by using different values of permeability in the initial blocks. They performed calculations to match the results of Blackwell et al. (1959) and obtained fingering as shown in Fig. 5.33 for one of their calculated concentration maps. The results are qualitatively similar to those of Blackwell et al. (1959). Garder et al. (1964) applied an approximate method based on the method of characteristics to solve Eqs. (5.110) and (5.111) using a stationary grid and a set of moving points. They simulated the data of Pozzi and Blackwell (1963) quite well. However, as the instability became more severe the calculated fingers were much thicker than those in the model.

Semiempirical methods for pure viscous fingers have been developed by Koval (1963), Dougherty (1963), and Perrine (1963). In these methods a conceptual model is envisioned with lateral symmetry. Fingering is considered as a contribution to longitudinal dispersion. Empirical parameters are introduced which correlate with the viscosity ratio. Koval (1963) and Dougherty (1963) apply the relative

(a) 0.1 Pore Volume Injected

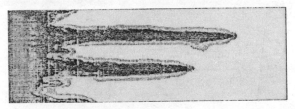

(b) 0.25 Pore Volume Injected

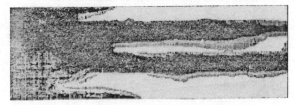

(c) 0.5 Pore Volume Injected

FIG. 5.33 Concentration maps for miscible displacement, mobility ratio 86. (From D. W. Peaceman and H. H. Rachford, Jr., Numerical Calculation of Multi-dimensional Miscible Displacement, Pet. Trans. AIME 225, SPEJ 327. © 1962 SPE-AIME.)

permeability model. The relative permeability is considered as the product of permeability and concentration in this model.

In summary, microscopic mixing (dispersion) in miscible unstable displacements has a minor effect on finger behavior. Sharp-front models can normally be used to predict fingering on the linear region. Mixing (dispersion) acts to soften the effect of viscosity ratio. However, viscous fingers dominate the flow to such an extent that the effects of dispersion only show up in long narrow systems. Heterogeneity and viscous forces dominate unstable miscible flow. Gravity forces may stabilize the flow up to a critical rate, then cause a single finger. Perturbation analysis gives some qualitative stability criteria. Numerical solutions of the dispersion model, Eqs. (5.110) and (5.111) give semiquantitative results—they show fingers. Scaled model studies provide the most complete information for estimating miscible unstable flow.

5.5 TRANSIENT FLOW

The equation of condition describing pressure behavior in multifluid miscible systems is the same as for single-fluid flow. However, in the case of multifluid

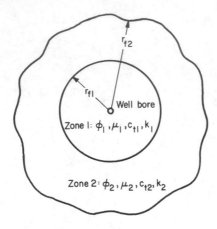

FIG. 5.34 Composite reservoir. (From Muskat, 1937.)

miscible flow, viscosity and compressibility are functions of concentration:

$$\phi\mu(c)c_t(c)\frac{\partial p}{\partial t} = \underline{\nabla}\cdot\underline{\underline{k}}\,\underline{\nabla}p \tag{5.112}$$

Consider two-fluid miscible flow with a mixed zone separating the two fluids. Also assume the mixed zone is small enough that it can be modeled as a sharp front. In this composite system the properties of the fluids $\mu(c)$ and $c_t(c)$ may vary stepwise. At an instant in time the same model applies to a system where permeability and porosity vary stepwise. We can model both situations by considering k/μ as a step

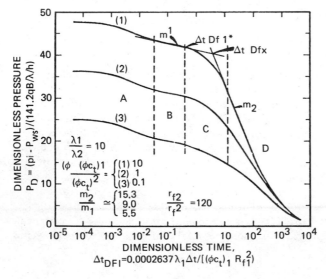

FIG. 5.35 Simulated pressure falloff data for a two-zone system. (From L. S. Merrill, Jr., H. Kazemi, and W. B. Gogarty, Pressure Falloff Analysis in Reservoirs with Fluid Banks, Pet. Trans. AIME 257, SPEJ 809. © 1974 SPE-AIME.)

and ϕc_t as a step. For the sake of simplicity let us discuss the case where $\mu(c)$ and $c_t(c)$ vary in a stepwise fashion. This problem has been considered for one dimension in radial systems analytically (Hurst, 1960; Larkin, 1963; Loucks and Guerrero, 1961; Carter, 1966; Odeh, 1969) and numerically (Bixel and van Poolen, 1967; Merrill et al., 1974).

Consider the radial situation sketched in Fig. 5.34 where the two fluids are miscible. A step change in k/μ (either by changing k, μ, or their combination) in the radial direction results in a change in slope of the semilog plot p_D versus log Δt_{Dfl} simulated in Fig. 5.35 by Merrill et al. (1974). Region A of Fig. 5.35 is dominated by fluid 1 being injected in the well (effect of well-bore storage). Region B of Fig. 5.35 is a semilog straight line reflecting the mobility of the injected (displacing) fluid. Region C of the figure is a transition zone. Region D is a second semilog straight line representing the mobility of the inplace (displaced) fluid. Merrill et al. (1974) propose two ways of estimating the fluid front, r_{fl}, from the plot of the falloff data, Fig. 5.35. Extrapolate the intersection time of the two semilog straight lines on the plot of the bottom hole pressure p_{ws} versus log Δt as in Fig. 5.36, then

$$r_{fl} = \sqrt{\frac{0.0002637(k/\mu)_1 \, \Delta t_{fx}}{(\phi c_t)_1 \, \Delta t_{Dfx}}} \qquad (5.113)$$

where Fig. 5.37 correlates Δt_{Dfx} with the ratio of the two semilog slopes from the falloff curve and the ratios $(\phi c_t)_1/(\phi c_t)_2$. The second method uses the point of deviation from the first straight line of Fig. 5.35 at Δt^*_{fi}.

$$r_{fl} = \sqrt{\frac{0.0002637(k/\mu)_1 \, \Delta t^*_{fl}}{(\phi c_t)_1 \, \Delta t^*_{Dfl}}} \qquad (5.114)$$

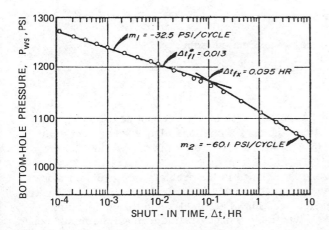

FIG. 5.36 Falloff test data. From L. S. Merrill, Jr., H. Kazemi, and W. B. Gogarty, Pressure Falloff Analysis in Reservoirs with Fluid Banks, Pet. Trans. AIME 257, SPEJ 809. © 1974 SPE–AIME.)

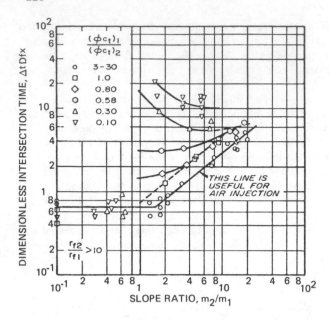

FIG. 5.37 Correlation of dimensionless intersection time, Δt_{Dfx}, for falloff data from a two-zone reservoir. (From L. S. Merrill, Jr., H. Kazemi, and W. B. Gogarty, Pressure Falloff Analysis in Reservoirs with Fluid Banks, Pet. Trans. AIME 257, SPEJ 809. © 1974 SPE-AIME.)

where $0.13 < \Delta t_{Dfl} < 1.39$ with an average of 0.389. The permeability of the injected fluid may be estimated from the slope (m_1) of the first semilog straight line.

$$k_1 = \frac{162.6q\,B\mu_1}{m_1 h} \qquad (5.115)$$

The mobility in the second zone may be estimated from

$$\left(\frac{k}{\mu}\right)_2 = \frac{(k/\mu)_1}{\lambda_1/\lambda_2} \qquad (5.116)$$

where λ_1/λ_2 may be found from Fig. 5.38. Note that m_2 cannot be used to get the mobility of zone 2. When $r_{f2} < 10r_{f1}$, the approximations discussed are not valid and numerical solutions of Eq. (5.112) with μ and c_t as a function of position must be used.

If the transition (mixed) zone is included in the problem, the problem must be simulated numerically since concentration must be first determined as a function of position by solving the dispersion equation, Eq. (5.11), and then determining viscosity and compressibility as a function of concentration through the mixed zone.

If the system is heterogeneous, the preceding solutions are not unique and the bank cannot be located; rather we can only locate changes of the ratios k/μ and the product ϕc_t.

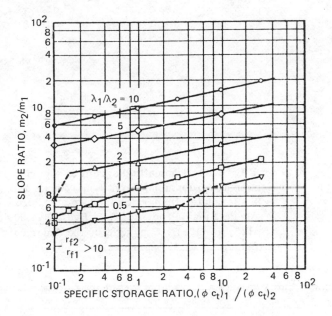

FIG. 5.38 Effect of specific storage ratio and mobility ratio on the slope ratio for falloff testing in a two-zone reservoir. (From L. S. Merrill, Jr., H. Kazemi, and W. B. Gogarty, Pressure Falloff Analysis in Reservoirs with Fluid Banks, Pet. Trans. AIME 257, SPEJ 809. © 1974 SPE-AIME.)

The mixed zone is a result of a concentration transient. However, transient flow, a variation of fluid velocity, will also affect the mixed zone concentrations. Banks and Jerasate (1962) considered the effect of a natural decrease in the pressure (head) on dispersion due to water draining from a porous medium under influence of gravity. The experiment is diagramed in Fig. 5.39. The column is initially filled with salt water. Fresh water fills the space above the porous medium. At time t = 0 the control valve is opened a set amount and the fluid drains through the column creating a mixed zone moving with a decreasing velocity. Figure 5.40 is a plot of dimensionless velocity $\zeta = v_x/v_0$, where v_x is interstitial velocity in the x direction and v_0 is v_x immediately after the control valve is opened versus mt, where m is defined as

$$m = \frac{\phi A^2 v_0}{gc^2 s^2} + \frac{L}{\phi k} \qquad (5.117)$$

c is the discharge coefficient of the control valve, s is the area of the control valve. Two forms of fitting the data are shown in Fig. 5.40. The following one was used in the analytical solution of the dispersion.

$$\zeta = \frac{v_x}{v_0} = e^{-mt} \qquad (5.118)$$

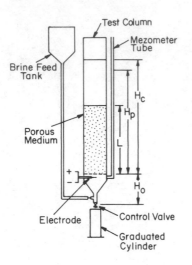

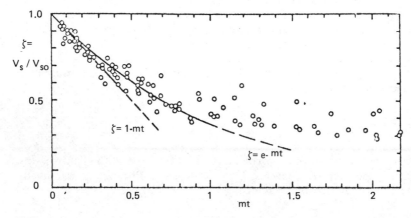

FIG. 5.39 Schematic diagram of experimental apparatus. [From R. B. Banks and S. Jerasate, Dispersion in Unsteady Porous-Media Flow, Proc. ASCE J. Hyd. Div. 90, HY5, 13 (1964). Copyright by the American Society of Civil Engineers.]

The partial differential equation in the z direction describing the concentration distribution of the mixed zone is

$$\frac{\partial c}{\partial t} + v_z \frac{\partial c}{\partial z} = D \frac{\partial^2 c}{\partial z^2} + r(c, t) \qquad (5.119)$$

where $r(c, t)$ is the rate of adsorption. The effect of $r(c, t)$ on dispersion is discussed in detail in Section 5.6. For the discussion here we assume an adsorption coefficient defined by linear equilibrium adsorption.

$$\Theta = 1 + \frac{k_1}{\phi} \qquad (5.120)$$

FIG. 5.40 Dimensionless plot of velocity versus time. [From R. B. Banks and S. Jerasate, Dispersion in Unsteady Porous-Media Flow, Proc. ASCE J. Hyd. Div. 90, HY 5 (3) (1964). Copyright by the American Society of Civil Engineers.]

where k_1 is the adsorption rate and rewrite Eq. (5.119) as

$$\Theta \frac{\partial c}{\partial t} + v_z \frac{\partial c}{\partial t} = D \frac{\partial^2 c}{\partial z^2} \tag{5.121}$$

Assume dispersion is related to the first power of velocity.

$$D = a v_z \tag{5.122}$$

Multiply Eq. (5.121) by ζ and substitute Eq. (5.122),

$$\frac{1}{\zeta} \frac{\partial c}{\partial t} + \frac{v_0}{\Theta} \frac{\partial c}{\partial z} = \frac{D_0}{\Theta} \frac{\partial^2 c}{\partial x^2} \tag{5.123}$$

where D_0 is defined at v_0. Defining

$$\tau = \int_0^t \zeta \, dt \tag{5.124}$$

Then

$$\frac{\partial t}{\partial \tau} = \frac{1}{\zeta} \tag{5.125}$$

And Eq. (5.123) written in terms of τ is

$$\frac{\partial c}{\partial \tau} + \frac{v_0}{\Theta} \frac{\partial c}{\partial z} = \frac{D_0}{\Theta} \frac{\partial^2 c}{\partial z^2} \tag{5.126}$$

The initial and boundary conditions for Eq. (5.126) assumed for the experiment are

$$c(z, 0) = c_0 \qquad z \geq 0 \tag{5.127}$$
$$c(0, \tau) = 0 \qquad \tau > 0 \tag{5.128}$$
$$c(\infty, \tau) = 0 \qquad \tau > 0 \tag{5.129}$$

The solution of Eq. (5.126) with the preceding conditions is

$$\frac{c}{c_0} = 1 - \frac{1}{2} \left(\text{erfc} \frac{1 - \xi}{2\sqrt{\xi \eta}} + e^{1/\eta} \text{erfc} \frac{1 + \xi}{2\sqrt{\xi \eta}} \right) \tag{5.130}$$

where

$$\xi = \frac{v_0 t}{\Theta z} \tag{5.131}$$

$$\eta = \frac{D_0}{v_0 z} \tag{5.132}$$

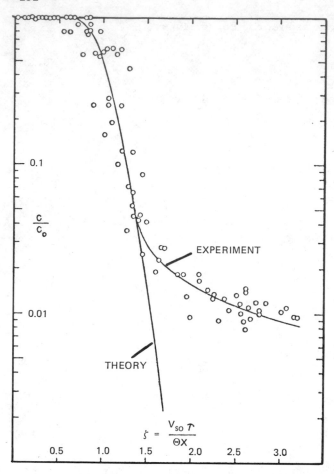

FIG. 5.41 Concentration ratio versus time parameter. [From R. B. Banks and
S. Jerasate, Dispersion in Unsteady Porous-Media Flow, Proc. ASCE J. Hyd. Div.
90, HY5, 13 (1964). Copyright by the American Society of Civil Engineers.]

If $\eta < 0.05$,

$$\frac{c}{c_0} = \frac{1}{2} \left(1 + \text{erf} \frac{1 - \xi}{2\sqrt{\xi\eta}} \right) \tag{5.133}$$

τ is determined using Eq. (5.118) to be

$$\tau = \frac{1}{m}(1 = e^{-mt}) \tag{5.134}$$

The experimental results and the results using Eq. (5.133) are shown in
Fig. 5.41. The disagreement between the calculations and the experiment are only

in the small concentration range. (The log scales exaggerate the effect.) When $c/c_0 \le 0.04$, Banks and Jerasate (1962) explain, the deviation is due to adsorption. It seems in this region that diffusion may enter the problem—using D_0 in this region may not explain the mixing.

Another practical problem involving a transient velocity is mentioned by Banks and Jerasate (1962), that of the effect of oscillating (sinusoidal) variation. Oscillating flow may model saltwater intrusion due to tides. Raats and Scotter (1968) discuss the implications of this motion. Scotter et al. (1967) performed exploratory experiments on dispersion resulting from sinusoidal flow. Scotter and Raats (1968) considered dispersion in oscillating flow to be described on the average, $\langle v \rangle = 0$, by

$$\frac{\partial c}{\partial t} = D \frac{\partial^2 c}{\partial x^2} \tag{5.135}$$

The experiment they describe is sketched in Fig. 5.42. They postulate that the dispersion depends on the amplitude and frequency of the oscillations. For slow oscillatory motion, from dimensional analysis

$$\frac{c}{c_0} = \frac{c}{c_0}\left(\frac{\ell}{\tilde{t}\tilde{v}}, \frac{\tilde{v}\ell}{\mathscr{D}}, \frac{a_0}{\ell}\right) \tag{5.136}$$

where ℓ is a microscopic length, $\tilde{t}$ a characteristic time, $\tilde{v}$ a characteristic speed of the oscillations. The initial and boundary conditions for the experiment are

$$c(x, 0) = 0 \qquad 0 < x < d + \hat{c} \tag{5.137}$$

$$c(0, t) = c_0 \qquad t > 0 \tag{5.138}$$

$$D \frac{\partial c}{\partial x} + \hat{c} \frac{\partial c}{\partial t} = 0 \qquad x = d, \ t > 0 \tag{5.139}$$

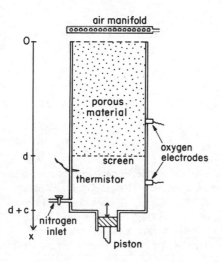

FIG. 5.42 Schematic diagram of apparatus. [From D. R. Scotter and P. A. C. Raats, Dispersion in Porous Mediums Due to Oscillating Flow, Water Resour. Res. 4 (6), 1201 (1968). Copyrighted by the American Geophysical Union.]

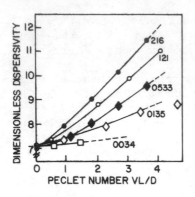

FIG. 5.43 Constant a_0/L curves for 3-mm spheres, and a theoretical curve for large values of a_0/L. The numbers on the curves are values of a_0/L. [From D. R. Scotter and P. A. C. Raats, Dispersion in Porous Mediums Due to Oscillating Flow, <u>Water Resour. Res</u>. <u>4</u> (6), 1201 (1968). Copyrighted by the American Geophysical Union.]

The solution of Eq. (5.135) for these boundary conditions is

$$\frac{c}{c_0} = 1 - \sum_{n=1}^{\infty} \frac{2(\alpha_n + h)^2 \exp(-D\alpha_n^2 t) \sin(\alpha_n x)}{\alpha_n[d(\alpha_n^2 + h^2) + h]} \tag{5.140}$$

where $h = \phi/\hat{c}$ and α_n are the solutions of

$$h = \alpha \tan \alpha d \tag{5.141}$$

For a sine wave oscillation,

$$a = a_0 \sin \omega t \tag{5.142}$$

and

$$\tilde{v} = \frac{2a_0 \omega}{\pi} \tag{5.143}$$

The Peclet number for oscillating flow is defined as

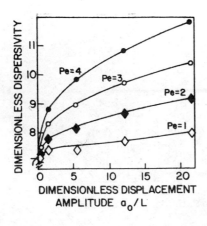

FIG. 5.44 Constant Peclet number curves for 3-mm spheres. [From D. R. Scotter and P. A. C. Raats, Dispersion in Porous Mediums Due to Oscillating Flow, <u>Water Resour. Res</u>. <u>4</u> (6), 1201 (1968). Copyrighted by the American Geophysical Union.]

$$\tilde{P}e = \frac{2a_0 \omega \ell}{\pi \mathcal{D}} \tag{5.144}$$

Figure 5.43 shows the experimental results where Eq. (5.139) was used to determine the dispersion. The results show that dispersion due to oscillatory flow depends on a_0/ℓ and $\tilde{P}e$. Figure 5.44 shows the dependence of dispersion on a_0/ℓ and $\tilde{P}e$. Further experiments using different size porous media packing were interpreted using Fig. 5.44 with good results. More complicated transient flows or flows involving adsorption require numerical solution.

5.6 DISPERSION WITH ADSORPTION

The movement of various chemicals during miscible flow in a porous medium with dispersion and adsorption is of interest in many situations:

> Chemical engineering. In adsorption and ion exchange columns, in reaction engineering, and in chromatography
> Petroleum engineering. In miscible displacement of oil with surfactant solutions and in interpretation of chemical tracers
> Hydrology. In groundwater flow and in saltwater intrusion
> Soil physics. In movement of nutrients and pollutants in the soil.

Initial studies of the effects of adsorption on mixing in adsorption columns and ion exchange resins neglected longitudinal dispersion. Lapidus and Amundson (1952) solved the dispersion model with linear equilibrium and finite rate adsorption. The solution for first-order finite rate adsorption is an integral form approximated in closed form by Ogata (1958). Lindstrom and Boersma (1970) considered the theory of chemical transport with simultaneous sorption in porous media; the convective term was not considered. Greenkorn (1962) used the bilinear model to interpret breakthrough curves for chemical tracers. Lai and Jurinak (1972) solved the problem of cation adsorption involving a general nonlinear exchange function numerically.

The dispersion model Eq. (5.11) is assumed to include the adsorption-desorption of chemical species in solution by addition of a rate term.

$$\frac{\partial c}{\partial t} + v_i \frac{\partial c}{\partial x_i} = \frac{\partial}{\partial x_i} D_{ij} \frac{\partial c}{\partial x_j} + r(c, t) \tag{5.145}$$

The term $r(c, t)$ can express a homogeneous reaction in the liquid, adsorption-desorption on the medium, or a heterogeneous reaction with the medium. Consider one-dimensional flow where the displacing fluid has the same properties as the displaced fluid and the fluids are miscible. The displacing fluid contains a solute, such as salt, surfactant, or a chemical tracer, that may adsorb and desorb on the medium. If F is the amount of adsorbate on the absorbent (chemical on the medium) per unit volume, then Eq. (5.145) is written in one dimension as

$$\frac{\partial c}{\partial t} = D \frac{\partial^2 c}{\partial x^2} - v_x \frac{\partial c}{\partial x} - \frac{1}{\phi} \frac{\partial F}{\partial t} \tag{5.146}$$

The adsorption mechanisms on porous media are usually nonlinear and follow a Langmuir isotherm or a Freundlich isotherm.

Linear adsorption is introduced in Section 5.5

$$F = k_1 c + k_2 \tag{5.147}$$

where k_1 and k_2 are kinetic rate constants. Equation (5.147) implies equilibrium is reached at each point in the medium. For linear adsorption Eq. (5.146) becomes

$$\left(1 + \frac{k_1}{\phi}\right) \frac{\partial c}{\partial t} = D \frac{\partial^2 c}{\partial x^2} - v_x \frac{\partial c}{\partial x} \tag{5.148}$$

Banks and Ali (1964) and Banks and Jerasate (1962) used the analytical solution of Eq. (5.130) to model this situation. The model does not reproduce the data well in the low concentration regions.

Nonlinear adsorption is described by a Langmuir isotherm or a Fruendlich isotherm. A Langmuir adsorption equilibrium isotherm is represented by

$$F = \frac{ac}{1 + bc} \tag{5.149}$$

Substituting Eq. (5.149) into Eq. (5.146) yields

$$\left(1 + \frac{a}{\phi(1 + bc)^2}\right) \frac{\partial c}{\partial t} = D \frac{\partial^2 c}{\partial x^2} - v_x \frac{\partial c}{\partial x} \tag{5.150}$$

A bilinear adsorption mechanism proposed for ion exchange and adsorption columns by Hiester and Vermuelen (1952) is appropriate for porous media containing clay. Bilinear adsorption is described by a Langmuir isotherm. The bilinear adsorption-desorption reaction can be represented by

$$A + \text{sorbent} \rightleftarrows A \cdot \text{sorbent} \tag{5.151}$$

Writing a second-order rate expression for the reaction of Eq. (5.151)

$$\frac{d(A \cdot \text{sorbent})}{dt} = k_{kin} \left[(A)(\text{sorbent}) - \frac{1}{K^{ad}} (A \cdot \text{sorbent}) \right] \tag{5.152}$$

Let $A \equiv c$, $A \cdot \text{sorbent} \equiv q$, and $\text{sorbent} \equiv Q - q$, where q is the amount of A adsorbed on the dry medium and Q is the ultimate capacity of the dry medium; then

$$\frac{dq}{dt} = k_{kin} \left[c(Q - q) - \frac{1}{K^{ad}} q \right] \tag{5.153}$$

For $dq/dt = 0$, $c = c_0$, $q = q_\infty$, and

$$K^{ad} = \frac{q_\infty}{c_0 (Q - q_\infty)} \tag{5.154}$$

where Eq. (5.154) is the form of a Langmuir isotherm. Writing Eq. (5.146) in terms of q

$$\frac{\partial c}{\partial t} = D \frac{\partial^2 c}{\partial x^2} - v_x \frac{\partial c}{\partial x} - \frac{\rho_b}{\phi} \frac{\partial q}{\partial t} \tag{5.155}$$

where ρ_b is the bulk density of the medium.

A Freundlich equilibrium isotherm is represented by

$$F = k c^n \tag{5.156}$$

Substituting Eq. (5.154) into Eq. (5.146)

$$\left(1 + \frac{nk}{\phi} c^{n-1}\right) \frac{\partial c}{\partial t} = D \frac{\partial^2 c}{\partial x^2} - v_x \frac{\partial c}{\partial x} \tag{5.157}$$

Gupta and Greenkorn (1973) solved the coupled partial differential equations. Eqs. (5.155) and (5.153), numerically and studied the effect of kinetic rate constant, fluid velocity, and dispersion on breakthrough curves. Writing Eqs. (5.155) and (5.153) in dimensionless terms $c^* = c/c_0$, $x^* = x/L$, $t^* = t/t_0$, and $q^* = q/q_\infty$, where t_0 is a characteristic time

$$\frac{1}{\eta \xi} \frac{\partial c^*}{\partial t^*} = \frac{\partial^2 c^*}{\partial x^{*2}} - \frac{1}{\eta} \frac{\partial c^*}{\partial \eta} - \frac{\beta}{\eta \xi} \frac{\partial q^*}{\partial t^*} \tag{5.158}$$

and

$$\frac{\partial q^*}{\partial t^*} = \gamma_1 \left[c^* \left(\frac{1 + A_1}{A_1} - q^* \right) - \frac{1}{A_1} q^* \right] \tag{5.159}$$

where

$$\eta = \frac{D}{v_x L} \tag{5.160}$$

$$\beta = \frac{\rho_b}{\phi} \frac{q_\infty}{c_0} \tag{5.161}$$

$$\xi = \frac{v_x t}{L} \tag{5.162}$$

$$\gamma_1 = k_{kin} t_0 c_0 \tag{5.163}$$

$$A_1 = K^{ad} c_0 \tag{5.164}$$

The initial and boundary conditions are

TABLE 5.8 Variables for Bilinear Rate of Adsorption[a]

	Velocity (cm min^{-1})	Dispersion coefficient (cm^2 min^{-1})	Rate constant (cm^2 (g mol)$^{-1}$ min^{-1})	Dimensionless dispersion coefficient	Remark
Figure 5.45a					
Curve 1	0.05	0.05	.002	60	Effect of kinetic rate constant
Curve 2	0.05	0.05	0.2	60	
Curve 3	0.05	0.05	2.0	60	
Figure 5.45b					
Curve 1	0.025	0.10	2.0	15	Effect of velocity of flow
Curve 2	0.05	0.10	2.0	30	
Curve 3	0.10	0.10	2.0	60	
Curve 4	0.20	0.10	2.0	120	
Figure 5.45c					
Curve 1	0.20	0.8	2.0	15	Effect of dispersion coefficent
Curve 2	0.20	0.1	2.0	120	

[a]The inlet concentration c_0 was 0.00002 g mol cm^{-3}, the ultimate capacity of the dry adsorbent was 0.00012 g mol g^{-1}, and the equilibrium adsorption constant was 540,000.0 cm^3 (g mol)$^{-1}$.

Source: S. P. Gupta and R. A. Greenkorn, Dispersion During Flow in Porous Media with Bilinear Adsorption, Water Resour. Res. 9, 5, 1357 (1973). Copyrighted by the American Geophysical Union.

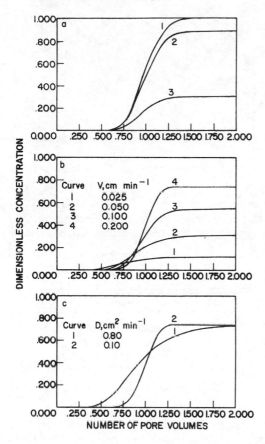

FIG. 5.45 Effect on breakthrough curves of (a) kinetic rate constant (V $= 0.05$ cm^2 min^{-1}, D $= 0.05$ cm min^{-1}, $c_0 = 0.00002$ g mol cm^{-3}, $K^{ad} = 540,000.0$ cm^3 g^{-1}, and Q $= 0.00012$ g mol g^{-1} adsorbent; for curves 1, 2, and 3, k_{sin} is 0.002, 0.200, and 2.000 cm^3 (g mol)$^{-1}$ min^{-1}, respectively; (b) fluid velocity (D $= 0.1$ cm^2 min^{-1}, $k_{sin} = 2.0$ cm^3 (g mol)$^{-1}$ min^{-1}, $c_0 = 0.00002$ g mol cm^{-3}, $K^{ad} = 540,000.0$ cm^3 (g mol)$^{-1}$, and Q $= 0.00013$ g mol g^{-1} adsorbent); and (c) dispersion coefficient (V $= 0.20$ cm min^{-1}, $k_{sin} = 2.0$ cm^3 (g mol)$^{-1}$ min^{-1}, $c_0 = 0.00002$ g mol cm^{-3}, $K^{ad} = 540,000$ cm^3 (g mol)$^{-1}$, and Q $= 0.00012$ g mol adsorbent). [From S. P. Gupta and R. A. Greenkorn, Dispersion During Flow in Porous Media with Bilinear Adsorption, Water Resour. Res. 9 (5), 1357 (1973). Copyrighted by the American Geophysical Union.]

$$c^*(x^*, 0) = 0 \qquad x^* > 0 \tag{5.165}$$

$$q^*(x^*, 0) = 1 \qquad t^* \geq 0 \tag{5.166}$$

$$c^*(0, t^*) = 1 \qquad t^* \geq 0 \tag{5.167}$$

$$c^*(\infty, t^*) = 0 \qquad t^* \geq 0 \tag{5.168}$$

To solve this system of equations numerically the semi-infinite boundary conditions, Eq. (5.168), were made finite using the transformation,

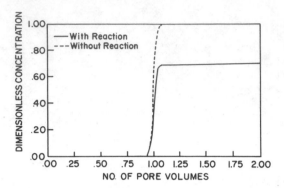

FIG. 5.46 Field example of a 100-ft-long system. [From S. P. Gupta and R. A. Greenkorn, Dispersion During Flow in Porous Media with Bilinear Adsorption, Water Resour. Res. 9 (5), 1357 (1973). Copyrighted by the American Geophysical Union.]

$$z^* = e^{-x^*} \qquad (5.169)$$

Rewriting Eqs. (5.158) and (5.165) to (5.168) using the transformation of Eq. (5.169)

$$\frac{1}{\eta\xi}\frac{\partial c^*}{\partial t^*} = z^{*2}\frac{\partial^2 c^*}{\partial z^{*2}} + \left(1 + \frac{1}{\eta}\right)z^*\frac{\partial c^*}{\partial z^*} - \frac{\beta}{\eta\xi}\frac{\partial q^*}{\partial t^*} \qquad (5.170)$$

$$c^*(z^*, 0) = 0 \qquad z^* < 1 \qquad (5.171)$$

$$q^*(z^*, 0) = 0 \qquad z^* < 1 \qquad (5.172)$$

$$c^*(1, t^*) = 1 \qquad t^* \geq 0 \qquad (5.173)$$

$$c^*(0, t^*) = 0 \qquad t^* \geq 0 \qquad (5.174)$$

Equation (5.159) retains the same form. Table 5.8 shows the variables used in the numerical calculations. (A discussion of the numerical solution and the finite difference approximation of the equations are given in the appendix to Chapter 6.) Figure 5.45 shows the effects of kinetic rate, fluid velocity, and dispersion on the breakthrough curves. Figure 5.46 shows the effect of adsorption on movement of phosphate ion in a field system 100 ft. long. The mass balance without adsorption is 0.9997. The mass balance with adsorption is 0.6836. Also, 31.46 percent of the solute is adsorbed in 10.6 days. The number of pore volumes needed for a 100 ft length at equilibrium saturation is 36.5. The time for flow of 36.5 pore volumes in the system is 193 days. Overman et al. (1976) use the dispersion with bilinear adsorption model to describe movement of phosphate ions in field systems. Trogus et al. (1977) show experimentally that adsorption of surfactants can be modeled with the bilinear mechanism.

Gupta and Greenkorn (1974) measured the dispersion with adsorption of phosphate and nitrate ions in close random pack models of Ottawa washed sand mixed with 0 to 7.5% kaolin clay. They compare their experimental results with the solution for dispersion and linear adsorption, Eq. (5.148), and dispersion and Freundlich adsorption, Eq. (5.157). The properties of their 12 in. long, 1.5-in.-inside-diameter models are given in Table 5.9. Using the dimensionless variables

TABLE 5.9 Pore Structure and Dispersion Parameters

Packing (% clay)	Total ϕ	k (95% confidence limit) (darcys)	V (cm min⁻¹)	D (cm² min⁻¹)	s.d. of curve fitting for D	Correlation of dispersion data with regression models[a]		
						$b_1 = b_2$	a_1	$a_2 \times 10^{-3}$
0.0	0.330	9.65 ± 0.05	1.65	0.641	0.033	1.16	0.35	3.81
			0.82	0.247	0.025			
			0.82	0.275	0.028			
			0.41	0.128	0.029			
2.5	0.352	3.67 ± 0.15	1.55	1.822	0.065	1.09	1.13	1.30
			0.77	0.854	0.063			
			0.77	0.831	0.070			
			0.39	0.410	0.061			
5.0 (column 1)	0.358	1.12 ± 0.02	1.52	2.198	0.057	0.98	1.32	10.74
			0.76	1.172	0.060			
			0.76	0.891	0.058			
			0.38	0.562	0.040			
5.0	0.347	1.11 ± 0.03	1.57	1.803	0.079	0.97	1.16	8.82
			0.78	0.954	0.073			
			0.78	0.937	0.072			
			0.39	0.468	0.068			
7.5	0.365	0.22 ± 0.01	0.74	1.557	0.050	1.00	2.07	11.31
			0.74	1.535	0.050			
			0.56	1.122	0.036			
			0.37	0.790	0.047			

[a]The two models are $D = a_1 V_1^b$ and $D = a_2 v R_2^b$.

Source: S. P. Gupta and R. A. Greenkorn, Determination of Dispersion and Nonlinear Adsorption Parameters for Flow in Porous Media, Water Resour. Res. 10, 4, 839 (1974). Copyrighted by the American Geophysical Union.

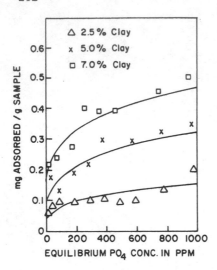

FIG. 5.47 Experimental points and theoretical curves for static adsorption of phosphate for the Freundlich isotherm. [From S. P. Gupta and R. A. Greenkorn, Determination of Dispersion and Nonlinear Adsorption Parameters for Flow in Porous Media, Water Resour. Res. 10 (4), 839 (1974). Copyrighted by the American Geophysical Union.]

$c^* = c/c_0$, $x^* = x/L$, and $t^* = t/t_0$, and the transformation of Eq. (5.169) to change the semi-infinite boundary conditions Eq. (5.157) becomes

$$(1 + rc^{*n-1}) \frac{1}{\xi \eta} \frac{\partial c^*}{\partial \eta} = z^{*2} \frac{\partial c^*}{\partial z^*} + \left(1 + \frac{1}{\eta}\right) z^* \frac{\partial c^*}{\partial z^*} \tag{5.175}$$

where ξ and η are given by Eqs. (5.160) and (5.162) and

$$\gamma = \frac{nk}{\phi} c_0^{n-1} \tag{5.176}$$

Equation (5.171) is solved numerically using the initial and boundary conditions

$$c^*(z^*, 0) = 0 \qquad z^* < 1 \tag{5.177}$$

$$c^*(1, t^*) = 1 \qquad t^* \geq 0 \tag{5.178}$$

$$c^*(0, t^*) = 0 \qquad t^* \geq 0 \tag{5.179}$$

(A discussion of the numerical solution and finite difference approximations are given in the appendix to Chapter 6.) Figure 5.47 shows the static adsorption of phosphate ion on the different mixtures of sand and clay, used in the flow models, is represented by a Freundlich isotherm. Table 5.10 shows the parameters determined by the static measurements and the dynamic measurements using the solution to Eq. (5.155). Figure 5.48 compares a typical experimental breakthrough curve with the results of the linear model, Eq. (5.133) and the numerical solution of

TABLE 5.10 Static and Dynamic Adsorption Parameters

Packing (% clay)	Static case $F = k_s c^{n_s}$			V (cm min^{-1})	Dynamic case					
					$F = k_1 c$			$F = k_2 c^{n_s}$ ($k_2 = k_s$)		
	k_s	n_s	s.e.[a]		k_1	s.e.[a]	$\langle k_1 \rangle$	n_2	s.e.[a]	$\langle n_2 \rangle$
2.5	0.024	0.266	0.026	1.55	0.068	0.030	0.0565	1.155	0.031	1.090
				0.77	0.056	0.040		1.089	0.042	
				0.77	0.057	0.043		1.089	0.047	
				0.39	0.045	0.029		1.027	0.031	
5.0	0.52	0.266	0.043	1.52	0.069	0.025	0.073	0.993	0.027	1.006
				0.76	0.068	0.032		0.984	0.033	
				0.76	0.078	0.028		1.028	0.031	
				0.38	0.077	0.025		1.022	0.025	
7.5	0.103	0.221	0.035	0.74	0.189	0.047	0.141	1.006	0.028	0.994
				0.74	0.126	0.036		1.006	0.036	
				0.56	0.116	0.043		0.990	0.043	
				0.37	0.133	0.036		0.973	0.038	

[a] Here s.e. denotes standard error.
Source: Gupta and Greenkorn (1974).

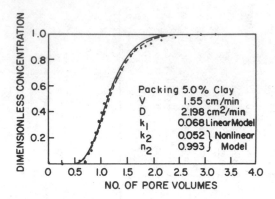

FIG. 5.48 Typical experimental and theoretical breakthrough curves, where the
solid curve indicates the nonlinear model, the broken curve indicates the linear
model, and the points indicate the experimental data. [From S. P. Gupta and
R. A. Greenkorn, Determination of Dispersion and Nonlinear Adsorption Parameters for Flow in Porous Media, Water Resour. Res. 10 (4), 839 (1974). Copyrighted by the American Geophysical Union.]

Eq. (5.176). In general, adsorption parameters can be determined from dynamic
experiments. Changing flow rates does not change the values of the adsorption
parameters. The data were fit best with the Freundlich model although the linear
model is good outside the low concentration range.

Gupta and Greenkorn (1976) solved the dispersion equation with Freundlich
adsorption for one-dimensional radial flow. Writing Eq. (5.157) in radial coordinates and dimensionless form using $c^* = c/c_0$, $r^* = r/R$, and $t^* = t/t_0$, where R
is the radius of the system

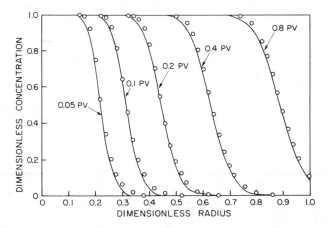

FIG. 5.49 Comparison of numerical and approximate analytical solution for radial
flow with dispersion. [From S. P. Gupta and R. A. Greenkorn, Solution for Radial
Flow with Nonlinear Adsorption, Proc. ASCE J. Env. Div. 102, EE1, 87 (1976).
Copyright by the American Society of Civil Engineers.]

$$\left(1 + \gamma c^{*^{n-1}}\right) \frac{\partial c^*}{\partial t^*} = \frac{\beta}{r^*} \frac{\partial^2 c^*}{\partial r^{*2}} - \frac{\alpha}{r^*} \frac{\partial c^*}{\partial z^*} \qquad (5.180)$$

where

$$A = \frac{Q}{2\pi h \phi} \qquad (5.181)$$

$$D = \frac{aA}{r} \qquad (5.182)$$

$$\alpha = \frac{At_0}{R^2} \qquad (5.183)$$

$$\beta = \frac{aAt_0}{R^3} \qquad (5.184)$$

The initial and boundary conditions are

$$c^*(r^*, 0) = 0 \qquad r^* \geq 0 \qquad (5.185)$$

$$c^*(0, t^*) = 1 \qquad t^* \geq 0 \qquad (5.186)$$

$$\frac{\partial}{\partial t} c^*(1, t^*) = 0 \qquad t^* \geq 0 \qquad (5.187)$$

The solution of Eq. (5.180) without adsorption is

$$c^* = \frac{1}{2} \operatorname{erfc} \left[\frac{r/2a - At/a^2}{(4r^3/3a^3)^{\frac{1}{2}}} \right] \qquad (5.188)$$

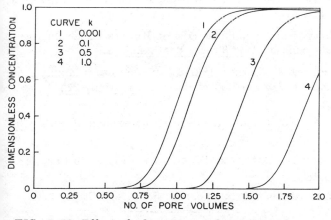

FIG. 5.50 Effect of adsorption on breakthrough curves: radial flow (Q – 5.0 cm^2 min^{-1}, R – 15.0 cm, b – 1.0 cm, $\phi = 0.35$, $a_1 = 0.15$ cm, $c_0 = 100$ ppm, and Freundlich isotherm $F = kc^{0.75}$). [From S. P. Gupta and R. A. Greenkorn, Solution for Radial Flow with Nonlinear Adsorption, Proc. ASCE J. Env. Div. 102, EE1, 87 (1976). Copyright by the American Society of Civil Engineers.]

(A discussion of the numerical solution and finite difference approximations are given in the appendix to Chapter 6.) Figure 5.49 compares the numerical solution of Eq. (5.180) without adsorption to Eq. (5.188). Figure 5.50 shows the effect of adsorption on the breakthrough curves for the numerical solution of Eq. (5.180).

REFERENCES

Aris, R., and N. R. Amundson (1957): Some Remarks on Longitudinal Mixing or Diffusion in Fixed Beds, A.I.Ch.E. J. 3, 280.

Bachmat, Y., and J. Bear (1964): The General Equations of Hydrodynamic Dispersion in Homogeneous Isotropic, Porous Mediums, J. Geophys. Res. 69, 2561.

Banks, R. B., and I. Ali (1964): Dispersion and Adsorption in Porous Media, Proc. ASCE J. Hyd. Div. 90, HY5, 13.

Banks, R. B., and S. Jerasate (1962): Dispersion in Unsteady Porous-Media Flow, Proc. ASCE J. Hyd. Div. HY3, 3109.

Bear, J. (1961): On the Tensor Form of Dispersion in Porous Media, J. Geophys. Res. 66, 1185.

Benham, A. L., and R. W. Olson (1963): A Model Study of Viscous Fingering, Pet. Trans. AIME 228, SPEJ 138.

Biggar, J. W., and D. R. Nielsen (1962): Some Comments on Molecular Diffusion and Hydrodynamic Dispersion in Porous Media, J. Geophys. Res. 27, 3636.

Bixel, H. C., and H. K. van Poolen (1967): Pressure Drawdown and Buildup in the Presence of Radial Discontinuities, Pet. Trans. AIME 240, SPEJ 301.

Blackwell, R. J. (1962): Laboratory Studies of Microscopic Dispersion Phenomena, Pet. Trans. AIME 225, SPEJ 51.

Blackwell, R. J., W. M. Terry, and J. R. Rayne (1959): Factors Influencing the Efficiency of Miscible Displacement, Pet. Trans. AIME 216, 1.

Brigham, W. E., P. W. Reed, and J. N. Dew (1961): Experiments on Mixing During Miscible Displacement in Porous Media, Pet. Trans. AIME 222, SPEJ 1.

Burch, J. C., and R. L. Street (1967): Two-Dimensional Dispersion, Proc. ASCE J. Sanitary Eng. SA6 6, 17.

Carberry, J. S., and R. H. Bretton (1958): Axial Dispersion of Mass in Flow through Fixed Beds, A.I.Ch.E. J. 4, 367.

Carman, P. C. (1939): Permeability of Saturated Sands, Soils, and Clays, J. Agri. Sci. 29, 262.

Carter, R. D. (1966): Pressure Behavior of a Limited Circular Composite Reservoir, Pet. Trans. AIME 237, SPEJ 328.

Coats, K. H., and B. D. Smith (1964): Dead-End Pore Volume and Dispersion in Porous Media, Pet. Trans. AIME 231, SPEJ 73.

Craig, F. F., Jr., J. L. Sanderlin, D. W. Moore, and T. M. Geffen (1957): A Laboratory Study of Gravity Segregation in Frontal Drives, Pet. Trans. AIME 210, 275.

Dagan, G. (1971): Perturbation Solutions of the Dispersion Equation in Porous Mediums, Water Resour. Res. 7(1), 135.

Deans, H. A. (1963): A Mathematical Model for Dispersion in the Direction of Flow in Porous Media, Pet. Trans. AIME 228, SPEJ 49.

deJosselin de Jong, G. (1958): Longitudinal and Transverse Diffusion in Granular Deposits, Trans. Amer. Geophys. Union 59, 67.

deJosselin de Jong, G., and M. J. Bossen (1961): Discussion of Paper by Jacob Bear, "On the Tensor Form of Dispersion in Porous Media," J. Geophys. Res. 66, 3623.

Dougherty, E. L. (1963): Mathematical Model of an Unstable Displacement, Pet. Trans. AIME 228, SPEJ 155.

Eldor, M., and G. Dagan (1972): Solution of Hydrodynamic Dispersion in Porous Media, Water Resour. Res. 8(3), 1316.

Garder, A. O., Jr., D. W. Peaceman, and A. L. Pozzi, Jr. (1964): Numerical Calculation of Multidimensional Miscible Displacement by the Method of Characteristics, Soc. Petrol. Eng. J. 4(1), 26.

Gelhar, L. W., and M. A. Collins (1971): General Analysis of Longitudinal Dispersion in Nonuniform Flow, Water Resour. Res. 7(6), 1511.

Goad, T. L. (1970): Permeability and Dispersion During Flow in Linear Heterogeneous Anisotropic Porous Media, M.S. Thesis, Purdue University.

Grane, F. E., and G. N. F. Gardner (1961): Measurements of Transverse Dispersion in Granular Media, J. Chem. Eng. Data 6, 283.

Greenkorn, R. A. (1962): Experimental Study of Waterflood Tracers, Pet. Trans. AIME 225, 87.

Greenkorn, R. A., C. R. Johnson, and R. E. Haring (1965): Miscible Displacement in a Controlled Natural System, Pet. Trans. AIME 234, 1329.

Greenkorn, R. A., and D. P. Kessler (1969): A Statistical Model for Flow in Nonuniform Porous Media, Proc. ASCE-ECM Specialty Conference on Probabilistic Methods and Concepts, p. 91.

Greenkorn, R. A., and D. P. Kessler (1970): Dispersion in Heterogeneous Nonuniform Anisotropic Porous Media, in Flow Through Porous Media, Amer. Chem. Soc., Washington, D.C., Chap. 8, p. 149.

Guin, J. A., D. P. Kessler, and R. A. Greenkorn (1972): The Dispersion Tensor in Anisotropic Porous Media, Ind. Eng. Chem. Fund. 11(4), 477.

Gupta, S. P., and R. A. Greenkorn (1973): Dispersion During Flow in Porous Media with Bilinear Adsorption, Water Resour. Res. 9(5), 1357.

Gupta, S. P., and R. A. Greenkorn (1974): Determination of Dispersion and Nonlinear Asorption Parameters for Flow in Porous Media, Water Resour. Res. 10(4), 839.

Gupta, S. P., and R. A. Greenkorn (1976): Solution for Radial Flow with Nonlinear Adsorption, Proc. ASCE J. Env. Div. 102, EE1, 87.

Hall, H. N., and T. M. Geffen (1957): A Laboratory Study of Solvent Flooding, Pet. Trans. AIME 210, 48.

Handy, L. L. (1959): An Evaluation of Diffusion Effects in Miscible Displacement, Pet. Trans. AIME 216, 382.

Haring, R. E., and R. A. Greenkorn (1970): A Statistical Model of a Porous Medium with Nonuniform Pores, A.I.Ch.E. J. 16, 477.

Harleman, D. R. F., P. F. Mehlhorn, and R. R. Rumer, Jr. (1963): Dispersion-Permeability Correlation in Porous Media, Proc. ASCE J. Hydr. Div. 67, 67.

Hawthorne, R. C. (1960): Two-Phase Flow in Two-Dimensional Systems—Effects of Rate Viscosity, and Density on Fluid Displacement in Porous Media, Pet. Trans. AIME 219, 81.

Heller, J. P. (1966): Onset of Instability Patterns between Miscible Fluids in Porous Media, J. Appl. Phys. 37, 1566.

Hiester, V. K., and T. Vermeulen (1952): Saturation Performance of Ion-Exchange and Adsorption Columns, Chem. Eng. Prog. 48(10), 505.

Hill, S. (1952): Channeling in Packed Columns, Chem. Eng. Sci. 1(6), 246.

Hurst, W. (1960): Interference between Oil Fields, Pet. Trans. AIME 219, 175.

Klinkenberg, L. J. (1951): Analog between Diffusion and Electrical Conductivity in Porous Rocks, Bull. GSA 62, 559.

Koval, E. J. (1963): A Method for Predicting the Performance of Unstable Miscible Displacement in Heterogeneous Media, Pet. Trans. AIME 228, SPEJ 145.

Kramers, H., and G. Albreda (1953): Frequency Response Analysis of Continuous Flow Systems, Chem. Eng. Sci. 2, 173.

Kyle, C. R., and R. L. Perrine (1965): Experimental Studies of Miscible Displacement Instability, Pet. Trans. AIME 234, SPEJ 189.

Lai, S. H., and J. J. Jurinak (1972): Cation Adsorption in One-Dimensional Flow Through Soils, Water Resour. Res. 8(1), 99.

Lantz, R. B. (1971): Quantitative Evaluation of Numerical Diffusion (Truncation Error), Soc. Petrol. Eng. J., 11(3), 315.

Lapidus, L., and N. R. Amundson (1952): Mathematics of Adsorption in Beds, J. Phys. Chem. 56, 584.

Larkin, B. K. (1963): Solutions to the Diffusion Equation for a Region Bounded by a Circular Discontinuity, Pet. Trans. AIME 228, SPEJ 113.

Lindstrom, F. T., and L. Boersma (1970): Theory of Chemical Transport with Simultaneous Sorption in Water Saturated Porous Media, Soil, Sci. 110, 1.

Loucks, T. L., and E. T. Guerrero (1961): Pressure Drop in a Composite Reservoir, Pet. Trans. AIME 222, SPEJ 170.

Merrill, L. S., Jr., H. Kazemi, and W. B. Gogarty (1974): Pressure Falloff Analysis in Reservoirs with Fluid Banks, Pet. Trans. AIME 257, 809.

Moranville, M. B., D. P. Kessler, and R. A. Greenkorn (1973): A Study of Dispersion in a Stochastic Model of a Nonuniform Porous Media, Proc. RILEM/IUPAC International Symposium on Pore Structure and Properties of Materials, Prague, pA-93.

Moranville, M. B., D. P. Kessler, and R. A. Greenkorn (1977a): Directional Dispersion Coefficients in Anisotropic Porous Media, Ind. Eng. Chem. Fund. 16(3), 327.

Moranville, M. B., D. P. Kessler, and R. A. Greenkorn (1977b): Dispersion in Layered Porous Media, A.I.Ch.E. J. 23, 786.

Muskat, M. Flow of Homogeneous Fluids, McGraw-Hill, New York.

Nieman, E. H. (1969): Dispersion During Flow in Non-uniform, Heterogeneous Porous Media, M.S. Thesis, Purdue University.

Nikolaevskii, V. N. (1959): Convective Diffusion in Porous Media, Prikl. Math. Mech. 23(6), 1042.

Nalluswami, M. (1971): Numerical Simulation of General Hydrodynamic Dispersion in Porous Media, Ph.D. Thesis, Colorado State University, Ft. Collins.

Odeh, A. S. (1969): Flow Test Analysis for a Well with Radial Discontinuity, Pet. Trans. AIME 246, 207.

Ogata, A. (1958): Dispersion in Porous Media, Ph.D. Thesis, Northwestern University, Evanston, Ill.

Ogata, A., and R. B. Banks (1961): A Solution of the Differential Equation of Longitudinal Dispersion in Porous Media, Geological Survey Professional Paper 411-A, U.S. Govt. Printing Office, Washington, D.C.

Overman, A. R., R. Chu, and W. G. Leseman (1976): Phosphorous Transport in a Packed Bed Reactor, J. Water Poll. Control 48(5), 881.

Pakula, R. J., and R. A. Greenkorn (1971): An Experimental Investigation of a Porous Media Model with Nonuniform Pores, A.I.Ch.E. J. 17, 1265.

Peaceman, D. W., and H. H. Rachford, Jr. (1962): Numerical Calculation of Multicomponent Miscible Displacement, Pet. Trans. AIME 219, SPEJ 327.

Perkins, T. K., and O. C. Johnston (1963): A Review of Diffusion and Dispersion in Porous Media, Pet. Trans. AIME 228, SPEJ 70.

Perkins, T. K., O. C. Johnston, and R. N. Hoffman (1965): Mechanics of Viscous Fingering in Miscible Systems, Pet. Trans. AIME 234, SPEJ 301.

Perrine, R. L. (1961): The Development of Stability Theory for Miscible Liquid-Liquid Displacement, Pet. Trans. AIME 222, SPEJ 17.

Perrine, R. L. (1963): A Unified Theory for Stable and Unstable Miscible Displacement, Pet. Trans. AIME 228, SPEJ 205.

Perrine, R. L., and G. M. Gay (1966): Unstable Miscible Flow in Heterogeneous Systems, Pet Trans. AIME 237, SPEJ 228.

Pfannkuch, H. O. (1962): Contribution à l'étude des deplacement de fluides miscible dans un milieu poreux, Rev. Inst. Fr. Petrol. 18(2), 215.

Pinder, G. F., and H. H. Cooper, Jr. (1970): A Numerical Technique for Calculating the Transient Position of a Saltwater Front, Water Resour. Res. 6, 875.

Poreh, M. (1965): The Dispersivity Tensor in Isotropic and Axisymmetric Mediums, J. Geophys. Res. 70, 3909.

Pozzi, A. L., and R. J. Blackwell (1963): Design of Laboratory Models for Study of Miscible Displacement, Pet. Trans. AIME 228, SPEJ 28.

Raats, P. A. C., and D. R. Scotter (1968): Dynamically Similar Motions of Two Miscible Constituents in Porous Mediums, Water Resour. Res. 4, 566.

Raimondi, P., G. H. F. Gardner, and C. B. Petrick (1959): Effect of Pore Structure and Molecular Diffusion on the Mixing of Miscible Liquids Flowing in Porous Media, AIChE-SPE Joint Symposium, A.I.Ch.E. Meeting San Francisco, Preprint 43.

Rifai, N. R. E., W. J. Kaufman, and D. K. Todd (1956): Dispersion Phenomena in Laminar Flow through Porous Media, Sanitary Engineering Research Laboratory and Division of Civil Engineering Report no. 3, University of California Berkeley.

Saffman, P. G. (1959): A Theory of Dispersion in a Porous Medium, Fluid Mech. 6, 21.

Scheidegger, A. E. (1954): Statistical Hydrodynamics in Porous Media, J. Appl. Phys. 25(8), 994.

Scheidegger, A. E. (1957): On the Theory of Flow of Miscible Phases in Porous Media, Compt. Rend. Assoc. Geo., Toronto Assoc. Intern. Hydrol. Sci. 2, 236.

Scheidegger, A. E. (1959): Statistical Approach to Miscible Displacement in Porous Media, Can. Min. Metallurgy Bull. 52, 26.

Scheidegger, A. E. (1961): General Theory of Dispersion in Porous Media, Geophys. Res. 66, 3273.

Schowalter, W. R. (1965): Stability Criteria for Miscible Displacement of Fluids from a Porous Medium, A.I.Ch.E. J. 11, 99.

Scotter, D. R., R. G. W. Thurtell, and P. A. C. Raats (1967): Dispersion Resulting from Sinusoidal Gas Flow in Porous Media, Soil Sci. 104.

Scotter, D. R., and P. A. C. Raats (1968): Dispersion in Porous Mediums Due to Oscillating Flow, Water Resour. Res. 4(6), 1201.

Shamir, U. Y., and D. R. F. Harleman (1967): Dispersion in Layered Porous Media, Proc. ASCE J. Hydraulic Div. HY5, 237.

Slobod, R. L., and R. A. Thomas (1963): Effect of Transverse Diffusion on Fingering in Miscible Displacement, Pet. Trans. AIME 228, SPEJ 9.

Taylor, G. (1953): Dispersion of Soluble Matter in Solvent Flowing Slowly through a Tube, Proc. Roy. Soc. London A219, 186.

Torelli, L., and A. E. Scheidegger (1972): Three-Dimensional Branching-Type Models of Flow through Porous Media, J. Hydrology 15, 23.

Trogus, F. J., T. Sophany, R. S. Schechter, and W. M. Wade (1977): Static and Dynamic Adsorption of Ionic and Nonionic Surfactants, Pet. Trans. AIME 263, SPEJ 337.

van der Poel, C. (1962): Effect of Lateral Diffusivity on Miscible Displacement in Horizontal Reservoirs, Pet. Trans. AIME 225, SPEJ 317.

Varnon, J. E., and R. A. Greenkorn (1969): Unstable Two-Fluid Flow in a Porous Medium, Soc. Pet. Eng. J. 9, 293.

von Rosenberg, D. V. (1969): Methods for the Numerical Solution of Partial Differential Equations, American Elsevier, New York.

Whitaker, S. (1967): Diffusion and Dispersion in Porous Media, A.I.Ch.E. J. 13, 420.

SUGGESTED READING

Aris, R., On the Dispersion of Solute in a Fluid Flowing through a Tube, Proc. Roy. Soc. London A235, 56 (1956).

Day, P. R., Dispersion of a Moving Salt-Water Boundary Advancing through a Saturated Sand, Trans. Amer. Geophys. A 37, 595 (1956).

Delhomme, J. P., Spatial Variability and Uncertainty in Groundwater Flow Parameters: A Geostatistical Approach, Water Resour. Res. 15(2), 269 (1979).

Ebach, E. A., and R. R. White, Mixing of Fluids Flowing through Beds of Packed Solids, A.I.Ch.E. J. 4, 161 (1958).

Freeze, R. A., A Stochastic-Conceptual Analysis of One-Dimensional Groundwater Flow in Nonuniform Homogeneous Media, Water Resour. Res. 11(5), 725 (1975).

Fried, J. J., Groundwater Pollution, Elsevier, New York, 1975.

Gelhar, L. W., A. L. Gutjahr, and R. L. Hoff, Stochastic Analysis of Macrodispersion in a Stratified Aquifer, Water Resour. Res. 15(6), 1387 (1979).

Gelhar, L. W., and C. L. Axness, Stochastic Analysis of Macrodispersion in Three-Dimensionally Heterogeneous Aquifers, Report No. H-8, Hydrology Research Program, Geophysical Research Center, Research and Development Division, New Mexico Institute of Mining and Technology, Socorro, N.M. (1981).

Matheron, G., and G. De Marsily, Is Transport in Porous Media Always Diffusive? A Counterexample, Water. Resour. Res. 16(5), 901 (1980).

Schwartz, F. W., Macroscopic Dispersion in Porous Media: The Controlling Factors, Water Resour. Res. 13(4), 743 (1977).

Smith, L., and F. W. Schwartz, Mass Transport: I. A Stochastic Analysis of Macroscopic Dispersion, Water Resour. Res. 16(2), 303 (1980).

Warren, J. E., and F. F. Skiba, Macroscopic Dispersion, Pet. Trans. AIME 23, SPEJ 215 (1964).

6

Phenomenological Behavior

Much of the discussion in previous chapters is concerned with modeling flow phenomena in porous media. The discussion centers around replacing the real world—an element of a porous medium and its contained fluids—with a mathematical model. In Chapters 3 to 5 we explained the real world problems with differential equations and with analytical solutions of the equations where possible. Such analytical solutions are usually not possible without simplifying assumptions.

In this chapter the real world is replaced with scaled physical models assuming a dimensionally homogeneous mathematical equation describes the real world. Also, the real world can be replaced with a physical analog described by a topologically equivalent mathematical equation. Solution of the real world problem results from observing homologous behavior in the scaled physical model or analogous behavior in the analog model.

Another approach when the equations describing the phenomena cannot be solved analytically is to simulate the problem numerically. In simulation the differential equations are expressed with finite differences or polynomial solutions that match the differential equations and explain phenomena with calculations via a computer. This technique is called <u>numerical modeling</u> or <u>numerical simulation.</u> Numerical simulation and computer calculations are complex. In contrast to physical homologs and analogs which usually apply to a specific problem, numerical simulations are general. It is easy to lose sight of the real world system. The idea of abstract manipulation of the simulator is fine as long as it reproduces the essential parts of the real world under study. An incomplete simulator will guarantee incomplete predictions regardless of how elegantly the simulator is manipulated.

6.1 SIMILARITY, SCALING, AND PHYSICAL MODELS

In this section scaling and dimensional analysis is reviewed, and the techniques are applied for determining scaling laws for constructing models and dimensionless correlations. Scaling relations are found for two-fluid miscible and immiscible flow.

Any scaling law or model law should be derived by dimensional analysis both from the general standpoint of deriving a complete set of dimensionless products and by inspection of the dimensionless equations of change. One can easily find a

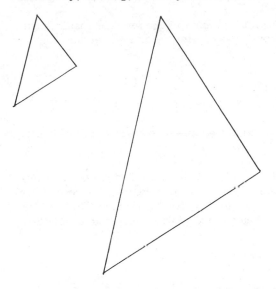

FIG. 6.1 Geometrically similar triangles.

complete set of dimensionless products for any given set of variables. The dimensionless groups that are not applicable can then be deleted so that as small a number of dimensionless products as possible remains. This step of deleting groups takes judgment and usually proves to be the most difficult part of the problem. When one writes the equations of change the flow regimes are assumed; thus some dimensionless groups may not show up which would be present if the variables alone were considered. By attacking the problem both ways the task of selecting correct groups usually becomes a little more obvious. In the end there is no easy way; it may be necessary to study various regimes before selecting the particular experimental conditions for a given analysis.

The use of scaling to design physical models to be used to obtain experimental data is directly related to dimensional analysis. The concept of similarity is described by a theory due to Buckingham (1915). Scaled models are designed using dimensionless parameters. Two plane figures, such as two triangles, are geometrically similar if corresponding angles are equal and corresponding sides are in constant ratio as in Fig. 6.1. Similarity is possible in other than a geometric sense. Two geometrically similar bodies are dynamically similar if similar points experience similar net forces. In general, two systems are similar if their relevant physical properties are in constant ratio.

The principle of similarity is expressed abstractly by Buckingham's theorem

$$\pi_1 = f(\pi_2, \pi_3, \ldots, \pi_{n-k}) \tag{6.1}$$

where the π's are dimensionless. Two or more similar systems undergoing a given physical phenomenon are all described by Eq. (6.1). Equation (6.1) is called the classical principle of similarity.

If we assume the function in Eq. (6.1) is a power function, the result is called the extended principle of similarity. Writing π_1 explicitly, as a power function equation

$$\pi_1 = C(\pi_2)^{x_2}(\pi_3)^{x_3} \cdots (\pi_{n-k})^{x_{n-k}} \tag{6.2}$$

the parameter C is related to the shape of the system—it is a shape factor—x_2, x_3, $\ldots$, x_{n-k} are empirical coefficients. The value of C and x_2, x_3, $\ldots$, x_{n-k} must be determined experimentally.

If we compare two systems of similar shape, the shape factor can be eliminated by dividing Eq. (6.2) for one system by Eq. (6.2) written for the second system. Designate one system as the model and identify it with a prime (') on the π's; the unprimed system is the prototype. Then the extrapolated principle of similarity is

$$\frac{\pi_1'}{\pi_1} = \left(\frac{\pi_2'}{\pi_2}\right)^{x_2}\left(\frac{\pi_3'}{\pi_3}\right)^{x_3} \cdots \left(\frac{\pi_{n-k}'}{\pi_{n-k}}\right)^{x_{n-k}} \tag{6.3}$$

Equation (6.3) is called the extrapolated principle because the exponents x_2, x_3, $\ldots$, x_{n-k} are assumed the same for both model and prototype. The values of the exponents still have to be determined from experiment.

The exponents in Eq. (6.3) can be eliminated by comparing geometrically similar systems so that

$$\pi_2' = \pi_2 \qquad \pi_3' = \pi_3 \qquad \cdots \qquad \pi_{n-k}' = \pi_{n-k} \tag{6.4}$$

then

$$\pi_1' = \pi_1$$

this approach—the most common in modeling—is called the principle of corresponding states. Most models are designed for geometrically similar systems, and experiments are run in such a way that relevant dimensionless groups are the same for the model and the prototype.

Physical models are used to interpret flow phenomena in porous media. Models are used because situations occur that cannot be simulated mathematically or numerically. We need a certain amount of information about the porous medium in order to model it. With reservoirs as the prototype, the models are scaled down to locate significant variables and compare effect of variables in a shorter time than would be possible in the prototype situation. Models normally have the same gross shape as corresponding parts of the prototype. One major problem in modeling, due to scale change, is the possibility of scale effects. For example, in the study of the force of waves on a harbor breakwater, the model will have a scale effect if it is made so small that the surface tension of the water in the model affects the wave formation. In scaling porous media the particles and pore sizes (that is, permeability) are usually not scaled. If we attempted to scale by scaling particle and pore sizes, the time required to run the model would be the same as that for the prototype, defeating the purpose of the model study. This may or may not cause a scale effect.

Most mathematical and physical models are written for geometrically similar situations. Experiments are designed so relevant physical quantities are the same for the prototype and its model. Homologous points in a prototype and its model are points that correspond in time and space. For example, homologus points in steady-state fully developed flow of a liquid in a pipe 1 ft in diameter and its 1/12 scale model, 1 in. in diameter, would be on the center line of each, half the distance to the center line one-quarter of the distance along the pipe, etc. Phenomena occurring at homologous points in a prototype, and its geometrically and dynamically similar model occur at homologous times. Similar systems behave identically at homologous points at homologous times.

Given two homologous coordinate systems (x, y, z, t) and x', y', z', t') describing the spatial and temporal components of a prototype and its model then scale factors are defined as

$$K_x = \frac{x'}{x} \tag{6.6}$$

$$K_y = \frac{y'}{y} \tag{6.7}$$

$$K_z = \frac{z'}{z} \tag{6.8}$$

$$K_t = \frac{t'}{t} \tag{6.9}$$

If the prototype and its model are geometrically similar, then

$$K_x = K_y = K_z = K_L \tag{6.10}$$

The functions describing a particular physical phenomenon f(x, y, z, t) and f'(x', y', z', t') are similar if

$$K_f = \frac{f'}{f} \tag{6.11}$$

and, further, if the functions are both evaluated at homologous points and times. For example, if the dependent variables represented by f and f' are the temperatures at homologous points and times, then the prototype and its model are thermally similar.

Kinematic similarity is expressed by

$$K_v = \frac{K_L}{K_t} \tag{6.12}$$

Dynamic similarity is expressed by

$$K_F = \frac{K_M K_L}{K_t^2}$$ (6.13)

The scaling relation for the drag on a body submerged in a stream of incompressible fluid is described by

$$\frac{F}{\rho v^2 L^2} = f\left(\frac{Lv\rho}{\mu}\right)$$ (6.14)

where F is the drag force and L is a characteristic length. The prototype and its model behave homologously if f = f', that is, if

$$\frac{Lv\rho}{\mu} = \left(\frac{Lv\rho}{\mu}\right)'$$ (6.15)

Therefore, if

$$K_L K_v K_\rho = K_\mu$$ (6.16)

then

$$K_F = K_\rho K_v^2 K_L^2$$ (6.17)

The scaling relation results from combining Eqs. (6.16) and (6.17) to yield

$$K_F = \frac{K_\mu^2}{K_\rho}$$ (6.18)

The scaling law in words: a geometrically similar model will behave like its prototype if the model is tested in the same fluid as the prototype.

Dimensionless similarity criteria are ratios of physical quantities which are functions of various forces and resistances. The number of dimensionless criteria depends on the number of controlling factors. The controlling process is called the regime of the systems considered, and where there are several controlling factors, there will be several dimensionless criteria; often these dimensionless criteria are incompatible. For example, viscous drag, gravitational forces, or surface tension may restrict the motion of a fluid. The corresponding dimensionless criteria are the Reynolds number, the Froude number, and the Weber number, and for homologous systems of different absolute size these three criteria are not compatible. Each group indicates a different fluid velocity dependence. (Velocity is predicted to be proportional to $1/d$, $\sqrt{d}$, $1/\sqrt{d}$, respectively.) Sometimes two of the groups may be satisfied by using analog fluids but seldom can all three be satisfied.

Scale-up or scale-down is based on homologous systems represented by dimensionless dependent variables, dimensionless independent variables, and dimensionless parameters for the system. Dimensions are a means of determining how the numerical value of a given quantity changes when its unit of measurement changes. An equation is dimensionally homogeneous if the form of the equation does not depend on the fundamental units of measurement. For example,

$$\text{Velocity} = \text{distance} \times \text{time} \tag{6.19}$$

is a dimensionally homogeneous equation. Similarly,

$$y = x + z + \cdots \tag{6.20}$$

is a dimensionally homogeneous equation if all the variables have the same dimensions.

A dimensionless product is the product of variables such that the result is dimensionless. For example, the Reynolds number is a dimensionless product.

$$\text{Re} = \frac{D\bar{v}\rho}{\mu} \tag{6.21}$$

where D is pipe diameter in feet, $\bar{v}$ is average velocity in feet per second, and ρ is fluid density in lb_m (ft sec)$^{-1}$.

Dimensionally homogeneous functions are a special class of functions, and the theory underlying dimensional analysis is the mathematical theory of this particular class of functions. Dimensional analysis is based on the principle that the mathematical model for a given natural phenomenon is represented as a dimensionally homogeneous equation.

Buckingham's theorem, Eq. (6.1), stated in words: If an equation is dimensionally homogeneous, it can be rewritten as a relationship among a complete set of dimensionless products. A complete set of dimensionless products is one such that each product in the set is independent of the others, and every other dimensionless product of the variables is a combination of the dimensionless products in the set.

For example, consider a physical situation (involving mass transfer to a fluid flowing in a pipe) described by the variables D, diameter in feet, $\bar{v}$, average velocity in feet per second, ρ, density in lb_m ft^{-3}, μ, viscosity in lb_m (ft-sec)$^{-1}$, k, mass transfer coefficient, in feet per second, and $\mathscr{D}$, diffusion coefficient, in ft^2 sec^{-1}. The general form of the mathematical model is

$$f(D, \bar{v}, \rho, \mu, k, \mathscr{D}) = 0 \tag{6.22}$$

A complete set of dimensionless products for this set is

$$g\left(\frac{kD}{\mathscr{D}}, \frac{D\bar{v}\rho}{\mu}, \frac{\mu}{\rho\mathscr{D}}\right) = 0 \tag{6.23}$$

Other combinations are possible

$$g'\left(\frac{\mu}{kD\rho}, \frac{\mathscr{D}\bar{v}\rho}{k\mu}, \frac{D\bar{v}}{\mathscr{D}}\right) = 0 \tag{6.24}$$

but

$$\frac{\mu}{KD\rho} = \frac{1}{kD/\mathscr{D}}\,\frac{\mu}{\rho\mathscr{D}} \tag{6.25}$$

and

$$\frac{\mathscr{D}\bar{v}\rho}{k\mu} = \frac{1}{kD/\mathscr{D}}\,\frac{D\bar{v}\rho}{\mu} \tag{6.26}$$

and

$$\frac{D\bar{v}}{\mathscr{D}} = \frac{D\bar{v}\rho}{\mu}\,\frac{\mu}{\rho\mathscr{D}} \tag{6.27}$$

so Eq. (6.24) can be generated from Eq. (6.23).

If we define a set of symbols Q_i to represent variables and another set of symbols π_i to represent dimensionless products, we can restate Buckingham's theorem symbolically: For n variables a function

$$f(Q_1,\ Q_2,\ \ldots,\ Q_n) = 0 \tag{6.28}$$

can be determined so that

$$g(\pi_1,\ \pi_2,\ \ldots,\ \pi_{n-r}) = 0 \tag{6.29}$$

The number of dimensionless products is fewer than the number of variables by a number r related to the number of fundamental units. The number r (which results from the theory of this class of functions) is the rank of the <u>dimensional matrix</u>— the matrix which contains the powers of the fundamental dimensions in the various variables. For example, for the variables

$$f(D,\ v,\ \rho,\ \mu) = 0 \tag{6.30}$$

Table 6.1 is the dimensional matrix. This matrix gives dimensions for each variable. For viscosity μ, the dimension is $L^{-1}t^{-1}M = M/Lt$ typically $[lb_m\ (ft\ sec)^{-1}]$. The rank of this matrix, r, is 3, the same number as the fundamental dimensions (L, t, M, or typically ft, sec, lb_m). In <u>most</u> cases $r = k$, where k is the number of fundamental dimensions. We will assume $r = k$ for the remaining discussion.

In many cases, as with Eqs. (6.22) and (6.23), the dimensionless products can be obtained easily by inspection, getting a complete set of dimensionless products. In general, however, we need a systematic way of going from the variables to the dimensionless products, from Eq. (6.28) to Eq. (6.29). The following is an algorithm for systematically obtaining a complete set of dimensionless products from a set of variables.

The problem is to map $f(Q_1,\ Q_2,\ \ldots,\ Q_n)$ into $g(\pi_1,\ \pi_2,\ \ldots,\ \pi_{n-k})$. The algorithm is

1. List the variables and their dimension, and transfer any dimensionless variables to g directly.
2. Determine the number of fundamental dimensions, k.

TABLE 6.1 Dimensional Matrix

	D	$\bar{v}$	ρ	μ
L	1	1	-3	-1
t	0	-1	0	-1
M	0	0	1	1

3. Select a subset of the dimensional variables identified as $\tilde{Q}_k$ equal in number to the number of fundamental dimensions such that
   a. None of the $\tilde{Q}_k$ is dimensionless.
   b. The set of $\tilde{Q}_k$ includes all the fundamental dimensions.
   c. No two of the $\tilde{Q}_k$ have the same dimensions.

The dimensionless products (π_i) are determined one at a time:

4. From the product of the set of $\tilde{Q}_k$'s, each raised to an unknown power, times one of the remaining $(n-k)$ Q's raised to a known power (usually 1),
5. Determine the exponents of $\tilde{Q}_k$ and $\tilde{Q}_i$ by the principle of dimensional homogeneity—the dimensions must "cancel" in each product.

Repeat steps 4 and 5 n - k times using a different Q_i each time.

Consider deriving scaling laws for the miscible flow of two fluids in a porous medium—displacement of oil by a miscible solvent in an oil reservoir. Assume that the porosity of the model and the reservoir are equal, and that the concentration of oil and solvent are the same in the model and the reservoir. The variables for the oil involved in this particular process are given as the equation

$$f(l, t, v, \mu, P, D, k, g, V, d, \rho) = 0 \qquad (6.31)$$

where l is length, t is time, v is velocity, μ is viscosity, P is pressure, D is dispersion, k is permeability, V is velocity of approach, d is a length perpendicular to flow, and ρ is density. Applying the algorithm where the system of fundamental dimensions is the mass M, length L, and time t system:

1. l, L   P, $\dfrac{M}{Lt^2}$ d, L

   t, t   D, $\dfrac{L^2}{t}$ ρ, $\dfrac{M}{L^3}$

   v, $\dfrac{L}{t}$ k, L^2 g, $\dfrac{L}{t^2}$

   μ, $\dfrac{M}{Lt}$ V, $\dfrac{L}{t}$

2. The number of variables n = 11, the number of fundamental dimensions k = 3, therefore there should be n - k = 8 dimensionless products or π's.

3. $\tilde{Q}_1 = V,\ \tilde{Q}_2 = d,\ \tilde{Q}_3 = \rho$

4. $\pi_1 = V^a d^b \rho^c l$ $\qquad$ $\pi_5 = V^m d^n \rho^o D$

$\qquad$ $\pi_2 = V^d d^e \rho^f t$ $\qquad$ $\pi_6 = V^p d^q \rho^r k$

$\qquad$ $\pi_3 = V^g d^h \rho^i \mu$ $\qquad$ $\pi_7 = V^s d^t \rho^u g$

$\qquad$ $\pi_4 = V^j d^k \rho^l P$ $\qquad$ $\pi_8 = V^v d^w \rho^x v$

π_1 for M	$c = 0$	$a = 0$	
for L	$a + b - 3c + 1 = 0$	$b = -1$	$\pi_1 = \dfrac{1}{d}$
for t	$-a = 0$	$c = 0$	

π_2 for M	$f = 0$	$d = 0$	
for L	$d + e - 3f = 0$	$d = -1$	$\pi_2 = \dfrac{tV}{d}$
for t	$-d + 1 = 0$	$f = 0$	

π_3 for M	$i + 1 = 0$	$g = -1$	
for L	$g + h - 3i - 1 = 0$	$h = -1$	$\pi_3 = \dfrac{\mu}{Vd\rho}$
for t	$-g - 1 = 0$	$i = -1$	

π_4 for M	$l + 1 = 0$	$j = -2$	
for L	$j + k - 3 - 1 = 0$	$k = 0$	$\pi_4 = \dfrac{P}{\rho V^2}$
for t	$-j - 2 = 0$	$l = -1$	

π_5 for M	$o = 0$	$o = 0$	
for L	$m + n - 3 + 2 = 0$	$n = -1$	$\pi_5 = \dfrac{D}{Vd}$
for t	$-m - 1 = 0$	$m = -1$	

π_6 for M	$r = 0$	$r = 0$	
for L	$p + q - 3r + 2 = 0$	$q = -2$	$\pi_6 = \dfrac{k}{d^2}$
for t	$-p = 0$	$p = 0$	

π_7 for M	$u = 0$	$s = -2$	
for L	$s + t - 3u + 1 = 0$	$l = 1$	$\pi_7 = \dfrac{dg}{V^2}$
for t	$-s - 2 = 0$	$u = 0$	

π_8 for M	$x = 0$	$v = -1$	
for L	$v + w - 3x + 1 = 0$	$w = 0$	$\pi_8 = \dfrac{v}{V}$
for t	$-v - 1 = 0$	$x = 0$	

Thus

$$F\left(\frac{1}{d}, \frac{tV}{d}, \frac{v}{V}, \frac{Vd\rho}{\mu}, \frac{p}{\rho V^2}, \frac{\mu}{\rho D}, \frac{k}{d^2}, \frac{V^2}{dg}\right) = 0 \qquad (6.32)$$

A limited set of dimensionless products can be found by writing the equation of change in dimensionless form. The set is limited to those terms that are included in the equations of change. The equations for the above situation are

$$\phi \frac{\partial c}{\partial t} + v_i \frac{\partial c}{\partial x_i} = \frac{\partial^2 c}{\partial x^2} \qquad (6.33)$$

and

$$\phi v_i = -\frac{k}{\mu} \frac{dp}{dx_i} - \frac{k\rho g}{\mu} \delta_z \qquad (6.34)$$

where δ_z is the unit vector in the z-direction (the direction parallel to gravity). Let $t^* = tV/d$, $x_i^* = x_i/d \equiv \ell/d$, $v_i^* = v_i/V$. Assume c is dimensionless and multiply Eq. (6.33) by d/v and Eq. (6.34) by $1/V$; then

$$\phi \frac{\partial c}{\partial t^*} + v_i^* \frac{\partial c}{\partial x_i^*} = \frac{D}{Vd} \frac{\partial^2 c}{\partial x_i^{*2}} \qquad (6.35)$$

and

$$\phi v_i^* = -\frac{\partial}{\partial x_i^*}\left(\frac{kp}{\mu Vd}\right) - \frac{k\rho g}{\mu V} \delta^2 \qquad (6.36)$$

Using the same assumptions as with Eq. (6.32) that porosity and concentration are the same in the model and the prototype Eqs. (6.35) and (6.36) are represented functionally by

$$F'\left(\frac{1}{d}, \frac{tV}{d}, \frac{v}{V}, \frac{kp}{\mu Vd}, \frac{k\rho g}{\mu V}, \frac{D}{Vd}\right) = 0 \qquad (6.37)$$

In applying the rule that dimensionless products can be multiplied and divided by other dimensionless groups it can be shown: that the group $p/\rho V^2$ in Eq. (6.32) can be transformed to $kp/\mu dV$; that the group $\mu/\rho D$ can be transformed to D/dV; that the group $V^2/\rho g$ can be transformed to $k\rho g/\mu V$. Thus the equations are the same except the groups $Vd\rho/\mu$ and k/d^2 do not occur in Eq. (6.37). They are missing because the equation of motion for flow in porous media, Eq. (6.34), is not complete. According to this analysis there should be two more terms in it, one to take into account the effect of boundaries enclosing the medium $(Vd\rho/\mu)$ and the other to take into account any relationship between microscopic and macroscopic flow (k/d^2).

The difference between Eqs. (6.32) and (6.37) points out an advantage to using both approaches in deriving a set of dimensionless products. The approach used in arriving at Eq. (6.32) gives the correct number of groups for the postulated variables, but does not relate them physically. The method of Eq. (6.37), which requires a mathematical model, does give more physical sense, but does not contain all the terms. It should be pointed out that Eq. (6.34), Darcy's law, is the one always used to calculate flow in porous media. Even though we model using Eqs. (6.34) and (6.37), we know from Eq. (6.32) that other variables exist. It may prove, as it does in this case, that these additional terms need not be taken into account if the models are operated in such a manner that contributions of these terms are minimized.

A possible further discrepancy in the analysis is that there may be more important variables than the ones shown in Eq. (6.31). Thus, it would seem more reasonable, especially in cases where little is known about the correct form of the equations of change, to get a complete set of variables and dimensionless products and then eliminate dimensionless products rather than variables. In this way one will gain a little insight into the physics of the problem simply by knowing what has been ignored.

Before obtaining scaling laws, any further terms due to two-phase flow and interaction with the medium must be included. Thus, any terms which include fluid properties must be duplicated, and we need three more coefficients in terms of the ratios of the two fluid properties and including c and ϕ

$$G\left(c, \frac{1}{d}, \frac{tV}{d}, \frac{v}{V}, \frac{Vd\rho}{\mu}, \frac{p}{\rho V^2}, \frac{V^2}{dg}, \frac{\mu}{\rho D}, \frac{d^2}{k}, \frac{\mu}{\mu_s}, \frac{\rho}{\rho_s}, \frac{D}{D_s}\right) = 0 \qquad (6.38)$$

where the subscript s refers to the solvent phase.

Since we are interested in gross effect, and since it is very difficult to scale permeability, the groups $Vd\rho/\mu$ and k/d^2 will not be scaled, and the final equation is

$$G'\left(c, \frac{1}{d}, \frac{tV}{d}, \frac{v}{V}, \frac{kp}{\mu dV}, \frac{k\rho g}{\mu V}, \frac{D}{dV}, \frac{\mu}{\mu_s}, \frac{\rho}{\rho_s}, \frac{D}{D_s}\right) = 0 \qquad (6.39)$$

Thus, the dependent variables are c, c/V, and $kp/\mu dV$, the independent variables are $1/d$ and tV/d, and the remaining groups are coefficients.

The scaling is geometric and down by the quantity a, and thus $K_1 = K_d = 1/a$ and length scales down by a. If the same fluids are used in the model and prototype

$$K_\mu = K_{\mu_s} = K_\rho = K_{\rho_s} = K_D = K_{D_s} \qquad (6.40$$

Since $K_D = K_d K_V$, $K_V = 1/K_d = a$, velocity scales up by a. From $K_k K_\rho K_g = K K_V$, then $K_k = K_V$ and permeability scales up by a. $K_t K_V = K_d$, thus $K_d = K_d/K_V$ and time scales down by a^2. Since $K_k K_P = K_\mu K_d K_V$, $K_P = 1/K_k$, pressure scales down by a.

The scaling law in words: The dimensionless concentrations, pressure and dimensionless velocity are the same functions of dimensionless length and time in geometrically scaled models if the same fluids are used in the model and prototype. This assumes dispersion in the laboratory and field are scaled correctly by the Bodenstein number. As we discussed earlier, this may or may not be the case.

Offeringa and van der Poel (1954) determined scaling laws for miscible flow by making the equations of change dimensionless as above. They tested the laws

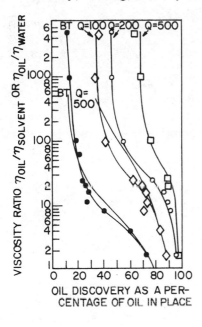

FIG. 6.2 Oil recovery with solvent flooding, as dependent on the viscosity ratio of oil and solvent compared with oil recovery from waterflooding tests according to Croes and Schwarz. Solvent used: kerosene, viscosity 1.2 cp. Q (oil + kerosene or water) expressed as a percentage of the oil originally in place. (From I. Offeringa and C. van der Poel, Displacement of Oil from Porous Media by Miscible Liquids, Pet. Trans. AIME 201, SPEJ 310. © 1954 SPE-AIME.)

successfully with laboratory models of different lengths (1.03, 1.60, and 3.00 meters); their results are summarized in Fig. 6.2.

Leverett et al. (1942), Rapoport (1955), and Geertsma et al. (1956) studied scaling immiscible flow requiring relative permeability to water, oil, and gas and dimensionless capillary pressure to be the same in both. It is almost impossible to meet these requirements unless the model is constructed of the same material as the prototype. If model fluids are selected to give equal viscosity ratios, different mobility ratios will result unless the model and prototype are constructed of the same material.

Perkins and Collins (1960) developed scaling criteria for immiscible flow which permit different relationships between saturation, relative permeability, or capillary pressure in the model and prototype. For incompressible oil-water flow in a porous medium the equations of condition are

$$\phi \frac{\partial S_w}{\partial t} = \frac{1}{\mu_w} \frac{\partial}{\partial x_i} \left(k_w \frac{\partial \Phi_w}{\partial x_i} \right) \tag{6.41}$$

$$\phi \frac{\partial S_o}{\partial t} = \frac{1}{\mu_o} \frac{\partial}{\partial x_i} \left(k_o \frac{\partial \Phi_o}{\partial x_i} \right) \tag{6.42}$$

where k_w and k_o are specific permeabilities to oil and water and

$$S_o + S_w = 1 \tag{6.43}$$

where

$$\Phi_w = p_w + \rho_w gz \tag{6.44}$$

$$\Phi_o = p_o + \rho_o gz \tag{6.45}$$

The pressures in the fluids are related by

$$p_c = p_o - p_w \tag{6.46}$$

Combining Eqs. (6.44) through (6.46),

$$\Phi_o = \Phi_w + p_c - \Delta\rho gz \tag{6.47}$$

Defining the following dimensionless variables:

$$x_i^* = \frac{x_i}{L_i} \tag{6.48}$$

$$\Phi_o^* = \frac{k_{wro}\Phi_o}{v\,\mu w\,L_{x_1}} \tag{6.49}$$

$$\Phi_w^* = \frac{k_{wro}\Phi_w}{v\mu_w\,L_{x_1}} \tag{6.50}$$

$$S_w^* = \frac{S_w - S_r}{1 - S_r - S_m} \tag{6.51}$$

$$J(S_w^*) = \frac{p_c}{\sigma}\left(\frac{k}{\phi}\right)^{\frac{1}{2}} \tag{6.52}$$

where k_{wro} is the specific permeability of water at residual oil saturation and v is injection velocity. (k_{ocw} is specific permeability to oil at residual water saturation.) Making Eqs. (6.41) to (6.43) and (6.47) dimensionless using the above definitions yields

$$\frac{\partial S_w^*}{\partial t^*} = \left(\frac{L_{x_1}}{L_{x_i}}\right)^2 \frac{\partial}{\partial x_i^*}\left(k_{rw}\frac{\partial \Phi_v^*}{\partial x_i^*}\right) \tag{6.53}$$

$$\frac{\partial S_o^*}{\partial t^*} = \left(\frac{\mu_w k_{ocw}}{\mu_o k_{wro}}\right)\left(\frac{L_{x_1}}{Lx_i}\right)^2 \frac{\partial}{\partial x_i^*}\left(k_{ro}\frac{\partial \Phi_o^*}{\partial x_i^*}\right) \tag{6.54}$$

$$S_o^* + S_w^* = 1 \tag{6.55}$$

$$\Phi_w^* + \frac{\sigma k_{wro}}{v\mu_w \frac{L}{x_1}} J(S_w^*) - \frac{k_{wro}\Delta\rho g L_{x_3}}{v\mu_w \frac{L}{x_1}} z^* = \Phi_o^* \tag{6.56}$$

Equations (6.53) to (6.56) with appropriate dimensionless initial and boundary conditions represent a description of the problem functionally:

$$f\left[S_w^*, S_o^*, \Phi_w^*, \Phi_o^*, t^*, x_i^*, \frac{L_{x_1}}{L_{x_i}}, \frac{L_{x_1}}{L_{x_3}}, \frac{k_{wro}\mu_o}{k_{ocw}\mu_w}, \frac{\sigma k_{wro}}{v\mu_w \frac{L}{x_1}}\left(\frac{\phi}{k}\right)^{\frac{1}{2}}, \frac{k_{wro}g\,\Delta\rho L_{x_3}}{v\mu_w \frac{L}{x_1}}\right] = 0 \tag{6.57}$$

The scaling laws for the flow are

Dimensionless initial and boundary conditions must be the same; k_{rw}, k_{ro}, and $J(S_W^*)$ must be the same functions of the dimensionless saturation S_W^* in the model and prototype where

$$k_{rw} = \frac{k_w}{k_{wro}} \tag{6.58}$$

$$k_{ro} = \frac{k_o}{k_{ocw}} \tag{6.59}$$

The dimensionless parameters in Eq. (6.57) must have the same function of the dimensionless saturation S_W^* in the model and prototype; then Φ_w^*, Φ_o^*, S_W^*, and S_o^* will be the same functions of x_i^* and t^* in model and prototype.

Corey (1977) summarizes the requirements for two-fluid immiscible flow as similarity of functional relationships using k_w, k_n, S, and p_c (where w is wetting, n is nonwetting, and S is saturation of wetting fluid); similarity of macroscopic

TABLE 6.2 Scale Factors of Several Investigators

Unit	Rapoport (1955)	Richardson (1961)	Miller & Miller (1956)	Corey (1965)
Length	$\dfrac{\sigma/\Delta\gamma}{\sqrt{k/\phi}}$	$\dfrac{\sigma f(\alpha)/\Delta\rho}{\sqrt{k/\phi}}$	$\dfrac{\sigma}{\Delta\rho\,d}$	$\dfrac{P_d}{\Delta\rho}$
Capillary pressure	$\dfrac{\sigma}{\sqrt{k/\phi}}$	$\dfrac{\sigma f(\alpha)}{\sqrt{k/\phi}}$	$\dfrac{\sigma}{d}$	P_d
Time	$\dfrac{\mu L\phi}{k\,\Delta\rho}$	$\dfrac{\mu L\phi}{k\,\Delta\rho}$	$\dfrac{\mu L}{\Delta\rho\,d^2}$	$\dfrac{\mu L\phi_e}{k\,\Delta\rho}$
Flux rate	$\dfrac{k\,\Delta\rho}{\mu}$	$\dfrac{k\,\Delta\rho}{\mu}$	$\dfrac{\Delta\rho\,d^2}{\mu}$	$\dfrac{k\,\Delta\rho}{\mu}$

Source: Corey, 1977.

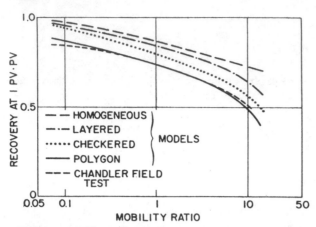

FIG. 6.3 Effect of mobility ratio on recovery for miscible floods. (From R. A. Greenkorn, C. R. Johnson, and R. E. Haring, Miscible Displacement in a Controlled Natural System, Pet. Trans. AIME 234. SPEJ 1329. © 1965 SPE-AIME.)

geometry and orientation with gravity; equal ratios of corresponding macroscopic dimensions to a scale factor for length which is inversely proportional to pore dimensions; equal viscosity ratios, equal initial conditions. Table 6.2 shows a comparison of the scale factors of Rapport (1955), Richardson (1961), Miller and Miller (1956), and Corey (1965).

Distorted models are occasionally used, that is, where $K_x \neq K_y \neq K_z$. van Daalen and van Domselaar (1972) discuss scaled models with geometries different from the prototype. Bear (1960) considers the problem of scaling in groundwater studies.

Much of the displacement information for two-dimensional flow in a porous medium with adverse mobility ratios has been obtained with scaled model studies. Greenkorn et al. (1965) suggest that in heterogeneous porous media, the spatial variation of heterogeneity is part of the prototype geometry and the degree and location must be preserved in the scaled model. The heterogeneity effect in miscible flow is the most significant variable. Heterogeneity further aggravates the effect of adverse mobility ratios. Figure 6.3 summarizes miscible flow experiments in a sandstone reservoir and its 1/13.5 scale model. The various models represent different approaches to scaling permeability heterogeneity. In the polygon model the degree and location of the heterogeneity were preserved in the model as in the prototype.

6.2 ANALOG MODELS

The models discussed in Section 6.1 are homologs; a prototype is modeled with a similar physical system scaled in such a way that phenomena observed in each occur at homologous points and times. The homolog is a scaled duplicate (ideally) in every way. An analog is a model which may have completely different physical components but the prototype and its model are described by the same mathematical equations. Several different kinds of analogs have been used to model flow phenomena in porous media. We will discuss a viscous flow analog—the Hele-Shaw model and two types of electrical analogs—an electrolytic or continuous electric analog, and a network or discrete analog. The Hele-Shaw analogy models flow in two dimensions with gravity either perpendicular or parallel to the flow. The electrical analogs can be used in two or three dimensions. Other analogs such as ionic and membranes have also been used.

Hele-Shaw (1898) showed experimentally that the streamline configuration for creeping flow around an obstacle located between two closely spaced parallel plates is the same as for two dimensional ideal flow about the same obstacle. Stokes (1898) verified these observations mathematically. A form of Darcy's law, within a multiplicative constant, is the same as the expression for the average velocity over the plate gap in the plane of a Hele-Shaw model. This analog may be used to describe flow in nonideal media.

A Hele-Shaw model is constructed by placing two plates, usually glass, very close together and allowing flow between them (see Fig. 6.4). Incompressible fluid flow through a Hele-Shaw model and a porous medium are analogous because the equations for the average velocity components in two dimensions are identical if

$$k = \frac{h^2}{12} \tag{6.60}$$

where h is the spacing between the two plates of the Hele-Shaw model. Streamlines in the model can be made visible by introducing a colored fluid into the space between the plates at a number of points across the model as in Fig. 6.4.

The average velocity components for two-dimensional flow in a porous medium are

$$q_x = -\frac{k}{\mu}\frac{dp}{dx} \tag{6.61}$$

$$q_y = -\frac{k}{\mu}\frac{dp}{dy} \tag{6.62}$$

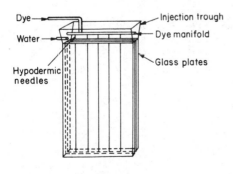

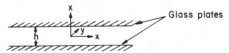

FIG. 6.4 Hele-Shaw model. (From Greenkorn et al., 1964.)

For two-dimensional flow in a Hele-Shaw model the Navier-Stokes equation for steady and compressible flow are

$$\rho\left(v_x \frac{\partial v_x}{\partial x} + v_y \frac{\partial v_x}{\partial y}\right) = -\frac{\partial p}{\partial x} + \mu\left(\frac{\partial^2 v_x}{\partial x^2} + \frac{\partial^2 v_x}{\partial y^2} + \frac{\partial^2 v_x}{\partial z^2}\right) \tag{6.63}$$

$$\rho\left(v_x \frac{\partial v_y}{\partial x} + v_y \frac{\partial v_y}{\partial y}\right) = -\frac{\partial p}{\partial y} + \mu\left(\frac{\partial^2 v_y}{\partial x^2} + \frac{\partial^2 v_y}{\partial y^2} + \frac{\partial^2 v_y}{\partial z^2}\right) \tag{6.64}$$

Assume the convective terms on the left are negligible when compared to the viscous forces. Assume the two plates are sufficiently close so the second derivative of the velocity with respect to x and y are small compared to the z direction, then

$$\frac{\partial p}{\partial x} = \mu \frac{\partial^2 v_x}{\partial z^2} \tag{6.65}$$

$$\frac{\partial p}{\partial y} = \mu \frac{\partial^2 v_y}{\partial z^2} \tag{6.66}$$

Integrate Eqs. (6.65) and (6.66) using $v_x = v_y = 0$ at the surface of the two plates.

$$v_x = \frac{1}{2\mu} \frac{\partial p}{\partial x}\left(z^2 - \frac{h^2}{4}\right) \tag{6.67}$$

$$v_y = \frac{1}{2\mu} \frac{\partial p}{\partial y}\left(z^2 - \frac{h^2}{4}\right) \tag{6.68}$$

There is a parabolic velocity profile between the two plates so that "planes" of fluid perpendicular to the z direction are moving at different velocities, so that for the x direction

$$v_x = \frac{\Delta p_x h^2}{4\mu L}\left[1 - \left(\frac{2z}{h}\right)^2\right] \tag{6.69}$$

According to Eqs. (6.67) and (6.68) streamlines in parallel planes perpendicular to the z direction are congruent, though each plane moves at a velocity given by Eq. (6.69). The streamlines are congruent when one "looks down" on the xy plane it appears there is one streamline resulting from the "stack" of the streamlines in each plane. The average velocities are

$$\langle v_x \rangle = \frac{1}{h} \int_{-h/2}^{h/2} v_x \, dz \tag{6.70}$$

$$\langle v_y \rangle = \frac{1}{h} \int_{-h/2}^{h/2} v_y \, dz \tag{6.71}$$

Thus over the plate gap

$$\langle v_x \rangle = -\frac{h^2}{12\mu} \frac{\partial p}{\partial x} \qquad (6.72)$$

$$\langle v_y \rangle = -\frac{h^2}{12\mu} \frac{\partial p}{\partial y} \qquad (6.73)$$

The velocity potential for Eqs. (6.61) and (6.62) is

$$\tilde{\Phi} = \frac{kp}{\mu} \qquad (6.74)$$

And for Eqs. (6.72) and (6.73)

$$\Phi = \frac{h^2 p}{12\mu} \qquad (6.75)$$

There is a direct analogy between two-dimensional flow in a Hele-Shaw model and a porous medium if

$$\tilde{\Phi} = \Phi \qquad (6.76)$$

which yields the relation in Eq. (6.60).

Gravity can be included in the definitions of the flow potential and the same condition Eq. (6.67) or (6.60) results. The analogy can be shown for other two-dimensional flow geometries—such as radial flow. If we take the derivatives of Eqs. (6.67) and (6.68) and add them together

$$\frac{\partial v_x}{\partial x} + \frac{\partial v_y}{\partial y} = \frac{1}{2\mu}\left(z^2 - \frac{h^2}{4}\right)\left(\frac{\partial^2 p}{\partial x^2} + \frac{\partial^2 p}{\partial y^2}\right) \qquad (6.77)$$

Assume $v_z = 0$; then from the continuity equation

$$\frac{\partial v_x}{\partial x} + \frac{\partial v_y}{\partial y} = 0 \qquad (6.78)$$

$$\frac{\partial^2 p}{\partial x^2} + \frac{\partial^2 p}{\partial y^2} = 0 \qquad (6.79)$$

This verifies that the two-dimensional flow was ideal.

The above derivation is not valid in the neighborhood of a total or partial obstruction or expansion in the Hele-Shaw model. Obstructions or expansions are used to model nonideal porous media. By placing a partial obstruction if the flow field of the model there is an abrupt change in the spacing or analogously an abrupt change in permeability—this models a heterogeneity. By causing gradual changes in spacing one can model a nonuniformity. By creating obstructions that have different spacings in different directions (striations), one can model anisotropy. Figure 6.5 is a sketch of a heterogeneous Hele-Shaw model.

When we introduce changes in plate spacing this violates the assumption that $v_z = 0$. If the change in h is gradual such that dh/dz is much less than 1, then v_z is much less than v_x, v_y and the analogy holds. If the change in h is abrupt, the

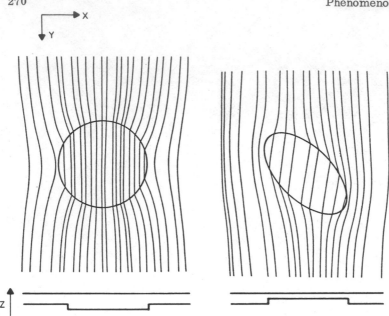

FIG. 6.5 Sketch of heterogeneity in Hele-Shaw model. (From Greenkorn et al., 1964.)

analog breaks down in the neighborhood of the discontinuity, and the width of the affected zone is of the order of h. Generally, deviations in the analogy are either small for small changes in h or restricted to small regions near the change as long as overall plate separation is small. Riegels (1938) included convection in the equations describing flow around a cylindrical obstruction and showed the velocity distribution depends on the Reynolds number, the plate spacing, and a dimension characteristic of the obstruction. A solution of Poisson's equation with the condition the flow rate into the obstacle is zero is used. Smith and Greenkorn (1969) investigated the effects of convective inertia on flow about obstructions in a Hele-Shaw model. Writing the Navier-Stokes equation in dimensionless form assuming two-dimensional steady-state incompressible flow perpendicular to gravity, where $v_x^* = v_x/v_\infty$, $v_y^* = v_y/v_\infty$, $x^* = x/2h$, $y^* = y/2h$, $p^* = p/\rho v_\infty^2$, $Re = v_\infty \rho h/2\mu$, we have

$$v_x^* \frac{\partial v_x^*}{\partial x^*} + v_y^* \frac{\partial v_x^*}{\partial y^*} = -\frac{\partial p^*}{\partial x^*} + \frac{1}{Re}\left(\frac{\partial^2 v_x^*}{\partial x^{*2}} + \frac{\partial^2 v_x^*}{\partial y^{*2}} + \frac{\partial^2 v_x^*}{\partial z^{*2}}\right) \qquad (6.80)$$

$$v_x^* \frac{\partial v_y^*}{\partial x^*} + v_y^* \frac{\partial v_y^*}{\partial y^*} = -\frac{\partial p^*}{\partial y^*} + \frac{1}{Re}\left(\frac{\partial^2 v_y^*}{\partial x^{*2}} + \frac{\partial^2 v_y^*}{\partial y^{*2}} + \frac{\partial v_y^*}{\partial z^{*2}}\right) \qquad (6.81)$$

Drop the asterisks. Assume a perturbation solution of Eqs. (6.80) and (6.81) is possible where

$$v_x = v_x^{(1)} + v_x^{(2)} \tag{6.82}$$

$$v_y = v_y^{(1)} + v_y^{(2)} \tag{6.83}$$

$$p = p^{(1)} + p^{(2)} \tag{6.84}$$

$v_x^{(1)}$ and $v_y^{(1)}$ are the velocities neglecting inertia. $v_x^{(2)}$ and $v_y^{(2)}$ are the perturbation velocities due to inertia. For small h the perturbation terms are small. Assume the second derivative of velocities with respect to x and y are small; then

$$v_x^{(1)} \frac{\partial v_x^{(1)}}{\partial x} + v_y^{(1)} \frac{\partial v_x^{(1)}}{\partial y} = - \frac{\partial p^{(1)}}{\partial x} + \frac{1}{Re} \frac{\partial^2 v_x^{(2)}}{\partial x^2} \tag{6.85}$$

$$v_x^{(1)} \frac{\partial v_y^{(1)}}{\partial x} + v_y^{(1)} \frac{\partial v_y^{(1)}}{\partial y} = - \frac{\partial p^{(1)}}{\partial y} + \frac{1}{Re} \frac{\partial^2 v_y^{(2)}}{\partial z^2} \tag{6.86}$$

The left side of Eqs. (6.85) and (6.86) can be determined from potential flow theory to be $I_1(x, y)[1 - (z/\tilde{h})^2]^2$ and $I_2(x, y)[1 - (z/\tilde{h})^2]^2$, where $\tilde{h} = h/2L$. L is the characteristic dimension of the obstacle perpendicular to the velocity of approach v_∞. Using the no-slip condition for the perturbation velocities yields a Poisson equation for the perturbation pressure in two dimensions.

$$\nabla^2 p^{(2)} = - \frac{24}{35} \left(\frac{\partial I_1}{\partial x} + \frac{\partial I_2}{\partial y} \right) \tag{6.87}$$

The solutions of Eq. (6.87) for the perturbation pressure gradients are

$$\frac{\partial p^{(2)}}{\partial x} = - \frac{24}{35} I_1(x, y) \tag{6.88}$$

$$\frac{\partial p^{(2)}}{\partial y} = - \frac{24}{35} I_2(x, y) \tag{6.89}$$

Substituting Eqs. (6.88) and (6.89) into Eqs. (6.85) and (6.86) yields

$$v_x^{(2)} = Re\, \tilde{h}^2 \left\{ I_1(x, y) \left[- \frac{1}{42} + \frac{11}{70} \left(\frac{z}{\tilde{h}} \right)^2 - \frac{1}{6} \left(\frac{z}{\tilde{h}} \right)^4 + \frac{1}{30} \left(\frac{z}{\tilde{h}} \right)^6 \right] \right\} \tag{6.90}$$

$$v_y^{(2)} = Re\, \tilde{h}^2 \left\{ I_2(x, y) \left[- \frac{1}{42} + \frac{11}{70} \left(\frac{z}{\tilde{h}} \right) - \frac{1}{6} \left(\frac{z}{\tilde{h}} \right)^4 + \frac{1}{30} \left(\frac{z}{\tilde{h}} \right)^6 \right] \right\} \tag{6.91}$$

Solutions may be found for arbitrary shapes by determining I_1 and I_2 for those shapes from the ideal flow problem. Figures 6.6 to 6.8 show the calculations for flow past a circle, square, and ellipse for $Re' = 1$, 3 and for $z/\tilde{h} = 0.75$, where

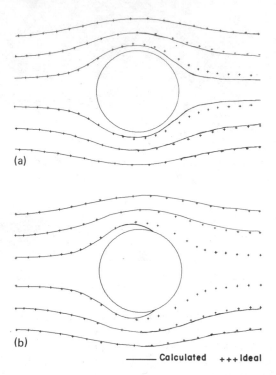

(a)

(b)

——— Calculated   +++Ideal

FIG. 6.6 Streamlines past a circular obstruction. (a) $N'_{Re} = 1$, $z/\tilde{h} = 0.75$;
(b) $N'_{Re} = 3$, $z/\tilde{h} = 0.75$. (From R. C. Smith and R. A. Greenkorn, An Investigation of the Flow Regime for Hele-Shaw Flow, Pet. Trans. AIME 264, SPEJ 434. © 1969 SPE-AIME.)

$$Re' = \frac{h}{2L} Re \tag{6.92}$$

For $Re' \leq 1$ the ideal flow lines and the perturbation results are close to each other. For $1 < Re' < 3$ there are noticeable distortions between the ideal and perturbation values. The perturbation velocities vanish at $z/\tilde{h} = 0.433$; in this plane the ideal flow lines are correct. As one looks down on the model the streamlines will begin to appear fuzzy since they are no longer congruent in the parallel planes perpendicular to z. However, an average value is close to the ideal flow. For $Re' > 3$ the perturbation solution requires higher order terms. For partial obstruction the speeding up or slowing down at the obstruction boundary cause additional inertial effects. For partial obstructions additional criteria are required (Polubrinova-Kochina, 1962)

$$\frac{\tan \alpha_1}{\tan \alpha_2} = \frac{h_1^3}{h_2^3} \tag{6.93}$$

where α_1 is the angle between the normal and tangent to the surface of the obstruction and where 1 is outside and 2 is inside the obstruction.

   Variations of the Hele-Shaw analogy can include semi-impervious layers and storativity. The layers can be modeled by putting a perforated plate in the middle of the gap. The storativity is modeled by gluing a network of short vertical tubes over holes in the upper plate of the model. Trapping models can be constructed by connecting the open ends of the tubes. Wettability may be changed by coating portions of the plates.

   Electrical analogs have been used to interpret flow in porous media for many years. Continuous electrical analogs such as a tank containing an electrolytic fluid or blotting paper soaked with a conducting fluid are described by Laplace's equations (as is flow in a porous medium). Darcy's law is

$$\underline{q} = -\frac{\underline{\underline{k}}}{\mu} \underline{\nabla} \Phi \qquad (6.94)$$

Ohm's law is

$$\underline{i} = \underline{\underline{\sigma}} \, \underline{\nabla} \, V \qquad (6.95)$$

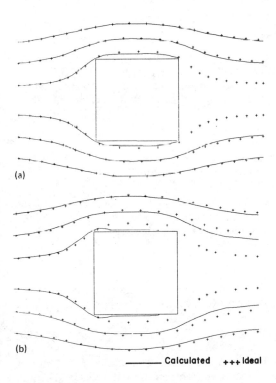

(a)

(b)

_____ Calculated   +++ Ideal

FIG. 6.7 Streamlines past a square obstruction. (a) $N'_{Re} = 1$, $z/\tilde{h} = 0.75$; (b) $N'_{Re} = 3$, $z/\tilde{h} = 0.75$. (From R. C. Smith and R. A. Greenkorn, An Investigation of the Flow Regime for Hele-Shaw Flow, Pet. Trans. AIME 264, SPEJ 434. © 1969 SPE-AIME.)

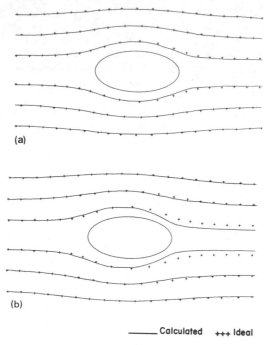

(a)

(b)

—— Calculated   +++ Ideal

FIG. 6.8 Streamlines pas an elliptical obstruction. (a) $N'_{Re} = 1$, $z/\tilde{h} = 0.75$; (b) $N'_{Re} = 3$, $z/\tilde{h} = 0.75$. (From R. C. Smith and R. A. Greenkorn, An Investigation of the Flow Regime for Hele-Shaw Flow, Pet. Trans. AIME 264, SPEJ 434. © 1969 SPE-AIME.)

where i is current, $\underline{\sigma}$ is conductivity, and V is electromotive potential. Combining Darcy's law and the equation of continuity for incompressible flow

$$\nabla \cdot \underline{\underline{k}} \Phi = 0 \tag{6.96}$$

Combining Ohm's law and the equation of current for electrical flow

$$\underline{\nabla} \cdot \underline{\underline{\sigma}} \, \underline{\nabla} \, V = 0 \tag{6.97}$$

The dimensionless flow of current under a dimensionless voltage potential is analogous to the dimensionless flow of fluid under a dimensionless fluid potential if

$$k^* = \sigma^* \tag{6.98}$$

where the asterisk indicates the terms are dimensionless. The analog problem is scaled much as the homolog problem with the scale factors for potential, flow, and size being

$$K_\Phi = \frac{V}{\Phi} \tag{6.99}$$

$$K_q = \frac{i}{q} \tag{6.100}$$

$$K_L = \frac{L'}{L} \tag{6.101}$$

Figure 6.9 shows the streamlines and potential lines determined for a 1/4 five-spot pattern, 1/4 sink, 1/4 source in a radial flow field using an electrolytic model by Wyckoff et al. (1933).

Resistance networks and resistance-capacitance networks are discrete electrical analogs of flow in porous media. The equation for electrical networks have the same analogy as above; however in the network the distributed parameters system is represented at discrete points where the network elements are connected. The discrete models have the advantage of more easily representing changes in media properties. Unsteady flow can also be modeled with a resistance-capacitance network. (With the electrolytic model the permeability is changed analogously by changing the depth of the electrolyte at a given point or over a given area.) Since the analog is discrete we are actually representing the medium with a finite-difference approximation; so in order to get decreased errors of approximation more resistances or more functions are required. A resistance-capacitance network is the analog of

$$S \frac{\partial \Phi}{\partial t} - (\underline{\nabla} \cdot \underline{\underline{T}} \, \Phi) \tag{6.102}$$

where

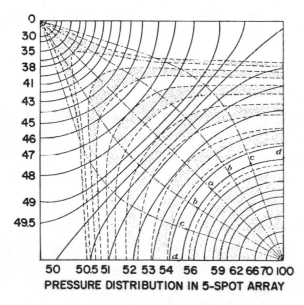

PRESSURE DISTRIBUTION IN 5-SPOT ARRAY

FIG. 6.9 Pressure distribution in five-spot array. (From R. D. Wyckoff, H. G. Botset, and A. N. Muskat, The Mechanics of Porous Flow Applied to Water Flooding Problems Pet. Trans. AIME 103, SPEJ 219. © 1933 SPE-AIME.)

$$S = \phi ch \tag{6.103}$$

$$T = \frac{kh}{\mu} \tag{6.104}$$

Consider a steady flow situation represented by a resistance network; then

$$\underline{\nabla} \cdot \underline{\underline{T}} \, \underline{\nabla}\Phi = 0 \tag{6.105}$$

If we are modeling flow in a two-dimensional isotropic porous medium with a phreatic surface assuming specific gravity of the fluid is 1

$$T\left(\frac{\partial^2 h}{\partial x^2} + \frac{\partial^2 h}{\partial y^2}\right) + \frac{\partial h}{\partial x}\frac{\partial T}{\partial x} + \frac{\partial h}{\partial y}\frac{\partial T}{\partial y} = 0 \tag{6.106}$$

where $T = kb/\mu$, b is saturated thickness, and h is fluid height above an arbitrary datum. Writing Eq. (6.106) in finite difference form where $\Delta x = \Delta y$

$$\frac{T_{p,n}(h_{p-1} + h_{p+1} + h_{n-1} + h_{n+1} - 4h_{p,n})}{\Delta x^2} + \frac{(h_{p+1} - h_{p-1})}{2\,\Delta x}\frac{(T_{p+1} - T_{p-1})}{2\,\Delta x}$$

$$+ \frac{(h_{n+1} - h_{n-1})}{2\,\Delta y}\frac{(T_{n+1} - T_{n-1})}{2\,\Delta y} = 0 \tag{6.107}$$

Figure 6.10 shows a small segment of the porous medium subdivided by a rectangular grid of spacing Δx and Δy. An element of the analog resistance network is also shown in Fig. 6.10. From Kirchhoff's law the equation of steady electrical flow to the junction (p, n) is

$$\frac{E_{p-1} - E_{p,n}}{R_{p-1}} + \frac{E_{p+1} - E_{p,n}}{R_{p+1}} + \frac{E_{n-1} - E_{p,n}}{R_{n-1}} + \frac{E_{n+1} - E_{p,n}}{R_{n+1}} = 0$$

where E is voltage. Rewriting Eq. (6.108),

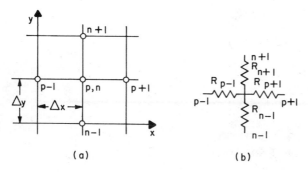

(a)   (b)

FIG. 6.10 Model relations at a point in a two-dimensional field of flow: (a) grid reference in the prototype system; (b) analogous resistance junctions of an electric model.

$$\frac{E_{p-1}}{R_{p-1}} + \frac{E_{p+1}}{R_{p+1}} + \frac{E_{n-1}}{R_{n-1}} + \frac{E_{n+1}}{R_{n+1}} - E_{p,n}\left(\frac{1}{R_{p-1}} + \frac{1}{R_{p+1}} + \frac{1}{R_{n-1}} + \frac{1}{R_{n+1}}\right) = 0 \quad (6.109)$$

Rewriting Eq. (6.107)

$$h_{p-1}\left(T_{p,n} - \frac{T_{p+1} - T_{p-1}}{4}\right) + h_{p+1}\left(T_{p,n} - \frac{T_{p+1} - T_{p-1}}{4}\right) + h_{n-1}\left(T_{p,n}\right.$$

$$\left. - \frac{T_{n+1} - T_{n-1}}{4}\right) + h_{n+1}\left(T_{p,n} + \frac{T_{n+1} - T_{n-1}}{4}\right) - 4h_{p,n}T_{p,n} = 0$$
$$(6.110)$$

Let

$$T_{p,n} - \frac{T_{p+1} - T_{p-1}}{4} = \frac{c}{R_{p-1}} \quad\quad\quad\quad\quad\quad (6.111)$$

$$T_{p,n} - \frac{T_{p+1} - T_{p-1}}{4} = \frac{c}{R_{p+1}} \quad\quad\quad\quad\quad\quad (6.112)$$

$$T_{p,n} - \frac{T_{n+1} - T_{n-1}}{4} = \frac{c}{R_{n-1}} \quad\quad\quad\quad\quad\quad (6.113)$$

$$T_{p,n} + \frac{T_{n+1} - T_{n-1}}{4} - \frac{c}{R_{n+1}} \quad\quad\quad\quad\quad\quad (6.114)$$

where c relates transmissivity to resistance by adding Eqs. (6.111) to (6.114)

$$T_{p,n} = \frac{c}{4}\left(\frac{1}{R_{p-1}} + \frac{1}{R_{p+1}} + \frac{1}{R_{n-1}} + \frac{1}{R_{n+1}}\right) \quad\quad\quad (6.115)$$

Relative values of the resistances can be determined. The scaling is

$$R_d = \frac{R}{R_p} = c \quad\quad\quad\quad\quad\quad\quad\quad\quad (6.116)$$

$$\phi_d = \frac{\Delta E}{\Delta \phi} \quad\quad\quad\quad\quad\quad\quad\quad\quad\quad (6.117)$$

$$Q_d = \frac{\Delta z}{\Delta Q} = \frac{\phi_d}{R_d} \quad\quad\quad\quad\quad\quad\quad\quad (6.118)$$

The unsteady-state situation may be modeled analogously by including a capacitance at each junction since on the right-hand side of Eq. (6.107) we have

TABLE 6.3 Applicability of Models and Analogs

Feature	Sand box model	Hele–Shaw analog		Electrolytic	Electric analogs		Membrane analog
		Vertical	Horizontal		RC network	Ion motion	
Dimensions of field	Two or three	Two	Two	Two or three	Two or three	Two (horizontal)	Two (horizontal)
Steady or unsteady flow	Both	Both	Both	Steady	Both	Steady	Steady
Simulation of elastic storage	Yes, for two dimensions	Yes	Yes	Yes, for two dimensions	Yes	No	No
Simulation of capillary fringe and capillary pressure	Yes	Yes	No	No	No	No	No
Simulation of phreatic surface	Yes[a]	Yes[a]	No	Yes[b]	No[c]	No	No
Simulation of anisotropic media[d]	Yes	Yes ($k_x \neq k_z$)	Yes ($k_x = k_y$)	Yes	Yes	Yes ($k_x \neq k_y$)	Yes ($k_x = k_y$)
Simulation of medium inhomogeneity	Yes	Yes[e]	Yes[e]	Yes	Yes	Yes	No
Simulation of leaky formations	Yes	Yes	Yes	Yes[e]	Yes	No	No
Simulation of accretion	Yes	Yes	Yes	Yes, for two dimensions	Yes	No	Yes
Flow of two liquids with an abrupt interface	Approximately	Yes	Yes (no gravity)	No[f]	No[f]	Yes (no gravity)	No
Hydrodynamic dispersion	Yes	No	No	No	No	No	No
Simultaneous flow of two immiscible fluids	Yes	No	No	No	No	No	No
Observation of streamlines and path lines	Yes, for two dimensions, near transparent walls for three dimensions	Yes	Yes	No	No	No	No

[a] Subject to restrictions because of the presence of a capillary fringe.
[b] By trial and error for steady flow.
[c] By trial and error for steady flow, or, as an approximation, for relatively small phreatic surface fluctuations.
[d] By scale distortion in all cases, except for the RC network and sometimes the Hele–Shaw analog where the hydraulic conductivity of the analog can be made anisotropic.
[e] With certain constraints.
[f] For a stationary interface by trial and error.
<u>Source</u>: Bear, 1972.

S $\partial\Phi/\partial t$ and on the right-hand side of Eq. (6.108) we have c' $\partial E/\partial t$. Bear summarizes the use of scaled homologs and analogs in Table 6.3.

6.3 NUMERICAL MODELS

The procedure for simulating flow phenomena in porous media is to discretize the element of medium under study into blocks or grids and integrate the equations of change in all the blocks or grids simultaneously or iteratively. Figure 6.11 is a schematic of overlaying a two-dimensional grid on a porous medium reservoir. The simulator representing the differential equations that describe flow may be used to solve many problems from the amount and type of fluids in the reservoir to interpreting flow data to describe the reservoir (the inverse problem). Simulators are used to study alternate methods of producing a reservoir as well as determining the parameters (rates, pressures, etc.) to produce them.

Chapters 3 to 5 were concerned with describing flow mechanisms for various types of fluid flow and summarizing this flow with mathematical models. We will not concern ourselves here with the mathematical models per se, but rather with methods of numerical simulation—rewriting the partial differential equations algebraically for computer solutions. Numerical simulation of differential equations is the topic of many books and papers. Numerical simulation of reservoirs is the full time occupation of many people. The objective here is to expose the basic techniques used in simulating flow phenomena in porous media, e.g., oil and water reservoirs and the soil. The book by Chrichlow (1977) is devoted entirely to numerical simulation of gas and oil reservoirs. There have been several symposia dealing directly with numerical simulation of reservoirs.

The numerical simulation of partial differential equations by finite differences in the space domain usually uses a rectangular grid system and then the time domain

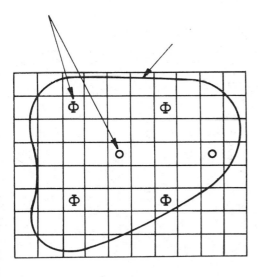

FIG. 6.11 Grid overlay. (From Chrichlow, 1977.)

is broken into a number of time steps. The derivations of various quantities are represented numerically by finite differences. Simulation of the first and second derivatives is needed for most of the equations. The first derivative of pressure in terms of forward differences is

$$\frac{\partial p}{\partial x} = \frac{p(x + \Delta x) - p(x)}{\Delta x} + 0(\Delta x) = \frac{p_{i+1} - p_i}{\Delta x} + 0(\Delta x)$$

(6.119)

where $0(\Delta x)$ represents truncation error. In terms of backward differences

$$\frac{\partial p}{\partial x} = \frac{p(x) - p(x - \Delta x)}{\Delta x} + 0(\Delta x) = \frac{p_i - p_{i-1}}{\Delta x} + 0(\Delta x)$$

(6.120)

In terms of central differences

$$\frac{\partial p}{\partial x} = \frac{p(x + \Delta x) - p(x - \Delta x)}{2 \Delta x} + 0(\Delta x)^2 = \frac{p_{i+1} - p_{i-1}}{2 \Delta x} + 0(\Delta x)^2$$

(6.121)

The second derivative is

$$\frac{\partial^2 p}{\partial x^2} = \frac{p(x + \Delta x) - 2p(x) + p(x - \Delta x)}{(\Delta x)^2} + 0(\Delta x)^2 = \frac{p_{i+1} - 2p_i + p_{i-1}}{(\Delta x)^2} + 0(\Delta x)^2$$

(6.122)

The simulation in time and space is formulated explicitly (in terms of calculated values at the previous time step) or implicitly (in terms of values to be calculated in the current time step). Suppose we are interested in solutions of the one-dimensional diffusion equation via numerical simulation.

$$\frac{\partial c}{\partial t} = \mathcal{D} \frac{\partial^2 c}{\partial x^2}$$

(6.123)

The analytical solution of Eq. (6.123), $c(x, t)$ is continuous in space and time. The numerical solution is discretized giving values of c at specific values of x and t. Assume the values at a given time, $t = 0$, are used to compute $c = (x, 1)$ at the next time, $t = 1$; then the values of $t = 1$ are used to compute $c(x, 2)$ at the next time, $t = 2$; etc. We proceed by calculating at each grid point from $x = 0$ to $x = L$ for a given time value—this is an explicit procedure. The result is a sequential solution

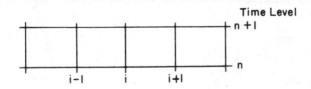

FIG. 6.12 Space grid at two time levels. (From Chrichlow, 1977.)

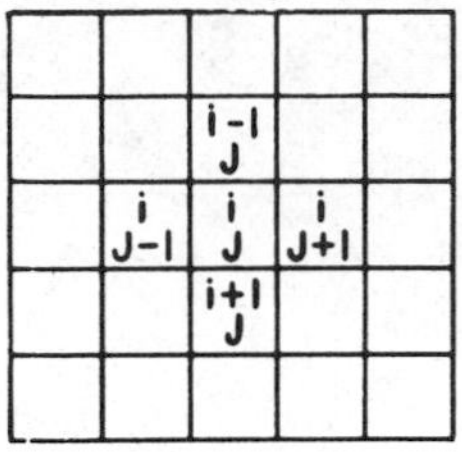

FIG. 6.13 Indices in two directions.
(From Crichlow, 1977.)

of one equation in one unknown. Figure 6.12 is a schematic of a space grid at two time levels. The finite difference form of Eq. (6.123) is

$$\frac{c_i^{n+1} - c_i^n}{\Delta t} = \varnothing \frac{c_{i+1}^n - 2c_i^n + c_{i-1}^n}{(\Delta x)^2} \tag{6.124}$$

Equation (6.124) can be solved explicitly for the value of the dependent variable at time step $n + 1$ as follows

$$c_i^{n+1} = c_i^n + \varnothing \, \Delta t \frac{c_{i+1}^n - 2c_i^n + c_{i-1}^n}{(\Delta x)^2} \tag{6.125}$$

Figure 6.13 is a space grid in two dimensions. Imagine a stack of these grids corresponding to time levels n, $n + 1$, ... Consider a two-dimensional equation for pressure of the form

$$\frac{1}{\xi} \frac{\partial p}{\partial t} = \frac{\partial^2 p}{\partial x^2} + \frac{\partial^2 p}{\partial y^2} \tag{6.126}$$

The explicit finite difference form of Eq. (6.126) is

$$\frac{1}{\xi} \frac{p_{i,j}^n - p_{i,j}^n}{\Delta t} = \frac{p_{i,j+1}^n - 2p_{i,j}^n + 2p_{i,j-1}^n}{(\Delta x)^2} + \frac{p_{i+1,j}^n - p_{i+1,j}^n + p_{i-1,j}^n}{(\Delta y)^2} \tag{6.127}$$

$$p_{i,j}^{n+1} = p_{i,j}^n + \xi \, \Delta t \left[\frac{p_{i,j+1}^n - 2p_{i,j}^n + 2p_{i,j-1}^n}{(\Delta x)^2} + \frac{p_{i+1,j}^n - 2p_{i,j}^n + p_{i-1,j}^n}{(\Delta y)^2} \right] \tag{6.128}$$

Equations (6.125) and (6.128) each have one unknown. One moves systematically through the grid calculating all of the value of c or p at time step $n + 1$ using the value of t at time step n.

In the implicit procedure the formulation of the space derivatives is in terms of time step $n + 1$ and the unknown values of the dependent variables must be determined simultaneously. Consider Fig. 6.12 again but formulate the difference equation for Eq. (6.123) implicitly

$$\frac{c_i^{n+1} - c_i^n}{\Delta t} = \wp \frac{c_{i+1}^{n+1} - 2c_i^{n+1} + c_{i-1}^{n+1}}{(\Delta x)^2} \qquad (6.129)$$

Rewriting Eq. (6.129) as

$$c_{i-1}^{n+1} - 2c_i^{n+1} + c_{i+1}^{n+1} = \frac{(\Delta x)^2}{\wp \Delta t}(c_i^{n+1} - c_i^n) \qquad (6.130)$$

or

$$c_{i-1}^{n+1} - \left[2 + \frac{(\Delta x)^2}{\wp \Delta t}\right]c_i^{n+1} + c_{i+1}^{n+1} = \frac{(\Delta x)^2}{\wp \Delta t} c_i^n \qquad (6.131)$$

Everything is known in Eq. (6.131) except c at n + 1 and at space points i − 1, i, and i + 1. Equation (6.131) can be written as

$$\alpha_i c_{i-1}^{n+1} + \beta_i c_i^{n+1} + \gamma_i c_{i+1}^{n+1} = \delta_i \qquad (6.132)$$

Writing Eq. (6.132) for N cells in the linear grid of Fig. 6.1 we obtain N equation and N unknowns as shown in Table 6.4. Writing the set of equations in Table 6.4 in matrix form

$$AC = D \qquad (6.133)$$

The matrix A of coefficients is tridiagonal, and the system of equations can be solved using a modified Gaussian elimination technique: the Thomas algorithm. For a two-dimensional problem we use the grid of Fig. 6.13 writing Eq. (6.126) in implicit finite difference form

$$\frac{1}{\xi}\frac{p_{i,j}^{n+1} - p_{i,j}^n}{\Delta t} = \frac{p_{i,j-1}^{n+1} - 2p_{i,j}^{n+1} + p_{i,j+1}^{n+1}}{(\Delta x)^2} + \frac{p_{i+1,j}^{n+1} - 2p_{i,j}^{n+1} + p_{i-1,j}^{n+1}}{(\Delta y)^2} \qquad (6.134)$$

TABLE 6.4

Cell		
1	$\alpha_1 c_0 + \beta_1 c_1 + \gamma_1 c_2$	$= \delta_1$
2	$\alpha_2 c_1 + \beta_2 c_2 + \gamma_2 c_3$	$= \delta_2$
3	$\alpha_3 c_2 + \beta_3 c_3 + \gamma_3 c_4$	$= \delta_3$
$\cdots$	$\cdots \cdots \cdots \cdots \cdots \cdots$	$\cdots$
N	$\alpha_N c_{n-1} + \beta_N c_n + \gamma_N c_{n+1}$	$= \delta_N$

It is less complex to use a grid such that $\Delta x = \Delta y$ and rewrite Eq. (6.134) similarly to Eq. (6.131):

$$p_{i,j-1}^{n+1} - p_{i-1,j}^{n+1} - \left(4 + \frac{(\Delta x)^2}{\zeta \Delta t}\right) p_{i,j}^{n+1} + p_{i+1,j}^{n+1} + p_{i,j+1}^{n+1} = \frac{(\Delta x)^2}{\xi \Delta t}\left(p_{i,j}^{n+1} - p_{i,j}^n\right) \qquad (6.135)$$

Rewriting Eq. (6.135) similar to Eq. (6.132)

$$\epsilon_i p_{i,j-1}^{n+1} + \alpha_i p_{i-1,j}^{n+1} + \beta_i p_{i,j}^{n+1} + \gamma_i p_{i-1,j}^{n+1} + \mu_{i,j+1}^{n+1} = \delta_i \qquad (6.136)$$

The matrix form of Eq. (6.136) is

$$AP = D \qquad (6.137)$$

If we wrote out each equation as we did in Table 6.4, we would find the matrix Λ is pentadiagonal. This equation system has N equations and N unknowns.

There is a half-way approach between explicit and implicit formulation. The Crank-Nicholson procedure combines both the n and n + 1 time steps in calculating the space derivative. For Eq. (6.123) with a grid of Fig. 6.12

$$\frac{c_i^{n+1} - c_i^n}{\Delta t} = \emptyset\, \frac{c_{i+1}^{n+1} - 2c_i^{n+1} + c_{i-1}^{n+1}}{2(\Delta x)^2} + \frac{c_{i+1}^n - 2c_i^n + c_{i-1}^n}{2(\Delta x)^2} \qquad (6.138)$$

This scheme obtains a space derivative at the n + 1/2 time step. Equation (6.123) is written in finite difference form in general as

$$\frac{1}{\emptyset}\frac{c_1^{n+1} - c_i^n}{\Delta t} = \ell\, \frac{c_{i+1}^{n+1} - 2c_1^{n+1} + c_{i-1}^{n+1}}{(\Delta x)^2} + (1 - \ell)\, \frac{c_{i+1}^n - 2c_i^n + c_{i-1}^n}{(\Delta x)^2} \qquad (6.139)$$

If $\ell = 0$, the procedure is explicit. If $\ell = 1/2$, the procedure is Crank-Nicholson. If $\ell = 1$ the procedure is implicit.

Explicit methods usually require more computer time than implicit methods. Both methods are stable with appropriate grid sizes. The finite difference formulation uses either a block-centered grid or a lattice network (see Fig. 6.14). With the block-centered grid the parameters needed in the quations (permeability,

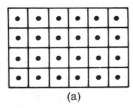   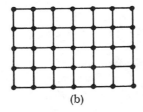

(a)   (b)

FIG. 6.14 (a) Block-centered grid; (b) lattice-centered grid. (From Chrichlow, 1977.)

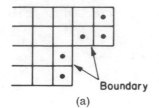   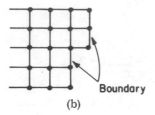

Boundary   Boundary
(a)   (b)

FIG. 6.15 (a) Neumann boundary; (b) pressure specified (p(i, j; t) = P_e). (From Crichlow, 1977.)

porosity, density, etc.) are calculated at the center of the block. With a lattice network the parameters are calculated at the intersections of the lattice. For the block-centered grid flow across a boundary is used as a boundary condition, and the equations for the boundary blocks include source or sink terms (see Fig. 6.15). With the lattice the boundary conditions are specified at the lattice points and the lines connecting them (see Fig. 6.15). Spacing in the grid may be irregular in that $\Delta x_i \neq \Delta x_{i+1}$ or $\Delta x \neq \Delta y$ or both. Spacing depends on the problem to be solved.

To include sources and sinks (wells) in the reservoir simulation problem, input or output must be added or subtracted in the appropriate grid blocks. The pressure for the block must also be adjusted for the buildup or drawdown due to the flow into or out of the block. Since blocks are usually square (or cubic) and wells are cylindrical an equivalent radius for the block is used

$$r = \sqrt{\frac{\Delta x\, \Delta y}{\pi}} \qquad\qquad (6.140)$$

For a well penetrating n blocks the total rate is the sum of the rates in each block accounting for the change in pressure using a radius of drainage based on steady flow [see, for example, Jenkins and Aronofsky (1953)]. For oil, water, and gas

$$q_o = \sum_{i=1}^{n}\left[\frac{7.07h}{\ln (r_e/r_w) - 3/4 + s}\frac{\lambda_o}{B_o}(p_e - p_{wi})\right]_j \qquad (6.141)$$

$$q_w = \sum_{i=1}^{n}\left[\frac{7.07h}{\ln (r_e/r_w) - 3/4 + s}\frac{\lambda_w}{B_w}(p_e - p_{wi})\right]_j \qquad (6.142)$$

$$q_g = \sum_{i=1}^{n}\left[\frac{7.07h}{\ln (r_e/r_w) - 3/4 + s}\frac{\lambda_o R_s + \lambda_g}{B_g}(p_e - p_{wi})\right]_j \qquad (6.143)$$

where there are n blocks penetrated and j wells. s is skin effect, r_e is the equivalent radius of the block containing the well, and p_e is the average pressure in the well. If the cells are small, the boundary pressure should include the effects of adjacent blocks: using a weighting of 0.6,

$$p_{ei} = \bar{p}_i + 0.6 \frac{\left[\sum_{j=1}^{4} (h\lambda)(\bar{p}_j - \bar{p}) \right]_i}{\sum_{j=1}^{4} (h\lambda)_i} \tag{6.144}$$

The total flow rate for a well using N blocks of the grid is

$$q_{oT} = \sum_{i=1}^{N} \left[\frac{7.07h}{\ln(r_e/r_w) - 3/4 + s} \frac{m_o}{B_o}(p_e - p_{wi}) \right] \tag{6.145}$$

Equations for water and gas are given by Eqs. (6.142) and (6.143). The production rates for each layer are needed to calculate the new pressure and saturations. We are usually given the total oil production q_o^{n+1}. Then

$$q_{oi}^{n+1} = q_{oT}^{n+1} \left[\frac{(\lambda_o/B_o)_i}{\sum_{j=1}^{N} (\lambda_o/B_o)_j} \right]^n \tag{6.146}$$

$$q_{wi}^{n+1} = \left(\frac{\lambda_w/B_w}{\lambda_o/B_o} \right)^n q_{oi}^{n+1} \tag{6.147}$$

$$q_{gi}^{n+1} = \left(\frac{\lambda_g/B_g}{\lambda_o/B_o} \right)^n q_{oi}^{n+1} + \left(\frac{R_s}{B_g} \right)^n q_{oi}^{n+1} \tag{6.148}$$

For the implicit procedure the ratios of mobility and formation volume factor are evaluated from the slope of the plot of λ/B versus saturation at the $n + 1/2$ time step.

The unknowns in simulation of oil reservoirs are usually pressure and saturation for oil, water, and gas. The simulator selected depends on the particular system modeled—whether it is areal, vertical (coning), contains single or multiple wells, and so on. Two approaches are used to simulate the equations of change. In the implicit pressure/explicit saturation procedure the pressure equations are solved implicitly; then saturations are found explicitly using the pressures. In the implicit pressure/implicit saturation procedure, saturation is expressed in terms of capillary pressure; then the pressures are found implicitly, capillary pressure is determined, and the saturation is computed implicitly.

Consider the simultaneous flow of oil, water, and gas in one dimension through a constant area. (See Chapter 4. If area varies, it would be included inside the first space derivative.)

$$V_r \frac{\partial}{\partial t}\left(\phi \frac{S_o}{B_o} \right) = A_x \frac{\partial}{\partial x}\left(\frac{k_o}{\mu_o B_o} \frac{\partial \Phi_o}{\partial x} \right) + q_o \tag{6.149}$$

$$V_r \frac{\partial}{\partial t}\left(\phi \frac{S_w}{B_w} \right) = A_x \frac{\partial}{\partial x}\left(\frac{k_w}{\mu_w B_w} \frac{\partial \Phi_w}{\partial x} \right) + p_w \tag{6.150}$$

$$V_r \frac{\partial}{\partial t}\left[\phi\left(\frac{S_g}{B_g} + \frac{R_{so}S_o}{B_o} + \frac{R_{sw}S_w}{B_w}\right)\right] = A_x \frac{\partial}{\partial x}\left[\left(\frac{k_g}{\mu_g R_g} + \frac{R_{so}k_o}{\mu_o B_o} + \frac{R_{rw}k_w}{\mu_w B_w}\right)\frac{\partial \Phi_q}{\partial x}\right] + q_g \tag{6.151}$$

where A_x is area perpendicular to the x direction and q_o, q_w, and q_g are source or sink terms. With the usual definition that $\Phi_i = p_i + \rho_i g_o$, $p_{cw} = p_o - p_w$, $p_{cg} = p_g - p_o$, and letting $\lambda = (k/\mu B)_i$, then substituting Eqs. (6.149) and (6.150) into Eq. (6.151),

$$\frac{R_T}{A_x B_T}\frac{\partial p_o}{\partial t} = \frac{\partial}{\partial x}\left(\lambda_T \frac{\partial p_o}{\partial x}\right) + \frac{\partial}{\partial x}\left(\lambda_g \frac{\partial p_{cg}}{\partial x} - \lambda_w \frac{\partial p_{cw}}{\partial x}\right)$$

$$+ \frac{\partial}{\partial x}\left(\lambda_g \frac{\partial(\rho_g gz)}{\partial x} + \lambda_o \frac{\partial(\rho_o gz)}{\partial x} + \lambda_w \frac{\partial(\rho_w gz)}{\partial x}\right) + \frac{q_T}{A_x} \tag{6.152}$$

Assume λ_i, p_{cg}, and p_{cw} are either evaluated at previous pressures and saturations or they are iteratively updated. Usually the particular application will determine which approach; for normal simulation evaluating at previous pressures and saturations is adequate. Equation (6.152) is written in finite difference form either for direct solution or by an iterative process. Direct solutions are simultaneous solutions of the pressure equations for each block of the grid by matrix inversion using some technique. Iterative solution involves continuous recalculation relaxing through the entire grid refining the result with each iteration. The form of the difference equation of Eq. (6.152) will depend on the procedure selected. Once the pressure distribution is determined the potential distribution can be calculated and the saturation at the new time level is calculated. The explicit saturation equation, Eq. (6.149) for oil is

$$\left(\frac{S_o}{B_o}\right)^{n+1} = \left(\frac{S_o}{B_o}\right)^n + \frac{\Delta t}{\phi}\left[\frac{\partial}{\partial x}\left(\frac{\lambda_o}{B_o}\frac{\partial \phi_o}{\partial x}\right)\right] + q_o \tag{6.153}$$

and likewise for water and gas.

The implicit pressure/implicit saturation method involves simultaneous solution for pressure and saturation (Douglas et al., 1959). For incompressible two-fluid immiscible flow in one dimension,

$$\phi \frac{\partial S_o}{\partial t} = \frac{\partial}{\partial x}\left(\lambda_o \frac{\partial \Phi_o}{\partial x}\right) \tag{6.154}$$

$$\phi \frac{\partial S_w}{\partial t} = \frac{\partial}{\partial x}\left(\lambda_w \frac{\partial \Phi_w}{\partial x}\right) \tag{6.155}$$

$$S_o = 1 - S_w \tag{6.156}$$

$$\frac{\partial S_o}{\partial t} = -\frac{\partial S_w}{\partial t} \tag{6.157}$$

For capillary pressure

$$p_c = \Phi_o - \Phi_w + \Delta\rho g z \qquad (6.158)$$

Since $S_w = S_w(p_c)$

$$\frac{dS_w}{dt} = \frac{\partial S_w}{\partial p_c} \frac{dp_c}{dt} = \tilde{S}_w \frac{dp_c}{dt} \qquad (6.159)$$

Writing Eq. (6.159) in terms of Φ_o and Φ_w

$$\frac{dS_w}{dt} = \tilde{S}_w \left(\frac{\partial \Phi_o}{\partial t} - \frac{\partial \Phi_w}{\partial t} \right) \qquad (6.160)$$

Rewriting Eqs. (6.154) and (6.155) using Eq. (6.160),

$$\phi \tilde{S}_w \left(\frac{\partial \Phi_w}{\partial t} - \frac{\partial \Phi_o}{\partial t} \right) = \frac{\partial}{\partial x} \left(\lambda_o \frac{\partial \Phi_o}{\partial x} \right) \qquad (6.161)$$

$$\phi \tilde{S}_w \left(\frac{\partial \Phi_o}{\partial t} - \frac{\partial \Phi_w}{\partial t} \right) = \frac{\partial}{\partial x} \left(\lambda_w \frac{\partial \Phi_w}{\partial x} \right) \qquad (6.162)$$

$\tilde{S}_w$ is approximated as

$$\tilde{S}_w = \frac{S_w^{n+1} - S_w^n}{\Phi_o^{n+1} - \Phi_w^n} \qquad (6.163)$$

Writing Eqs. (6.161) and (6.162) in implicit form,

$$\frac{\phi \tilde{S}_w}{\Delta t} \left\{ \left[(\Phi_w)_i^{n+1} - (\Phi_w)_i^n \right] - \left[(\Phi_o)_i^{n+1} - (\Phi_o)_i^n \right] \right\}$$

$$= \frac{1}{\Delta x} \left[(\lambda_o)_{i+1/2} \frac{(\Phi_o)_{i+1}^{n+1} - (\Phi_o)_i^{n+1}}{\Delta x} - (\lambda_o)_{i-1/2} \frac{(\Phi_o)_i^{n+1} - (\Phi_o)_{i-1}^{n+1}}{\Delta x} \right] \qquad (6.164)$$

$$\frac{\phi \tilde{S}_w}{\Delta t} \left\{ \left[(\Phi_o)_i^{n+1} - (\Phi_o)_i^n \right] - \left[(\Phi_w)_i^{n+1} - (\Phi_w)_i^n \right] \right\}$$

$$= \frac{1}{\Delta x} \left[(\lambda_w)_{i+1/2} \frac{(\Phi_w)_{i+1}^{n+1} - (\Phi_w)_i^{n+1}}{\Delta x} - (\lambda_w)_{i-1/2} \frac{(\Phi_w)_i^{n+1} - (\Phi_w)_{i-1}^{n+1}}{\Delta x} \right] \qquad (6.165)$$

Writing Eqs. (6.164) and (6.165) as linear algebraic equations

TABLE 6.5

Cell		

1
$$\alpha_1(\Phi_0)_0 + \beta_1(\Phi_0)_1 + \gamma_1(\delta_0)_2 + \delta(\Phi_w)_1 = (\epsilon_0)_1$$

$$\tilde{\alpha}_1(\Phi_w)_0 + \tilde{\beta}_1(\Phi_w)_1 + \tilde{\gamma}_1(\Phi_w)_2 + \tilde{\delta}(\Phi_0)_1 = (\epsilon_w)_1$$

2
$$\alpha_2(\Phi_0)_1 + \beta_2(\Phi_0)_2 + \gamma_2(\Phi_w)_3 + \delta_2(\Phi_w)_2 = (\epsilon_0)_2$$

$$\tilde{\alpha}_2(\Phi_w)_1 + \tilde{\beta}_2(\Phi_w)_2 + \tilde{\gamma}_2(\Phi_w)_3 + \tilde{\delta}_2(\Phi_0)_2 = (\epsilon_w)_2$$

. . . .

N
$$\alpha_N(\Phi_0)_{N-1} + \beta_N(\Phi_0)_N + \gamma_N(\Phi_0)_{N+1} + \delta_N(\Phi_w)_N = (\epsilon_0)_N$$

$$\tilde{\alpha}_N(\Phi_0)_{N-1} + \tilde{\beta}_N(\Phi_0)_N + \tilde{\gamma}_N(\Phi_0)_{N+1} + \tilde{\delta}_N(\Phi_w)_N = (\epsilon_w)_N$$

$$\alpha(\Phi_0)_{i-1}^{n+1} + \beta(\Phi_0)_i^{n+1} + \gamma(\Phi_0)_{i+1}^{n+1} + \delta(\Phi_w)_i^{n+1} = (\epsilon_0)_i \qquad (6.166)$$

and

$$\tilde{\alpha}(\Phi_w)_{i-1}^{n+1} + \tilde{\beta}(\Phi_w)_i^{n+1} + \tilde{\gamma}(\Phi_w)_{i+1}^{n+1} + \tilde{\delta}(\Phi_0)_i^{n+1} = (\epsilon_w)_i \qquad (6.167)$$

Writing Eqs. (6.166) and (6.167) for each of N cells in the grid results in a system of equations as in Table 6.5. Normally the boundary conditions are imposed so that for $i = 1$, $i - 1 = 0$ and for $i = N$, $i + 1 = N + 1 = 0$; so these terms are dropped from the above equations as in Table 6.5. The matrix solution of the system of the equations in Table 6.5 is solved simultaneously using a procedure similar to the Thomas algorithm. Once values of $(\Phi_w)_i$ and $(\Phi_0)_i$ are determined $(p_c)_i$ can be determined and saturation is found from the p_c's. In the computation flow is usually calculated as

$$q = \lambda_i(p_i - p_{i-1}) \qquad (6.168)$$

It is inferred in the above discussion that the pressure equations are solved directly or iteratively. Direct solution is usually by matrix solution of the equation. [See, for example, Varga (1962).] The actual method of inversion of the coefficient matrix depends on the particular problem. For diagonal matrices Gaussian elimination is used on the matrix equation

$$AP = D \qquad (6.169)$$

The Gauss–Jordon method is a modification of Gaussian elimination and is also used to solve Eq. (6.169). Matrix decomposition, that is, tearing the matrix into several smaller matrices before solving, speeds up solution.

$$AP = (BC)P = D \tag{6.170}$$

where A is torn into B and C.

In an iterative process Eq. (6.169) is written as

$$p_1 = \frac{1}{a_{11}}[d_1 - (a_{12}p_2 + a_{13}p_3 + \cdots + a_{1n}p_n] \tag{6.171}$$

$$p_2 = \frac{1}{a_{22}}[d_2 - (a_{21}p_1 + a_{23}p_3 + \cdots + a_{2n}p_n)] \tag{6.172}$$

$$\vdots$$

$$p_n = \frac{1}{a_{nn}}[d_n - (a_{n1}p_1 + a_{n2}p_2 + \cdots + a_{n,n-1}p_{n-1})] \tag{6.173}$$

The values of $p_i \equiv P$ are estimated iteratively and the equations are repeatedly solved iterating to a solution.

Jacobi and Gauss-Seidel methods are similar. The Jacobi method uses the value of p from the previous values. The Gauss-Seidel uses the current values of p. The Gauss-Seidel algorithm is

$$p_1^k = \frac{1}{a_{ii}}\left(d_1 - \sum_{\substack{j=1 \\ j \neq i}}^{n} a_{1j}p_j \left\{ \begin{matrix} k & j < i \\ k-1 & j > i \end{matrix} \right\} \right) \qquad i = 1, 2, \ldots, n \tag{6.174}$$

Equation (6.174) is used for the Jacobi method except the exponent of p_j is always $k - 1$.

Relaxation is a method where the current values are modified to speed convergence. The residual is defined as

$$r_i^{k-1} = d_i - \sum_{j=1}^{h} a_{ij}p_i^{k-1} \qquad i = 1, 2, \ldots, n \tag{6.175}$$

p_i is overcorrected by

$$p_i^k = p_i^{k-1} + w \frac{r_i^{k-1}}{a_{ii}} \qquad i = 1, 2, \ldots, n \tag{6.176}$$

The alternating-direction implicit procedure (ADIP) (Peaceman and Rachford, 1955; Douglas and Rachford, 1956), successive overrelaxation (SOR) (Young, 1954), and the strongly implicit procedure (SIP) (Stone, 1968; Weinstein et al., 1969) are iterative procedures used in solving the pressure equations. In the alternating-direction implicit procedure, the time steps are divided into two equal substeps. During the first substep the grid is solved in the x direction one row at a time for

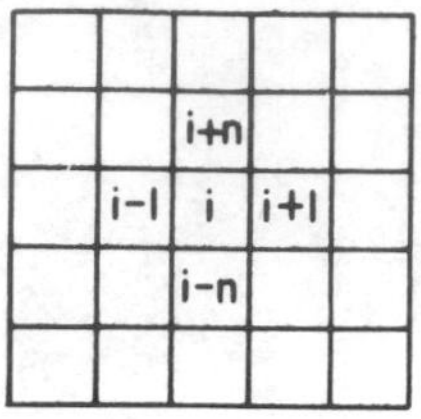

FIG. 6.16 Alternating direction implicit grid. (From Douglas et al., 1959.)

the unknown pressures. Then the grid is solved in the y direction one column at a time for the pressures. The appropriate equations describing the two-dimensional pressure distribution are Eqs. (6.126), (6.134), and (6.136). This is a system of N × N linear equations of pentadiagonal form. In ADIP for the first half-time step Eq. (6.136) is written for the x direction using the grid of Fig. 6.18.

$$\alpha_i p_{i-1}^{n+1/2} + \beta_i p_i^{n+1/2} + \gamma_i p_{i+1}^{n+1/2} = \delta_i - (\epsilon_i p_{i-\eta}^{n+1/2} - \mu_i p_{i+\eta}^{n}) \qquad (6.177)$$

Equation (6.177) is tridiagonal and can be solved using the Thomas algorithm. In the second half-time step p_{i-1}^{n+1} and p_{i+1}^{n+1} are assumed known from the previous half-time step calculation and Eq. (6.136) is written for the y direction using the grid of Fig. 6.16.

$$\epsilon_i p_{i+\eta}^{n+1} + \tilde{\beta}_i p_i^{n+1} + \mu_i p_{i-\eta}^{n+1} = \delta_i - (\alpha_i p_{i-1}^{n+1} + \gamma_i p_{i+1}^{n+1/2}) \qquad (6.178)$$

Equation (6.178) is a wide tridiagonal and can be calculated with a modified Thomas algorithm. At the end of the two half steps the new value of the pressure at time n + 1 is obtained. Solving the problem by this two-step process is much faster than solving the pentadiagonal equations of the original matrix. The ADIP procedure is used with an acceleration parameter with the Crank-Nicholson formulation. An acceleration parameter ω is added multiplicatively to the pressure term, resulting in $\omega(p^{k+1/2} - p^{k})$ and $\omega(p^{k-1} - p^{k+1/2})$. Optimum values of ω are determined sequentially as explained by Varga (1962).

Point or line successive overrelaxation (SOR) is applied sequentially on a grid. The fully implicit formulation of the pressure equation at the n + 1 time level of Eq. (6.126) is given by Eq. (6.134). Using the grid of Fig. 6.17, the pressure values are solved on a line assuming the i - η terms are known and the values of i + η are used from the previous iteration.

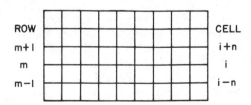

FIG. 6.17 Successive over relaxation. (From Young, 1954.)

$$p_{i+1}^{\substack{k+1 \\ n+1}} - (4 + C)p_i^{\substack{k+1 \\ n+1}} + p_{i-1}^{\substack{k+1 \\ n+1}} = p_{i-\eta}^{\substack{k+1 \\ n+1}} - p_{i+\eta}^{\substack{k+1 \\ n+1}} - Cp_i^n \qquad (6.179)$$

where k is iteration, n is time, i is grid, and

$$C = \frac{(\Delta x)^2}{\xi \, \Delta t} \qquad (6.180)$$

Equation (6.179) is tridiagonal. An iteration parameter (Varga, 1962) can be used to speed up the relaxation after each line computation by overrelaxing, so that

$$p_i^{k+1} = p_i^k + \omega (p_i^{k+1} - p_i^k) \qquad (6.181)$$

The strongly implicit procedure (SIP) has a fast convergence and is not sensitive to iteration parameters. In this procedure the simultaneous solution of the pressure equation, Eq. (6.136) is solved using a modified matrix approach.

$$AP - A'P - (BC)P = D \qquad (6.182)$$

Using matrix notation this time

$$A'P^{k+1} - A'P^k = AP^k - D^k \qquad (6.183)$$

$$\Delta^{k+1} = P^{k+1} - P^k \qquad (6.184)$$

$$A'\Delta^{R+1} = R^k \qquad (6.185)$$

Transferring A' to BC

$$HC \, \Delta^{k+1} = R^k \qquad (6.186)$$

$$BV = R^k \qquad (6.187)$$

then

$$C \, \Delta^{k+1} = V \qquad (6.188)$$

The new pressure increment is calculated from Eq. (6.188). A set of nine evenly spaced parameters is used as acceleration parameters. SIP is better than most iterative schemes except for simple homogeneous and isotropic problems.

Another numerical technique for solving the partial differential equations of flow in porous media is the finite element method (Mitchell and Wait, 1977). Essentially these procedures assume solutions of the partial differential equations are piecewise continuous polynomials. Consider the problem of approximating a real valued function f(x) over a finite interval of the x axis. Break the interval into a number of nonoverlapping subintervals and interpolate linearly between values of

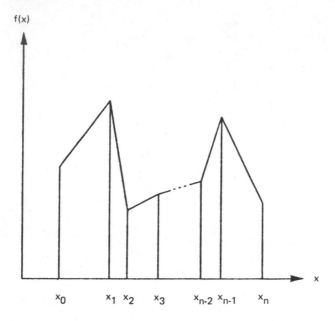

FIG. 6.18 Basis function. (From Mitchell and Wait, 1977.)

f(x) at the end points of each of the subinterval as in Fig. 6.18. If there are n sub-intervals $[x_i, x_{i+1}]$, $i = 0, 1, 2, \ldots, n - 1$, then the piecewise approximating func-tion depends only on the function values $f_i = f(x_i)$, $i = 0, 1, 2, \ldots, n$. If f(x) is given implicitly by a differential equation, the values of f_i are the unknown parame-ters of the problem. (For interpolation the f_i are known in advance.)

For the interval $[x_i, x_{i+1}]$ the linear approximating function is

$$p_1^{(i)}(x) = \alpha_i(x)f_i + \beta_{i+1}(x)f_{i+1} \qquad x_i < x < x_{i+1} \qquad (6.189)$$

where

$$\alpha_i(x) = \frac{x_{i+1} - x}{x_{i+1} - x_i} \qquad i = 0, 1, a, \ldots, n - 1 \qquad (6.190)$$

$$\beta_{i+1}(x) = \frac{x - x_i}{x_{i+1} - x_i} \qquad i = 0, 1, 2, \ldots, n - 1 \qquad (6.191)$$

The piecewise approximating function of $x_0 \le x \le x_n$ is

$$p_1(x) = \sum_{i=0}^{n} \phi_i(x)f_i \qquad (6.192)$$

where

$$\phi_0(x) = \begin{cases} \dfrac{x_i - x}{x_i - x_0} & x_0 \le x \le x_1 \\[2em] 0 & x_1 \le x \le x_n \end{cases}$$

(6.193)

$$\phi_i(x) = \begin{cases} 0 & x_0 \le x \le x_{i-1} \\[1em] \dfrac{x - x_{i-1}}{x_i - x_{i-1}} & x_{i-1} \le x \le x_i \\[2em] \dfrac{x_{i+1} - x}{x_{i+1} - x_i} & x_i \le x \le x_{i+1} \\[2em] 0 & x_{i+1} \le x \le x_n \end{cases}$$

(6.194)

$$\phi_n(x) = \begin{cases} 0 & x_0 \le x \le x_{n-1} \\[1em] \dfrac{x - x_{n-1}}{x_n - x_{n-1}} & x_{n-1} \le x \le x_n \end{cases}$$

(6.195)

The pyramid function of Eqs. (6.193) to (6.195) shown in Fig. 6.19 is a basis function. The first derivatives of the piecewise continuous approximating polynomial, Eq. (6.189) are not the same as for $f(x)$. To fit the nodes plus first- or higher order derivatives, We must construct a different approximating function. A quadratic spline basis function fits the node and its first derivative. A cubic spline basis function fits the node, first, and second derivatives.

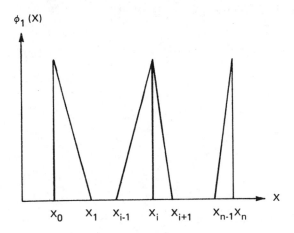

FIG. 6.19 Basis function. (From Mitchell and Wait, 1977.)

Price et al. (1968) used the variational method to simulate the dimensionless one-dimensional dispersion equation

$$\frac{\partial c}{\partial t} = \frac{\partial^2 c}{\partial x^2} - v \frac{\partial c}{\partial x} \qquad \alpha > 0 \qquad 0 \leq x \leq \ell \qquad t > 0 \tag{6.196}$$

where v is a dimensionless velocity, with initial and boundary conditions

$$c(x, 0) = 0 \qquad 0 < x < \ell \tag{6.197}$$

$$c(0, 1) = 1 \qquad t > 0 \tag{6.198}$$

$$\frac{\partial c(\ell, t)}{\partial x} = 0 \qquad t > 0 \tag{6.199}$$

For $0 < x < 1$, $0 < \ell < \infty$, and $R = \{x, t\}$,

$$L[c] = \frac{\partial u}{\partial t} - \frac{\partial^2 u}{\partial x^2} + v \frac{u}{x} = 0 \qquad (x, \ell) \in R \tag{6.200}$$

with initial and boundary conditions given by Eqs. (6.197) to (6.199) in terms of u. To start, the space variable is made discrete using a Galerkin process. The resulting ordinary differential equation is then discretized in time. Let S be the class of all real piecewise-continuous differentiable functions $w(x)$ on $(0, 1)$ such that $w(0) = 0$, $(dw/dx)_{x=1} = 0$. S_m is an m-dimensional subspace of S spanned by the m basis functions $\{w_i(x)\}_{i=1}^m$. The approximation of Eq. (6.196) and associated initial and boundary conditions is

$$u_m(x, t) = \sum_{R=1}^{m} C_{m,k}(t) w_k(x) + w_0(x) \tag{6.201}$$

where $C_{m,k}(t)$ are determined by the conditions that $L[u_m]$ is orthogonal to S_m for all $t > 0$ and u_m satisfies the initial and boundary conditions so that

$$\int_0^1 L[u_m] w_k \, dx = 0 \qquad k = 1, 2, \ldots, m \qquad t > 0 \tag{6.202}$$

$$w_0(0) = 1.0 \tag{6.203}$$

$$\frac{dw_0(1)}{dx} = 0 \tag{6.204}$$

$$\sum_{k=1}^{m} C_{m,k}(t) \left(\frac{dw_k(x)}{dx} \right)_{x=1} = 0 \tag{6.205}$$

The coefficients $C_{m,k}(0)$ are uniquely determined so that

$$\|w_0(x) + \sum_{k=1}^{m} C_{m,k}(t) w_R(x)\|_{L^2} = \min_{\alpha_i} \|w_0(k) + \sum_{k=1}^{m} \alpha_k w_k(x)\|_{L^2} \qquad (6.206)$$

Equation (6.202) has the form

$$\int_0^1 \sum_{j=1}^{m} [C'_{m,j}(t) w_j(x) - C_{m,j}(t) w''_m(x) + \alpha C_{m,j}(t) w'_j(x) \; w_k(x)] \, dx$$

$$= \int_0^1 w''_0(x) w_k(x) \, dx - \alpha \int_0^1 w'_0(x) w_k(x) \, dx \qquad (6.207)$$

Integrating by parts recognizing

$$w_k(0) = \frac{dw_k(1)}{dx} = 0 \qquad k = 1, 2, \ldots, m \qquad (6.208)$$

yields

$$\sum_{j=1}^{m} [\, C'_{m,j}(t) \int_0^1 w'_j(x) w_k(x) \, dx + C_{m,j}(t) \int_0^1 w'_j(x) w'_R(x) \, dx$$

$$+ \alpha C_{m,j}(t) \int_0^1 w'_j(x) w_k(x) \, dx]$$

$$= -\int_0^1 w'_0(x) [(\alpha w_R(x) + w'_R(x)] \, dx \qquad k = 1, 2, \ldots, m \qquad (6.209)$$

The interval $0 \leq x \leq \gamma$ is divided into m mesh blocks of length h and the chapeau basis functions are

$$w_0(x) = \begin{cases} \dfrac{-(x - h)}{h} & 0 \leq x \leq h \\[2mm] 0 & h \leq x \leq 1 \end{cases} \qquad (6.210)$$

$$w_i(x) = \begin{cases} \dfrac{[x - (i - 1)h]}{h} & (i - 1)h \leq x \leq ih \\[2mm] \dfrac{[(i + 1)h - x]}{h} & ih \leq x \leq (i + 1)h \qquad i = 1, 2, \ldots, m - 2 \qquad (6.211) \\[2mm] 0 & \text{elsewhere} \end{cases}$$

$$
w_{m-1}(x) = \begin{cases} \dfrac{[x - (m - 2)h]}{h} & (m - 2)h \le x \le (m - 1)h \\[2mm] \dfrac{(mh - x)^2}{h^2} & (m - 1)h \le x \le mh = 1 \\[2mm] 0 & 0 \le x \le (m - 2)h \end{cases} \qquad (6.212
$$

$$
w_m(x) = \begin{cases} \dfrac{-[(x - mh)^2 - h^2]}{h^2} & (m - 1)h \le x \le mh = 1 \\[2mm] 0 & 0 \le x \le (m - 1)h \end{cases} \qquad (6.213)
$$

Figure 6.20 is a schematic of the chapeau basis functions. Carrying out the integration of Eq. (6.209) we obtain, in matrix form

$$
B \frac{d \underline{Cm}}{dt} = -A \underline{Cm} + \underline{S} \qquad (6.214)
$$

where

$$
B = h \begin{bmatrix} \dfrac{2}{3} & \dfrac{1}{6} & & & & & \\[2mm] \dfrac{1}{6} & \dfrac{2}{3} & \dfrac{1}{6} & & & & \\[2mm] & \cdot & \cdot & \cdot & \cdot & \cdot & \\[2mm] & & \dfrac{1}{6} & \dfrac{2}{3} & \dfrac{1}{6} & & \\[2mm] & & & \dfrac{1}{6} & \dfrac{8}{15} & \dfrac{2}{15} & \\[2mm] & & & & \dfrac{2}{15} & \dfrac{8}{15} \end{bmatrix} \qquad (6.215)
$$

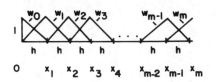

FIG. 6.20 Chapeau basis functions. (From Price et al., 1968.)

$$a = \frac{1}{h} \begin{bmatrix} 2 & -(1-\gamma) & & & & \\ -(1+\gamma) & 2 & -(1-\gamma) & & & \\ & -(1+\gamma) & 2 & -(1-\gamma) & & \\ & & -(1+\gamma) & \frac{7}{3} & -\left(\frac{4}{3}-\gamma\right) \\ & & & -\left(\frac{4}{3}-\gamma\right) & \frac{4}{3}+\gamma \end{bmatrix} \qquad (6.216)$$

$$\underline{S} = \frac{1}{h} \begin{bmatrix} 1+\gamma \\ 0 \\ \vdots \\ 0 \end{bmatrix} \qquad (6.217)$$

where $\gamma = \alpha h/2$. The Crank-Nicholson approximation is used for the time derivative. Price et al. (1968) determined solutions to the dispersion equation that compared favorably with the analytical solution and finite difference methods.

The method of collocation is similar to the Galerkin method. It involves selecting coefficients in the approximation

$$p_1(x) = \sum_{i=1}^{n} \phi_i(x) f_i \qquad (6.218)$$

such that the differential equation is satisfied exactly at certain points. The accuracy is similar to the Galerkin method. If we use Hermite piecewise polynomials of degree $7r-1$ as the basis, then the collocation points in each subinterval $[x_i, x_{i+1}]$ are the zeros of the Legendre polynomial

$$L(x) = Pr\left(\frac{2x - x_{i+1} - x_1}{x_{i+1} - x_i}\right) \qquad (6.219)$$

There are no inner products. The resulting algebraic equations have fewer terms. The disadvantage of collocation is it is necessary to use basis functions of at least degree $2k$ for differential equations of order $2k$.

Another approach to solving the equations numerically is the procedure of Higgins and Leighton (1962) which uses a stream tube model based on equipotential results and calculates displacement through the stream tubes via computer. Martin and Wegner (1978) used this method with the Buckley-Leverett procedure to calculate multiphase two-dimensional incompressible flow.

APPENDIX: NUMERICAL SOLUTIONS

Explicit or implicit methods for the one-dimensional convective diffusion equation
are unsatisfactory if the coefficient of the second space derivative is small compared to the coefficient of the first space derivative. Peaceman and Rachford (1962)
observed that when the one-dimensional convective diffusion equation is solved by
using the backward-in distance approximation, the numerical solution overshoots
in the neighborhood of the sharp change in concentration. This overshoot is the
result of distance truncation error.

Lantz (1971) quantitatively evaluated the value of the numerical dispersion
due to this truncation error. Over a wide range of spatial and time steps the truncation error expression is presented and can provide a guideline for choosing the
spatial and time increments such that the effect of numerical dispersion can be
minimized. Chaudhari (1971) added a negative dispersion term to the continuity
equation to account for the numerical dispersion. The value of the negative dispersion depends on the velocity, time, and spatial step size.

Shamir and Harleman (1976) listed the various methods used for the numerical
solution for dispersion in porous media and compared them. Garder et al. (1964)
used a method of characteristics in which moving points are used. Nalluswami
(1971) used the principle of calculus of variations for solving the two-dimensional
dispersion equation with mixed partial derivatives, which is the result of treating
the dispersion coefficients as second-order symmetric tensors. A functional is
developed. The minimization of the functional with the finite element method leads
to a system of simultaneous first-order linear ordinary differential equations.
Because of the severe stability criterion, a large value of the dispersion coefficient
was used.

The one-dimensional convective diffusion equation is

$$\frac{\partial^2 c}{\partial x^2} - a \frac{\partial c}{\partial x} = \frac{\partial c}{\partial t} \tag{6.220}$$

where the parameter a is the ratio of the velocity of flow to the dispersion coefficient. For a large value of a, i.e., for small dispersion coefficients, the second
term on the left-hand side of Eq. (6.220) is controlling. The first space derivative
over two increments is set equal to the first time derivative over one increment.
The value of $c_{i,n}$ has little influence from the spatial derivative in calculating
$c_{i,n+1}$. Thus the solution oscillates around the true curve.

For a stable method a higher order approximation is required for the term
$\partial c / \partial t$. The desired second-order approximation is called the Crank-Nicholson
method, which is the average of the explicit and implicit methods. In the Crank-
Nicholson method all the finite differences are written about the point x_i, $t_{n+(1/2)}$,
which is halfway between the known and unknown time levels. Although the Crank-
Nicholson method requires a little more computation per time step than the backward difference method, a larger time increment can be used, since its time derivative analog is second-order correct. The Crank-Nicholson equation is more effi-
cient than the backward difference method and is preferred for numerical solutions
to parabolic differential equations.

Price et al. (1968) have shown that for Eq. (6.220) the oscillations can be
eliminated by using a fine spatial mesh. With the Crank-Nicholson method there
will be no oscillations when a $\Delta x / 2 < 1$ for x between 0 and 1.

The coupled equations Eqs. (5.158) and (5.159) are solved by the Crank-Nicholson method for practical values of velocity and dispersion in soil, though for very small dispersion coefficients and large velocities the convective diffusion equation changes from a parabolic to a hyperbolic form and may cause numerical dispersion. In the Crank-Nicholson method the nonlinear coefficients can be approximated either by a forward projection of coefficients at the half level in time or by an iterative procedure using the old value. The Douglas method is used because it has less stringent restrictions on the time step size for stability and converges more rapidly than the iterative method. The method is basically a two-step process.

The finite difference approximations for various terms at i, n + 1/2, obtained by the analog, which is second-order correct, are given by (the asterisk has been dropped for convenience)

$$\left(\frac{\partial c}{\partial z}\right)_{i,\,n+1/2} \doteq \frac{1}{2}\left(\frac{c_{i+1,n} - c_{i-1,n}}{2\,\Delta z} + \frac{c_{i+1,n+1} - c_{i-1,n+1}}{2\,\Delta z}\right) \tag{6.221}$$

$$\left(\frac{\partial^2 c}{\partial z^2}\right)_{n,\,n+1/2} \doteq \frac{1}{2}\left(\frac{c_{i+1,n} - 2c_{i,n} + c_{i-1,n}}{(\Delta z)^2} + \frac{c_{i+1,n+1} - 2c_{i,n+1} + c_{i-1,n+1}}{(\Delta z)^2}\right)$$
$$\tag{6.222}$$

$$\left(\frac{\partial c}{\partial t}\right)_{i,\,n+1/2} \doteq \frac{c_{i,n+1} - c_{i,n}}{\Delta t} \tag{6.223}$$

$$\left(\frac{\partial q}{\partial t}\right)_{i,\,n+1/2} \doteq \frac{q_{i,n+1} - q_{i,n}}{\Delta t} \tag{6.224}$$

The nonlinear coefficients $c_{i,\,n+1/2}$ and $q_{i,\,n+1/2}$ are approximated by truncated Taylor series as

$$c_{i,\,n+1/2} \doteq c_{i,n} + \left(\frac{\partial c}{\partial t}\right)_{i,n} \frac{\Delta t}{2} \tag{6.225}$$

and

$$q_{i,\,n+1/2} \doteq q_{i,n} + \left(\frac{\partial q}{\partial t}\right)_{i,n} \frac{\Delta t}{2} \tag{6.226}$$

where $(\partial c/\partial t)_{i,n}$ and $(\partial q/\partial t)_{i,n}$ are substituted from the governing equations in which the approximations given by the following equations are also used.

$$\left(\frac{\partial c}{\partial z}\right)_{i,n} \doteq \frac{c_{i+1,n} - c_{i-1,n}}{2\,\Delta z} \tag{6.227}$$

and

$$\left(\frac{\partial^2 c}{\partial z^2}\right)_{i,n} \doteq \frac{c_{i+1,n} - 2c_{i,n} + c_{i-1,n}}{(\Delta z)^2} \tag{6.228}$$

Thus the governing differential equation is approximated at x_i, $t_{n+1/2}$ using Eqs. (6.220) to (6.224). The nonlinear coefficients are approximated from Eqs. (6.223) and (6.226). On simplication, we obtain a tridiagonal matrix of the form

$$P_i c_{i-1} + q_i c_i + r_i c_{i=1} = s_i \qquad 1 \le i \le R \qquad p_1 = r_R = 0 \qquad (6.229)$$

This algorithm is called the <u>Thomas algorithm</u> and can be solved for the concentration at the advanced time step. At the two extreme points, i.e., $i = 1$ and $i = R$ boundary conditions are used. The term s_i on the right hand side contains terms at the previous time step which are known.

Once $c_{i,n+1}$ and $q_{i,n+1}$ are known then the iterations can be performed by calculating the new values of $c'_{i,n+1/2}$ and $q'_{i,n+1/2}$ by

$$c'_{i,n+1/2} = \frac{c_{i,n+1} + c_{i,n}}{2} \qquad (6.230)$$

and

$$q'_{i,n+1/2} = \frac{q_{i,n+1} + q_{i,n}}{2} \qquad (6.231)$$

These improved values are used in place of the approximations given by Eqs. (6.225) and (6.226). This iterative procedure is repeated till the old and new values of $c_{i,n+1/2}$ and $q_{i,n+1/2}$ are comparable.

Von Rosenburg (1969) presents the Thomas algorithm as

$$P_i c_{i-1} + q_i c_i + r_i c_{i+1} = d_i \qquad 1 \le i \le R \qquad p_1 = r_R = 0 \qquad (6.232)$$

The algorithm is as follows: compute

$$\beta_i = q_i - \frac{p_i r_{i-1}}{\beta_{i-1}} \qquad \beta_1 = q_1$$

and

$$\gamma_i = \frac{d_i - p_i \gamma_i - 1}{\beta_i} \qquad \gamma_1 = \frac{d_1}{q_1} \qquad (6.234)$$

Finally, the values of the dependent variable are then calculated as

$$c_R = \gamma_R \qquad (6.235)$$

and

$$c_i = \gamma_i - \frac{r_i c_i + 1}{\beta_i} \qquad (6.236)$$

The difference approximation for bilinear adsorption is as follows: Approximating nonlinear coefficient $c_{i,n+1/2}$ from Eq. (6.225) by substituting $\partial c/\partial t$ from Eq. (5.168),

$$c_{i,n+1/2} = c_{i,n} + \frac{\Delta t}{2}\left[d_2\left(z^2\frac{\partial^2 c}{\partial z^2} + d_1 z\frac{\partial c}{\partial z} - d_3\frac{\partial q}{\partial t}\right)_{i,n}\right] \qquad (6.237)$$

where

$$d_1 = 1 + \frac{1}{\eta} \qquad (6.238)$$

$$d_2 = \frac{1}{\eta\xi} \qquad (6.239)$$

$$d_3 = \frac{\beta}{\eta\xi} \qquad (6.240)$$

From Eq. (5.159), we have

$$\left(\frac{\partial q}{\partial t}\right)_{i,n} = \gamma [c_{i,n}(d_4 - q_{i,n}) - d_5 q_{i,n}] \qquad (6.241)$$

where

$$d_4 - 1 + \frac{1}{A} \qquad (6.242)$$

$$d_5 = \frac{1}{\Lambda} \qquad (6.243)$$

Substituting approximations at i, n from Eqs. (6.227), (6.228), and (6.241) into Eq. (6.237), on simplification, one obtains

$$c_{i,n+1/2} = c_{i,n} + \frac{\Delta t}{2d_2}[p_i c_{i-1,n} - (2i^2 + d_3 d_4 \gamma)c_{i,n} + r_i c_{i+1,n} + d_3 \gamma c_{i,n} q_{i,n} - d_5 q_{i,n}]$$
$$(6.244)$$

where

$$p_i = i^2 - d_1\frac{i}{2} \qquad (6.245)$$

$$r_i = i^2 + d_1\frac{i}{2} \qquad (6.246)$$

Similarly, the nonlinear coefficient $q_{i,n+1/2}$ is evaluated from Eq. (6.226). On replacing $(\partial q/\partial t)_{i,n}$ from Eq. (6.241),

$$q_{i,n+1/2} = q_{i,n} + \frac{\gamma\Delta t}{2}[c_{i,n}(d_4 - q_{i,n}) - d_5 q_{i,n}] \qquad (6.247)$$

The governing differential equation, Eq. (5.168) is approximated at i, n + 1/2 using Eqs. (6.221) and (6.222). On substituting $\partial q/\partial t$ in Eq. (5.168) from Eq. (5.159), one obtains

$$p_i c_{i-1, n+1} + q_i c_{i, n+1} + r_i c_{i+1, n+1} = d_i \tag{6.248}$$

where

$$q_i = -2i^2 - 2\frac{d_2}{\Delta t} \tag{6.248}$$

$$s_i = 2i^2 - 2\frac{d_2}{\Delta t} \tag{6.250}$$

$$d_i = -p_i c_{i-1, n} + s_i c_{i, n} - r_i c_{i+1, n}$$
$$+ 2d_3 \gamma [c_{i, n+1/2}(d_4 - q_{i, n+1/2}) - d_5 q_{i, n+1/2}] \tag{6.251}$$

Thus, Eq. (6.248) gives a set of tridiagonal matrices which can be solved by Thomas algorithm using the boundary conditions at i = 1 and R.

The difference approximation for Freundlich equilibrium adsorption follows: The nonlinear coefficient $c_{i, n+1/2}$ is calculated from Eq. (6.225). Substituting $\partial c/\partial t$ from Eq. (5.171), we obtain

$$c_{i, n+1/2} = c_{i, n} + \frac{\xi \eta \Delta t}{2(1 + \gamma c_{i, n}^{n-1})}\left[z^2 \frac{\partial^2 c}{\partial z^2} + \left(1 + \frac{1}{\eta}\right) z \frac{\partial c}{\partial z}\right]_{i, n} \tag{6.252}$$

defining

$$d_1 = 1 + \frac{1}{\eta} \tag{6.253}$$

and

$$d_2 = \frac{1}{\eta \xi} \tag{6.254}$$

then

$$c_{i, n+1/2} = c_{i, n} + \frac{\Delta t}{2d_2(1 + \gamma c_{i, n}^{n-1})}\left(z^2 \frac{\partial^2 c}{\partial z^2} + d_1 z \frac{\partial c}{\partial z}\right)_{i, n} \tag{6.255}$$

Substituting approximations given by Eqs. (6.227) and (6.228), and on simplification,

$$c_{i,n+1/2} = c_{i,n} + \frac{\Delta t}{2d_2(1 + \gamma c_{i,n}^{n-1})} \left[c_{i-1,n}\left(i^2 - d_1\frac{i}{2}\right) \right.$$

$$\left. - 2.0\,i^2 c_{i,n} + c_{i+1,n}\left(i^2 + d_1\frac{i}{2}\right) \right] \qquad (6.256)$$

Approximating the governing differential equation, Eq. (6.170) at i, n + 1/2
Eq. (6.180) is approximated at i, n + 1/2. First the nonlinear term,

$$p_i c_{i-1,n+1} + q_i c_{i,n+1} + r_i c_{i+1,n+1} = d_i \qquad (6.257)$$

where

$$p_i = i^2 - d_1\frac{i}{2} \qquad (6.258)$$

$$q_i = -2i^2 - (1 + \gamma c_{i,n+1/2}^{n-1})\frac{2d_2}{\Delta t} \qquad (6.259)$$

$$r_i = i^2 + d_1\frac{i}{2} \qquad (6.260)$$

$$s_i = 2i^2 - (1 + \gamma c_{i,n+1/2}^{n-1})\frac{2d_2}{\Delta t} \qquad (6.261)$$

$$d_i = -p_i c_{i-1,n} + s_i c_{i,n} - r_i c_{i+1,n} \qquad (6.262)$$

Equation (6.225) can be written as

$$c_{i,n+1/2} = c_{i,n} + \frac{\Delta t}{2d_2(1 + \gamma c_{i,n}^{n-1})} [c_{i-1,n}p_i - 2i^2 c_{i,n} + c_{i+1,n}r_i] \qquad (6.263)$$

Thus the Eq. (6.257) can be solved by the Thomas algorithm using Eq. (6.263).
The difference approximation for the solution of the one-dimensional radial dispersion with Freundlich equilibrium adsorption follows: The governing differential
Eq. (5.180) is approximated at i, n + 1/2. First the nonlinear term, $c_{i,n+1/2}$ is
approximated by a Taylor's series expansion as

$$c_{i,n+(1/2)} = c_{i,n} + \frac{\Delta t}{2}\frac{\partial c}{\partial t} \qquad (6.264)$$

Substituting $\partial c/\partial t$ from Eq. (5.180), we get

$$c_{i,n+(1/2)} = c_{i,n} + \frac{\Delta t}{2(1 + \gamma c_{i,n}^{n-1})}\left(\frac{\beta}{r}\frac{\partial^2 c}{\partial r^2} - \frac{\alpha_1}{r}\frac{\partial c}{\partial r}\right)_{i,n} \qquad (6.265)$$

Using the approximations given by

$$\left(\frac{\partial c}{\partial r}\right)_{i,n} = \frac{c_{i+1,n} - c_{i-1,n}}{2\Delta r} \tag{6.266}$$

and

$$\left(\frac{\partial^2 c}{\partial r^2}\right)_{i,n} = \frac{c_{i+1,n} - 2c_{i,n} + c_{i-1,n}}{(\Delta r)^2} \tag{6.267}$$

we obtain

$$c_{i,n+(1/2)} = c_{i,n} + \frac{\Delta t}{2(1 + \gamma c_{i,n}^{n-1})}\left[p_i c_{i-1,n} - \frac{2\beta}{i(\Delta r)^3} c_{i,n} + r_i c_{i+1,n}\right] \tag{6.268}$$

in which

$$p_i = \frac{\beta}{i(\Delta r)^3} + \frac{\alpha_1}{2i(\Delta r)^2} \tag{6.269}$$

$$r_i = \frac{\beta}{i(\Delta r)^3} - \frac{\alpha_1}{2i(\Delta r)^2} \tag{6.270}$$

The following approximations are used for the governing differential equation Eq. (5.180):

$$\left(\frac{\partial c}{\partial t}\right)_{i,n+(1/2)} = \frac{c_{i,n+1} - c_{i,n}}{\Delta t} \tag{6.271}$$

$$\left(\frac{\partial c}{\partial r}\right)_{i,n+(1/2)} = \frac{1}{2}\left(\frac{c_{i+1,n} - c_{i-1,n}}{2(\Delta r)} + \frac{c_{i+1,n+1} - c_{i-1,n+1}}{2(\Delta r)}\right) \tag{6.272}$$

and

$$\left(\frac{\partial^2 c}{\partial r^2}\right)_{i,n+(1/2)} = \frac{1}{2}\left(\frac{c_{i+1,n} - 2c_{i,n} - c_{i-1,n}}{(\Delta r)^2} + \frac{c_{i+1,n+1} - 2c_{i,n+1} + c_{i-1,n+1}}{(\Delta r)^2}\right) \tag{6.273}$$

and on simplification we obtain

$$p_i c_{i-1,n+1} + q_i c_{i,n+1} + r_i c_{i+1,n+1} = d_i \tag{2.274}$$

in which

$$q_i = -\frac{2\beta}{i(\Delta r)^3} - \frac{2}{\Delta t}\left(1 + c_{n-(1/2)}^{n-1}\right) \tag{6.275}$$

$$d_i = -p_i c_{i-1,n} + s_i c_{i,n} - r_i c_{i+1,n} \tag{6.276}$$

and

$$s_i = \frac{2\beta}{i(\Delta r)^3} - \frac{2}{\Delta t}(1 + \gamma c_{i,\,n+(1/2)}^{n-1}) \tag{6.277}$$

Equation (6.274) is the Thomas algorithm and one can find the concentrations at new time step $n + 1$ for all i using Eq. (6.274) and boundary conditions at two end points. At $i = 1$, the inlet concentration is known. At the last spatial point, where the concentration gradient is known, the reflection boundary condition is used. Because of the concentration discontinuity at $r = 0$ for $t = 0$, the condition is modified as $c(0, 0) = 0.5$ and at any other time $c(0, 5) = 1.0$.

REFERENCES

Bear, J. (1961): Scales of Viscous Analogy Models for Ground Water Studies, Proc. ASCE J. Hyd. Div. 86, HY2, 11.

Bear, J. (1972): Dynamics of Fluids in Porous Media, American Elsevier, New York.

Buckingham, E. (1915): Model Experiments and the Forms of Empirical Equations, Trans. AIME 37.

Chaudhari, N. M. (1971): An Improved Numerical Technique for Solving Multidimensional Miscible Displacement Equations, Soc. Petrol Eng. J. 11(3), 277.

Crichlow, H. B. (1977): Modern Reservoir Engineering: A Simulation Approach, Prentice-Hall, Englewood Cliffs, N.J.

Corey, A. T. (1977): Mechanics of Heterogeneous Fluids in Porous Media, Water Resources Publication, Fort Collins, Colo.

Corey, G. L. (1965): Similitude for Non-Steady Drainage of Partially Saturated Soils, Ph.D. Thesis, Colorado State University, Fort Collins, Colo.

Douglas, J., Jr., and H. H. Rachford, Jr. (1956): On the Numerical Solution of Heat Conduction Problems in Two or Three Space Variables, Trans. Amer. Math. Soc. 82, 421.

Douglas, J., Jr., D. W. Peaceman, and H. H. Rachford, Jr. (1959): A Method for Calculating Multi-dimensional Immiscible Displacements, Pet. Trans. AIME 216, 297.

Garder, A. O., Jr., D. W. Peaceman, and A. L. Pozzi, Jr. (1964): Numerical Calculation of Multidimensional Miscible Displacement by the Method of Characteristics, Pet. Trans. AIME 231, SPEJ 26.

Geertsma, J. G. A. Croes, and N. Schwarz (1956): Theory of Dimensionally Scaled Models of Petroleum Reservoirs, Pet. Trans. AIME 207, 118.

Greenkorn, R. A. (1964):Flow Models and Scaling Laws for Flow Through Porous Media, Ind. Eng. Chem. 56(3), 31.

Greenkorn, R. A., R. E. Haring, H. O. Johns, and L. K. Shallenberger (1964): Flow in Heterogeneous Hele-Shaw Models, Pet. Trans. AIME 231, SPEJ 124.

Greenkorn, R. A., C. R. Johnson, and R. E. Haring (1965): Miscible Displacement in a Controlled Natural System, Pet. Trans. AIME 234, 1229.

Hele-Shaw, H. S. (1898): Investigation of the Nature of Surface Resistance of Water and of Streamline Motion under Certain Experimental Conditions, Trans. Inst. Nav. Arch. XI, 25.

Higgins, R. V., and A. J. Leighton (1962): A Computer Model to Calculate Two-Phase Flow in Any irregularly Bounded Porous Medium, Pet. Trans. AIME 225, 679.

Jenkins, R., and J. S. Aronofsky (1953): Unsteady Radial Flow of Gas Through Porous Media, J. Appl. Mech. 20, 210.

Lantz, R. B. (1971): Quantitative Evaluation of Numerical Diffusion (Truncation Error), Pet. Trans. AIME 251, SPEJ 315.

Leverett, M. C., W. B. Lewis, and M. E. True (1942): Dimensional-Model Studies of Oil-Field Behavior, Pet. Trans. AIME 146, 175.

Martin, J. C., and R. E. Wegner (1978): Numerical Solution of Multi-Phase Two-Dimensional Incompressible Flow Using Stream Tube Relations, Regional Meeting, SPE, San Francisco, Paper No. 7140.

Miller, E. E., and R. C. Miller, Physical Theory of Capillary Flow Phenomena, J. Appl. Phys. 27(4), 324.

Mitchell, A. R., and R. Wait (1977): The Finite Element Method in Partial Differential Equations, Wiley-Interscience, New York.

Nalluswami, M. (1971): Numerical Simulation of General Hydrodynamic Dispersion in Porous Media, Ph.D. Thesis, Colorado State University, Ft. Collins.

Offeringa, J., and C. van der Poel (1954): Displacement of Oil from Porous Media by Miscible Liquids, Pet. Trans. AIME 201, 310.

Peaceman, D. W., and H. H. Rachford, Jr. (1955): The Numerical Solution of Parabolic and Elliptical Differential Equations, J. Soc. Ind. Appl. Math. 3, 28.

Peaceman, D. W., and H. H. Rachford, Jr. (1962): Numerical Calculation of Multidimensional Miscible Displacement, Pet. Trans. AIME 225, SPEJ 327.

Perkins, F. M., Jr., and R. E. Collins (1960): Scaling Laws for Laboratory Flow Models of Oil Reservoirs, Pet. Trans. AIME 219, 383.

Price, H. S., J. C. Cavendish, and R. S. Varga (1968): Numerical Methods of Higher-Order Accuracy for Diffusion Convection Equations, Pet. Trans. AIME 243, SPEJ 243.

Polubarinova-Kochina, P. Ya (1962): Theory of Ground Water Movement, Princeton University Press, Princeton, N.J.

Rapport, L. A. (1955): Scaling Laws for Use in Design and Operation of Water-Oil Flow Models, Pet. Trans. AIME 204, 143.

Richardson, J. G. (1961): Flow Through Porous Media, in Handbook of Fluid Dynamics (V. I. Streeter, ed.), McGraw-Hill, New York, Section 16.

Riegels, F. (1938): To the Criticism of the Hele-Shaw Experiment, Zertschuft fer Ang. Math. Mech. 18, 951.

Shamir, U. Y., and D. R. F. Harleman (1967): Numerical Solution for Dispersion in Porous Mediums, Water Resour. Res. 3(2), 557.

Smith, R. C., and R. A. Greenkorn (1969): An Investigation of the Flow Regime for Hele-Shaw Flow, Pet. Trans. AIME 246, SPEJ 434.

Stokes, G. G. (1898): Mathematical Proof of the Identity of the Streamlines Obtained by Means of a Viscous Film with Those of a Perfect Fluid Moving in Two Dimensions, British Assoc. Advan. Sci. 78 Annual Mtg., Bristol, 143.

Stone, H. L. (1968): Iterative Solutions of Implicit Approximations of Multidimensional Partial Differential Equations, SIAM J. Numerical Analysis 5, 530.

van Daalen, F., and H. R. van Domselaar (1972): Scaled Fluid-Flow Models with Geometry Differing from that of Prototype, Pet. Trans. AIME 253, 220.

Varga, R. S. (1962): Matrix Iterative Analysis, Prentice-Hall, Englewood Cliffs, N.J.

von Rosenberg, D. V. (1969): Methods for Numerical Solution of Partial Differential Equations, American Elsevier, New York.

Weinstein, H. G., H. L. Stone, and T. V. Kwan (1969): Iterative Procedures for Solutions of Systems of Parabolic and Elliptical Equations in Three Dimensions, Ind. Eng. Chem. Fund. 2, 281.

Wyckoff, R. D., H. G. Botset, and M. Muskat (1933): The Mechanics of Porous Flow Applied to Water Flooding Problems, Pet. Trans. AIME 103, 219.

Young, D. M. (1954): Iterative Methods for Solving Partial Differential Equations of Elliptic Types, Trans. Amer. Math. Soc., 76, 92.

SUGGESTED READING

Javandel, J., and P. A. Witherspoon, Application of the Finite Element Method to Transient Flow in Porous Media, Pet. Trans. AIME 243, SPEJ 241 (1968).

McMichael, G. L., and G. W. Thomas, Reservoir Simulation by Galerkins' Method, Pet. Trans. AIME 255, SPEJ 125 (1973).

Settari, A., H. S. Price, and T. Dupont, Development and Application of Variational Methods for Simulation of Miscible Displacement in Porous Media, 4th SPE Symp. Numerical Simulation of Reservoir Performance, Los Angeles, Feb. Paper No. SPE 5721, 1976.

Spivak, A., H. S. Price, and A. Settari, Solution of the Equation of Multi-Dimensional Two-Phase Immiscible Flow by Variational Methods, 4th SPE Symp. Numerical Simulation of Reservoir Performance, Los Angeles, Feb. Paper No. SPE 5123, 1976.

Price, H. S., and K. H. Coats, Direct Methods in Reservoir Simulation, Pet. Trans. AIME 257, SPEJ 295.

Rachford, H. H., Jr., A Sampling of Variational Networks, 4th SPE Symp. Numerical Simulation of Reservoir Performance, Los Angeles, Feb. Paper No. SPE 5720, 1976.

7

Applications in Petroleum Engineering

In this chapter we consider reservoir description, well testing, and various methods of producing oil reservoirs. The term primary production usually refers to production of oil and/or gas from a reservoir by natural mechanisms to replace the volume of oil and/or gas produced. There are four mechanisms: expansion of the fluids (and matrix) accounts for some of the volume replaced; fluid displacement by invading gas, dissolution of gas, or water encroachment; gravity drainage particularly when oil is pumped from a reservoir (the driving force of encroaching side or bottom water drive is due to gravity forces); imbibition of water into an oil-bearing medium causing expulsion of the oil. These mechanisms can occur simultaneously, but normally one mechanism will dominate at any particular time during the depletion of a reservoir. In a gas reservoir, expansion and water drive mechanisms dominate. Generally gas drive or water drive are dominant during primary oil production.

When primary production ends, water flooding is used for secondary production. After secondary production tertiary methods use CO_2 and/or surface-active agents.

7.1 RESERVOIR DESCRIPTION

Oil and gas reservoirs are described to model performance of a reservoir analytically, physically, and numerically. A complete set of data for a reservoir study is very large and includes the physical properties of the fluids; a physical description of the reservoir size, shape, permeability, porosity, and compressibility; fluid matrix data, i.e., relative permeability, capillary pressure, and saturation; operating variables, e.g., flow rates and pressures; mechanical data such as casing sizes, tubing sizes, etc.; well data such as skin effect, fractures, boundaries; and economic data.

Reservoir description usually includes that part of the data associated with the geometry of a reservoir and its related properties—thickness, porosity, permeability—plus a geological description which correlates these properties in blobs or layers including fractures and boundaries. A broader definition includes capillary pressure, relative permeabilities, and saturations as a function of position.

Physical properties of the fluids are considered separately. The monograph by Standing (1977) provides a good explanation of the measurement and correlation

308

of reservoir fluid properties. The operational data are not considered part of the reservoir description.

The major characteristics of reservoir description are the geometric interpretation of area, thickness, porosity, and permeability. The aerial extent of the reservoir results from geologic interpretation of a variety of data including knowledge of the particular geologic formations, outcrops, geophysical data, well information, well logs, and pressure behavior of the reservoir. Formation elevations as well as gross and net thicknesses are usually extracted from well data such as drilling records, cores, and logs. Drilling records contain data on rates of penetration and interpretation of the cuttings returned to the surface during drilling. When cores of the reservoir are removed location of the top and the bottom of the reservoir can be determined from the cores. Well-logging measures various parameters of the reservoir with appropriate instruments by lowering the sensors into wells and making measurements adjacent to the well bores. [See Lynch (1962).] These tools include electric or resistivity logs which measure the electric potential or resistance of the fluid-filled formation. These logs can be interpreted to locate the top and the bottom of a formation and to infer properties of the contained fluids. Acoustic or sonic logs can be used to determine the porosity of the formation adjacent to the well bore. Other logs such as nuclear logs, temperature logs, etc., have been developed to aid in prospecting a given formation and to provide descriptive information. The geologist or reservoir engineer uses the available data to prepare an isopac map of elevations and formation thicknesses. The data from the wells (cores, logs, etc.) are used to correlate geologic markers between wells which are used to infer a three-dimensional description or a two-dimensional layered description of a reservoir.

An isoporosity map is prepared using core data and logging data. If dynamic tests, such as pressure buildup or falloff tests, are available and compressibility data is available, porosity is calibrated with the dynamic results. Reservoir volume estimates are made from geometry, net thickness, and porosity data.

An isopermeability map is prepared based on cores, logs, and dynamic tests. Vertical permeability is determined from cores or from individual well tests. Transmissivity may also be calculated from permeability and gross-thickness data and fluid-property data for direct comparison with dynamic tests results.

Oil and gas saturations are difficult to determine quantitatively. Saturation may be estimated from true fluid cores—cores captured in such a way that they contain the original reservoir fluids. In situ measurements are also used (Tomich et al., 1973). The gas/oil contact and the water/oil contact are located from core data, electric logs, or capillary-pressure data. Residual or connate water is determined from capillary-pressure data and subtracted from the oil.

Capillary pressures are determined from laboratory measurements of core samples or estimated from correlations. Relative permeabilities are determined from laboratory measurements of cores or from producing gas/oil ratios, or from correlations. Compressibilities are measured on cores in the laboratory or obtained from correlations. The combination of gross thickness, net thickness, porosity, permeability, and saturation by aerial contours and vertical location "describe" the reservoir. Averages are developed for these properties by certain calculations such as series flow with a distinct concentration front in layers using the Dykstra and Parsons (1950) method. From physical or numerical modeling, suitable grids or cell blocks are overlayed on the description, and the grids or blocks are assigned the values corresponding to the intersections with the overlay.

A major problem in practical reservoir description is determining when significant changes in thickness, porosity, permeability, and saturation occur and the geometrical location of such changes. The spatial locations of significant changes are necessary to preserve the geometric scaling of the reservoir and predict movement of the reservoir fluids.

Reservoirs can be modeled in several dimensions depending on the situation for which the model is to be used. If a reservoir is ideal, that is, homogeneous, uniform, and isotropic, only material balance equations are required. A knowledge of the property that describes each part—porosity, permeability, etc.—is all that is required.

One-dimensional models, linear radial, or horizontal to vertical, are often used in simple situations. Gas/oil contacts are maintained. Properties such as thickness, permeability, and porosity vary in one dimension. Two-dimensional layered models are frequently used. The rock properties vary in two dimensions. Multiple layers may be used to give a pseudo-three-dimensional model. Partial two-dimensional radial models r-z are used in well test analysis and single-well studies. Two-dimensional vertical models x-z are used to study gravity segregation. Three-dimensional models are used when rock properties vary significantly in three dimensions, when layered systems communicate, when reservoir discontinuities such as faults are important, and with large multiple-well systems.

CASE STUDY 1: A Reservoir Description

Pursley et al. (1973) report the results of a surfactant pilot test conducted in a watered-out portion of the Loudon field. Additional discussion of the description of the pilot area is reported by Harris (1975). The pilot site (Fig. 7.1) contains five wells drilled on a 0.625 acre five-spot pattern. Additional wells surround the pattern and an observation well was drilled halfway between the producing well and the northernmost injector.

Figure 7.2 is a flow chart of the reservoir description. The parameters generally required are thickness, porosity, permeability (or transmissivity), the pore volume, and initial fluid saturations. The rock information and framework studies use logs, core analysis, and well tests. Figure 7.3 is an example of well log-

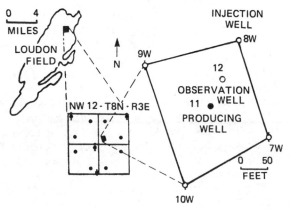

FIG. 7.1 Pilot test site. (From D. G. Harris, The Role of Geology in Reservoir Simulation Studies, Pet. Trans. AIME 259, SPEJ 625. © 1975 SPE-AIME.)

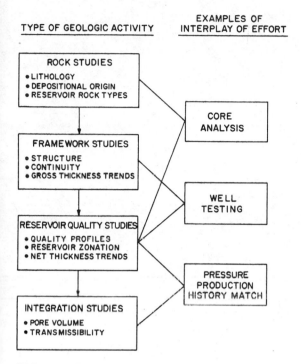

TYPE OF GEOLOGIC ACTIVITY | EXAMPLES OF INTERPLAY OF EFFORT

FIG. 7.2 Reservoir description flowchart. (From D. G. Harris, The Role of Geology in Reservoir Simulation Studies, Pet. Trans. AIME 259, SPEJ 625. © 1975 SPE-AIME.)

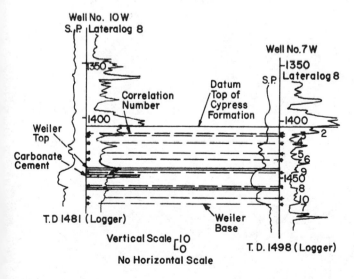

FIG. 7.3 Well-log correlation for sand continuity. (From D. G. Harris, The Role of Geology in Reservoir Simulation Studies, Pet. Trans. AIME 259, SPEJ 625. © 1975 SPE-AIME.)

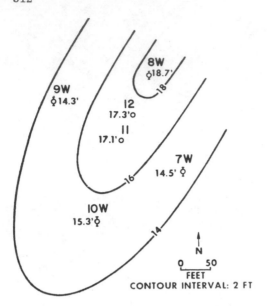

FIG. 7.4 Gross thickness map. (From D. G. Harris, The Role of Geology in Reservoir Simulation Studies, Pet. Trans. AIME 259, SPEJ 625. © 1975 SPE-AIME.)

correlation using wells 10W and 7W in the pilot area. Knowing the depositional environment is important since the distribution of reservoir rock is usually influenced by depositional conditions. The reservoir has a N70°E fracture pattern; however fractures were not detected in the pattern. The field is located on an anticline with

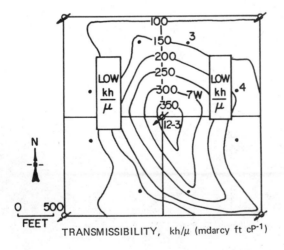

FIG. 7.5 Regional transmissivity distribution. (From S. A. Pursley, R. N. Healy, and E. I. Sandvik, A Field Test of Surfactant Flooding, Loudon, Illinois, Pet. Trans. AIME 255, SPEJ 793. © 1973 SPE-AIME.)

a northeast-southwest trend, a 165-ft closure, and a 1 to 2° dip. The logs and core description data showed the cored interval consists of six lithological units including shale, sandstone, and siltstone. The Loudon field is located on a flank of a large ancient river system. The Weiler sand, a subdivision of the Mississippian Chester Cypress formation is a stream mouth bar deposit. The sand is layered, is relatively uniform, and continues throughout the pilot area. Numerous thin shale beds of several millimeters thickness restrict vertical permeability. The areal extent of these shale laminations is small compared to the pilot area dimensions. The producing sandstone of the pilot is in the middle structural unit. Wells in the pilot and in the 160-acre area surrounding the pilot were correlated over the interval from the datum to the limestone marker in the Cypress formation.

A gross-thickness map was developed from logs and correlations for the pilot area shown in Fig. 7.4. A regional transmissivity (kh/μ) map of the 160-acre area surrounding the pilot was generated from a two-dimensional, areal steady-state simulation of waterflood rate-pressure behavior. This map is shown in Fig. 7.5.

As the pilot test wells were drilled pulse tests were performed between well 7W and the three nearest offsets (12-3, 3, and 4). Transmissivity for well 12-3 to well 7W and from well 3 to well 7W were 340 and 240 mdarcy cP^{-1}, respectively, and are in good agreement with Fig. 7.5. No response was detected between wells 4 and 7W so the pattern was located at maximum distance from this area of heterogeneity. The pattern was oriented to be parallel and perpendicular to the possible fracture orientation. A directional survey was run on the pattern wells and the wells are shown relative to their bottom hole locations.

Core analysis of the pattern wells show the Weiler sandstone at 1460-ft depth to be relatively uniform with a pay interval of about 14 ft. Table 7.1 is a summary of the core analysis for the six pattern wells. The porosity is nearly constant with an average of 30.6 percent. A net sand isopachous map (Fig. 7.6) was developed

TABLE 7.1 Core Analysis Results

Well	Net sand (ft)	Porosity (%)	Gas permeability (mdarcy)	Brine permeability (mdarcy)	Oil saturation (percent PV)
7W	13.7	20.7	154	129[a]	—
8W	17.8	20.9	—	117	30.2
9W	16.0	20.5	122	102	—
10W	11.9	20.5	170	104	27.9
11	16.6	21.1	112	94	—
12	17.0	20.0	98	69	—
Average	15.5	20.6		103	29.1

[a]Brine permeability not measured. Brine permeability calculated using average k_{brine}/k_{gas} value obtained from conventionally cored wells 9W and 11.

Source: S. A. Pursley, R. N. Nealy, and E. I. Sandvik, A Field Test Surfactant Flooding, Loudon, Illinois, Pet. Trans. SPE-AIME 255, 793. © 1973 SPE-AIME.

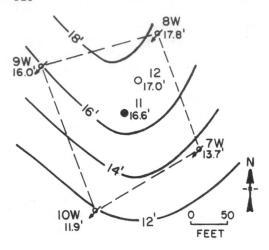

FIG. 7.6 Pilot net sand isopachous map. (From S. A. Pursley, R. N. Healy, and
E. I. Sandvik, A Field Test of Surfactant Flooding, Loudon, Illinois, Pet. Trans.
AIME 255, SPEJ 793. © 1973 SPE-AIME.)

from the gross-thickness map and core analysis. Intervals having less than 10
mdarcy permeability were ignored. A total pattern pore volume of 16,250 bbl was
derived by integrating a porosity-thickness map. The average residual oil saturation
measured on the cores from wells 8W and 10W is 29.1 percent which converts to a
residual oil volume within the pilot of 4730 bbl. ■

7.2 WELL-TEST ANALYSIS

Pressure transient testing of reservoirs via wells provides in situ information.
Pressure transient tests include pressure buildup, drawdown injectivity, falloff,
and interference tests. Transient tests are used in reservoir description to determine in situ estimates of permeability, porosity, and reservoir discontinuities.
Reservoir limits and fluid discontinuities are also estimated from transient tests.
Well damage and improvement are diagnosed with well tests. The material presented
in this section is introductory. There are many books and papers which discuss well
test analysis. There are several monographs directed specifically at well test analysis [see, for example, Earlougher (1977), Kruseman and deRidder (1970), Mathews
and Russel (1967), and Ramey et al. (1975)].
        Earlougher (1977) divides pressure-buildup, drawdown, injectivity, or falloff
behavior into three areas: front-end effects (well-bore storage, fractures, and
damage), the semilog straight-line region (the region where most procedures apply),
and boundary effects. Figure 7.7 is a composite curve that illustrates all of these
effects for a hypothetical buildup. Early-time behavior is affected by well-bore
storage effects and these effects may obscure early-time behavior of the reservoir.
Boundary conditions affect late-time pressure. Application of late-time analysis
methods such as the Muskat method to middle-time data may give misleading results.

There are several methods of single well testing. Muskat (1937) suggests plotting the logarithm of the average reservoir pressure minus the shut-in well pressure versus the shut-in time. This plot should yield a straight line. p_{avg} is unknown; so the procedure is trial and error. If p_{avg} is too high, the plot of the log $(p_{avg} - p_{ws})$ versus Δt will be concave upward. If p_{avg} is too low, the plot of log $(p_{avg} - p_{ws})$ will be concave downward. The effective region for a Muskat plot is shown in Fig. 7.7. The intercept of the plot for $\Delta t = 0$ is a function of kh and geometry. For a circular drainage volume,

$$kh = \frac{118.6 q \mu B}{(p_{avg} - p_{ws})_{\Delta t=0}} \tag{7.1}$$

For a square drainage area,

$$kh = \frac{94.6 q \mu B}{(p_{avg} - p_{ws})_{\Delta t=0}} \tag{7.2}$$

The slope m of the Muskat plot is a function of drainage volume. For a circular drainage volume,

$$\phi c_t \pi r_e^2 = \frac{0.00528 k}{m \mu} \tag{7.3}$$

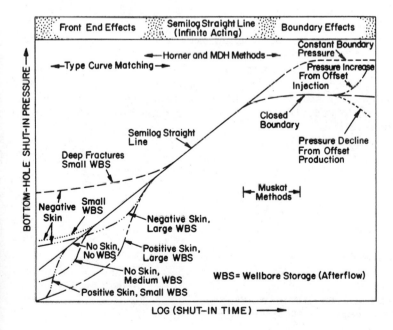

FIG. 7.7 Typical bottom-hole pressure-buildup curve shapes. For production at pseudosteady state before shut-in. (From Earlougher, 1977.)

For a square drainage area,

$$\phi c_t A = \frac{0.00471k}{m\mu} \qquad (7.4)$$

The skin effect is

$$s = 0.84 \frac{p_{avg} - p_{wf}}{(p_{avg} - p_{ws})_{\Delta t=0}} - \ln \frac{r_e}{r_w} + 0.75 \qquad (7.5)$$

CASE STUDY 2: Estimation of Drainage Volume from
Well-Test Analysis

In 1937, Muskat proposed plotting pressure-buildup data as log ($\bar{p}$ - p_{ws}) versus Δt. Subsequent theoretical studies by Ramey et al. (1975) indicate that this graph should be used with caution and only as a late-time analysis method. Since the time required to acquire the necessary data is usually much longer than that required for either a Horner or a Miller-Dyes-Hutchinson plot, practical considerations often limit the application of this method.

It is very important to remember that a buildup is always preceded by oil withdrawal and the buildup data are directly affected by this withdrawal. Ideally, the withdrawal starts from a stabilized reservoir condition represented by the stabilized reservoir pressure p_i. At a time t_p, the well is shut in, and the pressure buildup is continued for a time Δt.

Ramey and Cobb (1971) also showed that the production period does not necessarily have to extend into the pseudo-steady-state flow regime. However, the time required to reach the proper straight line does increase as production time decreases below the time required to reach pseudo-steady-state flow conditions.

The permeability of the formation immediately around the well can be damaged by the well-drilling process or improved by fracturing or acidizing the well on completion, and there will be an added pressure difference due to these effects. The skin effect s corresponds to an infinitesimal "skin" around the well causing additional or decreased resistance to flow.

The Muskat method uses a trial-and-error plot with several $\bar{p}$ estimates; a straight line is obtained for the correct $\bar{p}$. The intercept at Δt = 0, log ($\bar{p}$ - p_{ws}) of the correct line of an extended Muskat plot may be used to estimate reservoir permeability, k, from

$$k = \frac{141.2qB\mu p_D(t_{p_{DA}})}{h(\bar{p} - p_{ws})_{\Delta t=0}} \qquad (7.6)$$

The dimensionless pressure p_D, which is a function of dimensionless-producing time t_p, is given by Ramey and Cobb (1971) for a single unfractured well in the center of a closed-square drainage system. Figure 7.8 shows the data for this system. For the closed square system,

$$p_D(t_{p_{DA}} > 0.1) = 0.67$$

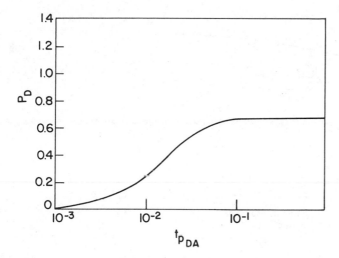

FIG. 7.8 Muskat dimensionless pressure for a well in the middle of a closed square. (From H. J. Ramey and W. M. Cobb, A General Buildup Theory for a Well in a Closed Drainage Area, Pet. Trans. AIME 251, SPEJ 1493. © 1971 SPE–AIME.)

if producing time exceeds the time to pseudo-steady state. For the closed circular system,

$$p_D(t_{p_{DA}} > 0.1) = 0.84$$

Under most circumstances, these two systems should behave identically.

The dimensionless-producing time, based on the total drainage area, is

$$t_{p_{DA}} = \frac{0.0002637kt_p}{\phi \mu c_t A} \qquad (7.7)$$

The slope m of the Muskat straight line may be used to estimate the total drainage area. For a closed square,

$$A = \frac{-0.00471k}{\phi \mu c_t m} \qquad (7.8)$$

The constant in Eq. (7.8) is −0.00528 for a closed circular system with the well at the center.

The beginning and the end of the Muskat straight line may be estimated from

$$\Delta t = \frac{\phi \mu c_t A(\Delta t_{DA})}{0.0002637k} \qquad (7.9)$$

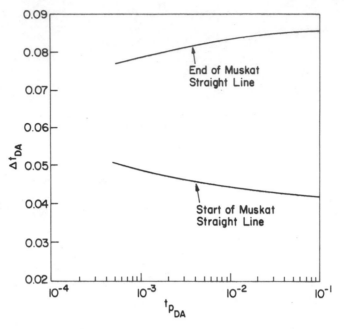

FIG. 7.9 Limits of Muskat line. (From Ramey et al., 1975.)

where Δt_{DA} is shown in Fig. 7.9. Data for both the start and the end of the straight line are given for the closed square.

The skin term can be obtained from a combination of terms,

$$s = 0.84 \frac{\bar{p} - p_{wf}}{(\bar{p} - p_{ws})_{\Delta t=0}} - \ln \frac{r_e}{r_w} + 0.75 \tag{7.10}$$

A positive value of s indicates a damaged well, and a negative value an improved well. In the case of negative skin, this concept, which is mathematically correct, leads to some difficulty in physical interpretation. Mathematically it is equivalent to superposing an injection well on top of the producing well.

The pressure drop at a damaged (or improved) well differs from that at an undamaged well by the additive amount

$$\Delta p_s = \frac{141.2qB\mu s}{kh} \tag{7.11}$$

This value can be used to calculate the flow efficiency of a well.

Consider a well which produces at a pressure of 3509 psig for 120.53 hr. The well is capped and the shut-in pressure recorded. Table 7.2 presents the reservoir and well data, while Table 7.3 presents the buildup data.

The Muskat graph of the well data is shown in Fig. 7.10. The pressure differences needed to make this graph are also given in Table 7.3. Three columns are

TABLE 7.2 Reservoir and Well Data[a]

P_{wf}	= 3509 psi
q	= 490 STB/D
μ	= 0.20 cP
c_t	= 22.6×10^{-6} psi
B	= 1.55 RB/STB
h	= 20 ft
r_w	– 0.29 ft
ϕ	= 0.10
Δt	= 49.87 hr
t_p	= 120.53 hr

[a]Drainage shape: square, no influx.
Well location: center.
Source: Theory and Practice of the Test-
ing of Gas Wells, Energy Resources Con-
servation Board, Calgary, Alberta,
Canada, 1975, pp. 5-12.

given for assumed pressures $\bar{p}$ of 3685, 3700, and 3715 psi. Inspection of Fig. 7.10
indicates the 3685-psi case bends sharply downward, indicating the estimated $\bar{p}$ is
too low. The line for 3700 psi appears straight after 30 hr. The line for 3715 psi
bends upward.

From Fig. 7.10,

$$(\bar{p} - p_{ws})_{\Delta t=0} = 43.5$$

and

$$m = -0.00885 \text{ cycles per hour}$$

Equation (7.6) is used to estimate permeability from the intercept, but $p_D(t_{p_{DA}})$
must be obtained from Fig. 7.8. Thus, it is assumed that there is a single well in
the center of a closed square. Combining Eqs. (7.7) and (7.8) yields

$$\Delta t_{DA} = -0.056 t_p m$$

$$= -0.056(120.53)(-0.00885)$$

$$= 0.06$$

Thus from Fig. 7.8, $p_D = 0.61$. Using Eq. (7.6),

TABLE 7.3 Buildup Data

Shut-in time Δt (hr)	p_{ws} (psig)	$\bar{p} - p_{ws}$ (psi) for		
		$\bar{p} = 3685$	$\bar{p} = 3700$	$\bar{p} = 3715$
0.53	3296	389	404	419
1.33	3296	389	404	419
1.60	3385	300	315	330
2.13	3521	164	179	194
2.67	3547	138	153	168
3.20	3562	123	138	153
3.73	3573	112	127	142
4.27	3582	103	118	133
4.80	3591	94	109	124
5.33	3599	86	101	116
5.87	3605	80	95	110
6.40	3605	80	95	110
6.93	3614	71	86	101
7.47	3619	66	81	96
8.00	3623	62	77	92
9.07	3630	55	70	85
9.87	3634	51	65	81
10.93	3640	45	60	75
12.00	3644	41	56	71
13.60	3650	35	50	65
14.67	3654	31	46	61
16.53	3660	25	40	55
18.67	3664	21	36	51
21.33	3668	17	32	47
24.53	3672	13	28	43
29.33	3676	9	24	39
35.73	3679	6	21	36
45.87	3683	2	17	32
49.87	3684	1	16	31

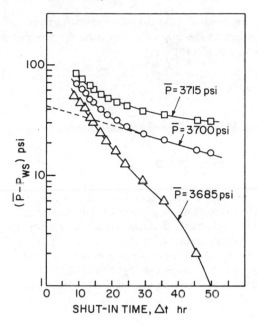

FIG. 7.10 Muskat graph.

$$k = \frac{141.2(490)(1.55)(0.20)(0.61)}{20(43.5)}$$

$$= 15.04 \ \text{mdarcy}$$

The area A is estimated from Eq. (7.8)

$$A = \frac{-0.00471(15.04)}{0.1(0.2)(22.6 \times 10^{-6})(-0.00885)}$$

$$= 17.71 \times 10^6 \ \text{ft}^2$$

This is approximately the size of a circular area with a radius

$$r_e = \sqrt{\frac{A}{\pi}} = \sqrt{\frac{17.71 \times 10^6}{3.1416}} = 2374.3 \ \text{ft}$$

Figure 7.9 indicates that for $t_{p_{DA}} = 0.06$, the straight line should meet the requirement

$$0.043 < \Delta t_{DA} < 0.085$$

Using Eq. (7.9) yields the result

$$87 < \Delta t < 171$$

This clearly indicates that even though the extended Muskat plot appears to have a straight line, that straight line does not occur at the right shut-in time for analysis.

The skin term can be calculated from Eq. (7.10)

$$s = (0.84) \frac{3700 - 3509}{43.5} - \ln \frac{2374}{0.29} + 0.75 = -4.57$$

The pressure drop from the skin effect can be calculated from Eq. (7.11)

$$\Delta p_s = \frac{141.2(490)(1.55)(0.2)(-4.57)}{15.04(20)} = -325.8 \text{ psi } \blacksquare$$

Miller et al. (1950) assumed a well produces long enough to reach a pseudo-steady state. If the flow boundaries are closed to flow, a pseudo-steady state will be reached when the pressure decline rate is constant. This is not a true steady state since pressure is changing continually with time everywhere in the drainage volume. Eventually the producing pressure reaches a limiting value and the production rate declines. A plot of the buildup pressure versus the logarithm of the buildup time is a straight line as identified in Fig. 7.11. The slope m of the straight line is inversely proportional to kh

$$kh = \frac{162.6q\mu B}{m} \qquad\qquad (7.12)$$

The skin effect using pressure at 1 hr is

$$s = 1.151 \frac{p_{1hr} - p_{wk}}{m} - \log \frac{k}{\phi\mu c_t r_w^2} + 3.23 \qquad\qquad (7.13)$$

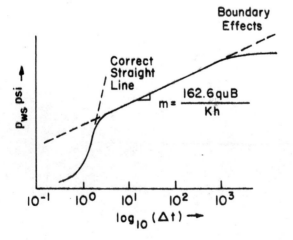

FIG. 7.11 Miller-Dyes-Hutchinson plot. (From Earlougher, 1977.)

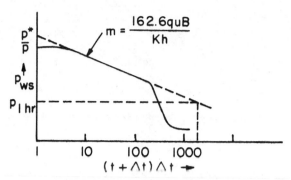

FIG. 7.12 Horner plot. (From Earlougher, 1977.)

Horner (1951) plots shut-in pressure versus logarithm of the time ratio $(t + \Delta t)/\Delta t$ as identified in Fig. 7.12. The slope of the straight line is inversely proportional to kh as in Eq. (7.12). The skin effect is given by Eq. (7.13). The flowing time t prior to shut-in is estimated by

$$t = \frac{Np}{q_{last}} \tag{7.14}$$

CASE STUDY 3: Comparison of Reservoir Properties and Skin Effect Determined by the Millers, Dyes, Hutchinson, and Horner Methods of Well-Test Analysis

Pressure-buildup tests involve shutting off a well being tested and then monitoring the well pressure. The Horner and the Miller-Dyes-Hutchinson are two commonly used methods to analyze pressure-buildup data. Each of these methods can be used to calculate the average reservoir permeability, the reservoir pressure, and the skin effect.

Horner (1951) reports the solution for the pressure equation in a circular reservoir with one well in the center as Eq. (7.15) below. He assumed that the reservoir was horizontal, homogeneous, and uniform and that the compressibility and viscosity of the fluid were relatively constant.

$$p_w = p_o - \frac{q\mu B}{kh}\left(162.6 \log \frac{t_o + \Delta t}{\Delta t} + 70.6y\left\{\frac{948.2r_e^2 \phi\mu c}{k(t_o + \Delta t)}\right\}\right.$$
$$\left. - 70.6y\left\{\frac{948.2r_e^2 \phi\mu c}{k(\Delta t)}\right\}\right) \tag{7.15}$$

where the function y is shown in Fig. 7.13. Equation (7.15) can be approximated, except at very large Δt, by the equation

$$p_w = p_o - \frac{70.6q\mu B}{kh} y\left\{\frac{948.2r_e^2 \phi c}{4kt_o}\right\} - \frac{162.6q\mu B}{kh} \log \frac{t_o + \Delta t}{\Delta t} \tag{7.16}$$

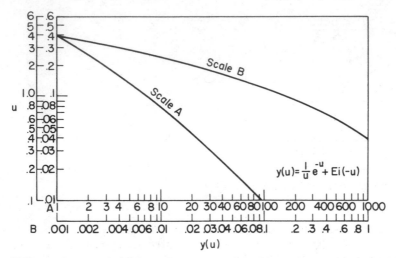

FIG. 7.13 y(u) to find reservoir pressure using the Horner method. (From Earlougher, 1977.)

which is of the form

$$p_w = b - m \log \frac{t_o + \Delta t}{\Delta t} \tag{7.17}$$

where

$$m = \frac{162.6 q \mu B}{kh} \tag{7.18}$$

so the permeability can be calculated from the slope of a p_w versus log $[(t_o + \Delta t)/\Delta t]$ curve.

To determine the reservoir pressure, Horner (1951) evaluated Eq. (7.15) at $\Delta t = \infty$ and got

$$p_e = p_o - \frac{70.6 q \mu B}{kh} \frac{kt_o}{948.2 r_e^2 \phi uc} \tag{7.19}$$

He also extrapolated Eq. (7.16) to $\Delta t = \infty$ to find p^*

$$p^* = P_o - \frac{70.6 q \mu B}{kh} y \left\{ \frac{948.2 r_e^2 \phi uc}{4kt_o} \right\} \tag{7.20}$$

Combining Eqs. (7.18) and (7.19), we can determine the reservoir pressure:

$$p_e = p^* + \frac{70.6q\mu B}{kh} \quad y\left\{\frac{\left(948.2r_e^2\,\phi uc\right)}{4kt_o}\right\} - \frac{kt_o}{948.2r_e^2\,\phi uc} \tag{7.21}$$

At very early times the flow into the well is from the area immediately around the wellbore. So the permeability calculated from the early part of the curve represents the permeability in the well–bore area, while the permeability from the later portion is for the entire reservoir. This permits an approximation to the skin factor.

$$s = 1.51 \frac{p_{1hr} - p_{wf}}{m} - \log \frac{k}{\phi ucr_w^2} + 3.23 \tag{7.22}$$

For the same reservoir situation, Miller et al. (1950) show that the pressure can be described by

$$\Delta\bar{p} = \ln \bar{r}_e - \frac{3}{4} + \frac{\pi^2}{2} \sum_{n=1}^{\infty} \frac{J_1^2(X_n)U^2(X_n)}{J_1(\Gamma_e X_n) - J_1(X_n)} e^{-\bar{r}_e X_n \bar{t}} \tag{7.23}$$

where

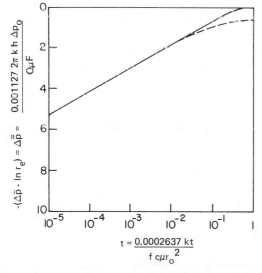

FIG. 7.14 Calculated rise in bottom–hole pressure in a shut–in well. Steady–state pressure distribution obtained prior to shut–in by constant production rate. Solid line for drainage area with constant pressure maintained at the radius of drainage. Dashed line for closed area with no influx over the radius of drainage. (From C. C. Miller, A. B. Dyes, and C. A. Hutchinson, Jr., The Estimation of Permeability and Reservoir Pressure from Bottom Hole Pressure Buildup Characteristics, Pet. Trans. AIME 189, SPEJ 91. © 1950 SPE–AIME.)

TABLE 7.4

Shut-in time, Δt (days)	Shut-in pressure, p_{ws} (psi)	
0	60.04	ϕ = 0.119
0.0001	93.14	
0.0002	117.69	K = 6.16 mdarcy
0.0003	136.18	P_o = 1500 psia
0.0004	150.20	
0.0005	170.95	r_e = 500 ft
0.001	229.13	h = 25 ft
0.005	355.62	
0.01	409.87	r_w = 0.5 ft
0.015	440.65	q = 25 STB day^{-1}
0.02	462.28	
0.03	493.82	μ = 1.93 cP
0.05	552.53	c = 4.168 × 10^{-4} psi
0.1	609.97	
0.15	639.49	B = 1.1
0.2	659.75	
0.25	675.24	t_o = 1240 days
0.3	687.94	
0.5	729.40	
0.8	769.73	
1	783.17	
2	824.49	
3	848.57	
5	878.18	
7	897.05	
8	911.97	
9	917.83	
10.6	929.99	
21	954.45	
30.6	970.10	
50.6	985.85	
70	990.08	
80	992.63	
90	994.31	
100	995.44	
140	997.27	

Source: R. Raghavan, Well Test Analysis: Wells Producing by Solution Gas Drive, Pet. Trans. SPE-AIME 261, SPEJ 196. © 1976 SPE-AIME.

$$\Delta \bar{p} = \frac{kh(p - p_f)}{162.6q\mu B}$$

$$\bar{r}_e = \frac{r_e}{r_w} \tag{7.24}$$

$$\bar{t} = \frac{0.0002637kt}{\phi c\mu r_e^2}$$

$-\Delta \bar{p} - \ln \bar{r}_e$ is plotted in Fig. (7.14).

Equation (7.23) gives a straight line when $\overline{\Delta p}$ is plotted versus $\log \bar{t}$; so a plot of p versus log t gives a straight line with slope

$$m = \frac{162.6q\mu B}{kh}$$

At infinite time Eq. (7.22) becomes

$$\Delta \bar{p}_s = \ln \bar{r}_e - \frac{3}{4} = \frac{kh(p_e - p_f)}{162.6q\mu B} \tag{7.25}$$

Subtracting Eq. (7.24) from Eq. (7.25)

$$\Delta \bar{p}_s - \Delta \bar{p} = \frac{kh}{162.6q\mu B}(p_e - p) \tag{7.26}$$

The reservoir pressure can be found from

$$p_e = (\Delta \bar{p}_s - \Delta \bar{p})\frac{162.6q\mu B}{kh} + p \tag{7.27}$$

The skin factor is determined by Eq. (7.22).

Raghavan (1976) reports data for a buildup test for a well in the center of a circular reservoir. The data are presented in Table 7.4. These data were used to generate Figs. 7.15 and 7.16, the Horner and Miller-Dyes-Hutchinson plots.

From the Horner plot (Fig. 7.15) to calculate permeability

$$m = 790 - 630 = 160$$

$$k = \frac{162.6q\mu B}{mh} = 2.158 \text{ mdarcy}$$

to determine reservoir pressure

From the plot at $(t_0 + \Delta t)/\Delta t = 1$

$$p^* = 1282$$

$$\frac{948.2r_e^2\phi uc}{kt_0} = \frac{948.2(500)^2(0.119)(1.93)(4.168 \times 10^{-4})}{2.158(1240)(24)} = 0.353$$

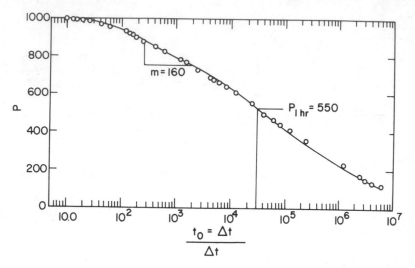

FIG. 7.15 Horner plot. (Data from Raghavan, 1976.)

From Fig. 7.13

$$y(0.353) = 0.7$$

From Eq. (7.21)

$$p_e = p^* + \frac{70.6q\mu B}{kh}\left[y(0.353) - \frac{kt_o}{948r_e^2\phi uc}\right]$$

$$= 1282 + \frac{70.6(25)(1.93)(1.1)}{(2.158)(25)}\left[.7 \frac{(2.158)(1240)(24)}{948.2(500)^2(.119)(1.93)(4.168 \times 10^{-4})}\right]$$

$$= 1134 \text{ psi}$$

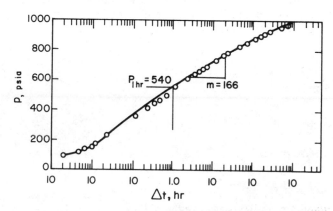

FIG. 7.16 Miller-Dyes-Hutchinson plot. (Data from Raghavan, 1976.)

and to calculate the skin factor

From Fig. 7.15

$$p_{1hr} = 550$$

Using Eq. (7.22)

$$s = \frac{1.15(p_{1hr} - p_{wf})}{m} - \log \frac{k}{\phi \mu c r_w^2} + 3.23$$

$$= \frac{1.15(550 - 60)}{160} - \log \frac{2.158}{0.119(1.93)(4.168 \times 10^{-4})(500)^2} + 3.23 = 7.80$$

From the Miller-Dyes-Hutchinson plot Fig. 7.16 to calculate permeability

$$m = 166$$

$$k = \frac{162.6q\mu B}{mh} = \frac{162.6(25)(1.93)(11)}{166(25)} = 2.079 \text{ mdarcy}$$

to determine the reservoir pressure

From Fig. 7.16

$$p = 610 \quad \text{at } \Delta t = 2.40 \text{ hr}$$

$$\overline{\Delta t} = \frac{0.000264K \Delta t}{\phi \mu c A} = \frac{0.000264(2.079)(2.40)}{(0.119)(4.168 \times 10^{-4})(1.93)\pi(500)^2} = 1.75 \times 10^{-5}$$

From Fig. 7.14

$$-(\Delta \bar{p} - \ln \bar{r}_e) = 5.1$$

$$\Delta \bar{p} = 1.81$$

$$\Delta \bar{p}_s = \ln \bar{r}_o - \frac{3}{4} = \ln 1000 - \frac{3}{4} = 6.15$$

$$p_e = p + (\Delta \bar{p}_S - \Delta p) \frac{Bq\mu}{0.00127\pi kh}$$

$$= 610 + (6.15 - 1.81) \frac{11(25)(1.93)}{0.001127\pi(2.079)(25)} = 1224 \text{ psi}$$

and to find the skin factor

From Fig. 7.16

$$p_{1hr} = 550$$

Using Eq. (7.22),

$$s = \frac{1.151(p_{1hr} - p_{wf})}{m} - \log \frac{k}{\phi u c r_w^2} + 3.23$$

$$= \frac{1.151(550 - 60)}{166} - \log \frac{2.079}{0.119(1.93)(4.168 \times 10^{-4})(500)^2} + 3.23 = 7.69$$

Each method gives approximately the same results for permeability and skin factor. The final pressure is different. The permeability here is smaller than that reported. ■

The type-curve analysis uses a match of the test-well data with an analytically derived solution. The type curve is an analytical solution of the unsteady-state pressure equation for the appropriate boundary conditions including well-bore storage effects. Buildup data are plotted in dimensionless terms as log p_D versus log t_D. The well data are superimposed on the type curve, and a coordinate translation of the ordinate and abscissas are used to determine the reservoir parameters. Dimensionless time, pressure, and storage are defined as

$$t_D = \frac{0.000264kt}{\phi\mu c_t r_w^2} \tag{7.28}$$

$$p_D = \frac{kh(p_i - p_{wf})}{141.3q\mu B} \tag{7.29}$$

$$c^* = \frac{5.615c}{2\pi h \phi c_t r_w^2} \tag{7.30}$$

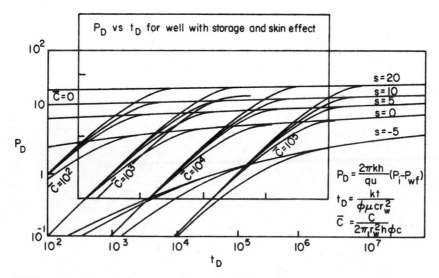

FIG. 7.17 Type curve analysis. (From Earlougher, 1977.)

TABLE 7.5 Data Required by Method for Analyzing Well Tests

		Muskat	M–D–H	Horner	Type curve
Pressure buildup data P_{ws} – Δt		Yes	Yes	Yes	Yes
Field data plot	Semilog	$\log(\bar{P} - P_{ws})$ vs. Δt	P_{ws} vs. $\log \Delta t$	P_{ws} vs. $\log \dfrac{t + \Delta t}{\Delta t}$	No
	Log-log	No	No	No	$\log (P_{ws} - P_{wf})$ vs. $\log \Delta t$
Producing time before shut–in, t		No	No	Yes	No
Trial–and–error method used		No	No	Yes	No
PVT data	Formation volume factor B	Yes	Yes	Yes	Yes
	Viscosity μ, cp	Yes	Yes	Yes	Yes
	Compressibility c(psi^{-1})	Yes	Yes	Yes	Yes
	P_{BP}	Yes	Yes	Yes	Yes
	R_s^*	No	No	No	No
	GOR*	No	No	No	No
Reservoir data	h	Yes	Yes	Yes	Yes
	ϕ	No	Yes	Yes	Yes
	$A(r_e)$	Yes	Yes	Yes	No
	Saturation[a]	No	No	No	No
	C_f^a	No	No	No	No

(continued)

TABLE 7.5 (continued)

		Muskat	M-D-H	Horner	Type curve
[Reservoir data]	T_R [a]	No	No	No	No
	Drainage shape and well location	Yes	Yes	Yes	No
Production data	q	Yes	Yes	Yes	Yes
	N_p	No	No	Yes	No
	r_w	Yes	Yes	Yes	Yes
	P_{wf}	Yes	Yes	Yes	Yes
Dimensionless chart used		No	ΔP_{DMDH} vs. Δt_{DA} chart	ΔP_{DMBH} vs. t_{DA} chart	P_D vs. t_D analytical type curve
Pseudo-steady state needed before shut-in		Yes	Yes	Yes	No

[a] Data required only for multiphase well or gas well. R_s = solution gas. GOR = gas/oil ratio.

TABLE 7.6 Parameters Obtained by Methods for Analyzing Well Tests

	Muskat	M-D-H	Horner	Type curve
Slope of straight line	$m = \dfrac{0.000264 k a_1}{2.303 \phi \mu c r_e^2}$	$m = \dfrac{162.6 q \mu B}{kh}$	$m = \dfrac{162.6 q \mu B}{kh}$	Unit slope for early buildup points
Intercept	$(\bar{P} - P_{ws})_{\Delta t=0} = \dfrac{141.2 q \mu B}{khb_1}$	—	$P^* = P_{ws}$ at $\log \dfrac{t + \Delta t}{\Delta t} = 1$	—
P_{1hr}	—	Extend from the straight line on field data plot	Extend from the straight line on field data plot	—
k	$k = \dfrac{94.6 q \mu B}{(\bar{P} - P_{ws})_{\Delta t=0}}$ (For square)	$k = \dfrac{162.6 q \mu B}{mh}$	$k = \dfrac{162.6 q \mu B}{mh}$	$k = \dfrac{141.4 q \mu B P_D}{h(P_{ws} - P_{wf})}$
$t_{DA} = \dfrac{0.000264 kt}{\phi \mu c A}$		Calculate t_{DA}	Calculate $\forall t_{DA}$	—
$\Delta t_{DA} = \dfrac{0.000264 k \, \Delta t}{\phi \mu c A}$	—	Calculate Δt_{DA}	—	—
$\Delta t_{De} = \Delta t_{DA} \times \pi$	—	Calculate Δt_{De}	—	—
ΔP_D	—	ΔP_{DMDH}	ΔP_{DMBH}	—
$\bar{P}$	By trial and error	$\bar{P} = \dfrac{m}{1.151} \Delta P_{DMDH} + P_{ws}$	$\bar{P} = P^* - \dfrac{m}{2.303} \Delta P_{DMBH}$	—

(continued)

TABLE 7.6 (continued)

	Muskat	M-D-H	Horner	Type curve
S	$S = 0.84 \dfrac{\bar{P} - P_{wf}}{(\bar{P} - P_{ws})_{\Delta t=0}}$ $- \ln \dfrac{r_e}{r_w} + 0.75$	$S = 1.151 \left(\dfrac{P_{1hr} - P_{wf}}{m} \right)$ $- \log_{10} \dfrac{k}{\phi \mu r_w^2 c} + 3.23)$	$S = 1.151 \left(\dfrac{P_{1hr} - P_{wf}}{m} \right)$ $- \log_{10} \dfrac{k}{\phi c r_w^2 \mu} + 3.23)$	Read directly from the analytical type curve

The unit storage constant is

$$c = \frac{qB \, \Delta t}{(p_{ws} - p_{wf})} \tag{7.31}$$

Taking the logarithm of Eqs. (7.28) and (7.29) and expanding them

$$\log t_D = \log \frac{0.000264k}{\phi \mu c_t r_w^2} + \log t \tag{7.32}$$

$$\log p_D = \log \frac{kh}{141.3q\mu\beta} + \log (p_i - p_{wf}) \tag{7.33}$$

The difference between the real and the dimensionless time and pressure are translations of both axes as shown in Fig. 7.17. Table 7.5 compares the four methods of transient analysis discussed above and Table 7.6 shows parameters obtained by the four methods.

Multiple-well transient tests include interference and pulse tests. In an interference test the flow rate of one well is changed for an extended period which creates a pressure interference at the other wells. Type-curve matching is usually applied to interference tests as with a single-well test. To analyze the data from an inter-

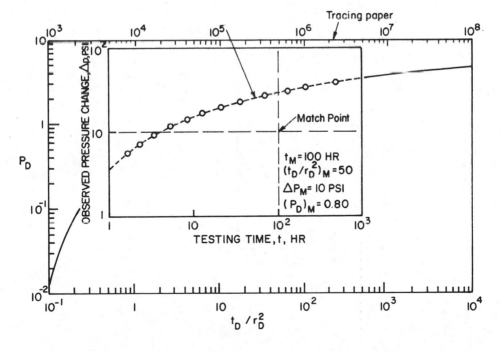

FIG. 7.18 Illustration of type-curve matching for an interference test using the type curve. (From R. C. Earlougher, Jr., Advances in Well Test Analysis, Monograph Series, Vol. 5, Soc. Pet. Eng., Dallas. © 1977 SPE–AIME.)

ference test the pressure data from the observation well is plotted as log Δp versus Δt and overlayed as in Fig. 7.18. When the data are matched to the type curve

$$kh = 141.2\,q\mu B\,\frac{(\Delta p_D)_m}{\Delta p_m} \qquad (7.34)$$

$$\phi c_t = \frac{0.0002637}{r^2}\frac{k}{\mu}\frac{t_m}{(t_D/r_D^2)_m} \qquad (7.35)$$

The type curve is usually the exponential integral solution to the unsteady-state pressure equation and is good if

$$\frac{r}{r_w} > 20 \qquad (7.36)$$

and

$$\frac{t_D}{r_D} > 0.5 \qquad (7.37)$$

A pulse test (Johnson et al., 1966) uses a series of short rate pulses at one well. The pulses are generated by alternately shutting the well in and letting it flow. The pressure response to the pulsing is measured at observation wells. Figure 7.19 is a schematic of a pulse test. Advantages of a pulse test are that the pulses are so short that the infinite system solutions of the unsteady-state pressure equation, the

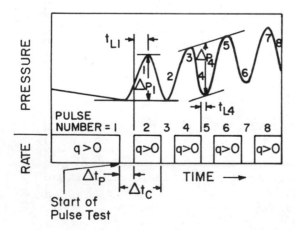

FIG. 7.19 Schematic pulse-test rate and pressure history showing definition of time lag (t_ℓ) and pulse-response amplitude (Δp). (From M. Kamal and W. E. Brigham, Pulse Testing Response for Unequal Pulse and Shut-in Periods, Pet. Trans. AIME 259, SPEJ 399. © 1975 SPE-AIME.)

exponential integral, apply. Reservoir trends and noise are automatically removed. The pulses are of such short duration that there is virtually no interruption to field operations. Two characteristics of the pressure response are the time lag and amplitude of the pressure. The ratio of pulse length to total cycle length is

$$F' = \frac{\Delta t_p}{\Delta t_c} \tag{7.38}$$

The dimensionless time lag is

$$(\Delta t_L)_D = \frac{0.0002637 k t_L}{\phi \mu c_t r_w^2} \tag{7.39}$$

The dimensionless distance between the pulsed and observation wells is

$$r_D = \frac{r}{r_w} \tag{7.40}$$

The dimensionless pressure response amplitude in

$$(\Delta p)_D = \frac{k h \, \Delta p}{141.2 q \mu \beta} \tag{7.41}$$

where q is the rate at the pulsed well when active. The time lag and pressure response amplitude from one or more pulse responses are used to estimate reservoir properties

$$k h = \frac{141.2 q \mu B \left[(\Delta p)_D (t_L / \Delta t_c)^2 \right]_{Fig}}{\Delta p (t_L / \Delta t_c)^2} \tag{7.42}$$

where Δp and t_L are from the observation well response and Δt_c is the cycling time of the pulsing well. $\left[(\Delta p)_D (t_L / \Delta t_c)^2 \right]_{Fig}$ is from Figs. 7.20a–d for the appropriate value of $t_L / \Delta t_c$.

$$\phi c_t = \frac{0.0002637 k t_L}{\mu r^2 \left[(t_L)_D / r_D^2 \right]_{Fig}} \tag{7.43}$$

$\left[(t_L)_D / r_D^2 \right]_{Fig}$ is from Figs. 7.20e–h.

FIG. 7.20 Pulse testing: relation between time lag and response amplitude. (From M. Kamal and W. E. Brigham, Pulse Testing Response for Unequal Pulse and Shut-in Periods, Pet. Trans. AIME 259, SPEJ 399. © 1975 SPE-AIME.) (Parts a through h are on pp. 338–341.)

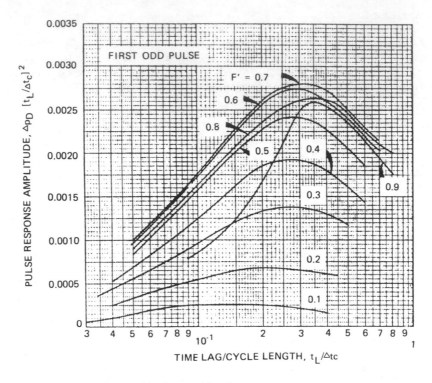

FIG. 7.20a Relation for first odd pulse.

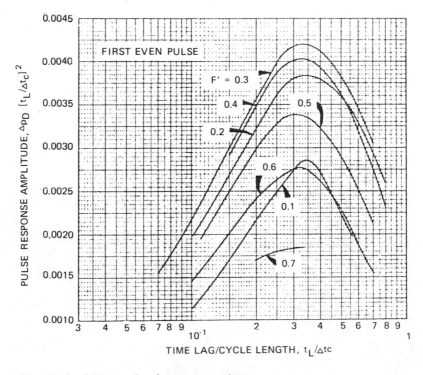

FIG. 7.20b Relation for first even pulse.

338

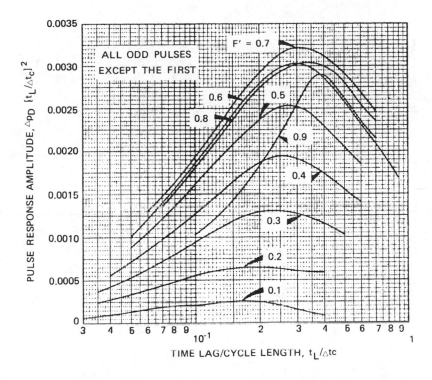

FIG. 7.20c Relation for all odd pulses after the first.

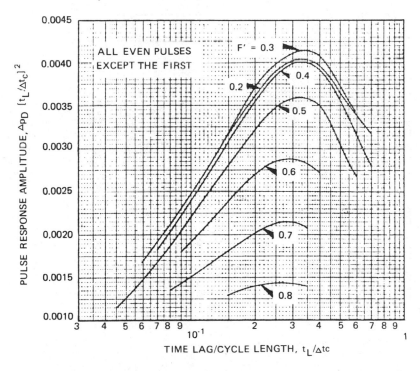

FIG. 7.20d Relation for all even pulses after the first.

339

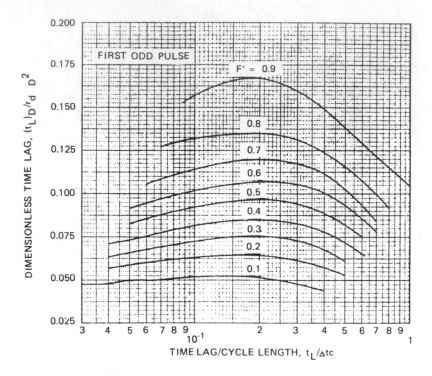

FIG. 7.20e Relation for first odd pulse.

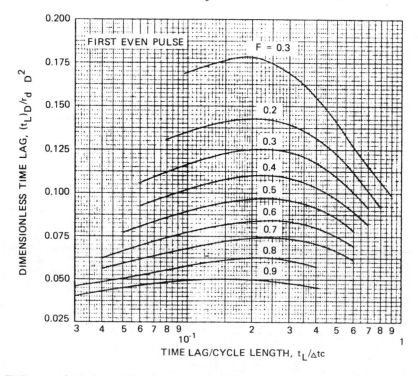

FIG. 7.20f Relation for first even pulse.

340

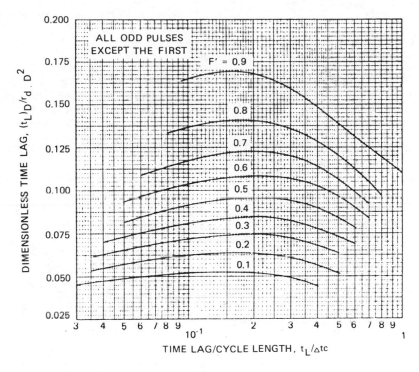

FIG. 7.20g Relation for all odd pulses after the first.

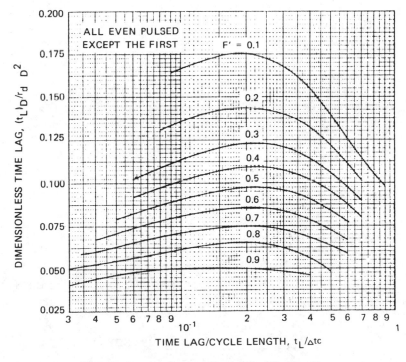

FIG. 7.20h Relation for all even pulses after the first.

341

CASE STUDY 4: Comparison of Standard Interference Test
and a Pulse Test in Analyzing Reservoir Properties

McKinley et al. (1968) performed pulse tests and interference tests in a producing
oil field. The reservoir is a structural trap because of a fault along the east side of
a north-south trending anticline. The reservoir has a natural water drive due to a
down-structure aquifer. The reservoir is dolomite limestone with mainly vugular
permeability. The contained oil has a gravity of 20° API with negligible dissolved
gas. The field has 19 wells on 40-acre spacing (the well spacing is 1320 ft). Figure
7.21 shows the location of all but two of the wells.

 Consider wells A8 and A9, which are in the watered-out part of the field. Well
A8 was pumped at a continuous rate of 550 STB D^{-1} for 70 hr. The pressure drawdown at well A9 due to the production at well A8 is given in Table 7.7. In order to
estimate the reservoir properties from the interference test we will use type-curve
matching. The data of Table 7.7 are plotting in on Fig. 7.22 using the same grid as
the type-curve solution Fig. 7.23. We then overlay Fig. 7.22 on Fig. 7.23 moving
Fig 7.22 vertically and horizontally to give the best match of the data and the type-
curve solution. Fig. 7.24 shows this overlay with the match point and match point
values for calculating the transmissivity using Eq. (7.34) assuming B = 1 RB/STB^{-1}

$$\frac{kh}{\mu} = 141.2qB \frac{(P_D)_m}{\Delta P_m}$$

$$= 141.2(550)(1) \frac{6.2}{10} = 48 \times 10^3 \text{ mdarcy ft cP}^{-1}$$

The storativity is calculated using Eq. (7.35)

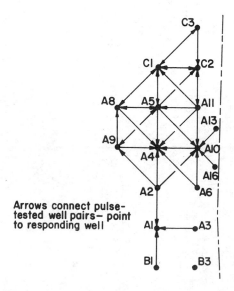

Arrows connect pulse-
tested well pairs— point
to responding well

FIG. 7.21 Well locations. (From R. M.
McKinley, S. Vela, and Y. A. Carlton,
A Field Application of Pulse Testing for
Detailed Reservoir Description, Pet.
Trans. AIME 243, SPEJ 313. © 1968
SPE-AIME.)

TABLE 7.7 Pressure Drawdown at Well A9
from Pumping A8

t (hr)	Δp (psi)
0	0
2	2.5
2.8	3.7
3	5.5
5	7.5
5.9	10
9	12
12	13.5
14	15.5
19	17
25	19.5
30	21.5
40	24.5
50	25.5
60	25.7
70	26

Source: R. M. McKinley, S. Vela, and L. A.
Carlton, A Field Application of Pulse Testing
for Detailed Reservoir Description, Pet. Trans.
AIME 243, SPEJ 313. © 1968 SPE–AIME.

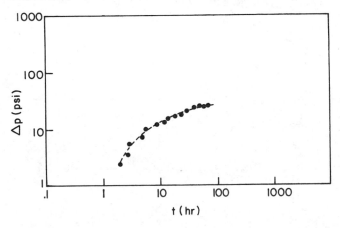

FIG. 7.22 Pressure drawdown at well A9 from pumping A8. (Data from McKinley
et al., 1968.)

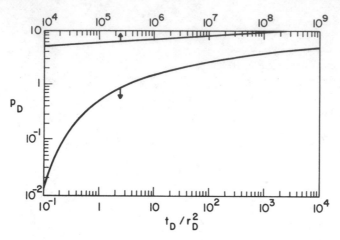

FIG. 7.23 Dimensionless pressure for a single well in an infinite system, no well-bore storage, no skin. Exponential-integral solution. (From Earlougher, 1977.)

$$\phi c_t h = \frac{0.0002637}{r_D^2} \frac{kh}{\mu} \frac{t_m}{(t_D/r_D^2)_m}$$

$$\phi c_t h = \frac{0.0002637}{(1320)^2} (48 \times 10^3) \frac{10}{2.2} = 33 \times 10^{-6} \text{ ft psi}^{-1}$$

The pulse test data for wells A8 and A9 are shown in Fig. 7.25. Well A8 was pulsed for one hour at 550 BPD and the response measured at well A9. Later well

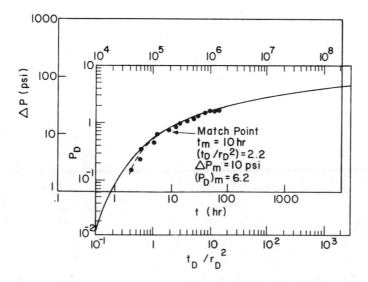

FIG. 7.24 Overlay of Fig. 7.22 on Fig. 7.23.

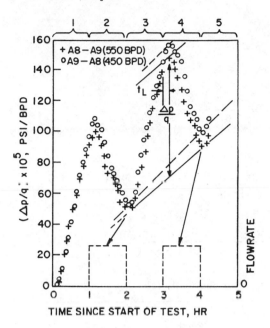

FIG. 7.25 Pulse test data. (From R. M. McKinley, S. Vela, and Y. A. Carlton, A Field Application of Pulse Testing for Detailed Reservoir Description, Pet. Trans. AIME 243, SPEJ 313. © 1968 SPE–AIME.)

well A9 was pulsed at 450 BPD and the response measured at A8. For the purpose of calculation the second peak (third pulse response) is analyzed for well A8 pulsed with A9 responding using Eqs. (7.42) with Fig. 7.20c. From Fig. 7.25 the difference between the value of $\Delta p/q$ at the third peak means the value at the reservoir time (the tangent line to the second and fourth peaks)

$$\frac{\Delta p}{q} = 146 - 70 = 7.6 \times 10^{-4} \text{ psi STB D}^{-1}$$

The time lag between the start of the third peak at well A8 and the response at A9 from Fig. 7.25 is $t_L = 0.15$ hr. The pulse length $\Delta t_p = 1$ hr. The cycle length $\Delta t_c = 2$ hr. From Fig. 7.20c with

$$\frac{t_L}{\Delta t_c} = \frac{0.15}{2} = 0.075$$

and

$$F' = \frac{\Delta t_p}{\Delta t_c} = \frac{1}{2} = 0.5$$

read a value

$$[\Delta p_D (t_c / \Delta t_c)^2]_{Fig} = 0.00128$$

Then from Eq. (7.42)

$$\frac{kh}{\mu} = \frac{141.2qB[\Delta p_D (t_L / \Delta t_c)^2]_{Fig}}{\Delta p (t_c / \Delta t_c)^2} = \frac{141.2(1)(0.00128)}{(7.6 \times 10^{-4})(0.075)^2}$$

$$= 42.3 \times 10^3 \text{ mdarcy ft } cP^{-1}$$

To estimate $\phi c_t h$ we use Eq. (7.43) and Fig. 7.20g. From Fig. 7.20g $[(t_c)_D / r_D^2]_{Fig} = 0.0876$. Then from Eq. (7.43)

$$\phi c_t h = \frac{0.0002637 t_c}{r^2 [(t_c)_D / r_D^2]_{Fig}} \frac{kh}{\mu} = \frac{0.0002637(42.3 \times 10^3)(0.15)}{(1320)^2 (0.0876)} = 11 \times 10^{-6} \text{ ft psi}^{-1}$$

The values using the data with well A9 the pulsing well yield

$$\frac{kh}{\mu} = 46 \times 10^3 \text{ mdarcy ft } cP^{-1}$$

and

$$\phi c_t h = 12.7 \times 10^{-6} \text{ ft psi}^{-1}$$

The values of transmissivity by both methods are within the probable error of measurement. The values of storativity are outside the measurement error. The values of transmissivity and storativity in the watered out region based on core data are reported as 50×10^3 mdarcy ft cP^{-1} and 10×10^{-6} ft psi^{-1}. ■

In detailed well test analysis the effects of heterogeneity, anisotropy, fracturing, and well bore storage must also be considered (see Earlougher, 1977).

7.3 GAS RESERVOIRS

A single-fluid model is used for gas reservoirs without a water drive. The same equations apply as for flow of single compressible fluids. However with gases the properties of the gas—density and viscosity—vary with temperature. There may be slip flow; so permeability is a function of pressure. Gas flow may be turbulent. Deviations from Darcy's law begin because of inertial effects and eventually as velocity increases turbulence dominates the flow.

For horizontal steady flow in one dimension (Forchheimer, 1901)

$$-\frac{dp}{dx} = \frac{\mu}{k}q + \beta \rho q^2 \tag{7.44}$$

where β is a turbulence factor. Rearranging

$$q = -\delta \frac{k}{\mu} \frac{dp}{dx} \qquad (7.45)$$

where

$$\delta = \frac{1}{1 + \beta\rho kq/\mu} \qquad (7.46)$$

when $\delta = 1$, Eq. (7.45) is Darcy's law. In three dimensions for an isotropic medium

$$q = -\frac{k\delta}{\mu} \nabla p \qquad (7.47)$$

At low pressure for gas flow the Klinkenberg effect exists

$$\tilde{k} = k\left(1 + \frac{b}{p}\right) \qquad (7.48)$$

where k is the absolute permeability.

In terms of the flow potential the equation of change is

$$\frac{\partial(\phi\rho)}{\partial t} = \underline{\nabla} \cdot \frac{\rho\tilde{k}\delta}{\mu} \underline{\nabla} \Phi \qquad (7.49)$$

An equation of state is used to write Eq. (7.49) in terms of the pressure. For a real gas

$$\rho = \frac{Mp}{RTz} \qquad (7.50)$$

Assume porosity is independent of time and substitute Eq. (7.50) into Eq. (7.49).

$$\phi \frac{\partial}{\partial t}\left(\frac{Mp}{RTz}\right) = \underline{\nabla} \cdot \frac{Mpk\tilde{}\delta}{RTz\mu} \underline{\nabla}\Phi \qquad (7.51)$$

At constant temperature

$$c = \frac{1}{\rho}\left(\frac{\partial\rho}{\partial p}\right)_T \qquad (7.51)$$

From Eq. (7.50)

$$\left(\frac{\partial\rho}{\partial p}\right)_T = \frac{M}{RTz} + \frac{Mp}{RT}\left(\frac{\partial^{1/z}}{\partial p}\right)_T \qquad (7.53)$$

Substituting Eq. (7.53) into Eq. (7.52)

$$c = \frac{1}{p} - z\left(\frac{\partial z}{\partial p}\right)_T \qquad (7.54)$$

Combining Eq. (7.54) with Eq. (7.51) gives

$$\phi \frac{\partial (p/z)}{\partial t} = \underline{\nabla} \cdot \frac{p k \tilde{\delta}}{z \mu} \underline{\nabla} \Phi \tag{7.55}$$

Assume isothermal, single-fluid, laminar flow with no gravity in an ideal medium, and $\tilde{k} = k$

$$\phi \frac{\partial (p/z)}{\partial t} = k (\underline{\nabla} \cdot \frac{p}{z \mu} \underline{\nabla} p) \tag{7.56}$$

$$\frac{\partial (p/z)}{\partial t} = \frac{p}{z} \frac{\partial p}{\partial t} \left[\frac{1}{p} - \frac{1}{z} \left(\frac{\partial z}{\partial p} \right)_t \right] \tag{7.57}$$

Combine Eqs. (7.54), (7.56), and (7.57)

$$\frac{\partial p}{\partial t} = \frac{k}{\phi \mu c} \left\{ \nabla^2 p - \frac{d}{dp} \left[\ln \left(\frac{z \mu}{p} \right) \right] (\underline{\nabla} p)^2 \right\} \tag{7.58}$$

If pressure gradients are small, $(\underline{\nabla} p)^2$ approaches zero or if $p/z\mu$ is constant, then

$$\frac{\partial p}{\partial t} = \frac{k}{\phi \mu c} \nabla^2 p \tag{7.59}$$

which is the equation for a slightly compressible fluid. Since

$$p \underline{\nabla} p = \frac{1}{2} \nabla p^2 \tag{7.60}$$

and

$$p \partial p = \frac{1}{2} \partial p^2 \tag{7.61}$$

Eq. (7.58) can be rearranged to

$$\frac{\partial p^2}{\partial t} = \frac{k}{\phi \mu c} \nabla^2 p^2 - \frac{d}{dp^2} (\ln z \mu)(\underline{\nabla} p^2)^2 \tag{7.62}$$

If $z\mu$ is constant or $(d^2 p)^2 \to 0$, then

$$\phi \mu c \frac{\partial p^2}{\partial t} = k \nabla^2 p^2 \tag{7.63}$$

For an ideal gas $c = 1/p$ and

$$\phi \mu \frac{\partial p}{\partial t} = k \nabla^2 p^2 \tag{7.64}$$

Al-Hussainy (1965) defined a pseudopressure Ψ to account for variation of z and μ with pressure

$$\Psi = \int_{p_0}^{p} \frac{p}{z\mu} \, dp \qquad (7.65)$$

where p_0 is a specified reference pressure. Since

$$\nabla \underline{\Psi} = \frac{\partial \Psi}{\partial p} \underline{\nabla} p = \frac{2p}{z\mu} \underline{\nabla} p \qquad (7.66)$$

and

$$\frac{\partial \Psi}{\partial t} = \frac{2p}{z\mu} \frac{\partial p}{\partial t} \qquad (7.67)$$

$$\frac{\partial (p/z)}{\partial t} = \frac{cp}{z} \frac{\partial p}{\partial t} \qquad (7.68)$$

Substituting Eqs. (7.66) to (7.68) into Eq. (7.56) yields

$$\phi\mu c \, \frac{\partial \Psi}{\partial t} = (\nabla \cdot k \, \underline{\nabla} \Psi) \qquad (7.69)$$

If pressures are low and permeability varies with pressure,

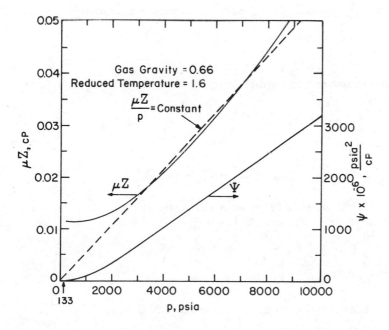

FIG. 7.26 Variation of Ψ and μZ with pressure. (From Wattenbarger, 1967.)

TABLE 7.8 $\tilde{\Phi}$ and κ for Eq. (7.72)

	$\tilde{\Phi}$	κ
Pressure	p	$\dfrac{k}{\phi\mu_{avg}c_{avg}}$
Pressure squared	p^2	$\dfrac{k}{\phi\mu_{avg}c_{avg}}$
Pseudopressure	ψ	$\dfrac{k}{\phi\mu_i c_i}$

$$\Psi' = 2 \int_{p_0}^{p} \frac{kp}{z\mu}\,dp \qquad (7.70)$$

and

$$\phi\mu c \,\frac{\partial \Psi}{\partial t} = \nabla^2 \Psi' \qquad (7.71)$$

Figure 7.26 shows the variation of Ψ and $z\mu$ with pressure. Tables of reduced pseudo-pressure in terms of reduced pressure and reduced temperature are available (Wattenbarger, 1967).

The equations of p, p , and ψ can be combined in a general equation

$$\frac{\partial \tilde{\Phi}}{\partial t} = \kappa \, \nabla^2 \tilde{\Phi} \qquad (7.72)$$

where Table 7.8 gives values of $\tilde{\Phi}$ and κ for each case. The dimensionless form of Eq. (7.72)

$$\frac{\partial(\Delta p_D)}{\partial t} = \nabla^2 (\Delta p_D) \qquad (7.73)$$

Table 7.9 gives the dimensionless definitions. Table 7.10 gives the dimensionless coefficients.

Analytical solutions of the radial flow case are available for certain boundary conditions. Equation (7.73) in radial coordinates is

$$\frac{\partial(\Delta p_D)}{\partial t} = \frac{1}{r_D}\frac{\partial}{\partial r_D}\left[r_D \frac{\partial}{\partial r_D}(\Delta p_D)\right] \qquad (7.74)$$

Initial and boundary conditions for a reservoir of infinite extent are

$$(\Delta p_D)\{r_D, 0\} = 0 \qquad \text{all } r_D \tag{7.75}$$

$$\frac{\partial(\Delta p_D)}{\partial r_D}\{1, t_D\} = -1 \qquad t_D > 0 \tag{7.76}$$

$$(\Delta p_D)\{\infty, t_D\} = 0 \qquad t_D > 0 \tag{7.77}$$

The solution of Eq. (7.74) for these conditions is

$$(\Delta p_D) = -\frac{1}{2} \text{Ei}\left(-\frac{r_D^2}{4t_D}\right) \tag{7.78}$$

Figure 7.27 shows values of (Δp_D) for Eq. (7.78). If the reservoir is finite, then the boundary condition of Eq. (7.77) is replaced by

TABLE 7.9 Definition of Dimensionless Variables in Terms of p, p^2, and ψ

Dimensionless variable	Flow geometry	Gas			Liquid
		p	p^2	ψ	p
x_D	Linear	$\dfrac{x}{x_f}$	$\dfrac{x}{x_f}$	$\dfrac{x}{x_f}$	$\dfrac{x}{x_f}$
r_D	Radial-cylindrical Radial-spherical	$\dfrac{r}{r_w}$	$\dfrac{r}{r_w}$	$\dfrac{r}{r_w}$	$\dfrac{r}{r_w}$
t_D	Linear	$\dfrac{\lambda kt}{\phi\bar{\mu}\bar{c}x_f^2}$	$\dfrac{\lambda kt}{\phi\bar{\mu}\bar{c}x_f^2}$	$\dfrac{\lambda kt}{\phi\mu_i c_i x_f^2}$	$\dfrac{\lambda kt}{\phi\mu c x_f^2}$
t_D	Radial-cylindrical Radial-spherical	$\dfrac{\lambda kt}{\phi\bar{\mu}\bar{c}r_w^2}$	$\dfrac{\lambda kt}{\phi\bar{\mu}\bar{c}r_w^2}$	$\dfrac{\lambda kt}{\phi\mu_i c_i r_w^2}$	$\dfrac{\lambda kt}{\phi\mu c r_w^2}$
p_D		$\dfrac{p}{p_i q_D}$	$\dfrac{p^2}{p_i^2 q_D}$	$\dfrac{\psi}{\psi_i q_D}$	$\dfrac{p}{p_i q_D}$
Δp_D		$\dfrac{p_i - p}{p_i q_D}$	$\dfrac{p_i^2 - p^2}{p_i^2 q_D}$	$\dfrac{\psi_i - \psi}{\psi_i q_D}$	$\dfrac{p_i - p}{p_i q_D}$
q_D	Linear Radial-cylindrical	$\dfrac{\gamma\bar{Z}Tq_{sc}\bar{\mu}}{\bar{p}khp_i}$	$\dfrac{\gamma\bar{Z}Tq_{sc}\bar{\mu}}{khp_i^2}$	$\dfrac{\gamma Tq_{sc}}{kh\psi_i}$	$\dfrac{\gamma Bq_{sc}\mu}{khp_i}$
q_D	Radial-spherical	$\dfrac{\gamma\bar{Z}Tq_{sc}\bar{\mu}}{\bar{p}krp_i}$	$\dfrac{\gamma\bar{Z}Tq_{sc}\bar{\mu}}{krp_i^2}$	$\dfrac{\gamma Tq_{sc}}{kr\psi_i}$	$\dfrac{\gamma Bq_{sc}\mu}{krp_i}$

Source: Energy Conservation Board, 1975.

TABLE 7.10 Values of Coefficients Used in the Dimensionless Terms

	Flow geometry	Gas			Liquid
		p	p^2	ψ	p
Darcy units					
λ	Linear/radial-cylindrical/ radial-spherical	1	1	1	1
γ	Linear	$\dfrac{p_{sc}}{T_{sc}}$	$2\dfrac{p_{sc}}{T_{sc}}$	$2\dfrac{p_{sc}}{T_{sc}}$	1
γ	Radial-cylindrical/ radial-spherical	$\dfrac{1}{2\pi}\dfrac{p_{sc}}{T_{sc}}$	$\dfrac{1}{\pi}\dfrac{p_{sc}}{T_{sc}}$	$\dfrac{1}{\pi}\dfrac{p_{sc}}{T_{sc}}$	$\dfrac{1}{2\pi}$
Field units					
λ	Linear/radial-cylindrical/ radial-spherical	2.637×10^{-4}	2.637×10^{-4}	2.637×10^{-4}	2.637×10^{-4}
γ	Linear (14.65 psia, 520°R)	4.452×10^6	8.903×10^6	8.903×10^6	887.3
	Radial-cylindrical/ radial-spherical (14.65/520)	7.085×10^5	1.417×10^6	1.417×10^6	141.2
γ	Linear (14.7 psia, 520°R)	4.467×10^6	8.933×10^6	8.933×10^6	887.3
	Radial-cylindrical/ radial-spherical (14.7/520)	7.110×10^5	1.422×10^6	1.422×10^6	141.2
Metric (SI) units					
λ	Linear/radial-cylindrical/ radial-spherical	3.601	3.601	3.601	3.601×10^{-3}
γ	Linear (101.325 kPa, 288 K)	9.624×10^{-2}	1.925×10^{-1}	1.925×10^{-1}	11.57
γ	Radial-cylindrical/ radial-spherical (101.325/288)	1.532×10^{-2}	3.064×10^{-2}	3.064×10^{-2}	1.84

Source: Energy Conservation Board, 1975.

$$\frac{\partial(\Delta p_D)}{\partial r_D}\{r_{eD}, t_D\} = 0 \qquad t_D > 0 \tag{7.79}$$

where

$$r_{eD} = \frac{r_e}{r_w} \tag{7.80}$$

The solution of Eq. (7.74) with initial and boundary conditions given by Eqs. (7.75), (7.76), and (7.79) is a combination of Bessel functions. In most practical situations

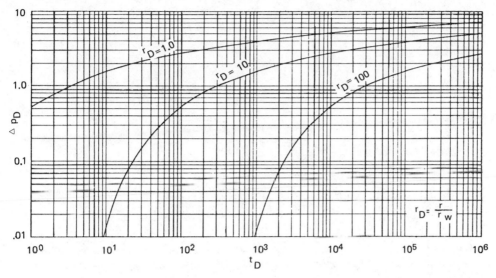

FIG. 7.27 Values of Δp_D at various dimensionless radii, for an infinite reservoir: Eq. (7.77). (From the Energy Conservation Board, 1975.)

$r_e \gg r_w$ and solutions in terms of p_t (values of Δp_D at the well excluding inertial, turbulent, and skin effects) is

$$p_t = \frac{2t_D}{r_{eD}^2} + \ln r_{eD} - \frac{3}{4} \qquad (7.81)$$

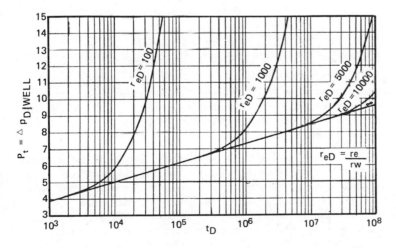

FIG. 7.28 Values of P_t ($= \Delta p_{D|\,well}$) for various finite circular reservoirs with no flow at the external boundary. (From the Energy Conservation Board, 1975.)

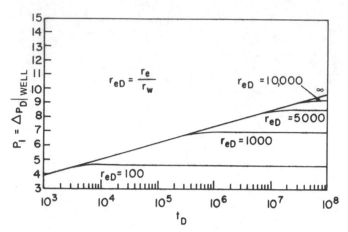

FIG. 7.29 Values of P_t (= $\Delta p_{D| well}$) for various finite circular reservoirs with constant pressure at the external boundary. (From the Energy Conservation Board, 1975.)

Figure 7.28 shows the solution for a circular reservoir.

 If the reservoir is finite but pressure is maintained constant at the outer boundary by a natural water drive or artificially by a pressure maintenance scheme, the boundary condition of Eq. (7.77) is replaced by

$$(\Delta p_D)\{r_{eD}, \; t_D\} = 0 \qquad t_D > 0 \tag{7.82}$$

The solution of Eq. (7.77) with initial and boundary conditions given by Eqs. (7.75), (7.76), and (7.82) is again a combination of Bessel functions. The solution is shown in Fig. 7.29, where

$$p_t = \ln r_{eD} \qquad t_D > r_{eD}^2 \tag{7.83}$$

CASE STUDY 5: Comparison of Flow and Pressure of a Gas Well
in an Infinite Circular Reservoir, a Finite Circular Reservoir
with No-Flow Boundaries, and a Finite Circular Reservoir with
a Boundary at Constant Pressure

This case study details the calculations of the pressure drop for the three boundary conditions listed above for a constant flow rate from a natural gas reservoir. The inverse problem of the flow rates for constant pressure drop is analogous. In addition, for the infinite cylindrical case, the three methodologies to account for physical property changes of the gas (pressure, pressure squared, and pseudopressure) are demonstrated. The reservoir to be considered has the following properties:

$h = 45$ ft $k = 25$ mdarcy $p_i = 2500$ psi $r_w = 0.5$ ft

  $T = 500\,°R$ ($120\,°F$) $Q = 10$ MMscf $\phi = 0.12$

The properties of the gas are evaluated from the "pseudocritical" properties, that is properties of the mixture determined from the critical properties of the components. The natural gas considered has the following composition, in terms of mole percent: 2.01% N_2, 1.2% CO_2, 4.3% H_2S, and only very minor hydrocarbon impurities. The correlations of Brown et al. (1948) and Carr et al. (1954)* were used to find the pseudocritical temperature and pressure to be, for a gravity of 0.612 that of air:

$$T_{pc} = 361 - 5 - 2 + 6 = 360 °R$$

$$P_{pc} = 672 - 9 + 5 + 28 = 696 \text{ psi}$$

and

$$M = 0.9249(16.04) + 0.0201(28.01) + 0.012(44.01) + 0.043(34.00)$$

$$= 17.39$$

The compressibility of the gas is then determined from the pseudoreduced properties,

$$p_{pr} = \frac{p}{p_{pc}} = 3.59 \qquad\qquad (7.84)$$

$$T_{pr} = \frac{T}{T_{pc}} = 1.61 \qquad\qquad (7.85)$$

utilizing the compressibility chart of Standing and Katz (1942)

$$\bar{z} = 0.818$$

The viscosity at standard conditions μ_1, 1 atm, and the temperature of operations is determined from Carr et al. (1954) with corrections for N_2, CO_2, and H_2S to be

$$\mu_1 = 0.0115 + 0.00015 + 0.00008 + 0.00009$$

$$= 0.01182 \text{ cP (at 1 atm)}$$

correcting for the pressure,

$$\frac{\mu}{\mu_1} = 1.5$$

$$\mu_{2500} = 0.01773$$

This is the viscosity at the initial pressure, 2500 psi. The isothermal compressi-

---

*The correlations of these two references and of Standing and Katz (1942) and Mattar et al. (1975) are conveniently reproduced in a work of the Energy Resources Conservation Board (1975).

bility c is defined by Eq. (7.52). In terms of the real gas correction, the compressibility z is defined by Eq. (7.54). Define for convenience a reduced compressibility, given as

$$c_r = cp_c = \frac{1}{p_r} - \frac{1}{z}\left(\frac{\partial z}{\partial p_r}\right) T_r \tag{7.86}$$

From the correlations of Mattar et al. (1975)

$$c_r T_r = 0.75 \qquad c = 0.000669$$

We have initial values for all the physical properties of the gases. However the pressure and pressure-squared solutions to the equations require physical properties evaluated at the "average" pressure; for example,

$$\frac{\mu_i - \mu_{wf}}{2} = \bar{\mu} \tag{7.87}$$

thus it is necessary to calculate new values for each value of p_{wf}. For systematic, frequent, usage a digital computer is employed, utilizing analytical expressions or trial and error procedures for determination of the properties.

For the first set of boundary conditions, the infinite radial reservoir, the sandface pressure p_{wf} is determined by all three methods. The equation solved in this case is Eq. (7.74) with boundary conditions (7.75-7.77). For all three boundary conditions assume laminar flow with negligible skin and inertial effects.

Pressure Method. Initially the average physical properties and pressure are that of the initial state. Solve the system at a time 50 hr into production as an example:

First calculate t_D from Tables 7.9 and 7.10

$$t_D = \frac{\lambda k t}{\phi \mu c r_w^2} = \frac{(2.637 \times 10^{-4})(25)(50)}{0.12(0.01773)(0.000669)(0.5)^2} = 926,326$$

for the sandface pressure, $r_D = 1.0$, defined in Table 7.9. From Fig. 7.27 we get $p_D = 7.0$. From Tables 7.9 and 7.10

$$q_D = \frac{\gamma \bar{z} T q_{sc} \bar{\mu}}{\bar{p} k h p_i} = \frac{(7.085 \times 10^5)(0.818)(580)(10)(0.01773)}{2500(25)(45)(2500)}$$

$$= 0.00848$$

From Table 7.9

$$\Delta p_D\Big|_{well} = \frac{p_i - p}{p_i q_D}$$

which gives

$$p = p_i - \Delta p_D\big|_{well} \; p_i q_D$$

so

$$p = 2500 - 7.0(2500)(0.00848) = 2352 \text{ psi}$$

Use this value to calculate a new p, μ, z, and c, then recalculate the p until no further change is observed. If, for this case, the physical properties are constant, only q_D change.

$$q_{D(new)} = 0.00873$$

so p = 2347 psi, a negligible change.

The Pressure-Squared Method. From Tables 7.9 and 7.10 t_D remains the same, however q_D is different. Again, let the physical properties remain constant.

$$q_D = \frac{(1.417 \times 10^6)(0.818)(580)(10)(0.01773)}{25(45)(2500)^2} = 0.0170$$

Now

$$\Delta p_D\big|_{well} = \frac{p_i^2 - p^2}{p_i^2 q_D}$$

$$p = (p_i^2 - p_i^2 \Delta p_D q_D)^{0.5} = [(2500)^2 - (2500)^2(7.0)(0.017)]^{0.5}$$

$$= 2347 \text{ psi}$$

which is the same result as the pressure method.

The Pseudopressure Method. This method requires no assumptions about the physical properties; however the ψ-p curve must be constructed. For our gas, this is shown in Fig. 7.30. Figure 7.30 was constructed by numerical integration (Standing and Katz, 1942). The program is capable of handling sweet and sour gases with N_2 and CO_2 impurities, but with no provision for the effects of hydrocarbon impurities. Tables constructed from numerical computations are available for sweet (Standing and Katz, 1942) and sour (Zana and Thomas, 1970) gases for the conversion to pseudo pressures. For our problem with the infinite-acting reservoir, find from Tables 7.9 and 7.10

$$t_D = \frac{(2.637 \times 10^{-4})kt}{\phi \mu_i c_i r_w^2} = \frac{(2.637 \times 10^{-4})(25)(50)}{0.12(0.01773)(0.000669)(0.25)} = 926,326$$

(Note that μ_i and c_i, the values at the initial pressure are now used.)

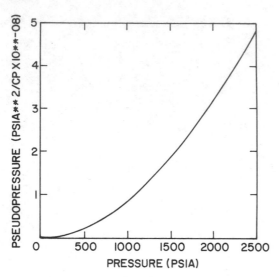

FIG. 7.30 P-psi plot.

From Fig. 7.27 $\Delta p_D = 7.0$. From Tables 7.9 and 7.10 and Fig. 7.30,

$$q_D = \frac{1.417 \times 10^6 \, T q_{sc}}{kh\psi_i} = \frac{(1.417 \times 10^6)(580)(10)}{25(45)(4.7977 \times 10^8)}$$

$$= 0.0152$$

Again from Table 7.9

$$\Delta p_D\Big|_{well} = \frac{\psi_i - \psi_{wf}}{\psi_i q_D}$$

so

$$\psi_{wf} = \psi_i - \psi_i \Delta p_D\Big|_{well} q_D = (4.798 \times 10^8) - (4.798 \times 10^8)(7)(0.0152)$$

$$\psi_{wf} = 4.287 \times 10^8$$

Converting back from Fig. 7.30

$$p = 2345 \text{ psi}$$

The p and p^2 methods give the same result. This is due to the use of the same average values for $\bar{z}$ and $\bar{\mu}$. The deviations between the three methods are due to the differences in the way the properties of the fluid are averaged, as well as the parameters of the reservoir. For large pressure drops one expects the ψ method to provide a more accurate result.

For the other two sets of boundary conditions use only the ψ method. For the finite circular reservoir with no-flow boundary solving Eq. (7.74) with boundary conditions (7.75), (7.76), and (7.79). This is solved as before except now Fig. 7.28 is used. Since

$$t_D = 926,326$$

$$q_D = 0.0152$$

let r_e = 2500 ft with r_{eD} defined by Eq. (7.80). Then from Fig. 7.28, $\Delta p_D\big|_{well} - 7.1$ and

$$\psi_{wf} = \psi_i - \psi_i \Delta p_D\big|_{well} q_D = (4.798 \times 10^8) - (4.798 \times 10^8)(7.1)(0.0152)$$

$$= 4.28 \times 10^8$$

and from Fig. 7.30,

$$p = 2343 \text{ psi}$$

For the constant pressure boundary condition, Eq. (7.82), is substituted in place of Eq. (7.79(with the other boundary conditions remaining the same to solve Eq. (7.74). Figure 7.29 provides the $\Delta p_D\big|_{well}$ for various r_{eD}. For $r_{eD}\big|_{well}$ = 5000, we find $\Delta p_D\big|_{well}$ = 7.2; so ψ_{wf} = 4.273×10^8 and p = 23.50 psi. From Fig. 7.29 for sufficiently large t_D there is a true steady-state solution, i.e., $\Delta p_D\big|_{well}$ = constant.

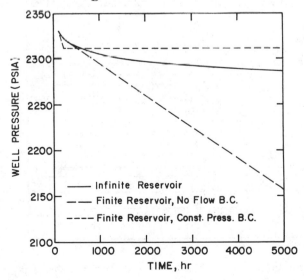

FIG. 7.31 Well pressure versus time.

The most important result is the behavior of each of these solutions as a function of time. At small times, all the solutions are nearly the same; however as time becomes large, the solutions differ significantly. Figure 7.31 shows the solutions as a function of time for all three conditions. It is easy to see the difference between the different assumptions. Assuming an infinite reservoir causes a slow steady pressure drop, while the finite, no-flow reservoir rapidly starts to see the effect of the boundary. The constant pressure case has a similar initial pressure drop as the others, but then reaches a true steady state. ∎

Figure 7.32 is a plot produced by superimposing Figs. 7.29, 7.30, and 7.31. It illustrates the behavior of the three types of reservoirs. For sufficiently small t_D [$t_D < 0.25r_{eD}^2$ (or r_D^2)] the solution is the same, forming the line running from lower left to upper right. For t_D greater than $0.25r_{eD}^2$ the solution changes, as has been demonstrated from the previous examples. The solution assumes a steady state for the constant pressure boundary; becomes increasingly smaller ($\Delta p_D|_{well}$ gets bigger) for the no flow boundary and remains along the original line for the infinite reservoir case. For t_D sufficiently small, the solutions for the two finite reservoirs are the same as the infinite-acting one, i.e., the well pressure has not yet seen the effect of the reservoir boundary.

Solutions for the three sets of boundary conditions discussed above are summarized by Aziz and Flock (1963) in Fig. 7.32. The effective drainage radius of a reservoir r_d—the radius of which the steady-state solution represents the transient solution—is shown if Fig. 7.33. The equations for the effective drainage radius r_d

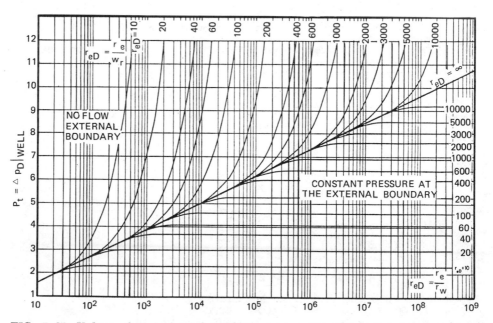

FIG. 7.32 Values of P_t (= Δp_D) for infinite reservoirs, for finite circular reservoirs with no flow at the external boundary and for finite circular reservoirs with constant pressure at the external boundary. (From Aziz and Flock, 1963.)

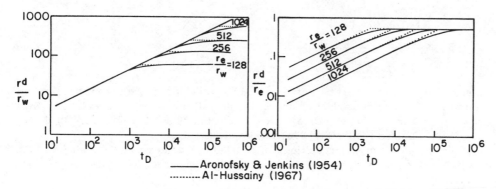

FIG. 7.33 Effective drainage radius as a function of time. (From J. S. Aronofsky and R. Jenkins, A Simplified Analysis of Unsteady Radial Gas Flow, Pet. Trans. AIME 201, SPEJ 149. © 1954 SPE-AIME; and Al-Hussainy, 1967.)

of various reservoir types are summarized in Table 7.11. The radius on investigation r_{inv} is the distance from the well where there is a significant influence on the production of the well

$$r_{inv} = r_w \sqrt{4t_D} \tag{7.88}$$

When the radius of investigation reaches the outer boundary, the well is said to be stabilized. The time to stabilization can be computed for Eq. (7.88), where $r_{inv} = r_{eD}$. In production situations wells are produced against a fixed back pressure. The flow rate declines continuously and the cumulative production is

$$Q_t = 2\pi\phi c r_w^2 h \frac{T_{sc}}{T} \frac{p_i}{p_{sc}} (p_i - p_{wf}) Q_t \tag{7.89}$$

where

TABLE 7.11 Effective Drainage Radius

Reservoir type	Time limit	Drainage radius
Infinite	$t_D > 25$	
Closed outer boundary	$25 < t_D < 0.25 r_{eD}^2$	$\ln \dfrac{r_d}{r_w} = \dfrac{1}{2}(\ln t_D + 0.809$
Constant pressure outer boundary	$25 < t_D < 0.25 r_{eD}^2$	
Closed outer boundary	$t_D > 0.25 r_{eD}^2$	$r_d = 0.472 r_e$
Constant pressure outer boundary	$t_D > 1.0 r_{eD}^2$	$r_d = r_e$

$$Q_t = \frac{\int_D^{t_D} (\partial p/\partial r_D)\, \partial t_D}{p_i - p_{wf}} \tag{7.90}$$

Values of Q_T are tabulated by Katz et al. (1959). The pressure distribution in an infinite reservoir with production at a constant well pressure is given by Van Everdingen and Hurst (1949) as an integral of Bessel functions. The pressure in the linear and spherical cases is also given by Katz et al. (1959).

The principle of superposition can be used to extend results for single wells in a radial system to multiple wells and polygonal geometries. The principle of superposition is

$$p_i - p_{wf} = p_i q_{D1} P_t\{t_D\} + p_i (q_2 - q_1)_D P_t\{t - t_1\}_D + \cdots \tag{7.91}$$

Figure 7.34 shows superposition schematically. The principle of superposition in space is

$$\Delta p\big|_{point} = \Delta p\big|_{point\ due\ to\ q_A\ in\ well\ A} + \Delta p\big|_{due\ to\ q_B\ in\ well\ B}$$

$$= p_i q_{AD}\left[-\frac{1}{2}E_i\left(-\frac{r_{AD}^2}{4t_D}\right)\right] + p_i q_{BD}\left[-\frac{1}{2}E_i\left(-\frac{r_{BD}^2}{4t_D}\right)\right] + \cdots \tag{7.92}$$

The method of images is an application of the principle of superposition. When two wells are producing at the same rate in an infinite reservoir, halfway between them there is a no-flow boundary. If one of the wells is producing the fluid is injected at the other, at a point halfway between them the pressure is constant. Thus these boundaries can be simulated using image wells. The superposition of the real and the image well gives the boundary condition: no flow or constant pressure. If $\Phi(x, y, t)$ is the solution to the flow of equation in two dimensions satisfying the boundary condition

$$\Phi(x, o, t) = 0 \tag{7.93}$$

then for $y < 0$

$$\Phi(x, y, t) = -\Phi(x, -y, t) \tag{7.94}$$

A producing well in a semi-infinite reservoir with a constant pressure boundary can be simulated by superimposing an image well at an equal distance on the other side of the boundary injecting at the same flow rate as the producing well. This is called an odd image.

If

$$\frac{\partial}{\partial x}\Phi(x, o, t) = 0 \tag{7.95}$$

then for $y < 0$

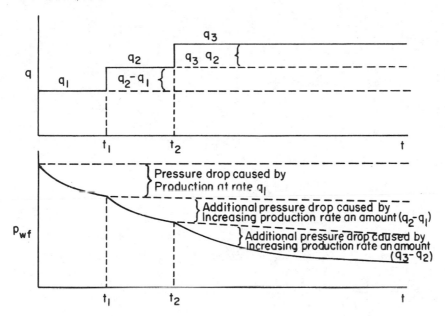

FIG. 7.34 Superposition. (From J. S. Aronofsky and R. Jenkins, A Simplified Analysis of Unsteady Radial Gas Flow, Pet. Trans. AIME 201, SPEJ 149. © 1954 SPE-AIME.)

$$\Phi(x, y, t) = \Phi(x, -y, t) \tag{7.96}$$

A producing well in a reservoir with a no-flow boundary can be simulated by a well producing in an infinite reservoir by superimposing an image well at an equal distance on the other side of the boundary producing at the same flow rate as the producing well. This is called an even image.

A well with a constant production rate in a rectangular reservoir with a no-flow boundary can be simulated with an infinite number of images. Practically the simulation is good after a few images. All the images are superimposed so that

$$(\Delta p)_{well} = p_i q_d \left[\frac{1}{2}(\ln t_D + 0.809) + \sum_{n=1}^{\infty} -\frac{1}{2}E1\left(-\frac{\phi \mu c d_n^2}{4\lambda k t}\right) \right] \tag{7.97}$$

where d_n is the distance from the real well to the nth well. Eq. (7.97) applies in any polygonal drainage area. Define a dimensionless time based on a drainage area A

$$t_{DA} = t_D \frac{r_w^2}{A} \tag{7.98}$$

in the transient region t_{DA} is small and the well behaves as if it is an infinite reservoir; so

$$p_t = \frac{1}{2} \ln \left(\frac{4At_{DA}}{1.781r_w^2} \right) \qquad\qquad (7.99)$$

In the region between transient and pseudo-steady state

$$p_i - p_{wf} = p_i q_D \frac{1}{2} \left(\ln \frac{4At_{DA}}{1.781r_w^2} + 4\pi t_{DA} - F \right) \qquad\qquad (7.100)$$

where

$$F = 4\pi t_{DA} + \sum_{n=1}^{\infty} Ei\left(-\frac{d_n^2}{4t_{DA}A} \right) \qquad\qquad (7.101)$$

In the pseudo-steady-state region t_{DA} is large; all the boundaries reflect and

$$p_t = \frac{1}{2} \ln \frac{4A}{1.781r_w^2 C_A} + 2\pi t_{DA} \qquad\qquad (7.102)$$

Values of F were determined for various closed rectangular reservoirs by Earlougher et al. (1968) and values of C_A for various reservoirs by Dietz (1965).

CASE STUDY 6: Extensions of Cylindrical Solutions of Flow
Equations in a Gas Reservoir to Rectangular Shapes

The same reservoir and gas properties from Case Study 5 are used here. For a well with a constant production rate, enclosed in a rectangular reservoir with no flow across the boundaries, there will be an infinite number of images because each image reflects off all the boundaries. Superimposing the effect of each of the image wells, the pressure drop at the real well is given by Eq. (7.97), where d_N is the distance of the Nth image well from the real well. For a regular pattern of wells d_N is easy to calculate and Eq. (7.97) converges rapidly.

Equation (7.97) applies for a well in any polygonal drainage area and the well does not have to be in the center. A dimensionless time based on the drainage area A is given by Eq. (7.98). The general equation for all t_{DA} and thus flow states are given by Eqs. (7.100) and (7.101). When t_{DA} is small in the transient flow region $F = 4\pi t_{DA}$; so

$$p_t = \frac{1}{2} \ln \frac{4At_{DA}}{1.781r_w^2} \qquad\qquad (7.103)$$

When t_{DA} is large, flow is in the pseudo-steady-state region and all boundaries reflect; so

$$p_t = \frac{1}{2} \ln \frac{4A}{1.781r_w^2 C_A} + 2\pi t_{DA} \qquad\qquad (7.104)$$

where C_A is a shape factor. F values for various rectangles have been calculated by Earlougher et al. (1968) and C_A values by Dietz (1965). For a 2-by-1 center drain rectangular area $\boxed{\;\cdot\;}\,1$, F = 3.02 (Earlougher et al., 1968),

$$\Delta p\big|_w = p_i q_D \frac{1}{2} \ln\left(\frac{4At_{DA}}{1.781r_w^2} + 4\pi t_{DA} - 3.02\right)$$

$$= 2500(0.00873)(0.5) \ln \frac{4 \times 10^6 \times 0.92}{1.781(0.5)^2} + 4\pi(0.92) - 3.02$$

$$= 174$$

$$p = 2500 - 174 = 2341 \text{ psi}$$

This is in the pseudo-steady-state region. The value for a similar condition in a finite circular well was 2312 psi at r_e = 2500 ft or A = 2×10^7 ft^2.

For a rectangular no-flow boundary for A = 2×10^7 ft^2 and t = 50 hr and 2-by-1 center drain rectangular area

$$t_{DA} = 926326 \frac{(0.5)^2}{2 \times 10^7} = 0.012$$

$$F = 0.125$$

$$\Delta p\big|_w = 159$$

p = 2500 - 159 = 2340 psi which is closer to the value in a circular area.

Since the first result was found to be in the pseudo-steady-state region, it can also be done using the shape factor C_A. For a 2-by-1 center drain rectangular area C_A = 22.6 (Dietz, 1965)

$$p_t = \frac{1}{2} \ln \frac{4A}{1.781r_w^2 C_A} + 2\pi t_{DA}$$

$$= \frac{1}{2} \ln \frac{4 \times 10^6}{1.781(0.5)^2(22.6)} + 2\pi(0.92)$$

$$= 12.2$$

$$p_t = \frac{p_i - p}{p_i q_D} \quad \text{or} \quad p = p_i - p_t p_i q_D$$

p = 2500 - 12.2(2500)(0.00873) = 2232 psi ∎

If the pressure equations and boundary conditions cannot be linearized, then the principle of superposition does not apply. In these complex situations the solution of the equations is numerical. A useful simulation model for gas reservoirs with water drive is a two-dimensional r-z geometry with a two-phase model.

CASE STUDY 7: Numerical Simulation of Pressure in a Gas
Reservoir with Water Influx

Givens (1968) presents an approximate simulation procedure for a gas reservoir
with water influx using the numerical solution for two-dimensional unsteady-flow of
a real gas coupled with a water influx calculation. The equation to be represented
numerically is

$$\frac{\partial}{\partial x}\left(\frac{k_x h}{\mu z}\frac{\partial p^2}{\partial x}\right) + \frac{\partial}{\partial y}\left(\frac{k_y h}{\mu z}\frac{\partial p^2}{\partial y}\right) = \frac{\phi hc}{z}\frac{\partial p^2}{\partial t} + 2q(x, y, t) \qquad (7.105)$$

where

$$c = \frac{1}{p} - \frac{1}{z}\frac{\partial z}{\partial p} \qquad (7.106)$$

The grid network of Fig. 7.35 is superimposed over a map of the reservoir. A modified ADIP procedure is used to calculate the pressure distribution with z and μ
determined at reservoir conditions. Matrices for $k_x h/\mu z$, $k_y h/\mu z$, and h/z were
determined. The pressure distribution at the last time step is used in evaluating
these matrices. The values are not changed during the alternate sweeps of the next
time step. The size of the time steps are controlled by the growth of the material
balance error.

The accumulative water influx is calculated from the unsteady-state equation
of van Everdingen and Hurst (1949).

$$W_e = C \sum q(t_D)\Delta p \qquad (7.107)$$

where

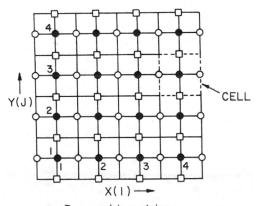

- ● P, q, and h matrices
- ○ kxh matrix
- □ kyh matrix

FIG. 7.35 Calculation grid. (From J. W. Givens, A Practical Two-Dimensional
Model for Simulating Dry Gas Reservoirs with Bottom Water Drive, J. Pet. Tech.
20, 1229. © November 1968 SPE-AIME.)

$$C = 1.191\,h\phi c_w r_e^2 \tag{7.108}$$

$$t_D = \frac{0.00634 k_w t}{\phi\mu c_w r_e^2} \tag{7.109}$$

Time increments corresponding to the calculated pressures at 10, 20, 30, 60, 90, 180, ... days are used. The cumulative water influx is computed at these times and the net influx is determined by coupling to the previous time step. The net water influx is distributed to the cells of Fig. 7.35 according to their pressure and thickness using

$$(\Delta W_e)_C = \frac{(\Delta W_e)_N [1 - (z_1/z_2)(p_2/p_1)]_{i,j}\, h_{i,j}}{\sum_{i=1}^{N_x}\sum_{j=1}^{N_y} [1 - (z_1/z_2)(p_2/p_1)]_{i,j}\, h_{i,j}} \tag{7.110}$$

TABLE 7.12 Reservoir Data

Initial reservoir pressure, psia	1,500
Reservoir temperature, °F	140
Average porosity, %	10
Average water saturation in gas zone, %	25
Standard temperature, °F	60
Standard pressure, psia	14.65
Gas z-factor (constant)	0.92
Gas viscosity (constant), cP	0.0121
Backpressure on wells (caused by pipeline pressure), psia	100
Water influx constant, bbl/psi	568
Dimensionless time constant, day^{-1}	0.04
Residual gas saturation in water zone, %	15
Number of producing wells	5
Grid size (Δx), ft	400
Initial gas in place, Bscf	7.16962

Source: J. W. Givens, A Practical Two-Dimensional Model for Simulating Dry Gas Reservoirs with Bottom Water Drive, J. Pet. Tech. 20, 1229, November 1968. © 1968, SPE-AIME.

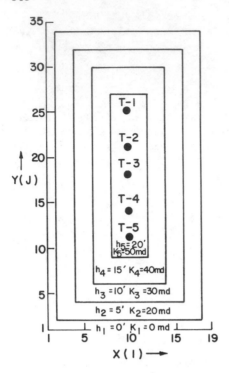

FIG. 7.36 Hypothetical reservoir. (From
J. W. Givens, A Practical Two-Dimen-
sional Model for Simulating Dry Gas
Reservoirs with Bottom Water Drive,
J. Pet. Tech. 20, 1229. © November 1968
SPE–AIME.)

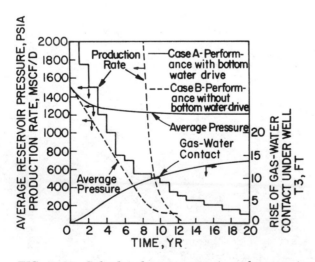

FIG. 7.37 Calculated pressure rate and gas–water contact. (From J. W. Givens,
A Practical Two-Dimensional Model for Simulating Dry Gas Reservoirs with Bottom
Water Drive, J. Pet. Tech. 20, 1229. © November 1968 SPE–AIME.)

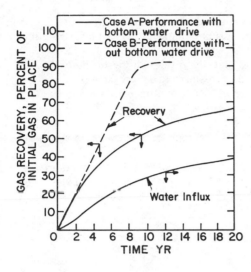

FIG. 7.38 Calculated recovery and water
influx. (From J. W. Givens, A Practical
Two-Dimensional Model for Simulating
Dry Gas Reservoirs with Bottom Water
Drive, J. Pet. Tech. 20, 1229. © November
1968 SPE-AIME.)

This equation is based on the volume of water influx with each cell under complete pressure maintenance. The water enrichment is assumed to change the thickness. The adjusted thickness is

$$h_{n+1} = h_n - \frac{(\Delta W_e)_x}{\Delta x^2 \phi (1 - S_w - S_{gr})} \tag{7.111}$$

The flow capacity of each cell is adjusted by

$$(kh)_{n+1} = \frac{(kh)_n h_{n+1}}{h_n} \tag{7.112}$$

The adjusted pressure is a function of the thickness and the pressure. This is determined iteratively:

$$p_{n+1} = p_n \frac{z_{n+1}}{z_n} \frac{h_n}{h_{n+1}} \tag{7.113}$$

To insure stability in the calculation the next step is kept small.

The calculation procedure is a sequence of unsteady-state pressure depletion calculations followed by a water influx correction. The procedure is as follows:

1. Set up initial values of pressure, rate, flow capacity in the x direction, flow capacity in the y direction, and the thickness matrices.
2. Select a time step (Δt) for the pressure calculation.
3. Calculate $k_x h/\mu z$, $k_y h/\mu z$, and h/z using the pressure at the previous time step to determine μ and z.
4. Calculate the new pressure distribution by ADIP.

5. Calculate the material balance error and repeat steps 2 to 4 until μ and z.
6. At the time interval for the water influx calculation, calculate the cumulative water influx.
7. Calculate the net water influx and distribute the volume to the cells. Change the thickness and adjust the pressure and flow capacity.
8. Calculate the material balance errors. If it is within limits and final prediction is not reached, repeat 2 through 6.

To prevent water coning, the prediction interval is kept small enough so the rate is less than the water coning rate.

The calculation is applied to a hypothetical reservoir with properties given by Table 7.12 and reservoir data according to Fig. 7.36. The results of the calculation are shown in Figs. 7.37 and 7.38. The results show that ultimate recovery is decreased by encroachment of water.

7.4 OIL RESERVOIRS

Solution gas-drive and water-drive oil reservoirs are considered in this section. If reservoir oil pressure, temperature, and composition are such that the oil is at or above its bubble point, primary production results from expansion of the oil and dissolved gases. If a large aquifer is in contact with the oil reservoir either at the bottom or side and its influx replaces the fluids withdrawn, primary production is by water drive. Muskat (1949) describes the effect of the fluid and reservoir properties for solution gas drive in terms of saturations, formation volume factors, and gas oil ratios by

$$\left(1 + \frac{\mu_o k_g}{\mu_g k_o}\right)\frac{dS_o}{dp} = S_o\left(\frac{B_g}{B_o}\frac{dR_s}{dp}\right) + (1 - S_o - S_w)\left(B_g\frac{d(1/B_g)}{dp}\right) + S_o\frac{k_g}{k_o}\left(\frac{\mu_o}{\mu_g B_o}\frac{dB_o}{dp}\right)$$

$$(7.114)$$

Equation (7.114) is solved graphically, by trial and error, or numerically. The material balance for oil and gas separately are for oil

$$\Delta N_d = V_P\left(\frac{S_{o1}}{B_{od_1}} - \frac{S_{o2}}{B_{od_2}}\right) = \frac{V_p}{B_{od_2}}\left[\Delta S_o - \Delta S_{o1}\left(\frac{B_{od_1} - B_{od_2}}{B_{od_1}}\right)\right] \qquad (7.115)$$

and for gas

$$\Delta G_d = V_p\left[S_{o2}\left(\frac{1}{B_g} - \frac{R_{sd}}{B_{od}}\right)_2 - S_{o1}\left(\frac{1}{B_g} - \frac{R_{sa}}{B_{od}}\right)_1 + S_{o2}\left(\frac{1}{B_{g1}} - \frac{1}{B_{g2}}\right)\right] = R_d\Delta N_d$$

$$(7.116)$$

The gas/oil ratio is

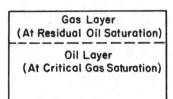

FIG. 7.39 Vertical saturation distribution for complete segregation.

$$R = R_s + \left(\frac{B_o \mu_o k_g}{B_g' \mu_g k_o} \right)$$ (7.117)

The assumption is made that the evolved gas and oil separate into two horizontal layers as in Fig. 7.39. The lower oil layer is at the gas saturation at which the evolved gas flows. The upper gas layer is at residual oil saturation. Recovery is assumed independent of rate. Differential pVT properties are used in the material balance equations. Reservoir hydrocarbon pore volume and temperature are assumed constant and the fluids are at equilibrium. Relative permeabilities are assumed uniform throughout the reservoir. Initial fluid saturations are determined from core data or from well log analysis. Volumetric estimates of the in-place oil are also required. The estimates of the in-place oil are determined from either production history or estimates from the material balancing equations.

The vertical segregation of oil and gas is assumed to take place at equal flow rates q_v (in barrels per day):

$$q_v = \frac{0.00112 k_v A (\rho_o - \rho_g)}{(\mu_o / k_o) + (\mu_g / k_g)}$$ (7.118)

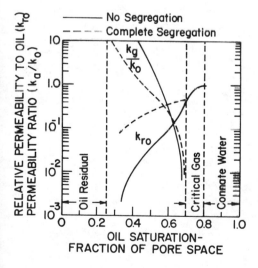

FIG. 7.40 Comparison of no-segregation and complete-segregation relative permeability data. (From Frick and Taylor, 1962.)

Calculation procedures are different whether the gas and oil are segregated or not. The minimum time required for segregation can be estimated from a Buckley-Leverett calculation or from empirical expressions. The relative permeabilities used depend on whether the fluids are segregated or not. Figure 7.40 shows the differences in relative permeability data between segregated and nonsegregated systems.

CASE STUDY 8: Calculation of Oil Production
from a Solution Gas-Drive Reservoir

In a solution gas-drive reservoir when the pressure of a producing well falls below the bubble point, oil is no longer produced by the expansion of reservoir fluids. Instead, production proceeds by solution gas drive. Two methods for predicting the performance of solution gas drive reservoirs from their rock and fluid properties are due to Muskat (1949) and to Tarner (1944). Because the gas originates within the oil, the mechanism for solution gas drive is complex. A number of simplifying assumptions are used in both Muskat's and Tarner's methods. These assumptions do not appreciably reduce the accuracy of the determinations. The assumptions are

1. Uniformity of the reservoir at all times regarding porosity, fluid saturations, and relative permeabilities
2. Uniform pressure throughout the reservoir in both the gas and oil zones
3. Negligible gravity separation forces
4. Equilibrium at all times between the gas and the oil phases
5. No water encroachment

In the Muskat method, incremental gas and oil productions are calculated for a small pressure drop. Then the reservoir properties are recalculated at the new pressure. This scheme is repeated until any desired abandonment pressure has been reached.

To derive the Muskat equation, let V_p be the reservoir pore volume in barrels. Then the stock tank barrels of oil remaining N_r at any pressure is given by

$$N_r = \frac{S_o V_p}{B_o} \qquad (7.119)$$

Differentiating with respect to pressure

$$\frac{dN_r}{dp} = V_p \left(\frac{1}{B_o} \frac{dS_o}{dp} - \frac{S_o}{B_o^2} \frac{dB_o}{dp} \right) \qquad (7.120)$$

The gas remaining in the reservoir, both free and dissolved, at the same pressure in standard cubic feet is

$$G_r = \frac{R_s V_p S_o}{B_o} + (1 - S_o - S_w) B_g V_p \qquad (7.121)$$

$$\frac{dG_r}{dp} = V_p \left[\frac{R_s}{B_o} \frac{dS_o}{dp} + \frac{S_o}{B_o} \frac{dR_s}{dp} - \frac{R_s S_o}{B_o^2} \frac{dB_o}{dp} + (1 - S_o - S_w) \frac{dB_g}{dp} - \frac{B_g dS_o}{dp} \right]$$

$$(7.122)$$

If the reservoir pressure is dropping at the rate dp/dt, then the producing gas/oil ratio (GOR) at this pressure is

$$R = \frac{\Delta G_p}{\Delta N_p} = \frac{\Delta G_r}{\Delta N_r} = \frac{\Delta G_r / \Delta p}{\Delta N_r / \Delta p} = \frac{dG_r / dp}{dN_r / dp}$$

$$(7.123)$$

Substituting Eqs. (7.120) and (7.122) into (7.123)

$$R = \frac{\dfrac{R_s}{B_o} \dfrac{dS_o}{dP} + \dfrac{S_o}{B_o} \dfrac{dR_s}{dp} - \dfrac{R_s S_o}{B_o^2} \dfrac{dB_o}{dp} + (1 - S_o - S_w) \dfrac{dB_g}{dp} - B_g \dfrac{dS_o}{dp}}{\dfrac{1}{B_o} \dfrac{dS_o}{dp} - \dfrac{S_o}{B_o^2} \dfrac{dB_o}{dp}}$$

$$(7.124)$$

The producing GOR can also be developed from the flow equations of oil and gas. The flow q_g is expressed at the mean reservoir pressure and temperature

$$q_g = \frac{7.08 k_g h (p_e - p_w)}{\mu_g \ln (r_e / r_w)}$$

$$(7.125)$$

and for oil

$$q_o = \frac{7.08 k_o h (p_e - p_w)}{\mu_o \ln (r_e / r_w)}$$

$$(7.126)$$

The flowing GOR is

$$\frac{q_g / B_g}{q_o / B_o} = B_o B_g \frac{k_g}{k_o} \frac{u_o}{u_g} \text{ SCF STB}^{-1}$$

$$(7.127)$$

Since Eq. (7.127) pertains only to the free flowing gas, another term must be added to account for the solution gas which flows to the well bore in the oil.

$$R = B_o B_g \frac{k_g}{k_o} \frac{u_o}{u_g} + R_s$$

$$(7.128)$$

Equations (7.128) and (7.124) are equated and dS/dp is solved

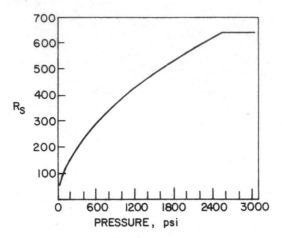

FIG. 7.41 R_s versus P, X(p).

$$\frac{dS_o}{dp} = \frac{\dfrac{S_o}{B_o B_g}\dfrac{dR_s}{dp} + \dfrac{S_o}{B_o}\dfrac{k_g}{k_o}\dfrac{u_o}{u_g}\dfrac{dB_o}{dp} + (1 - S_o - S_w)\dfrac{1}{B_g}\dfrac{dB_g}{dp}}{1 + \dfrac{k_g}{k_o}\dfrac{u_o}{u_g}} \qquad (7.129)$$

To simplify Eq. (7.129), the terms in the numerator which are functions of the pressure only may be grouped together

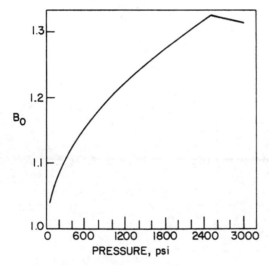

FIG. 7.42 B_o versus P, Y(p).

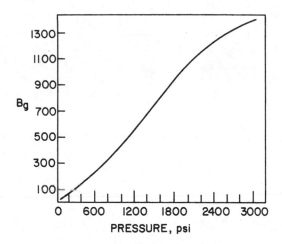

FIG. 7.43 B_g versus P, Z(p).

$$X(p) = \frac{dR_s}{dp} \qquad Y(p) = \frac{dB_o}{dp} \qquad Z(p) = \frac{dB_g}{dp}$$

and Eq. (7.129) is simplified to

$$\Delta S_o = \Delta p \, \frac{\dfrac{S_o}{B_o B_g} X(p) + \dfrac{S_o}{B_o} \dfrac{k_g}{k_o} \dfrac{u_o}{u_g} Y(p) + (1 - S_o - S_w) \dfrac{1}{B_g} Z(p)}{1 + \dfrac{k_g}{k_o} \dfrac{u_o}{u_g}} \qquad (7.130)$$

Equation (7.130) gives the change in oil saturation which accompanies a pressure drop Δp. The derivatives dR_s/dp, dB_o/dp, and dB_g/dp are found graphically from plots of R_s, B_o, and B_g versus pressure. The plots are shown in Figs. 7.41, 7.42, and 7.43. Typical reservoir data are shown in Table 7.13 and Figs. 7.44 and 7.45. Figure 7.44 shows the oil and gas viscosity versus pressure. Figure 7.45 shows the relative permeability as a function of total liquid saturation.

To determine a solution gas-drive performance curve using the Muskat model, Eq. (7.130) is used to solve for S_o at each new pressure drop. Then using Figs. 7.44

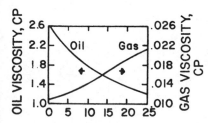

FIG. 7.44 Oil and gas viscosities. (From Craft/Hawkins, <u>Petroleum Reservoir Engineering</u>, © 1959, pp. 381, 382. Reprinted by permission of Prentice-Hall, Inc., Englewood Cliffs, N.J.)

TABLE 7.13 Typical Rock and Fluid Data

Pressure (psia)	Oil volume factor Bo (bbl STB^{-1})	Solution GOR Rs (SCF STB^{-1})	Gas volume factor Bg (SCF bbl^{-1})
2500	1.325	650	1256
2300	1.311	618	1186
2100	1.296	586	1102
1900	1.281	553	999
1700	1.266	520	880
1500	1.250	486	750
1300	1.233	450	619
1100	1.215	412	501
900	1.195	369	381
700	1.172	320	287
500	1.143	264	195
300	1.108	194	111
100	1.057	94	35
50	1.041	55	17

Source: B. C. Craft and M. F. Hawkins, Applied Petroleum Reservoir Engineering. © 1959, pp. 381, 387. Reprinted by permission of Prentice-Hall, Englewood Cliffs, New Jersey.

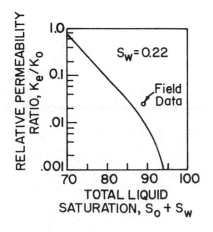

FIG. 7.45 Relative permeability ratio. (From Craft/Hawkins, Petroleum Reservoir Engineering, © 1959, pp. 381, 382. Reprinted by permission of Prentice-Hall, Inc., Englewood Cliffs, N.J.)

and 7.45, the viscosities and relative permeability are calculated at that pressure and saturation. R is calculated from Eq. (7.128).

As an example calculate the ΔS_o, S_o, and R for 1700 psi. The slopes X(p), Y(p), and Z(p) are found graphically at 1900 psi.

$$X(p) = 1.6 \times 10^{-1}$$
$$Y(p) = 9.5 \times 10^{-5}$$
$$Z(p) = 5.5 \times 10^{-1}$$

The connate water is constant at 22 percent. The reservoir properties are calculated at 1900 psi

$$\frac{k_g}{k_o} = 0.01 \qquad B_o = 1.28$$

$$\frac{u_o}{u_g} = 81.1 \qquad B_g = 999$$

$$S_w = 0.22 \qquad R_s = 553$$

$$S_o = 0.700$$

and the change in oil saturation is calculated using the Muskat equation

$$\Delta S_o = 9.86 \times 10^{-5}$$

The new saturation at 1700 psi is

$$S_o = 0.700 - 9.86 \times 10^{-5} = 0.699$$

and the gas/oil ratio is

$$R = 1475 \text{ SCF STB}^{-1}$$

According to Tracy (1955) the material balance for the case of a volumetric, undersaturated reservoir can be simply expressed

$$N = N_p \Phi_n + G_p \Phi_g \qquad (7.131)$$

where

$$\Phi_n = \frac{B_o - R_s B_g}{B_o - B_{o_i} + (R_{s_i} - R_s)B_g} \qquad (7.132)$$

$$\Phi_g = \frac{B_g}{B_o - B_{o_i} + (R_{s_i} - R_s)B_g} \qquad (7.133)$$

Φ_g and Φ_n group together all the pressure variant fluid properties. With the Tarner method assume an initial reservoir content of one stock barrel.

$$1 = N_p \Phi_n + G_p \Phi_g \qquad (7.134)$$

Tracy (1955) suggests that in going from pressure p_j to a lower pressure p_k, the producing gas/oil ratio R_k' should be estimated. Then the estimated average GOR between the two pressures is

$$R'_{ave} = \frac{R_j + R_k}{2} \qquad (7.135)$$

The estimated production ΔN_p is calculated using Eq. (7.134)

$$1 = (N_{pj} + \Delta N_p)\Phi_{nj} + (G_{pj} + R_{ave}\Delta N_p)\Phi_{gk} \qquad (7.136)$$

$$N_{pk} = N_{pj} + \Delta N_p \qquad (7.137)$$

The oil saturation is the oil volume divided by the pore volume

$$S_o = \frac{(N - N_p)B_o(1 - S_w)}{NB_{oi}} \qquad (7.138)$$

On the basis of $N = 1$,

$$S_o = (1 - N_p)\frac{B_o}{B_{o_i}}(1 - S_w) \qquad (7.139)$$

The values of the gas and oil viscosities and the relative permeability ratio are again found graphically. Then the producing GOR is calculated using Eq. (7.128). If this value of R_k is very close to the R_k' guessed, the calculations can proceed. If not, a new guess must be made. As a further check, the value of R_{ave} is recalculated using R_k, and this R_{ave} is placed in Eq. (7.136). If the answer is correct to 1 part in 1000, the process is continued.

As an example the GOR at 1300 psi will be calculated from the data at 1500 psi. From the previous calculation of R at higher pressures, R at 1300 is estimated to be 2085 SCF STB^{-1}. Then the R_{ave} is calculated using Eq. (7.135).

$$R_{ave} = \frac{2085 + 1361}{2} = 1725 \text{ SCF STB}^{-1}$$

At 1300 psi, Eq. (7.136) predicts

$$1 - (0.0947 + \Delta N_p)2.097 + (72.56 + 1723\,\Delta N_p)0.0067$$

$$\Delta N_p = 0.0231\,\text{STB}$$

$$N_p = 0.0947 + 0.0231 = 0.1178\,\text{STB}$$

The estimated oil saturation, Eq. (7.138), is

$$S_o = (1 - 0.1178)\frac{1.233}{1.315}(1 - 0.22) = 0.6452$$

The water saturation is 22 percent; so the total liquid saturation is 86.52 percent. Referring to Fig. 7.44,

$$\frac{k_g}{k_o} = 0.021$$

and by Eq. (7.128)

$$R_k = 1.233\,\frac{1}{0.001616}\,0.021(102.61) + 450 = 2079 \qquad (7.140)$$

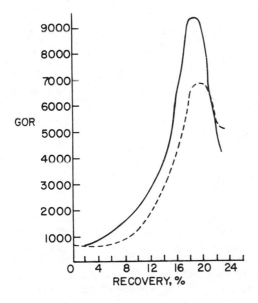

FIG. 7.46 Solution gas-drive performance curve: (——) Muskat and (- - -) Tarner.

This value is quite close to the assumed value of 2085 SCF STB^{-1}. As a check, R_k will be substituted in the material balance. The cumulative produced gas at 1300 psi is 72.56 SCF. Resubstituting into Eq. (7.136)

$$G_p = 72.56 + R_{avg} \Delta N_p = 112.29\,SCF$$

$$1 = 0.1178(2.1) + 112.27(0.0067) = 0.9995$$

Figure 7.46 shows the producing GOR plotted versus cumulative production down to 50 psi from 2500 psi for the reservoir whose data are given in Table 7.13. These are typical curves for solution gas drive performance. While the Muskat method is simpler in that it requires no trial and error calculation, it is not self-checking and cumulative errors may enter the calculations unless the pressure increment is sufficiently small. Figure 7.46 indicates that a $\Delta p = 200$ psi was too large. ∎

Aquifer performance associated with a water-drive reservoir either is determined from wells which penetrate the aquifer or is inferred from pressure production data from the oil reservoir. Solutions of the unsteady-state pressure equation are used to infer aquifer properties. Water-drive reservoirs are treated much as with gas-drive reservoirs with water influx. The oil-gas-water contacts are assumed equipotential lines. The problem becomes one of estimating pressure decline. In the oil region it is assumed water is at residual or connate water. Electric analogs have been used extensively to determine potentials in water-drive reservoirs.

CASE STUDY 9: Calculation of Oil Production from a
Water-Drive Reservoir

Oil removed by water displacement gives the reservoir a slow pressure decline. The slower the pressure decline the smaller the amount of oil that is left at abandonment. In predicting the performance of a water-drive reservoir Schilthuis (1936) developed an early approximation for predicting water influx

$$W_e = k \int_0^t (p_i - p)\, dt \qquad\qquad (7.141)$$

where W_e = gross water influx

p_i = initial boundary pressure

p = boundary pressure at some later time

k = water influx constant

This approximation is a steady-state equation and does not take into account magnitudes of pressures or the length of time.

Van Everdingen and Hurst (1949) developed another method for determining the water influx into a reservoir based on the continuity equation. The development of this method is shown for radial flow systems. The time rate of change of a volume element of radius r is

$$\frac{\partial V}{\partial t} = 2\pi rh\phi c(dr)\frac{\partial p}{\partial t} \tag{7.142}$$

or in terms of flow rate

$$\frac{\partial q}{\partial r} = 2\pi rh\phi c\left(\frac{dp}{dt}\right) \tag{7.143}$$

For radial flow Darcy's law is

$$q = \frac{2\pi khr}{\mu}\frac{dp}{dr} \tag{7.144}$$

Differentiating Eq. (7.144) with respect to r and combining with Eq. (7.143) gives,

$$\frac{\partial^2 P}{\partial r^2} + \frac{1}{r}\frac{\partial p}{\partial r} = \frac{\mu\phi c}{k}\frac{\partial p}{\partial t} \tag{7.145}$$

Let $N = k/\mu\phi c$ and $t_D = kt/\mu\phi cr^2$, then

$$\frac{\partial^2 P}{\partial r^2} + \frac{1}{r}\frac{\partial p}{\partial r} = \frac{\partial p}{\partial t_d}\left(\frac{1}{r^2}\right) \tag{7.146}$$

The cumulative water influx is found to be

$$W_e - 2\pi\phi cr^2 h\,\Delta PQ_t \tag{7.147}$$

where Q_t = dimensionless water influx. Writing Eq. (7.147)

$$W_e = B\,\Delta PQ_t \tag{7.148}$$

where

$$B = 2\pi\phi cr^2 h \tag{7.149}$$

The solution to Eq. (7.148) provides values of Q(t) and is given by van Everdingen and Hurst (1949). The solution assumes that water is encroaching from all sides. To modify this solution to eliminate this assumption the following equation is used for B:

$$B = 2\pi\phi cr^2 hf \tag{7.150}$$

where f = fraction of reservoir periphery into which water is encroaching.

The unsteady-state water influx equation is used in determining future performance of a water-drive reservoir. The equation is used with the material balance equation to predict performance. The following example illustrates this method of predicting water-drive reservoir performance.

TABLE 7.14 Discovery = (Year)$_0$

Initial oil in place	42,000,000 ST bbl
Initial pressure	2432 psig
Reservoir saturation pressure	1145 psig
Average porosity in aquifer	20%
Average permeability in aquifer	100 mdarcy
Average radius of field	5000 ft
Ratio of aquifer radius to field radius	7.0
Effective compressibility of reservoir water	1.2×10^{-6}
Water viscosity at initial reservoir conditions	0.72

It is desired to calculate oil production by water drive from a reservoir that is described in Table 7.14. Figure 7.47 shows the pressure decline for the first four years of performance. Figure 7.48 gives the initial performance of the reservoir. van Everdingen and Hurst (1949) provide tables of dimensionless water influx which enter the reservoir in response to unit pressure drop for any value of dimensionless time

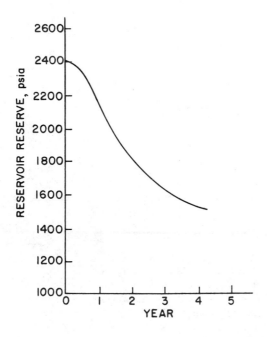

FIG. 7.47 Pressure versus time, water-drive reservoir.

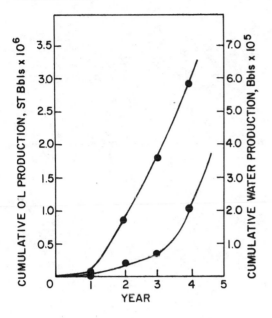

FIG. 7.48 Production data water-drive reservoir.

$$t_D = 6.323 \quad 10^{-3} \frac{kt}{\phi \mu c_e r_w^2} \qquad (7.151)$$

For the reservoir considered

$$t_D = \frac{6.323 \times 10^{-3} \times 100t}{0.2(0.72)(1.2 \times 10^{-6})(5000)^2} = 0.1464t$$

From Fig. 7.47 the pressure at the end of year 1 is 2150 psi. The material balance for year 1 is

$$W_e = N_p B_t + W_p - N(B_t - B_{ti}) \qquad (7.152)$$

where W_p is cumulative water produced, N_p is cumulative oil produced, and B_t is the total formation volume factor. Using Fig. 7.48 at the end of year 1

$$W_e = 180,000(1.102) + 1000 - 42,000,000(1.102 - 1.101)$$

$$= 157,360 \text{ bbl}$$

We now construct Table 7.15 and find the cumulative influx. The constant B is found for Eq. (7.148) and for the first 4 years B = 21.2, 79.9, 20.8, 63.3 for an average of 47. This value of B is used to calculate future performance. The future performance calculations are

TABLE 7.15

Period	t	t_d	Q_t	P	ΔP	$\Delta P Q_t$
Year 1						
1	365	53.4	19.67	2380	26	511.4
2	182.5	26.7	14.15	2190	121	1712.1
						Σ 2223.6
Year 2						
1	730	106.8	23.13	2380	26	601.4
2	547.5	80.1	22.15	2190	121	2680.1
3	365	53.4	19.67	1950	241	2740.5
4	182.5	26.7	14.15	1830	301	4259.5
						= 10281.2
Year 3						
1	1095	160.2	23.85	2380	26	620.1
2	912.5	133.6	23.66	2190	121	2862.8
3	730	106.8	23.13	1950	241	5574.3
4	547.5	80.1	22.15	1830	301	6667.3
5	365	53.4	19.67	1710	361	7100.9
6	182.5	26.7	14.15	1640	396	5603.4
						= 88428.7
Year 4						
1	1460	213.7	23.97	2380	26	623.2
2	1277.5	187.0	23.89	2190	121	2890.7
3	1095	160.2	23.85	1950	241	5747.8
4	912.5	133.6	23.66	1830	301	7121.7
5	730	106.8	23.13	1710	361	8349.9
6	547.5	80.1	22.15	1640	396	8771.4
7	365	53.4	19.67	1560	436	8576.1
8	182.5	26.7	14.15	1520	456	6452.4
						= 48533.2

TABLE 7.16 Trial 1

Period	t	t_d	Q_t	P	ΔP	ΔPQ_t
1	1642.5	240.4		2380	26	623.5
2	1460	213.7	23.97	2190	121	2900.4
3	1277.5	187.0	23.89	1950	241	5757.5
4	1095	160.2	23.85	1830	301	7178.8
5	912.5	133.6	23.66	1710	361	8541.3
6	730	100.8	23.13	1640	396	9159.5
7	547.5	80.1	22.15	1560	436	9657.4
8	365	53.4	19.67	1520	456	8969.5
9	182.5	26.7	14.15	1470	481	6806.1
						= 59,594

Trial 1: Assume $P_9 = 1470$,

$$W_e = 3,150,111(1.1090) + 180,000 - 42,000,000(1.1090 - 1.1010)$$

$$= 3,337,473$$

$$PQ_t = 59,594$$

$$W_e = 47(59,594) = 2,800,918$$

Tables 7.15 and 7.16 summarize the calculations. ■

7.5 CONDENSATE RESERVOIRS

A condensate reservoir is intermediate between a gas and an oil reservoir. A gas reservoir primarily produces a dry gas; the oil reservoir primarily produces oil (with accompanying water in later stages or in water influx situations). Gas condensate production is a wet gas which when brought to the surface where it reaches surface temperatures and pressures separates into gas and condensed liquid. The fluids exist in the reservoir as gases above the critical temperature; as temperatures and pressures of the gas are lowered, the gas condenses partially into light hydrocarbon liquids. A knowledge of the pVT properties of the fluids in the gas condensate reservoir is important. Section 4.7 contains the essential discussion associated with flow with phase change in gas condensate systems. Condensate reservoirs are produced by pressure depletion and/or pressure maintenance.

Calculations of condensate reservoir behavior during pressure depletion are normally made using laboratory pVT data. Liquid that condenses in the reservoir

TABLE 7.17 Laboratory Pressure-Depletion Study at 250°F (Total cell volume: 985.5 cm³)

Pressure (psia)	Operation	Avg pressure during gas removal (psia)	Volume of residual liquid in cell (cm³)	Volume % liquid	Residual liquid (bbl MMscf⁻¹ of reservoir gas)[a]	Cumulative gas removed from cell (% of original)[b]
4265	Observed	...	0	0	0	0
4265-3595	Removed gas	3930				
3595	Observed	...	128.1	13.0	95.47	10.77
3595-2865	Removed gas	3230				
2865	Observed	...	165.7	16.8	123.51	25.04
2865-2250	Removed gas	2558				
2250	Observed	...	166.6	16.9	124.18	38.90
2250-1675	Removed gas	1963				
1675	Observed	...	157.7	16.0	117.50	52.32
1675-1185	Removed gas	1430				
1185	Observed	...	147.8	15.0	110.16	63.86
1185-665	Removed gas	925				
665	Observed	...	133.9	13.6	99.81	75.96

[a]Based on entire original fluid composition as a gas, corrected to 60°F and 14.7 psia: barrels of liquid at reservoir temperature and indicated pressure, per million cubic feet of original reservoir gas corrected to standard conditions.

[b]Arbitrarily taken as A ÷ B, where A = summation of separate volumes removed at each step, measured at or corrected to standard conditions of 60°F and 14.7 psia, B = entire original fluid (vapor) in cell, corrected to standard conditions of 60°F and 14.7 psia. Note that B represents the original composition, while the composition of A varies continuously as pressure declines.

Source: Petroleum Production Handbook, Vol. II by Frick and Taylor. Copyright © 1962 McGraw-Hill. Used by permission of McGraw-Hill Book Company.

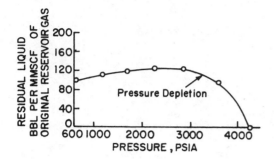

FIG. 7.49 Effect of pressure on residual liquid during laboratory pressure-depletion test of a gas condensate system at 250°F. (From Petroleum Production Handbook Vol. II by Frick and Taylor. Copyright © 1962 McGraw-Hill. Used by permission of McGraw-Hill Book Company.)

TABLE 7.18 Analysis of Gas Samples Removed from Cell During Laboratory Pressure-Depletion Study of a Gas Condensate System at 250°F (mol %)

Component	At 4,265 psia	At 3,930[a] psia	At 3,230[a] psia	At 2,558[a] psia	At 1,963[a] psia	At 1,430[a] psia	At 925[a] psia	At 665 psia
Nitrogen	3.56	1.68	2.27	1.69	1.52	2.03	2.29	1.20
Hydrogen sulfide	0	0	0	0	0	0	0	0
Carbon dioxide	2.43	2.86	2.86	2.86	2.86	2.86	2.86	2.86
Methane	65.01	69.12	70.94	73.34	74.46	74.17	72.57	71.39
Ethane	10.07	10.00	9.92	10.08	10.35	10.23	10.66	11.68
Propane	5.05	4.78	4.85	4.75	4.63	4.92	5.23	6.67
Isobutane	1.05	0.89	0.95	0.92	0.85	0.88	0.95	0.93
Normal butane	2.08	1.76	1.75	1.66	1.68	1.72	1.87	2.00
Isopentane	0.79	0.69	0.52	0.62	0.54	0.67	0.63	0.67
Normal pentane	1.03	0.93	0.96	0.67	0.65	0.53	0.59	0.78
Hexanes	1.47	1.44	1.11	1.00	0.71	0.60	0.87	0.68
Heptanes-plus	7.46	5.85	3.87	2.41	1.75	1.39	1.48	1.14
	100.00	100.00	100.00	100.00	100.00	100.00	100.00	100.00

[a]Average pressure during gas-removal step. Composition represents that for all gas removed during pressure decrements shown in Table 7.17.
Source: Petroleum Production Handbook, Vol. II by Frick and Taylor. Copyright © 1962 McGraw-Hill. Used by permission of McGraw-Hill Book Company.

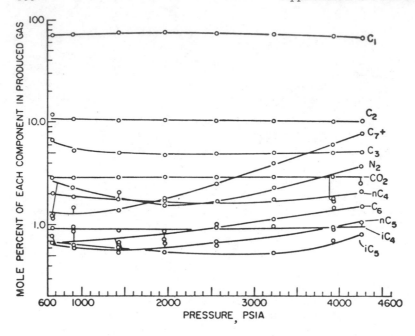

FIG. 7.50 Change in composition of produced gas during laboratory pressure-depletion test of a gas condensate system at 250°F. (From Petroleum Production Handbook, Vol. II by Frick and Taylor. Copyright © 1962 McGraw-Hill. Used by permission of McGraw-Hill Book Company.)

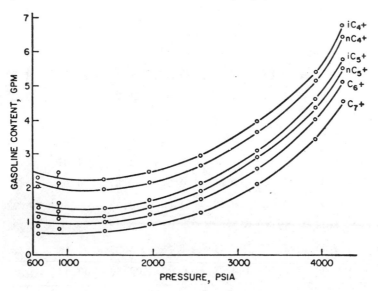

FIG. 7.51 Effect of pressure on gasoline content of produced gas during laboratory pressure–depletion test of a gas condensate system at 250°F. (From Petroleum Production Handbook, Vol. II by Frick and Taylor. Copyright © 1962 McGraw-Hill. Used by permission of McGraw-Hill Book Company.)

during pressure depletion is normally discontinuous and does not flow. Pressure depletion tests are made in laboratory pVT cells to simulate the change in properties of the fluid as the temperature and pressure decline. The laboratory procedure for simulating pressure depletion is to place a sample of recombined reservoir fluid of known composition in a visual pVT cell and raise its temperature and pressure to that of the reservoir. The pressure is then reduced stepwise by removing samples of gas. The volume of remaining gas and dew point liquid is determined. The gas removed at each step is analyzed for its composition. The fractional analysis gives composition data which is used to calculate condensable liquid contact of the gas that is produced at all stages of the pressure depletion. Table 7.17 shows results of a laboratory depletion study (Frick and Taylor, 1962). Figure 7.49 is a plot of the volume of the liquid in the cell after each step of the pressure reduction. Table 7.18 is the analysis of the gas samples from the depletion study of Table 7.17. Figure 7.50 is a plot of the produced gas composition and is a function of pressure. Figure 7.51 is the compressibility z of the produced gas as a function of depletion pressure.

CASE STUDY 10: Calculation for a Gas Condensate Reservoir
During Pressure Depletion

To calculate performance of a gas condensate reservoir, field sampling of the reservoir must be accomplished. These data involve determining original reservoir pressure, average reservoir temperature, composition of the reservoir contents, pore space, and compressibility factors of the produced fluid (both at the original reservoir conditions and at standard conditions).

The pore space can be calculated using the following formula:

$$V_p = 43,560 Ah\phi(1 - S_w) \qquad (7.153)$$

Knowing the reservoir fluid composition one can simulate the gas condensate system under pressure depletion.

TABLE 7.19 Gas Condensate Reservoir Data

Original reservoir	4900 psia
Dew point pressure	4500 psia
Assumed abandonment pressure	500 psia
Average reservoir temperature	250°F
Pore space	400,000 Mscf
Condensate content (butane "+") at sc[a]	160.7 bbl MMscf^{-1}
Compressibility factor of product gas:	
At original reservoir conditions	0.931
At standard conditions	0.995
Molecular weight of butane "+" (mole weighted average)	96.0
Density of butane "+" (mole weighted average)	5.78 lb gal^{-1}

[a]rc = reservoir conditions; sc = standard conditions.
Source: Frick and Taylor, 1962.

TABLE 7.20 Cell Volume = 1000 cm^3

Pressure (psia)	Operation	Volume of liquid in cell (cm^3)	Volume % liquid	Residual liquid (bbl MMscf^{-1} of reservoir gas)[a]	Sum of percent of original gas removed	Deviation factor
4500	Analysis	0	0	0	0	0.920
4500–3850	Gas removal					
3850	Analysis	130.6	13.06	97.46	11.79	0.875
3850–3200	Gas removal					
3200	Analysis	160.2	16.82	125.52	26.06	0.853
3200–2550	Gas removal					
2550	Analysis	169.1	16.91	126.19	39.92	0.846
2550–1900	Gas removal					
1900	Analysis	160.2	16.02	119.55	53.34	0.860
1900–1250	Gas removal					
1250	Analysis	150.3	15.03	112.16	64.88	0.890
1250–600	Gas removal					
600	Analysis	136.4	13.64	101.79	76.98	0.925

[a]Corrected to standard conditions (60°F and 14.7 psia).
Source: Frick and Taylor, 1962.

TABLE 7.21

Component	At 4500 psia	At 4175 psia	At 3525 psia	At 2875 psia	At 2225 psia	At 1575 psia	At 925 psia
Methane	65.01	69.12	71.03	73.43	74.55	72.66	71.48
Ethane	10.97	9.10	9.83	10.0	10.26	10.57	11.59
Propane	4.15	5.68	4.94	4.84	4.72	5.32	6.76
i-Butane	1.95	1.79	0.86	0.83	0.76	0.86	0.84
n-Butane	1.18	0.86	1.84	1.75	1.77	1.96	2.09
i-Pentane	1.69	0.99	0.43	0.53	0.45	0.54	0.58
n-Pentane	0.13	0.12	1.05	0.76	0.74	0.68	0.87
Hexanes	2.37	2.01	1.02	0.91	0.61	0.78	0.59
Heptanes "+"	6.56	3.64	3.96	2.50	1.84	1.57	1.23
Nitrogen	4.46	4.01	2.18	1.60	1.43	2.20	1.11
CO_2	1.53	2.68	2.86	2.86	2.86	2.86	2.86

Source: Frick and Taylor, 1962.

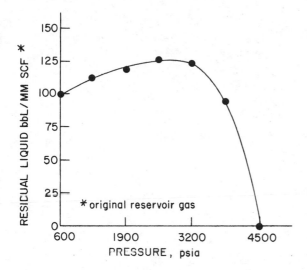

FIG. 7.52 Residual liquid versus pressure. (Data from Frick and Taylor, 1962.)

Table 7.19 shows data from Frick and Taylor (1962) for a gas condensate reservoir. A simulation is performed in a pVT cell, where the original reservoir fluid composition and conditions are reconstructed assuming the retrograde liquid is immobile in the reservoir. The pressure of the cell is then reduced by consecutive removal of gas samples. The gas samples from each removal are analyzed to obtain the composition, as shown in Table 7.20. The volume of the residual liquid in the cell is also obtained for each reduction, as shown in Table 7.21.

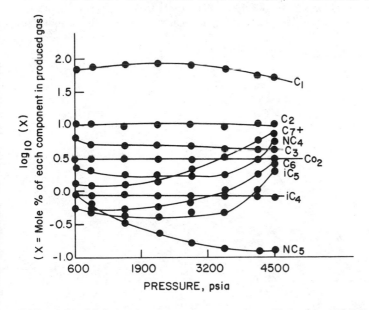

FIG. 7.53 Mole percent versus pressure. (Data from Frick and Taylor, 1962.)

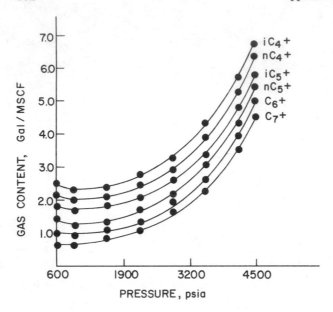

FIG. 7.54 Gas content versus pressure. (Data from Frick and Taylor, 1962.)

Having performed the simulation, one is able to construct graphical representations of the behavior of the reservoir during pressure depletion. Figure 7.52
shows how the residual liquid (bbl $MMscf^{-1}$ of original reservoir gas) varies with
pressure. Figure 7.53 shows the (mol %) composition of the product gas during
pressure depletion. Figure 7.54 shows how the gas content of the product gas varies
with pressure depletion. Figures 7.55 and 7.56 show the variance, with pressure,
of the deviation factor and the volume percent liquid phase, in the cell, respectively.

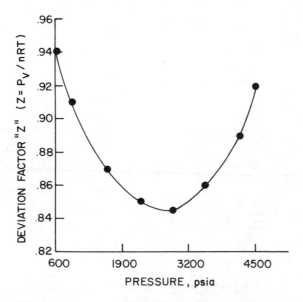

FIG. 7.55 Compressibility versus pressure. (Data from Frick and Taylor, 1962.)

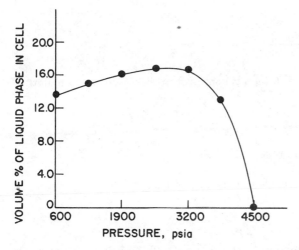

FIG. 7.56 Volume percent of liquid versus pressure. (Data from Frick and Taylor, 1962.)

The wet gas in place is calculated at standard conditions.

$$\text{Wet gas in place} = \frac{(\text{pore space})p_{rc}T_{sc}z_{sc}}{p_{sc}T_{rc}z_{rc}} \qquad (7.154)$$

TABLE 7.22

Pressure (psia)	Cumulative wet gas produced (MMscf)	Butane "+" content	
		(bbl MMscf^{-1})	(gal Mscf^{-1})
4900	0	160.7	6.75
4500	7,374	160.7	6.75
3850	17,116	115.5	4.85
3200	29,976	88.6	3.72
2550	44,596	71.4	3.00
1900	60,557	59.8	2.51
1250	76,516	55.5	2.33
600	91,504	59.5	2.50
500	97,000[a]	73.8[a]	

[a]Extrapolated value.
Source: Calculated from data from Frick and Taylor, 1962.

Original wet gas in place:

$$\frac{(400,000 \times 10^3)(4900)(520)(0.995)}{710(0.931)(14.7)} = 104366 \text{ MMscf}$$

The total wet gas produced up to the present pressure is found by calculating

$$104366 - \frac{(400,000 \times 10^3)(4500)(520)(0.995)}{710(14.7)(0.920)} = 7374 \text{ MMscf}$$

The original butane "+" in the reservoir at sc is

(Wet gas in place at sc) $\times$ (condensate content (butane "+" of original fluid at sc))

104366 MMscf $\times$ 160.7 bbl MMscf^{-1} = 16,772 Mbbl

Table 7.22 shows the cumulated wet gas produced and the butane "+" content. Figure 7.57 is a plot of the cumulated wet gas produced versus the pressure. From this plot one can determine by extrapolation the total wet gas produced when the pressure reaches the abandonment pressure (in this case the assumed abandonment pressure is 500 psia). The extrapolation gives a value for the total wet gas produced as 97 MMscf.

$$\text{Wet gas recovered} = \text{volume wet gas recovered} \times \frac{1}{379}$$

97000 MMscf $\frac{1}{379} \frac{\text{lb mol}}{\text{scf}} = 255937 \times 10^3 \text{ lb mol}$

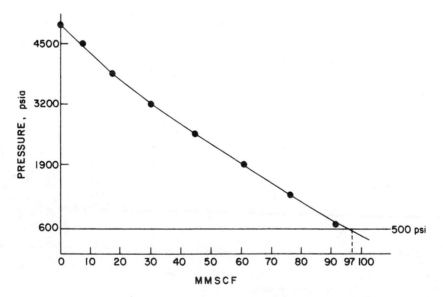

FIG. 7.57 Pressure versus volume. (Calculated from data from Frick and Taylor, 1962.)

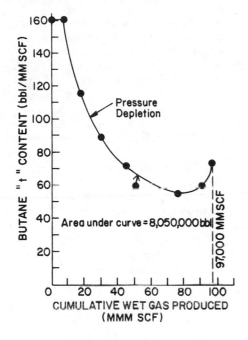

FIG. 7.58 Butane content versus gas
produced. (Calculated from data from
Frick and Taylor, 1962.)

The volume of butane "+" can be determined graphically as in Fig. 7.58 by
plotting butane "+" content (bbl MMscf^{-1}) against cumulative wet gas produced
(MMMscf) and determining the area under the pressure depletion curve (8,050,
bbl).

Butane "+" recovered:

$$\frac{\text{bbl} \times 42 \times \text{lb gal}^{-1}}{\text{mol wt}} = \frac{8.05 \times 10^6 \times 42 \times 5.78}{96.0} = 20356 \times 10^3 \text{ lb mol}$$

The residual gas is

(Wet gas recovered – butane "+" recovered) × 379

= (255937 – 20356) × 10^3 × 379 = 89,285 MMscf

The recoveries by pressure depletion are:

Wet gas recovery:

$$\frac{97000 \times 10^6 \times 100}{104366 \times 10^6} = 92.9\%$$

Butane "+" recovery:

$$\frac{8.05 \times 10^6 \times 100}{16.802 \times 10^6} = 48.0\% \blacksquare$$

The pressure of a gas condensate reservoir may be maintained by natural or by artificial water drive or by injection of dry gas into the reservoir. The objective of cycling the dry gas back into the reservoir is to maintain reservoir pressure above the dew point of the reservoir gas to minimize retrograde condensation. Hydrocarbons that condense in the reservoir will not be mobile and will be lost production. Dry gas, mostly methane, is miscible with the wet reservoir gas and the pressure cycling creates a miscible displacement. The efficiency of a gas cycling displacement depends on the well pattern, relative fluid properties, and reservoir properties. Since the mobility ratio of the displacement is close to 1, the pattern efficiency depends largely on the well locations and rates. Aerial sweep efficiency—the area contacted by the injected dry gas during displacement—has been calculated using electrical analogs. Pattern efficiency is an aerial efficiency weighted with the product of thickness, porosity, and saturation. The invasion efficiency represents the relative amount of hydrocarbon pore space contacted by the dry gas. Displacement efficiency is the relative volume of gas that is displaced. The reservoir cycling efficiency is the product of pattern, invasion, and displacement efficiency. The horizontal and vertical permeabilities strongly affect the reservoir recovery efficiency. In calculating displacement a discrete interface is assumed between the wet and dry gas.

CASE STUDY 11: Numerical Simulation of a Gas Cycling Project

Cycling of dry gas back into a reservoir is done to maintain reservoir pressure above the dew point of the reservoir gas in order to minimize retrograde condensation. This method of gas cycling is especially suited to wells with a large gas cap.

One such reservoir is the Bonnie Glen D-3A pool in Alberta, Canada. Thompson and Thachuk (1974) developed a compositional simulation for gas cycling this reservoir.

The Bonnie Glen D-3A pool is part of the Leduc chain reef and is one of the most capable producing fields in Canada. The original oil in place is estimated to be 657,138,000 STB with an original gas cap of 44,900 MMcf. Table 7.23 gives the reservoir properties of the Bonnie Glen D-3A pool.

The process of gas cycling in the Bonnie Glen pool consists of gas production from both the original gas cap and the gas-flushed portion of the original oil column at a rate of 140 MMcf D^{-1}. The produced gas is processed to remove liquids and the remaining residue gas is reinjected into both the gas cap and the gas flushed zone. Cycling of this reservoir results in liquid hydrocarbon recovery that would otherwise have been delayed some 24 years.

A preliminary study, using two one-dimensional linear compositional models to simulate gas cycling in both the gas cap and gas-flushed zones, was used to determine the liquid recovery from each zone. These one-dimensional models used the Benedict-Webb-Rubin equation of state to predict compositions. Results of the study were checked by a vaporization study on a sample of Bonnie Glen crude as shown in Fig. 7.59.

A three-dimensional three-phase black oil simulator was used to conduct the history match of the Bonnie Glen pool. Details of the simulator are as follows:

1. Hydrocarbon properties were handled by conventional laboratory pVT data.
2. All flow was governed by Darcy's law.
3. Gravity and capillary pressure were included in the mathematical technique.
4. The model was capable of handling three-phase relative permeability and variable geometry.

TABLE 7.23 Reservoir Properties

Original oil in place, STB	756,138,000
Original gas in place, MMcf	444,900
Average porosity, %	
Gas cap	8.89
Oil column	9.55
Connate water saturation, %	6.0
Critical gas saturation, %	10.0
Residual oil saturation, %	
Gas-flushed zone	24.5
Water-flushed zone	35.4
Average weighted arithmetic average permeability, mdarcy	
Horizontal	1,271
Vertical	115
Average weighted geometric average permeability, mdarcy	
Horizontal	252
Vertical	5.9
Critical reservoir datum pressure (4100 ft subsea)	2,477
Critical reservoir datum temperature, °F	172
Saturation pressure of oil column, psig	2,444
Dew point pressure of gas cap, psig	2,421
Initial oil formation volume factor, RB STB^{-1}	1.4580
Initial gas cap formation factor, RB Mcf^{-1}	1.044
Initial solution GOR, Scf STB^{-1}	732
Gravity, °API	42

Source: F. R. Thompson and R. Thachuk, Compositional Simulation of a Gas Cycling Project, Bonnie Glen D-3A Pool, Alberta, Canada, J. Pet. Tech. 1285, November (1974). © 1974 SPE-AIME.)

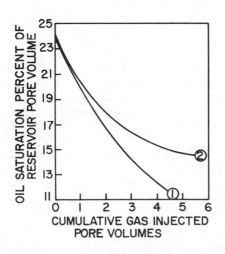

FIG. 7.59 Calculated and measured residual oil saturations. (1) Average residual saturation, cells 1 and 2. (2) Vaporization study. (From F. R. Thompson and R. Thackuk, Compositional Simulation of a Gas Cycling Project, Bonnie Glen D-3A Pool, Alberta, Canada, J. Pet. Tech. 1285. © November 1974 SPE-AIME.)

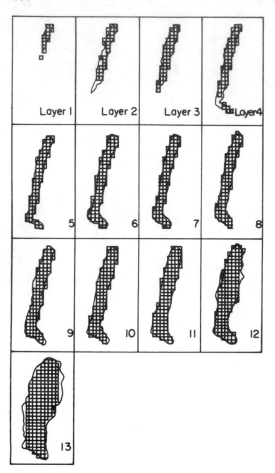

FIG. 7.60 Grid construction. (From F. R. Thompson and R. Thachuk, Compositional Simulation of a Gas Cycling Project, Bonnie Glen D-3A Pool, Alberta, Canada
J. Pet. Tech. 1285. © November 1974 SPE-AIME.)

A good history match was needed to ensure that the basic reservoir parameters
were defined correctly. The reservoir was divided into 13 layers to simulate vertical
variations in permeability, porosity, saturation, and pVT properties. Figure 7.60
shows the grid locations. Historical oil production rates calculated by material balance were entered into the simulator, with oil production taken from layer 12 and
water influx simulated by water injection into layer 13. A good historical match was
achieved, shown in Figs. 7.61 and 7.62. The historical match made possible the
conversion to the compositional model after several years of historical performance,
resulting in a substantial cost savings.

The final prediction of the gas cycling scheme was based on a combination of
a two-dimensional longitudinal compositional and the three-dimensional compositional
models. The initial production injection configuration simulated the scheme as outlined in Fig. 7.63, with production and injection from the reef top to the existing

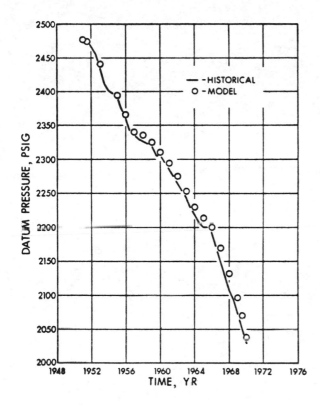

FIG. 7.61 Historical pressure match. (From F. R. Thompson and R. Thachuk, Compositional Simulation of a Gas Cycling Project, Bonnie Glen D-3A Pool, Alberta, Canada. J. Pet. Tech. 1285. © November 1974 SPE-AIME.)

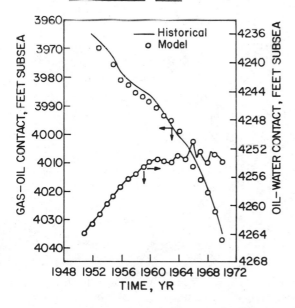

FIG. 7.62 Historical contact match. (From F. R. Thompson and R. Thachuk, Compositional Simulation of a Gas Cycling Project, Bonnie Glen D-3A Pool, Alberta, Canada, J. Pet. Tech. 1285. © November 1974 SPE-AIME.)

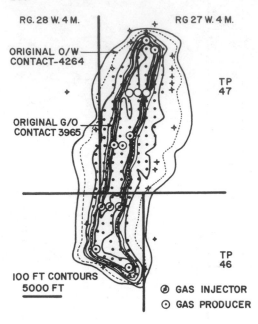

FIG. 7.63 Initial well configuration (1973 to 1980). (From F. R. Thompson and R. Thachuk, Compositional Simulation of a Gas Cycling Project, Bonnie Glen D-3A Pool, Alberta, Canada, J. Pet. Tech. 1285. © November 1974 SPE-AIME.)

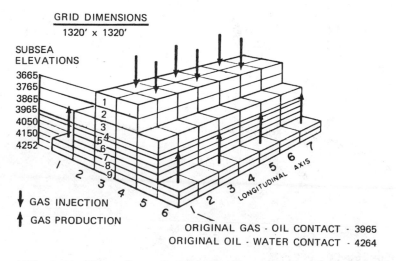

FIG. 7.64 Three dimensional grid. (From F. R. Thompson and R. Thachuk, Compositional Simulation of a Gas Cycling Project, Bonnie Glen D-3A Pool, Alberta, Canada, J. Pet. Tech. 1285. © November 1974 SPE-AIME.)

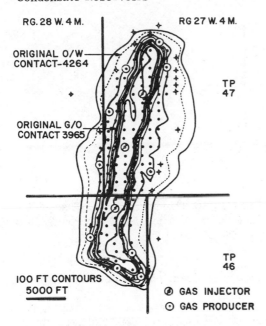

FIG. 7.65 Well configuration 1980-1996. (From F. R. Thompson and R. Thachuk, Compositional Simulation of a Gas Cycling Project, Connie Glen D-3A Pool, Alberta, Canada, J. Pet. Tech. 1285. © November 1974 SPE-AIME.)

gas-oil contact. The first portion of the cycling scheme was essentially a longitudinal displacement and required simulation with only a two-dimensional cross-sectional model of the three-dimensional grid of Fig. 7.64. After the gas-oil contact reached layer 9, the cycling scheme was converted to a vertical displacement, with gas being injected into layer 1 and production coming from flank wells in layer 8.

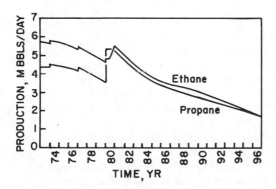

FIG. 7.66 Predicted production rate. (From F. R. Thompson and R. Thachuk, Compositional Simulation of a Gas Cycling Project, Bonnie Glen D-3A Pool, Alberta, Canada, J. Pet. Tech. 1285. © November 1974 SPE-AIME.)

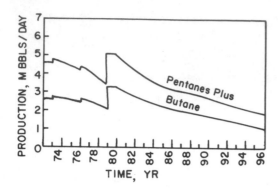

FIG. 7.67 Predicted production rate. (From F. R. Thompson and R. Thachuk, Compositional Simulation of a Gas Cycling Project, Bonnie Glen D-3A Pool, Alberta, Canada, J. Pet. Tech. 1285. © November 1974 SPE-AIME.)

The well configuration of Fig. 7.65 used a three-dimensional model and the grid of Fig. 7.64.

Predictions from Jan. 1, 1968, to Jan. 1, 1980, were made using the two-dimensional longitudinal compositional model. All pressure, compositions, and saturations were then entered into the three-dimensional compositional model. Cycling predictions with a vertical gas displacement were then continued until Dec. 31, 1996, at which time oil production was terminated.

The predicted production rates for each component are shown in Figs. 7.66 and 7.67. The large increase in production rate for each component in 1980 is due to relocating the gas producers to the flanks of the reservoir and thus producing a much richer gas. An independent calculation indicated that approximately 50 percent of the 113,745,000 bbl recovered during cycling is attributed to the original gas cap and the remaining 50 percent to products originally contained in the oil column. ∎

7.6 SECONDARY RECOVERY

Secondary recovery is a term applied to displacement of oil by injection of a fluid into an oil reservoir after the reservoir no longer produces oil economically by primary recovery. The definition of secondary recovery is arbitrary since good practice may be to inject gas or water prior to reaching the primary production economic limit. Gas cycling discussed in Section 7.5 in gas or water injection pressure maintenance transcend this definition. From the point of view of calculations associated with gas and/or water injection the techniques are the same whether the primary economic limit is reached or not.

Gas injection can result in immiscible or miscible displacement depending on pressure, temperature, and the properties of the reservoir oil and the injected gas. When the injection of the gas results in immiscible behavior, the calculations for gas and water injection are similar. When the injection of gas results in miscible behavior, the calculations are guided by the results of scaled laboratory experiments. The efficiency of gas injection depends on well patterns, reservoir heterogeneity, gravity, and the mobility ratio between the displaced and displacing fluids.

   Water injection results in an immiscible displacement and is calculated using
a sharp front, the Buckley-Leverett approach, or a numerical simulator. Systems
with distinct layering are calculated by sharp fronts adjusting for aerial sweep and
vertical conformance. Immiscible displacements are affected by mobility ratio.
Correlations determined from scaled laboratory models are used to include the
effect of mobility ratio.

   Gas injection is dispersed or external. Dispersed gas injection or pattern
injection utilizes a geometric array of wells to distribute gas throughout a reservoir.
The pattern may be random or symmetric such as five spots, seven spots, or nine
spots. Injection geometry may be determined by reservoir properties and existing
wells. There are some disadvantages of dispersed gas injection. Structure or gravity
do not help recovery. Unstable flow and resulting fingering decrease aerial
efficiency.

   External gas injection, or gas cap injection, utilizes wells completed in the
gas cap. Since the gas cap is structurally higher in the reservoir, gravity drainage
improves recovery.

   Buckley-Leverett fractional flow calculations are used for immiscible displacement to determine unit displacement efficiency: the fraction of oil recovered
within the totally swept volume. The fractional flow equation describes the fraction
of gas flowing in terms of the properties of a unit element of the reservoir

$$f_g = \frac{1 + 1.127(k_o A/\mu_o q)[(\partial p_c/\partial x) - 0.433(\rho_o - \rho_g)\sin\alpha]}{1 + k_o \mu_g / k_g \mu_o} \qquad (7.155)$$

and

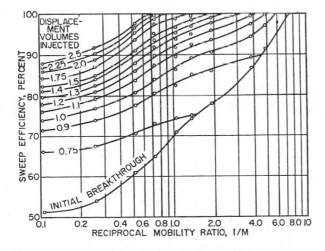

FIG. 7.68 Sweep efficiency as a function of mobility ratio. (From A. B. Dyes,
B. H. Caudle, and R. A. Erickson, Oil Production After Breakthrough as Influenced
by Mobility Ratio, Pet. Trans. AIME 201, 81. © 1954 by the Society of Petroleum
Engineers.)

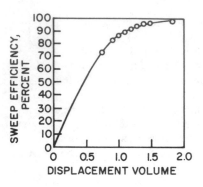

FIG. 7.69 Areal sweep efficiency as a function of injection fluid volume for a mobility ratio of unity. (From A. B. Dyes, B. H. Caudle, and R. A. Erickson, Oil Production After Breakthrough as Influenced by Mobility Ratio, <u>Pet. Trans. AIME</u> <u>201</u>, SPEJ 81. © 1954 SPE-AIME.)

$$x = \frac{5.61qt}{\phi A}\left(\frac{\partial f_g}{\partial S_g}\right) S_g \qquad (7.156)$$

The conformance—the fraction of the reservoir within the swept area contacted by the displacing fluid—is usually determined from laboratory measurements.

Conformance includes vertical efficiency, gravity, and mobility effects. The aerial efficiency—the fraction of the reservoir within the swept area—is also determined in the laboratory using scaled models. Figure 7.68 shows aerial sweep efficiency as a function of mobility ratio for a staggered line drive. Figure 7.69 shows aerial sweep efficiency as a function of injected volume for unit mobility ratio for a five-spot pattern.

CASE STUDY 12: Calculation for Pressure Maintenance

Stanley (1960) reports calculations for pressure maintenance in the Abqaiq field of Saudi Arabia. The reservoir was undersaturated prior to pressure maintenance,

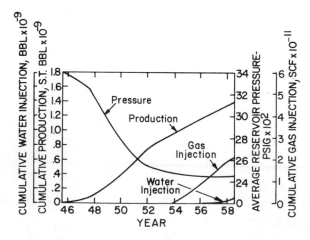

FIG. 7.70 Production history. (From L. T. Stanley, Approximation of Gas-Drive Recovery and Front Movement in the Abqaiq Field, Saudi Arabia, <u>Pet. Trans. AIME</u> <u>219</u>, SPEJ 273. © 1960 SPE-AIME.)

TABLE 7.24 Zone Properties

Zone	ϕ (%)	k (mdarcy)	kh (d ft)	S_w (%)	Rel. Inj
A	22.6	577	46.2	7.0	0.472
B	22.6	550	33.0	7.0	0.337
C	16.4	311	18.7	8.0	0.191

which created a free-gas saturation at the crest of the structure. Gas injection was begun in 1954 with an average injection rate of 122 MMscf D^{-1} into two wells near the crest of the structure maintaining the pressure at 2450 psig. The reservoir is a carbonate section about 200 ft thick with average porosity of 22 percent and average permeability of 500 mdarcy. The injected gas is first-stage separation gas taken at a pressure of 500 psig. The gas is partially miscible with the reservoir oil. Peripheral water injection was begun in 1957 along with an additional well on gas injection. The production history of the field is shown in Fig. 7.70.

The gas volume factor varied as a function of time but as the water influx increased it approached equilibrium behavior. The reservoir was divided into three zones: A, 80 ft thick; B and C, 60 ft thick each. Table 7.24 shows the properties of each zone. The relative injectivity results from the permeability thickness product.

In zone A the gas bubble is assumed to expand at a rate of $\Delta p/\Delta t$ directly proportional to the gas injection rate i_g, inversely proportional to displacement efficiency E_d and porosity ϕ, and directly proportional to the original gas saturation S_{gi}. In zones B and C the gas bubble is assumed to expand at a rate $\Delta A/\Delta t$ directly proportional to $i_g - q_{gv}$ inversely proportional to displacement efficiency and porosity, and directly proportional to original gas saturation.

Since all vertical migration is controlled by the oil drainage rate

$$(q_{gv})B \to A = (q_{gv})A \to B \qquad (q_{gv})C \to B = (q_{gv})B \to C \qquad (7.157)$$

The rate of oil change from Darcy's law is

$$q_{ov} = 0.488\,k_{ov}\frac{k_{ro}/\mu_o(k_{rg}/\mu_g}{k_{ro}/\mu_o + k_{rg}/\mu_g}A(\rho_o - \rho_g) \qquad (7.158)$$

Above critical gas saturation,

$$q_{ov} = \frac{0.488\,k_{ov}\,k_{ro}\,A(\rho_o - \rho_g)}{\mu_o} \qquad (7.159)$$

The gravity drainage rate from A to B is obtained by differentiating Eq. (7.159):

$$\frac{dq_{ov}}{dS_{gA1}} = C'A_B\frac{dk_{ro}}{dS_{gA1}} + C'k_{ro}\frac{dA_B}{dS_{gA1}} \qquad (7.160)$$

and from B to C

$$\frac{dq_{ov}}{dS_{gB1}} = C'A_C \frac{dk_{ro}}{dS_{gB1}} + C'k_{ro} \frac{dA}{dS_{gB1}} \qquad (7.161)$$

The gas saturations are

$$S_{gA} = E_d + \frac{\int_0^1 q_{ovA} \, dtb}{A_{B(1 - S_{wA})}\phi_A h_A} \qquad (7.162)$$

$$S_b = E_D + \frac{\int_0^1 q_{ovB} \, dt}{A_{C(1 - S_{wB})}\phi_B h_B} \qquad (7.163)$$

$$S_{gC} = E_d + \frac{\int_0^1 (i_{gC} - q_{ovB}) \, dt}{A_{C(1 - S_{wC})}\phi_c h_c} \qquad (7.164)$$

$$S_{gA2} = S_{gB2} = E_d \qquad (7.165)$$

Figure 7.71 shows the schematic for the calculations. The area swept by the gas bubble in each zone is

$$dA_A = \frac{i_{gA} \, dt}{h_A \phi_A (1 - S_{wA})(E_d - S_{gi})} \qquad (7.166)$$

$$dA_B = \frac{i_{gB} \, dt - [0.488k_{ov}(k_{ro}/\mu_o)(\rho_o - \rho_g)A_B] \, dt}{h_B \phi_B (1 - S_{wB})(E_d - S_{gi})} \qquad (7.167)$$

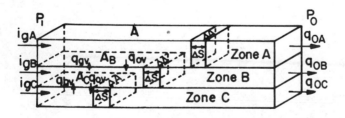

FIG. 7.71 Schematic of areas swept. (From L. T. Stanley, Approximation of Gas-Drive Recovery and Front Movement in the Abqaiq Field, Saudi Arabia, Pet. Trans. AIME 219, SPEJ 273. © 1960 SPE-AIME.)

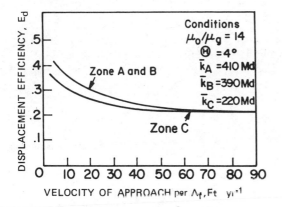

FIG. 7.72 Displacement efficiency. (From L. T. Stanley, Approximation of Gas-Drive Recovery and Front Movement in the Abqaiq Field, Saudi Arabia, Pet. Trans. AIME 219, SPEJ 273. © 1960 SPE-AIME.)

$$dA_C = \frac{i_{gC}\, dt - [0.488k_{ov}(k_{ro}/\mu_v)(\rho_o - \rho_g)A_C]\, dt}{h_C \phi_C (1 - S_{wC})(E_d - S_{gi})} \qquad (7.168)$$

The displacement efficiency was calculated following Welge (1957) and the result is shown in Figure 7.72. The average injection rates are given in Table 7.25. The ΔA for each zone is calculated as follows

$t = 1005$ days (end of 1956)

$\Delta t = 365$ days

In zone A,

TABLE 7.25 Average Injection Rates

Year	Zone A		Zone B		Zone C	
	AW-6	AW-50	AW-6	AW-50	AW-6	AW-50
1954	17.05	22.80	12.20	16.25	6.87	9.17
1955	25.10	21.90	17.90	15.60	10.15	8.82
1956	29.55	33.80	21.10	24.10	11.97	13.70
1957	33.10	40.00	23.50	28.60	13.37	16.22
1958	29.60	32.70	21.00	23.30	11.96	13.25

Source: L. T. Stanley, Approximation of Gas-Drive Recovery and Front Movement in the Abquiq Field, Saudi Arabia, Pet. Trans. AIME 219, SPEJ 273. © 1960 SPE-AIME.

$$A_A \text{ (at } t - \Delta t) = 56.18 \times 10^6 \text{ ft}^2$$

(The bubbles had coalesced in zone A in 1955; therefore, after this date wells AW-6 and AW-50 were treated as a unit in zone A.)

$$\text{Est. } A_A \text{ (at } t - \tfrac{1}{2}\Delta t) = 80 \times 10^6 \text{ ft}^2$$

$$i_{gA} = 63.35 \times 10^3 \text{ reservoir B/D}$$

$$\text{Est. } A \text{ (at } t - \tfrac{1}{2}\Delta t) = 3.61 \times 10^6 \text{ ft}^2$$

(from perimeter formed by 80×10^6 ft^2 on contour pattern). Then

$$V_o = \frac{(63.35 \times 10^3)(5.61)(365)}{3.61 \times 10^6} = 36 \text{ ft year}^{-1}$$

and from Fig. 7.72

$$E_{dA} = 0.251$$

$$\begin{aligned}
\Delta A_A &= \frac{i_{gA}\,\Delta t}{h_A \phi_A (1 - S_{wA})(E_d - S_{gt})A} \\[2mm]
&= \frac{(63.35 \times 10^3)(365)(5.61)}{80(0.226)(1 - 0.07)(0.251 - 0.078)} \\[2mm]
&= 44.60 \times 10^6 \text{ ft}^2
\end{aligned}$$

$$\sum_{j=1} \Delta A = 56.18 + 44.60 = 100.78 \times 10^6 \text{ ft}^2$$

$$\begin{aligned}
\frac{\Delta A_A}{\Delta t} \text{ (at } t - \tfrac{1}{2}\Delta t) &= \frac{44.60 \times 10^6}{365} \\[2mm]
&= 122 \times 10^3 \text{ ft}^2 \text{ day}^{-1}
\end{aligned}$$

In zone B, Well 6:

$$A_B \text{ (at } t - \Delta t) = 6.17 \times 10^6 \text{ ft}^2$$

Average displacement efficiency of zone A at t,

$$\begin{aligned}
\bar{E}_{dA} &= \frac{(\Delta A_A E_{dA})_1 + \cdots + (\Delta A_A E_{dA})_{n-1} + (\Delta A_A E_{dA})_n}{\Sigma_{j=1} \Delta t} \\[2mm]
&= \frac{[23.58(0.233) + 32.60(0.253) + 44.60(0.251)] \times 10^6}{100.78 \times 10^6} = 0.248
\end{aligned}$$

Average gas-injection rate into zone B to t,

$$\bar{i}_{gR} = \frac{(\Delta ti_g)_1 + \cdots + (\Delta ti_g)_{n-1} + (\Delta ti_g)_n}{\Sigma_{j=1} \Delta t}$$

$$= \frac{[275(12.20) + 365(17.90) + 365(17.90) + 365(21.10)] \times 10^3}{1005}$$

$$= 17.50 \times 10^3 \text{ reservoir B/D}$$

Est. $V_o = 16.5$ ft year^{-1} (at $t - \frac{1}{2}\Delta t$)

then

$$E_{dB} = 0.319 \text{ during } \Delta t$$

Try

$$\Delta A_B = 3.70 \times 10^6 \text{ ft}^2$$

then total average displacement efficiency to t in zone B,

$$\bar{E}_{dB} = \frac{(\Delta A_B E_{dB})_1 + \cdots + (\Delta A_B E_{dB})_{n-1} + (\Delta A_B E_{dB})_n}{\Sigma_{j=1}^n \Delta A_B}$$

$$= \frac{[6.17(0.280) + 3.70(0.319)] \times 10^6}{9.87 \times 10^6} = 0.295$$

During the interval t = 0 to t,

$$\bar{i}_{gR} - \bar{q}_{ovA} = \frac{A_B \phi_B h_B (1 - S_{wB})(\bar{E}_{dB} - S_{gi})}{5.61 \, \Sigma_{j=1}^n \Delta t}$$

$$= \frac{[(6.17 + 3.70) \times 10^6] (0.266)(60)(1 - 0.07)(0.295 - 0.077)}{5.61(1005)}$$

$$= 4.81 \times 10^3 \text{ reservoir B/D}$$

then

$$q_{ovA} = (17.50 - 4.81)(10^3) = 12.69 \times 10^3 \text{ reservoir B/D}$$

But, at the same time, S_{gA1} must be such that the relative permeability to oil will satisfy the above drainage rate; so

$$(S_{gA1})_t = \bar{E}_{dA} + \frac{\bar{q}_{ovA} \Sigma_{j=1}^{n} \Delta t}{A_B \phi_A h_A (1 - S_{wA})}$$

$$= 0.248 + \frac{(12.69 \times 10^3)(1005)(5.61)}{(9.87 \times 10^6)(0.226)(80)(1 - 0.07)} = 0.678$$

Average S_{gA1} during total time interval is

$$\bar{E}_{dA} + \frac{1}{2}\Delta S_{gA1} = 0.248 + \frac{0.430}{3} = 0.463$$

The relative permeability to oil of the typical Abqaiq reservoir rock may be expressed in terms of saturation by

$$k_{ro} = \frac{1.35}{10^{5.13 S_g}}$$

then

$$\bar{q}_{ovA} = 0.488 \frac{k_{ov}}{\mu_o} \frac{1.35}{10^{5.13 S_g}} (\rho_o - \rho_g)A_B$$

$$= \frac{0.488(0.298)(1.35)(0.433)(0.87 \times 10^6)}{(0.28) 10^{(5.13)(0.463)}}$$

$$= 12.65 \times 10^3 \text{ reservoir B/D}$$

which checks the 12.69×10^3 reservoir B/D calculated by material balance. Then,

$$\sum_{j=1}^{n} \Delta A_B = 9.87 \times 10^6 \text{ ft}^2$$

and

$$\frac{\Delta A_B}{\Delta t} = \frac{3.70 \times 10^6}{365} = 10.15 \times 10^3 \text{ ft}^2 \text{ day}^{-1} \text{ (at } t - \frac{1}{2}\Delta t)$$

A similar procedure was followed to compute the area swept about well 50. In zone C, Well 6:

$$A_C \text{ (at } t - \Delta t) = 5.43 \times 10^6 \text{ ft}^2$$

The method described for zone B was followed in computing the following zone C values.

$$\bar{E}_{dB} = 0.295$$

$$E_{dC} = 0.306$$

$$\bar{i}_{gC} = 9.91 \times 10^3 \text{ reservoir B/D}$$

Try

$$\Delta A_C = 3.00 \times 10^3 \text{ ft}^2$$

Then

$$\bar{E}_{dC} = 0.293$$

and

$$\bar{i}_{gC} - \bar{q}_{OvB} = \frac{A_C \phi_C h_C (1 - S_{wC})(\bar{E}_{dC} - S_{gi})}{5.61 \ \Sigma_{j=1}^{n} \ \Delta t}$$

$$= \frac{(5.43 + 3.00) \times 10^6 (0.164)(60)(1 - 0.08)(0.293 - 0.076)}{(5.61)(1005)}$$

$$= 2.94 \times 10^3 \text{ reservoir B/D}$$

$$\bar{q}_{ovB} = 9.91 \times 10^3 - 2.94 \times 10^3$$

$$= 6.97 \times 10^3 \text{ reservoir B/D}$$

$$(S_{gR1})_t = \bar{E}_{dB} + \frac{\overline{q_{ovB}} \ \Sigma_{j=1}^{n} \ \Delta t}{A_C \phi_B h_B (1 - S_{wB})}$$

$$= 0.295 + \frac{(6.97 \times 10^3)(1005)(5.61)}{(8.43 \times 10^6)(0.226)(60)(1 - 0.07)}$$

$$= 0.664$$

Average S_{gR1} during total time interval is

$$\bar{E}_{dB} + \frac{1}{2} \Delta S_{gR1} = 0.295 + \frac{0.369}{2} = 0.479$$

$$\bar{q}_{orB} = 0.488 \frac{k_{ov}}{11_o} \frac{1.35}{5.13 S_{gb1}} (\rho_o - \rho_g)(A_C)$$

$$= \frac{0.488(0.228)(1.35)(0.433)(8.43 \times 10^6)}{(0.28)10^{(5.13)(0.479)}}$$

$$= 6.90 \times 10^3 \text{ reservoir B/D}$$

which checks the value of 6.97×10^3 calculated by material balance. Then

$$\sum_{j=1}^{n} \Delta A_C = 8.43 \times 10^6 \text{ ft}$$

and

$$\frac{\Delta A_C}{\Delta t} = \frac{3.00 \times 10^6}{365}$$

$$= 8.22 \times 10^3 \text{ ft}^2 \text{ day}^{-1} \text{ (at } t - \frac{1}{2}\Delta t)$$

A similar procedure was followed in computing the area swept about well 50.

Using the preceding calculations the recovery and position of the gas oil contact as a function of time may be computed. ■

Miscible displacement results from high-pressure gas injection, condensing-gas injection, and a miscible slug drive. The phase relations for high-pressure gas injection are shown schematically in Fig. 7.73. A lean gas of composition C is injected into the reservoir containing fluid A. At the start A and C are immiscible—a line connecting A and C on the diagram of Fig. 7.73 passes through the two-phase region. As gas C moves through the reservoir it gets richer because of vaporization until the mixture reaches the critical composition B where it is miscible with A. Once miscibility is reached, any oil contacted by the gas is displaced so that after that point recovery is affected only by conformance and aerial sweep efficiency.

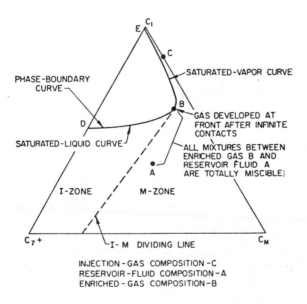

FIG. 7.73 Illustration of type M (miscible) displacement (high-pressure gas injection). (From Slobod and Koch, 1953.)

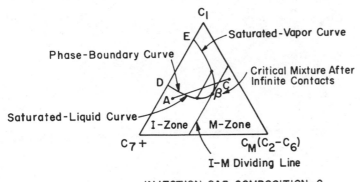

INJECTION-GAS COMPOSITION- C
RESERVOIR- FLUID COMPOSITION-A

FIG. 7.74 Illustration of type M (miscible) displacement (condensing-gas drive). (From Frick and Taylor, 1962.)

Laboratory studies are necessary to determine the phase behavior of the reservoir fluid and the injected gas. The reservoir must be deep enough to use high pressure (greater than 3000 psi). The reservoir must be undersaturated and contain enough light hydrocarbons (C_2-C_4) for vaporization to create miscibility.

The phase relations for condensing-gas drive are shown in Fig. 7.74. In this process the components of the injected gas condense while displacing the reservoir fluid. A rich gas composition C is injected into the reservoir containing oil A. A and C are immiscible at the start. As injection continues, components condense from C until at B miscibility with A is reached. The composition of gas C at the start must be to the right of the line IM for the process to work. The number of intermediate hydrocarbons required in gas C depends on the pressure. The higher the pressure, the lower the number of intermediates.

A miscible slug process is one where a slug of 5 to 10 percent of a pore volume of liquefied petroleum gas (LPG), alcohol, or carbon dioxide is injected into the reservoir chased with a dry gas. The slug must be miscible with both the reservoir fluid and the chaser. Design of the size of the slug depends on mobility ratio, conformance, and reservoir heterogeneity.

CASE STUDY 13: Calculations for a Miscible Displacement

Lantz (1970) presents an analogy to calculate miscible displacement using an immiscible reservoir simulator. Displacement is calculated involving two components within a single phase using relative permeabilities and capillary pressure to be special functions of saturation. If these functions are chosen properly, the equations for immiscible and miscible displacements are analogous. If two miscible fluids are assumed incompressible, the overall volume balance and component balance are respectively

$$\underline{\nabla} \cdot v = q \tag{7.169}$$

and

$$\underline{\nabla} \cdot \phi D \ \underline{\nabla} C - \underline{\nabla} \cdot vC = \phi \ \frac{\partial c}{\partial t} - c_i q \qquad (7.170)$$

The equations for each phase are

$$\underline{\nabla} \cdot \lambda_n \ \underline{\nabla} \Phi_n = -\phi \ \frac{\partial S_w}{\partial t} - q_n \qquad (7.171)$$

and

$$\underline{\nabla} \cdot \lambda_w \ \underline{\nabla} \Phi_w = -\phi \ \frac{\partial S_w}{\partial t} - q_w \qquad (7.172)$$

where $\lambda = kk_r/\mu$.

Adding Eq. (7.171) and (7.172) and subtracting Eq. (7.172) from Eq. (7.171) and rearranging

$$\underline{\nabla} \cdot (v_n + v_w) = q_n + q_w = q \qquad (7.173)$$

and

$$-\underline{\nabla} \cdot (v_n - v_w) = -2\phi \ \frac{\partial S_w}{\partial t} - (q_n - q_w) \qquad (7.174)$$

If we assume the following for the form of the relative permeabilities,

$$k_{rw} = \frac{\mu_w S_w}{\mu(S_w)} \qquad (7.175)$$

and

$$k_{rn} = \frac{\mu_n S_n}{\mu(S_w)} \qquad (7.176)$$

Substituting Eqs. (7.175) and (7.176) into Eq. (7.173)

$$\underline{\nabla} \cdot \frac{-k}{\mu(S_w)} [S_w \ \nabla p_w + S_n \ \nabla p_n + (\rho_w S_w + \rho_n S_n)] g \ \Delta h = q \qquad (7.177)$$

The concentration dependence of density is

$$\rho(c) = c\rho_1 + (1 - c)\rho_2 \qquad (7.178)$$

and of viscosity is

$$\mu(c) = \mu_1^c \mu_2^{(1-c)} \qquad (7.179)$$

then $v_n + v_w$ of Eq. (7.173) or Eq. (7.177) is the same as v of Eq. (7.169). Rewrite Eq. (7.173) and (7.174) as

$$\underline{\nabla} \cdot \frac{4\lambda_w \lambda_n}{\lambda_w + \lambda_n} \underline{\nabla} \frac{(\Phi_n - \Phi_w)}{2} - \underline{\nabla} \cdot \frac{\lambda_n - \lambda_w}{\lambda_w + \lambda_n} (v_n + v_w) = -2\phi \frac{\partial S_w}{\partial t} - (q_n - q_w)$$

$$(7.180)$$

where $v_n + v_w = -\lambda_n \underline{\nabla} \Phi_n - \lambda_w \Phi_w$

Also

$$\nabla p = S_w \nabla p_w + S_n \nabla p_n \tag{7.181}$$

Assuming capillary pressure has the following form

$$p_c = \frac{D\phi}{k} \int \frac{\mu(S_w)(\lambda_w + \lambda_n)}{\lambda_w \lambda_n} dS_w - (\rho_n - \rho_w)g \int dh \tag{7.182}$$

Substituting Eqs. (7.175), (7.176), and (7.182) into Eq. (7.180) gives

$$\underline{\nabla} \cdot D\phi \nabla S_w - (v_n + v_w) \cdot \nabla S_w = \phi \frac{\partial S_w}{\partial t} - (q_w - S_w q) \tag{7.183}$$

Equation (7.183) is of the same form as Eq. (7.170) if S_w is interpreted as a component volume fraction and $q_w = S_{wi}q$. The relative permeability functions for the miscible analogy are given in Fig. (7.75) for the logarithmic mixing model. The assumptions restricting the analogy are that mixing is due to molecular difference or a constant dispersion.

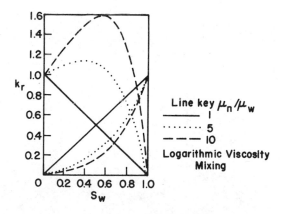

FIG. 7.75 Relative permeability functions. (From R. B. Lantz, Rigorous Calculations of Miscible Displacement Using Immiscible Reservoir Simulation, Pet. Trans. AIME 249, SPEJ 192. © 1970 SPE-AIME.)

TABLE 7.26 Properties

Difference type	CD
W, cm × L, cm	50 × 50
Number blocks	35 × 35
Time step, PV	Variable
$D\mu/K$, dyne/cm^2	1.25
Average (v/ϕ) cm sec^{-1}	0.04
Input $D\phi/vR$	0.0625

Source: R. B. Lantz, Rigorous Calculations of Miscible Displacement Using Immiscible Reservoir Simulation, Pet. Trans.
AIME 249, SPEJ 192. © 1970 SPE-AIME.

       Lantz (1970) presents a two-dimensional example for equal density and viscos-
ity with properties given in Table 7.26. The calculation represents one quarter of
a circle and is compared to an analytical solution for the radial case. The variable
time step was based upon the saturation change occurring during the first iteration.
Twenty percent saturation change per block was the criterion used to select the time
step. Since each block corresponds to 1/1000 of the pore space the resulting dimensionless numerical diffusion was of the order of 0.0002 which is negligible compared
to the input value of 0.0625. Figure 7.76 shows the comparison of the calculation
and analytical results. Lantz (1970) calculated several examples of using the
method. ∎

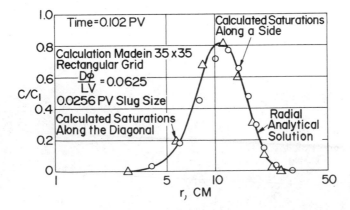

FIG. 7.76 Comparison of calculation and analytical results. (From R. B. Lantz,
Rigorous Calculations of Miscible Displacement Using Immiscible Reservoir Simulation, Pet. Trans. AIME 249, SPEJ 192. © 1970 SPE-AIME.)

Waterflooding is a common secondary recovery process and/or pressure maintenance technique. As with all displacement processes a complete reservoir description is necessary to design a waterflood. To determine the effect of a waterflood, it is necessary to estimate residual oil saturation after waterflooding. The estimated residual oil after waterflood is subtracted from the oil in place at the start of the waterflood to estimate expected recovery. Usually the residuals are estimated from laboratory floods on reservoir cores. In the process relative permeability data are determined for use in the waterflood calculations. An important consideration in determining data in the laboratory is that the wettability of the core material is preserved in the same state as the reservoir. Conformance for waterflooding is the vertical sweep efficiency. There is less chance of unstable flow in water flooding than with gas drives. However, the viscosity of the water may be raised by addition of soluble polymers or polysaccharides to improve the mobility ratio—to make the flood more favorable.

Two methods are used to calculate water flood behavior. The Stiles (1949) method for waterflood calculations used in layered systems assumes a sharp saturation front—the linear flood advance in a given layer is proportional to the permeability—thickness of the layer. After breakthrough, oil production rates are determined from the overall mobility ratio. The data needed for the calculations are permeability, thickness, water-oil mobility ratios, and formation volume factors. Once the calculations are complete, the data are corrected for sweep efficiency (product of aerial sweep and conformance).

Dykstra and Parsons (1950) concluded that recovery by water flooding is a function of mobility ratio, permeability variation, initial oil saturation, and initial water saturation from radial flow and pattern floods. Areal efficiency for pattern floods has been determined by electrical analogs (Muskat and Wyckoff, 1934; Wyckoff et al., 1933) and by x-ray model studies (Caudle et al., 1955; Dyes et al., 1954). The Buckley-Leverett fractional flow equation is given as Eq. (7.155) for gas and oil displacement. For waterfloods the effect of capillary pressure gradient and gravity are neglected so that

$$f_w = \frac{1}{1 + (k_{ro}\mu_w / k_{rw}\mu_o)} \qquad (7.184)$$

and

$$x = \frac{5.615qt \ \partial f_w}{\phi A \ \partial S_w} \qquad (7.185)$$

CASE STUDY 14: Calculations for a Waterflood in a Layered System

Oil produced from the Prudhoe Bay Field has passed the billion-barrel mark. Studies have been undertaken to determine the profitability of a waterflood project and/or reinjection of the produced gas (as opposed to selling the gas) (Anon. 1980a and b). Waterflooding would be conducted at Prudhoe Bay in portions of the reservoir from which ultimate recovery by the primary drive mechanisms is forecast to be low. Studies indicate that water injections would boost oil recovery by 5 to 9 percent. This increase in oil recovery by water injection would be about 1.5 billion bbl.

The optimum time for producing gas for sale is an issue being studied. Four commodities must be considered to determine the best time for overall energy recovery—oil, gas, condensate, and natural-gas liquids. When you take all four of these into consideration, the studies performed by the operating companies indicated that 1985-1986 would probably be the best time for maximum energy recovery. This date is as early as the gas sale can start because the gas pipeline cannot be completed before then.

Using a simple model, the Buckley-Leverett model, and some information about the reservoir, some prediction of the behavior of the reservoir during waterflooding can be determined. The following are calculations that allow following the waterfront with time and to determine the recovery at breakthrough.

Table 7.27 presents data for the calculation. The set of wells used is a quarter of a nine-spot pattern. Each well is on a 160-acre block. The distance between producer and injector is 2640 ft.

Some of the simplifying assumptions made in the calculations presented here are as follows:

Incompressible flow in one direction
No capillary pressure across the fluid-fluid interface
Linear displacement
No interfacial tension or density difference between fluids
Homogeneous, uniform formation

The calculational procedure is that outlined in an example in the Reservoir Engineering Manual by Cole (1969).

1. Calculate F_w, fraction of water flowing (assume a horizontal reservoir)

$$F_w = \frac{1}{1 + (k_o/k_w)(\mu_w/\mu_o)}$$

The results of this calculation are in Table 7.28, column 4, for each value of S_w.
2. Plot F_w versus S_w (see Fig. 7.77).
3. From Fig. 7.77, obtain the slope (dF_w/dS_w) at various values of S_w. See Table 7.28, column 5.
4. Plot dF_w/dS_w versus S_w (see Fig. 7.78).
5. For a given volume of water injected Q_t, calculate the distance of flood front travel Δx for a range of S_w. The continuity equation is used:

$$\Delta x = \frac{Q_t}{\phi A} \frac{dF_w}{dS_w} = \frac{q_t n}{\phi A} \frac{dF_w}{dS_w}$$

Results are in columns 6 to 10 of Table 7.27, each column representing the water profile at a different number of days after starting injection.
6. Plot S_w versus Δx (see Fig. 7.79). The area under the curve equals $S_w \Delta x$. Since $Q_t = \phi A \Delta x S_w$, then $S_w \Delta x = Q_t/\phi A$ = the area under the curve. To find the location of the discontinuity for each time interval, calculate

TABLE 7.27 Data

Layer (of model)	Porosity ϕ, porosity fraction	h, thickness (ft)
1	0.0	0.01
2	0.204	38.83
3	0.206	75.00
4	0.136	50.
5	0.175	6.8
6	0.199	6.8
7	0.245	54.4
8	0.252	27.2
9	0.227	27.2
10	0.094	27.19
11	0.105	54.38
12	0.212	27.19
13	0.156	25.
14	0.200	43.14
15	0.115	0.29

Viscosity of formation water	μ_w	= 0.312 cP
Average value for viscosity of oil	μ_o	= 1.2 cP
Cross-sectional area across which flow occurs	Λ	= 278058 ft^2
Injection rate	q_t	= 7000 bbl day^{-1}
Initial oil saturation	S_{oi}	= 0.81
Connate water saturation	S_{cw}	= 0.19
Distance between injector and producer wells	Δx_{max}	= 2640 ft

the corresponding value of $Q_t \phi A$. By trial and error, locate the Δx which causes the area to be equal to that quantity (see Fig. 7.79).

To find the time required for breakthrough, a similar procedure is followed, except that Δx is known (2640 ft, the distance between the producing and injecting wells) and Q_t is the only unknown. Draw in the breakthrough curve, calculate the area under it, determine Q_t:

$$Q_t = (\text{area}) \, \phi A = 6.09 \times 10^7 \text{ ft}^3$$

TABLE 7.28 Calculated Intermediates

SW	k_o	k_w	F_w	dF_w/dS_w	Δx (ft) n=100	Δx n=500	Δx n=1000	Δx n=5000	Δx n=10,000
0.19	1.0	0							
0.22	0.88	0.00003	0.00013						
0.25	0.74	0.00007	0.00036						
0.30	0.52	0.00014	0.00103	0	0	0	0	0	0
0.35	0.36	0.00027	0.00288	0.035	3.17	15.85	31.7	158.5	317
0.40	0.25	0.00052	0.00794	0.22	19.93	99.65	199.3	996.5	1993
0.45	0.17	0.0010	0.0221	0.60	54.37	271.85	543.7	2718.5	5437
0.50	0.10	0.0018	0.0647	1.24	112.4	562.0	1124	56.20	11240
0.55	0.062	0.0034	0.1742	3.40	308.08	1540.4	3081	15404	30808
0.60	0.034	0.0064	0.4199	6.33	573.58	2867.9	5736	28680	57358
0.65	0.017	0.012	0.7308	4.78	433.13	2166	4331	21656	43313
0.70	0.0084	0.023	0.9133	1.92	173.98	870	1740	8699	17398
0.75	0.0035	0.047	0.9810	0.88	79.74	399	797.4	3987	7974
0.80	0.0009	0.102	0.9977	0.12	10.87	54.4	108.7	543.5	1087
0.819	0.00028	0.150	0.9995	0	0	0	0	0	0
0.85	0.00005	0.280	1.0						
0.90	0.00003	0.560							
0.95	0.00001	0.890							
0.97	0	1.0							
1									

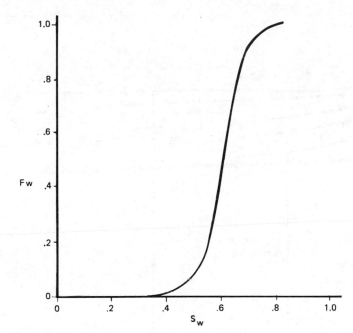

FIG. 7.77 F_w versus S_w.

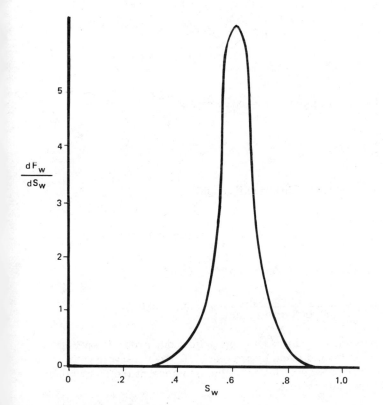

FIG. 7.78 dF_w/dS_w versus S_w.

421

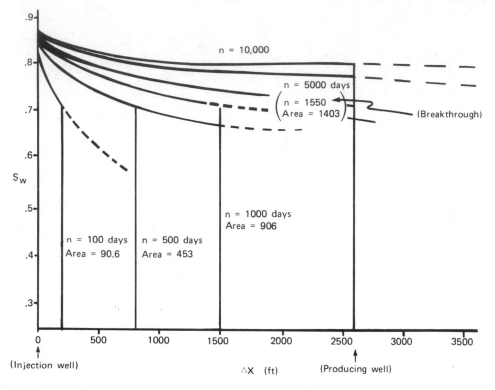

FIG. 7.79 S_w versus Δx.

Then calculate the time required for breakthrough:

$$\Delta t = \frac{Q_t}{q_t} = 1550 \text{ days}$$

7. Calculate the percent recovery at breakthrough:

Cumulative oil produced
at breakthrough   = (area under the curve of 1550 days)
                         + 6.087×10^7 ft^3

Original oil in place = $S_{oi} \phi A \Delta x = 0.276 \times 10^7$ ft^3

$$\text{Percent recovery} = \frac{\text{oil produced}}{\text{original oil in place}} \times 100\% = 65\%$$

This value of 65 percent recovery is an approximation, probably optimistic, much rougher than estimates made by representing the reservoir with three-dimensional computer simulation taking into account the many other details that affect the fluid flow. ∎

CASE STUDY 15: Numerical Simulation of a Waterflood

McCarty and Barfield (1958) present discussion of the use of a two-dimensional numerical simulation to predict flood front movements in a waterflood, assuming a distinct oil-water interface is formed, either water or oil is flowing in the predominantly water-saturated or oil-saturated regions of the reservoir. For the single-phase flow regions on each side of the interface,

$$\frac{\partial^2 p}{\partial x^2} + \frac{\partial^2 p}{\partial y^2} = \frac{c\phi\mu}{k}\frac{\partial p}{\partial t} \qquad\qquad (7.186)$$

With a varying fluid mobility,

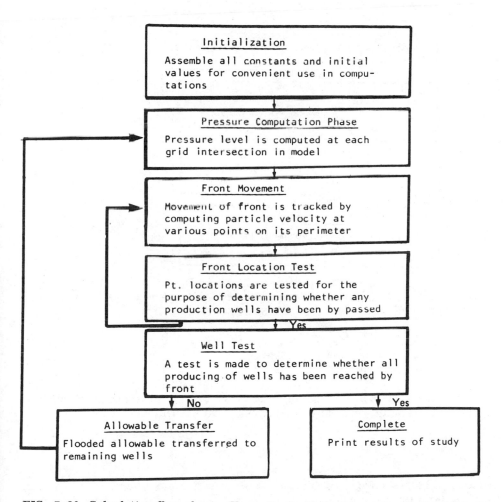

FIG. 7.80 Calculation flow sheet. (From D. G. McCarty and F. C. Barfield, The Use of High-Speed Computers for Predicting Flood-Out Patterns, Pet. Trans. AIME 213, SPEJ 139. © 1958 SPE-AIME.)

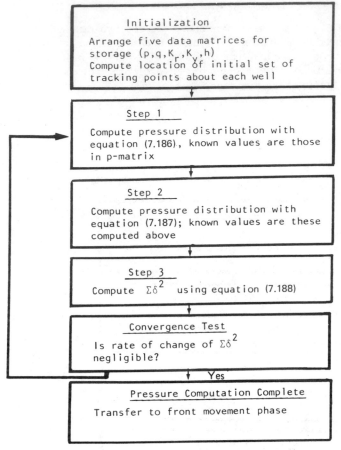

FIG. 7.81 Pressure computation flow sheet. (From D. G. McCarty and F. C. Barfield, The Use of High-Speed Computers for Predicting Flood-Out Patterns, Pet. Trans. AIME 213, SPEJ 139. © 1958 SPE-AIME.)

$$\frac{\partial}{\partial x}\left(\frac{kh}{\mu}\frac{\partial p}{\partial x}\right) + \frac{\partial}{\partial y}\left(\frac{kh}{\mu}\frac{\partial p}{\partial y}\right) = c\phi h\frac{\partial p}{\partial t} \qquad (7.187)$$

The basic steady-state differential equation describing the pressure distribution in the reservoir assuming incompressibility and equal mobility is

$$\frac{\partial}{\partial y}\left(kh\frac{\partial p}{\partial x}\right) + \frac{\partial}{\partial y}\left(kh\frac{\partial p}{\partial y}\right) = 0 \qquad (7.188)$$

Equation (7.188) is solved by replacing derivatives by finite differences and solving the resulting difference equation.

In defining a reservoir for the finite difference technique, a scale map of the reservoir is used. A grid is superimposed onto this map. Appropriate values of permeability and thickness are assigned to each grid point. These points form the basic information used in the reservoir analysis. Another necessary requirement is the initial and boundary conditions. The numerical technique used for the pressure calculations is the ADIP procedure discussed in Chapter 6.

The procedure for computation has three main parts: the pressure computations, the front movement, and the adjustment of the production-injection schedule. Each time a production well is flooded and removed from the pattern an adjustment is made to the production-injection schedule. Figure 7.80 is a flow diagram of the calculation procedure. Figure 7.81 summarizes the essential steps from the start of the problem through completion of the pressure computation phase. The final result is a steady-state pressure distribution for a particular schedule of products and injection rates. Figure 7.82 is a flow diagram for calculating the front movement.

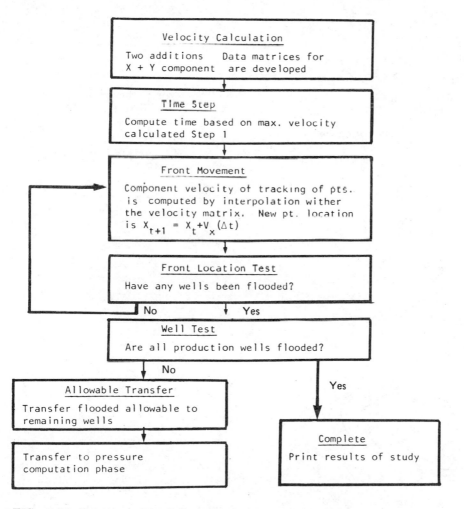

FIG. 7.82 Front movement flow sheet. (From D. G. McCarty and F. C. Barfield, The Use of High-Speed Computers for Predicting Flood-Out Patterns, Pet. Trans. AIME 213, SPEJ 139. © 1958 SPE-AIME.)

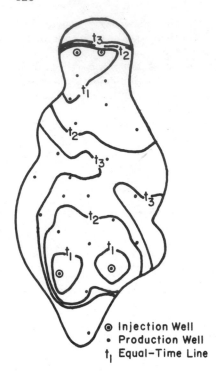

● Injection Well
• Production Well
t_1 Equal-Time Line

FIG. 7.83 Typical flood pattern resulting
from field study. (From D. G. McCarty
and F. C. Barfield, The Use of High-Speed
Computers for Predicting Flood-Out Patterns, Pet. Trans. AIME 213, SPEJ 139.
© 1958 SPE-AIME.)

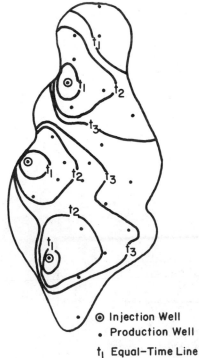

● Injection Well
• Production Well
t_1 Equal-Time Line

FIG. 7.84 Typical flood pattern resulting
from field study. (From D. G. McCarty
and F. C. Barfield, The Use of High-Speed
Computers for Predicting Flood-Out Patterns, Pet. Trans. AIME 213, SPEJ 139.
© 1958 SPE-AIME.)

This technique can be used to select optimum injection patterns. Figures 7.83 and 7.84 represent two different injection patterns for a reservoir. The reservoir was represented with a rectangular grid with 1400 grid points. There are 25 injection and production wells in the reservoir. When scaled to reservoir dimensions, the interval between grid points is 315 ft. In Fig. 7.83 an end-to-end injection is simulated. In this simulation a large "pocket" of oil is trapped in one area of the field where flood fronts do not coalesce. In Fig. 7.84 a peripheral-type injection is simulated. This type of pattern shows that a lesser volume of the reservoir is invaded before flooding all wells. It also indicates early flooding of some of the production wells. ∎

7.7 TERTIARY RECOVERY

Tertiary recovery is a term applied to methods or processes of obtaining the residual oil from a reservoir after it no longer produces oil economically by a secondary recovery process. As with the term secondary recovery, this definition is arbitrary. It may be desirable to enhance recovery using waterflooding or water additives to improve mobility ratios or lower interfacial tension in the primary or secondary stages of depletion. In a broader sense enhanced recovery includes most nonnatural methods for recovery of oil. Usually, tertiary recovery refers to some process of removing residual oil trapped discontinuously in the pores of the reservoir material. Normally this is after a reservoir has been waterflooded to its economic limit.

Carbon dioxide flooding is a method of enhancing recovery of oil by combination of solution gas drive, swelling of the oil, and subsequent reduction of viscosity, and miscible effects through extraction of hydrocarbons from the oil in reservoir. The extraction of the hydrocarbons, the C_5 to C_{30} hydrocarbons, from crude oil in place by CO_2 promotes a high displacement efficiency. Unlike LPG CO_2 does not achieve direct miscible displacement at practical reservoir pressures. Also unlike high-pressure gas-miscible processes the displacement of oil by CO_2 does not depend on the presence of light hydrocarbons in reservoir oil.

The CO_2 process is applicable to reservoirs in which the oil has been depleted of its gas and LPG components. The extraction of hydrocarbons by CO_2 at flooding temperatures below about 200°F from reservoir oils above 30°API occurs in the range of 1000 to 2000 lb in.$^{-2}$ The flooding pressure at which most of the contacted oil in place is recovered is several hundred psi above the pressure in which the extraction mechanism first takes place.

The use of carbon dioxide as an oil recovery agent has been investigated for many years. Holm and Josendal (1974) published both laboratory and field studies using CO_2 as an oil displacement agent. The characteristics of carbon dioxide that make it useful as an oil displacement agent in addition to swelling, reduction of oil viscosity and extraction of oil, are increase of oil density, and the acidic effect of CO_2 in the rock increasing the injectivity of water. With CO_2 flooding, a bank of light hydrocarbons forms at the leading edge of the CO_2 slug. Formation of this bank is probable due to a selective extraction by the CO_2 and probably accounts for high oil recoveries. In addition one obtains a favorable mobility ratio from the CO_2 waterflood behind the oil bank.

CASE STUDY 16: Interpretation of a CO_2 Pilot Flood

Holm and O'Brien (1971) report the results of a carbon dioxide field pilot to test the effectiveness of CO_2 as an oil recovery agent.

A small slug of CO_2 (4 percent PV) was injected followed by a slug of carbonated water (12 percent pV) and then brine. Prior to CO_2 injection, water was injected to raise the reservoir pressure in the test area from about 115 to 850 psi; the objective was to maintain the average reservoir pressure at a minimum of 850 psig throughout the test to ensure maximum effectiveness of the CO_2.

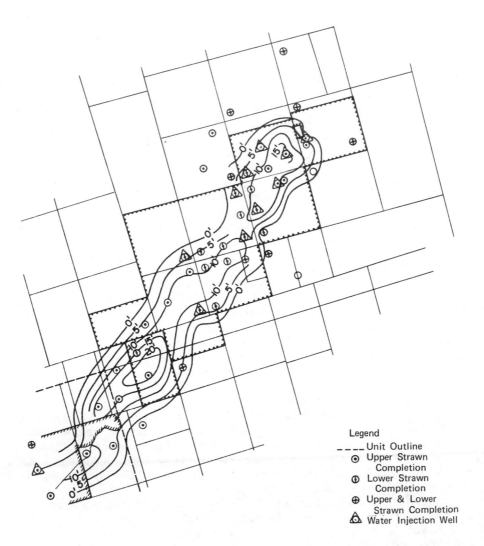

Legend
- - - - Unit Outline
⊙ Upper Strawn Completion
⊕ Lower Strawn Completion
⊕ Upper & Lower Strawn Completion
△ Water Injection Well

FIG. 7.85 Isopachous map of upper zone of Mead-Strawn sand. (From L. W. Holm and L. S. O'Brien, Carbon Dioxide Test at the Mead-Strawn Field, J. Pet. Tech. 431. © April 1977 SPE-AIME.)

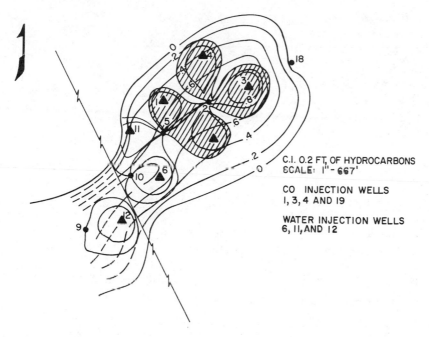

C.I. 0.2 FT. OF HYDROCARBONS
SCALE: 1" - 667'

CO INJECTION WELLS
1, 3, 4 AND 19

WATER INJECTION WELLS
6, 11, AND 12

FIG. 7.86 Mead–Strawn upper zone test area: predicted areal sweep. (From L. W. Holm and L. S. O'Brien, Carbon Dioxide Test at the Mead–Strawn Field, J. Pet. Tech. 431. © April 1977 SPE-AIME.)

The test was run in a small area of the Upper Stawn sand in the Mead field located near Abilene, Texas. Oil is produced from two zones in the Strawn sands; these zones are separated by about 400 ft of shale and lime. Approximately 2000 acre ft to the south and west of the CO_2 pilot in the Upper zone, and all

TABLE 7.29

Average thickness, ft	10.3
Average porosity, percent	11
Average permeability to air, mdarcy	9
Reservoir oil in place, percent PV	39
Water saturation, percent PV	40
Oil recovered by primary, bbl (acre ft)$^{-1}$	
From area around wells 1 through 5	105
From area around wells 7, 9, 10	114

Source: L. W. Holm and L. S. O'Brien, Carbon Dioxide Test at the Mead–Strawn Field, J. Pet. Tech. 431 (April 1971). © 1971 SPE-AIME.

TABLE 7.30 Properties of CO_2 and CO_2-saturated Mead-
Strawn Oil and Injection Water

Properties of CO_2

  Surface conditions, 0°F and 300 psi
    Density, lb ft^{-3} 63.69
    Viscosity, cP   0.14

  Reservoir conditions, 135°F
    Density, lb ft^{-3}
      at 1000 psi   9.8
      at 1500 psi   20.8
      at 2000 psi   36
    Viscosity, cP
      at 1000 psi   0.02
      at 1500 psi   0.02
    Solubility in injection water at 135°F, scf bbl^{-1}
      at 1000 psi   104
      at 1500 psi   118

Properties of CO_2-saturated Mead-Strawn crude oil

  Solution GOR, scf bbl^{-1} at 60°F
    at 1000 psi   640
    at 1500 psi   1,325

  Oil viscosity, cP
    at 1000 psi   0.58
    at 1500 psi   0.38

  Oil density, g cm^{-3}
    at 1000 psi   0.797
    at 1500 psi   0.806

  Relative oil volume
    at 1000 psi   1.25
    at 1500 psi   1.5

Source: L. W. Holm and L. S. O'Brien, Carbon Dioxide Test
at the Mead-Strawn Field, J. Pet. Tech. 431 (April 1971).
© 1971 SPE-AIME.

2000 acre ft of the Lower zone, were waterflooded conventionally at the same time
the CO_2 flood was being conducted.

    The Upper and Lower Strawn sands are Pennsylvanian in age and are at depths
of 4500 and 4900 ft, respectively. There is no known oil-water contact in the Mead-
Strawn sands. Structural attitude is monoclinal, with a dip of 80 to 100 ft mile^{-1} in
a northwesterly direction. The hydrocarbon traps occur independently of structure.

    Figure 7.85 is an isopachous map of the Upper Strawn sand. The area for the
CO_2 test is shown in Fig. 7.86. The pilot is a regular five-spot of 33.5 acres with
an adjacent area of about 9 acres. This total area assigned to the CO_2 test included
four injection wells (nos. 1, 3, 4, and 19) and two production wells (nos. 2 and 5).

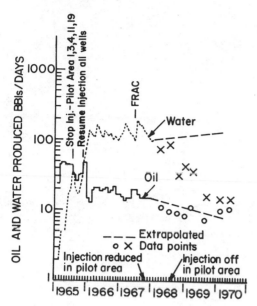

FIG. 7.87 Production history: Mead-Strawn CO₂ flood, well 2. (From L. W. Holm and L. S. O'Brien, Carbon dioxide Test at the Mead-Strawn Field, J. Pet. Tech. 431. © April 1977 SPE-AIME.)

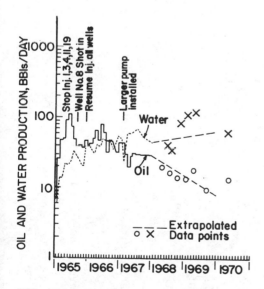

FIG. 7.88 Production history: Mead-Strawn CO₂ flood, well 10. (From L. W. Holm and L. S. O'Brien, Carbon Dioxide Test at the Mead-Strawn Field, J. Pet. Tech. 431. © April 1977 SPE-AIME.)

TABLE 7.31 Fluid Production from Mead–Strawn Upper and Lower Sands

Area in field		Number	Liquid recovered to Jan. 1, 1968		1969 Water cut (%)	Oil recovery extrapolated to 95% cut (bbl)	Drainage area[b] around well (acre ft)	Oil recovered	
			Oil (bbl)	Water (bbl)				To Jan. 1, 1968 [bbl(acre ft)$^{-1}$]	To 95% cut [bbl(acre ft)$^{-1}$]
CO$_2$ test pattern		2	27,098	98,968	86	35,098	306	89	115
		5	17,016	99,570	87	22,846	333	51	70
		10	48,460	57,500	65	69,460	158	307	438
Total			92,574	256,038		127,404	797	Avg. 116	160
Entire waterflood	9U		11,200	34,721	89	12,660	158	71	80
	14U&L		6,400	61,727	97+	6,400	135	47	47
	7L		23,123	168,205	97+	23,123	263	88	88
	8U&L		5,092[b]	11,800[a]	50	9,700[b]	—	—	92
	10L		17,592	84,637	88	23,322	254	70	45
	19L		7,025	17,126	91	7,935	176	40	75
Total			65,320	366,416		73,440	986	Avg. 66	
Best water-flood wells	9U								
	7L		51,915	287,563	—	59,105	675	76	88
	10L								

Oil recovery: CO$_2$ test area compared with total waterflood area
To Jan. 1, 1968 (116·66)/66 = 75 percent increase
To 95 percent cut (160·75)/75 = 113 percent increase

Oil recovery: CO$_2$ test area compared with best three waterflood wells
To Jan. 1, 1968 (116·76)/76 = 53 percent increase
To 95 percent cut (160·88)/88 = 82 percent increase

[a]Not included in totals because of lack of information to estimate drainage area.
[b]Calculations are based on potentiometric and mathematical studies and data from cores and logs from wells in area.
Source: L. W. Holm and L. S. O'Brien, Carbon Dioxide Test at the Mead–Strawn Field, J. Pet. Tech. 431 (April 1971). 1971 SPE-AIME.

The average distance between injection and production wells is about 800 ft. The test area was surrounded by dry holes to the east, north, and west and by two water injection wells to the south and southwest. These water injection wells (nos. 11 and 12) served as backup wells to help confine the injected fluids within the test area. Table 7.29 gives the properties of the pilot area. The properties of CO_2 and of CO_2-saturated Mead–Strawn oil and injection water are given in Table 7.30.

A minimum reservoir pressure of 800 psi was required for effective CO_2 flooding; therefore it was necessary to raise formation pressure in the pilot area by first injecting flood water. Water injection to raise pressure was started in mid-1964 and continued for about 5 months; 153,000 bbl of water—about 20 percent of the conforming pore volume (CPV)—was injected.

On December 1, 1964, CO_2 injection was started. Five thousand tons of CO_2 (about 4 percent CPV) was injected into the four pilot injection wells (no. 1, 3, 4, and 19). The amount injected into each well was in proportion to the reservoir pore volume surrounding the well. CO_2 injection was continued for 3 months at a rate of about 55 tons day^{-1}. On March 1, 1965, carbonated water injection was started; actually, alternating slugs of CO_2 and brine were injected. An additional 500 tons of CO_2 was injected in this manner. Regular flood water (7 percent brine) was used to push the CO_2 and carbonated water slugs through the formation. Brine injection into wells 6, 11, and 12 was begun at about the same time that the CO_2 flood was started. The waterflood of the remaining Upper and Lower sands in the unit was started early in 1965.

The production wells affected by the CO_2 flood did not respond dramatically. However, production from these wells was characterized by higher oil-flow rates that did not decline as rapidly as the rates of wells affected only by the waterflood. Other production data showed that the waterflood was not so successful as anticipated. There is evidence that this poor performance was due to reservoir heterogeneity, high initial water saturation, and inadequate flooding patterns. Figures 7.87 and 7.88 plot the oil and water production data for wells 2 and 10. Neither of these wells produced at the anticipated rate, nor did any of the waterflood producers.

The effectiveness of CO_2 flooding and of water flooding is illustrated by the data in Table 7.31. The CO_2 flood swept more than the five-spot test pattern originally selected for evaluation. Instead, the flood area was more like a line drive pattern. The remaining Upper zone also was swept by the waterflood in this manner. The oil recovered from each well was expressed as barrels per acre foot in order to compare the production from wells in the CO_2 test area with the production from wells in the waterflood area. These results were compared as of January 1, 1968 (the end of the waterflood), and also for the end of the CO_2 flood (obtained by extrapolation to a 95 percent cut). The data indicate that 53 to 82 percent more oil was produced by the CO_2 flood than was produced by water in the best areas of the waterflood.

The production history of CO_2 from the test area is an indication of the effectiveness of CO_2 as a flooding agent. There was no channeling of CO_2. The small amount of CO_2 produced (less than 10 percent of that injected considering the CO_2 in the produced oil, water, and gas) was produced at a rate similar to that observed in long laboratory core experiments, confirming that CO_2 was transported chromatographically through the porous medium by the oil and the injected water. Over 50 percent more oil was produced by the CO_2 carbonated flood than by conventional flooding. ■

The methods to lower interfacial tension are addition of surface-active agents to a bank of oil field brine, addition of a slug of alkaline material that reacts with the oil creating surface-active agents in place, and addition of materials which change the wettability of the reservoir medium. All of the methods are directed at lowering interfacial tension in the water-oil-solid system so that residual trapped oil becomes mobilized and forms a bank of oil which can be moved through the reservoir. The processes are complicated in that they require multiple banks of materials from a preflush which conditions the reservoir, surface-active agent bank, a bank of water with thickener for mobility control, to the driving water behind the first three banks.

Macroscopically there are several aspects to consider in the design of a surfactant flood. Several dimensionless groups should be considered. The reduced capillary pressure—a ratio of viscous to inertial forces—is represented by the capillary number or the J function. The capillary number

$$Ca = \frac{\mu q}{\phi \sigma} \tag{7.189}$$

is usually of the order of 10^{-6} during waterflooding. For a surfactant flood to approach values of residual oil near zero in laboratory floods, the capillary number must approach 10^{-2}. A capillary number of 10^{-2} can be obtained if interfacial tension between oil and water is of the order 10^{-3} dynes cm^{-1}. Figure 7.89 shows a plot of capillary number versus residual oil. Region A is the data of Foster (1973). Region B is the data of duPrey (1969). The scatter in Fig. 7.89 is attributed to experimental error; however as shown in Chapter 4 the nonuniformity of the media may explain some of this scatter. The viscosity ratio between water and oil is of

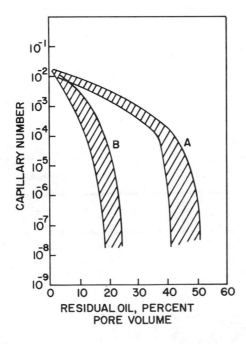

FIG. 7.89 Dependence of residual oil on capillary number. (From W. R. Foster, A Low-Tension Water Flooding Process, Pet. Trans. AIME 225, SPEJ 205. © 1973 SPE-AIME.)

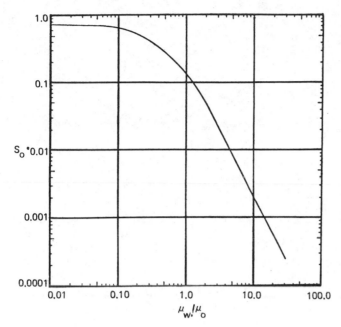

FIG. 7.90 Oil saturation following low-tension displacement. (From W. R. Foster, A Low-Tension Water Flooding Process, Pet. Trans. AIME 225, SPEJ 205. © 1973 SPE-AIME.)

primary importance as the region of zero residual oil propagates through the medium, since the oil moves with velocities that increase with increasing oil saturation. Depending on the viscosity ratio the movement of the low-surface-tension region and the faster-moving oil form an oil bank. The saturation of the low-tension region is called a critical oil isosaturation S_O^*. Figure 7.90 is a plot of S_O^* versus μ_W/μ_O. In a normal water flood the viscosity is about 10^{-1} to 10^0. From Fig. 7.90 if the viscosity ratio is greater than 4, the critical oil saturation is less than 0.01.

The Bond number—the ratio of gravitational to viscous forces—is

$$B_n = \frac{(\rho_w - \rho_o)\,gk}{\phi\sigma} \qquad\qquad (7.190)$$

In a normal water flood the Bond number has a value of about 10^{-6}. For a capillary number of 10^{-2} the Bond number also approaches 10^{-2}. At values of 10^{-2} gravity segregation may cause underrunning of the water. Raising the viscosity of the water behind the surfactant prevents this segregation. Dispersion and absorption of the surfactant become important since the concentration level of the surface-active agent is important in designing the process. Stability of the fluid movement in the various zones is also important since fingering could destroy the integrity of the process. The use of the thickened water bank is important for this aspect of the process.

Addition of a surface-active agent to an oil-water system, where the water is saline, creates a multiphase system which can be represented by a ternary diagram

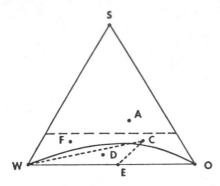

FIG. 7.91 Ternary representation of
water-oil-surfactant system. (From R. N.
Healy, R. L. Reed, and C. W. Carpenter,
Jr., A Laboratory Study of Microemulsion
Flooding, Pet. Trans. AIME 259, SPEJ
87. © 1975 SPE-AIME.)

such as in Fig. 7.91 (Healy et al., 1975). Points A and C are in the single-phase
miscible region. Composition D is in a two-phase immiscible region. Pairs of composition points are miscible, partially immiscible, or immiscible if a straight line
joining two points on the diagram is in the single-phase region, crosses from the
two-phase to the single-phase region, or is in the two-phase region, respectively.
A and C are miscible. C and W are partially miscible. D and F are immiscible.

The total set of compositions represented by Fig. 7.91 above the line separating the single- and two-phase regions (the binodal curve) are microemulsions.
Microemulsions are clear, thermodynamically stable dispersions of two immiscible
liquids (in this case oil and water) stabilized by the addition of the surface-active
agent (emulsifier). A microemulsion consists of droplets in the size range of 30 to
400 Å. Oil-continuous microemulsions absorb water until an equilibrium state is
reached. Water-continuous microemulsions absorb oil until an equilibrium state is
reached.

For simple systems a ternary diagram may have as many as four regions
as shown in Fig. 7.92. Above the binodal curves of Fig. 7.92 there is a "single-
phase" microemulsion. The micellar structure—the manner of arrangement of the
oil, water, and surfactant molecules in the interfacial region—may be different in
different parts of the diagram above the binodal curve. In the two- and three-phase
regions microemulsions containing excess water or excess oil exist in several
phases. These "phases" are usually distinguished by their being opaque in contrast
to the optically clear single-phase region. Real systems of saline water, oil, and
surface-active agents are similar to the idealized diagram of Fig. 7.92.

The mechanism of the displacement of residual discontinuous oil using a
microemulsion in a porous medium is complex. As the displacement process proceeds composition changes throughout the history of the flood. A displacement is
considered miscible if all composition pairs are miscible. In Fig. 7.91 composition
F can be miscibly displaced by composition C if the medium initially has a saturation represented by F. Ion exchange and absorption of the surface-active agents
create further complications to the displacement process. Once residual oil is
mobilized so that an oil bank forms ahead of the low-tension region, the composition
being displaced is represented by E in Fig. 7.91. As the displacement progresses,
the composition of the microemulsion is continually changing. Gradually as the flood
progresses the microemulsion-oil-water interface becomes immiscible. The dashed
line on Fig. 7.91 indicates the upper economic limit. This limit depends on the bank
size and concentration needed for a given flood.

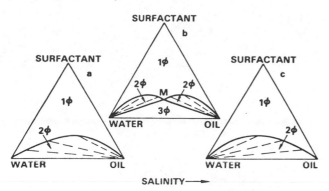

FIG. 7.92 Simple phase behavior. (From R. N. Healy, R. L. Reed, and D. G.
Stenmark, Multiphase Microemulsion Systems, Pet. Trans. AIME 261, SPEJ 147.
© 1976 SPE-AIME.)

Healy and Reed (1974) describe the use of surface-active agents in tertiary
recovery. Figure 7.93 is a schematic representation of their concept of microemulsion flooding. Starting at (a) and going to (c): (a) Microemulsion bank is injected into the porous medium (in this case a linear core) that contains a continuous
bring and a discontinuous oil. (b) The microemulsion bank is displaced by a thickened water made by adding a soluble polymer. (c) As the displacement proceeds
the oil is miscibly displaced as long as the composition of the microemulsion bank
remains in the single-phase region. (d) eventually the mixing of brine, oil, and
microemulsion change the composition such that the bank is in the two-phase region.

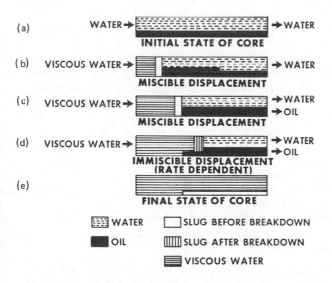

FIG. 7.93 Microemulsion flooding concepts. (From R. N. Healy and R. L. Reed,
Physicochemical Aspects of Microemulsion Flooding, Pet. Trans. AIME 257,
SPEJ 491. © 1974 SPE-AIME.)

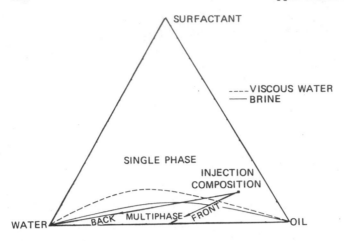

FIG. 7.94 Dilution path for microemulsion flooding. (From R. N. Healy and R. L. Reed, Physicochemical Aspects of Microemulsion Flooding, Pet. Trans. AIME 257, SPEJ 491. © 1974 SPE-AIME.)

The flood is immiscible and some oil is again trapped and left behind. (e) The viscous bank is followed by water and the various components are recovered until the core reaches its final state. Figure 7.94 shows the same displacement represented on a ternary diagram. The dashed line on this diagram indicates the effects of the viscous water on the phases.

The criteria for using a surface-active agent in a tertiary recovery process that will effectively recover oil are (1) the multiphase region of the phase diagram should be as small as possible. The binodal curve should be "low," and (2) interfacial tension in the multiphase region should be as low as possible to enhance the immiscible displacement.

The differences in the various tertiary processes using surface-active agents represent variations in the injection concentration and size of the surface-active agent bank. Absorption of the surface-active agent is an important consideration since loss of the surface-active agent will more rapidly drive the concentration of the microemulsion bank into the multiphase region. Mobility control is important to several aspects of the process. The integrity of the surface-active bank must be maintained as long as possible. The effects of gravity should be minimized. Since the process involves recovery of a small amount of residual oil, the volumetric sweep efficiency should be as high as possible so the oil is contacted.

A knowledge of residual oil saturation and possible distribution of oil saturation is important since if the amount of residual oil is low or in places difficult to reach the process may fail economically even though it is technically successful.

CASE STUDY 17: Interpretation of a Surfactant Pilot Flood

Danielson et al. (1976) report a tertiary recovery pilot in the Bradford field of Pennsylvania. The Bingham field test is a 0.75-acre pilot located in the Bradford field. Figure 7.95 is a map of the Bingham pilot area, and Table 7.32 is a summary of the reservoir properties of the pilot. As with many watered-out fields the oil saturation is not firmly established. However, from a variety of methods

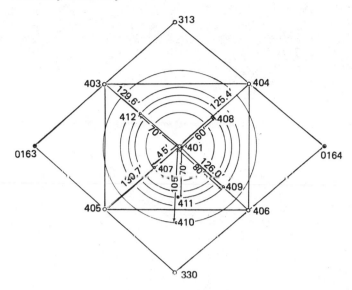

FIG. 7.95 Pilot area. (From H. H. Danielson, W. T. Paynter, and H. W. Milton, Jr., Tertiary Recovery by the Maraflood Process in the Bradford Field, J. Pet. Tech. 129. © February 1976 SPE-AIME.)

TABLE 7.32 Properties of Pilot Area

Property	Average data
Depth, ft	1860
Thickness, ft	23.7
Porosity, %	18
Oil saturation (after waterflood)	
Pilot area range, % PV	27 to 35
Expansion area range, % PV	36 to 43
Permeability, mdarcy	82
Permeability capacity, mdarcy ft	2207
Permeability maximum, mdarcy	311
Permeability minimum, mdarcy	8.4
Temperature, °F	68
Oil viscosity, cP	5

Source: H. H. Danielson, W. T. Paynter, and H. W. Milton, Jr., Tertiary Recovery by the Maraflood Process in the Bradford Field, J. Pet. Trans. 129, February 1976. © 1976 SPE-AIME.

TABLE 7.33 Slug Composition

Component	Volume (%)	
	Pilot	Expansion
Petroleum sulfonate	13	12
Petroleum distillate	39	47
Water	46	40
Cosurfactant	2	1

Source: H. H. Danielson, W. T. Paynter, and
H. W. Milton, Jr., Tertiary Recovery by the
Maraflood Process in the Bradford Field, J. Pet.
Tech. 129, February 1976. © 1976 SPE-AIME.

including core analysis, transient tests, and logs the saturation in the pilot area is
estimated in the range 27 to 35 percent. Laboratory studies were carried out to
determine the micellar solution slug size and the mobility buffer injection volume.
Design mobility of displacing fluids required for stable displacement was determined
by the minimum combined oil and water relative mobility ($k_{ro}/\mu_o + k_{rw}/\mu_r$) ahead
of the slug. A slug viscosity of 25 to 35 cP is required for stable displacement. The
primary surfactant in the micellar solution was a petroleum sulfonate. The composition of the slugs used is given in Table 7.33.

TABLE 7.34 Displacement Efficiencies

Cylinder or annular ring (wells)	Incremental oil in place (bbl)	Incremental oil displaced (bbl)	Incremental percent efficiency	Distance from injection well 401 (ft)
401–407	2,090	1,990	0.95	45 (407)
407–408	1,310	967	0.74	60 (408)
408–(411/412)	980	426	0.43	70 (411/412)
409–(411/412)[a]	1,067	424	0.40	80 (409)
409–410[a]	3,004	1,006	0.33	105 (410)
Totals	8,451	4,813	0.57	

Average S_{oi} = 50% upper zone
        = 40% middle zone

[a]Middle zone only affected. Total PV affected = 20,800 bbl.
Source: H. H. Danielson, W. T. Paynter, and H. W. Milton, Jr., Tertiary Recovery by the Maraflood Process in the Bradford Field, J. Pet. Tech. 129 (February
1976). © 1976 SPE-AIME.

The field test began with an inverted five-spot pattern on 0.75-acre spacing. The pattern was later expanded to 1.5 acres using old wells. A 6 percent PV slug was used with a total of 1800 bbl injected. Polymer was injected until 41,000 bbl was injected and then water was injected. Table 7.34 shows displacement efficiencies for the upper and middle zones with assumed oil saturation of 50 and 40 percent, respectively. This resulted in a 57 percent displacement efficiency. The incremental displacement of 4800 bbl compared with an extrapolation of the oil production indicating an oil displacement of 5100 bbl. The reversion to water injection brought a decline in oil production from the pilot area. ■

REFERENCES

Al-Hussainy, R. (1965): The Flow of Real Gases through Porous Media, M.S. Thesis Texas A&M University.

Al-Hussainy, R. (1967): Transient Flow of Ideal and Real Gases through Porous Media, Ph.D. Thesis, Texas A&M University.

Anon. (1980a): Prudhoe Bay Operators Lay Plans for Mammoth Seawater Injection Project, Oil Gas J. Feb. 25.

Anon. (1980b): Source Water Injection Seen Hiking Prudhoe Oil, Oil Gas. J. April 14.

Aronfsky, J. S., and R. Jenkins (1954): A Simplified Analysis of Unsteady Radial Gas Flow, Pet. Trans. AIME 201, 149.

Aziz, K., and D. L. Flock (1963): Unsteady State Gas Flow: Use of Drawdown Data in the Prediction of Gas Well Behavior, J. Can. Pet. Tech. 2(1), 9.

Brown, G. G., D. L. Katz, G. G. Oberfell, and R. C. Alden (1948): Natural Gasoline and the Volatile Hydrocarbons, NGAA (now NGPA), Tulsa, Okla.

Calhoun, J. C., Jr. (1953): Fundamentals of Reservoir Engineering, University of Oklahoma Press, Norman, Okla.

Carr, N. L., R. Kobyashi, and D. B. Burrows (1954): Viscosity of Hydrocarbon Gases Under Pressure, Pet. Trans. AIME 201, 264. (Note: The correlations of these two references are conveniently reproduced in Theory and Practice of the Testing of Gas Wells, 3rd ed., Energy Resources Board, Calgary, 1975.)

Caudle, B. H., R. A. Erickson, and R. L. Slobod (1955): The Encroachment of Injected Fluids beyond the Normal Well Patterns, Pet. Trans. AIME 204, 79.

Cole, F. W. (1969): Reservoir Engineering Manual, Gulf Pub. Co., Houston.

Danielson, H. H., W. T. Paynter, and H. W. Milton, Jr. (1976): Tertiary Recovery by the Maraflood Process in the Bradford Field, J. Pet. Tech., 129, February.

Dietz, D. N. (1965): Determination of Average Reservoir Pressure from Build-up Surveys, Pet. Trans. AIME 234, 955.

duPrey, E. L. (1969): Displacements Non Miscible Daus Les Milieux Poreaux:

Influence Des Parametres Interfaciaux Sur Les Permeabilities Relatives, Compte render de l'A.R.T.F.D. Ed. Technip, Paris, 251.

Dyes, A. B., B. H. Caudle, and R. A. Erickson (1954): Oil Production after Breakthrough as Influenced by Mobility Ratio, Pet. Trans. AIME 201, 81.

Dykstra, H., and R. L. Parsons (1950): The Prediction of Oil Recovery by Water Flood, Secondary Recovery of Oil in the United States, 2d ed. API, New York.

Earlougher, R. C., Jr., H. J. Ramey, Jr., F. S. Miller, and T. D. Mueller (1968): Pressure Distributions in Rectangular Reservoirs, Pet. Trans. AIME 243, 199.

Earlougher, R. C., Jr. (1977): Advances in Well Test Analysis, Monograph Series, Vol. 5, Soc. Pet. Eng., Dallas.

Energy Resources Conservation Board, Theory and Practice of the Testing of Gas Wells, 3d ed., Calgary.

Forchheimer, P. (1901): Wasserbewegung durch Boden, Z. Deutsch. Ing. 45, 1781.

Foster, W. R. (1973): A Low-Tension Waterflooding Process, Pet. Trans. AIME 255, 205.

Frick, T. C. (ed.), and R. W. Taylor (assoc. ed.) (1962): Petroleum Production Handbook Vol. 11. Reservoir Engineering. McGraw-Hill, New York, 1962.

Givens, J. W. (1968): Practical Two-Dimensional Model for Simulating Dry Gas Reservoirs with Bottom Water Drive, J. Pet. Tech. 20, 1229, November.

Harris, D. G. (1975): The Role of Geology in Reservoir Simulation Studies, Pet. Trans. AIME 259, 625.

Healy, R. N., and R. L. Reed (1974): Physicochemical Aspects of Microemulsion Flooding, Pet. Trans. AIME 257, SPEJ 491.

Healy, R. N., R. L. Reed, and C. W. Carpenter, Jr. (1975): Laboratory Study of Microemulsion Flooding, Pet. Trans. AIME 259, SPEJ 87.

Healy, R. N., R. L. Reed, and D. G. Stenmark (1976): Multiphase Microemulsion Systems, Pet. Trans. AIME 261, SPEJ 147.

Holm, L. W., and L. S. O'Brien (1971): Carbon Dioxide Test at the Mead-Strawn Field, J. Pet. Tech., 431, April.

Holm, L. W., and V. A. Josendal (1974): Mechanisms of Oil Displacement by Carbon Dioxide, Pet. Trans. AIME 257, 1427.

Horner, D. R. (1951): Pressure Buildup in Wells, Proc. Third World Pet. Congress 2, 503.

Johnson, C. R., R. A. Greenkorn, and E. G. Woods (1966): Pulse-Testing; A New Method for Describing Reservoir Flow Properties between Wells, Pet. Trans. AIME 237, 1599.

Kamal, M., and W. E. Brigham (1975): Pulse-Testing Response for Unequal Pulse and Shut-in Periods, Pet. Trans. AIME 259, SPEJ 399.

Katz, D. L., D. Cornell, R. Kobayoshi, F. H. Poettman, J. A. Vary, J. R. Elenboss, and C. F. Weinaug (1950): Handbook of Natural Gas Engineering, McGraw-Hill, New York.

Kruseman, G. D., and N. A. deRidder (1970): Analysis and Evaluation of Pumping Test Data, International Institute for Land Reclamation and Improvement, Wageningen, The Netherlands.

Lantz, R. B. (1970): Rigorous Calculation of Miscible Displacement Using Immiscible Reservoir Simulations, Pet. Trans. AIME 249, SPEJ 192.

Lynch, E. J. (1962): Formation Evaluation, Harper and Row, New York.

Mattar, L., G. S. Brar, and K. Aziz (1975): Compressibility of Natural Gases, J. Can. Pet. Tech. 14.

Mathews, C. S., and D. C. Russel (1967): Pressure Buildup and Flow Tests in Wells, Monograph Series Vol. 1, Society of Petroleum Engineering, Dallas.

McCarty, D. G., and F. C. Barfield (1958): The Use of High-Speed Computers for Predicting Flood-Out Patterns, Pet. Trans. AIME 213, 139.

McKinley, R. M., S. Vela, and L. A. Carlton (1968): A Field Application of Pulse-Testing for Detailed Reservoir Description, Pet. Trans. AIME 243, SPEJ 313.

Miller, C. C., A. B. Dyes, and C. A. Hutchinson, Jr. (1950): The Estimation of Permeability and Reservoir Pressure from Bottom Hole Pressure Buildup Characteristics, Pet. Trans. AIME 189, 91.

Muskat, M. (1937): Use of Data on Buildup of Bottom Hole Pressures, Pet. Trans. AIME 123, 44.

Muskat, M. (1946): Flow of Homogeneous Fluids, J. W. Edwards, Inc., Ann Arbor, Mich.

Muskat, M. (1949): Physical Principles of Oil Production, McGraw-Hill, New York.

Muskat, M., and R. C. Wyckoff (1934): A Theoretical Analysis of Waterflooding Networks, Pet. Trans. AIME 107, 62.

Pursley, S. A., R. N. Healy, and E. I. Sandvik (1973): A Field Test of Surfactant Flooding, Loudon, Illinois, Pet. Trans. AIME 255, 793.

Raghavan, R. (1976): Well Test Analysis: Wells Producing by Solution Gas Drive, Pet. Trans. AIME 203, 196.

Ramey, H. J., Jr., and Cobb, W. M. (1971): A General Buildup Theory for a Well in a Closed Drainage Area, Pet. Trans. AIME 251, 1493.

Ramey, H. J., Jr., A. Kumar, and M. Gulati (1975): Gas Well Test Analysis under Water-Drive Conditions, American Gas Association, Arlington, Va.

Schilthuis, R. J. (1936): Active Oil and Reservoir Energy, Pet. Trans. AIME 118, 33.

Slobod, R. L., and H. A. Koch, Jr. (1953): High Pressure Gas Injection: Mechanism of Recovery Increase, Drill and Prod. Proc. API, p. 82.

Standing, M. B. (1977): Volumetric and Phase Behavior of Oil Field Hydrocarbon System, Society of Petroleum Engineering, Dallas.

Standing, M. B., and K. L. Katz (1942): Density of Natural Gases, Pet. Trans. AIME 146, 140.

Stanley, L. T. (1960): Approximation of Gas-Drive Recovery and Front Movement in the Abqaiq Field, Saudi Arabia, Pet. Trans. AIME 219, 273.

Stiles, W. E. (1949): Use of Permeability Distribution in Water Flood Calculations, Pet. Trans. AIME 186, 9.

Tarner, J. (1944): How Different Size Gas Caps and Pressure Maintenance Programs Affect Amount of Recoverable Oil, Oil Weekly 144, 32.

Thompson, F. R., and R. Thachuk (1974): Compositional Simulation of a Gas Cycling Project, Bonnie Glen D-3A Pool, Alberta, Canada, J. Pet. Tech. 1285, November.

Tomich, J. F., R. L. Dalton, Jr., H. A. Deans, and L. K. Shallenberger (1973): Single-Well Tracer Method to Measure Residual Oil Saturation, Pet. Trans. AIME 255, 211.

Tracy, G. W. (1955): Simplified Form of the Material Balance Equation, Pet. Trans. AIME 204, 243.

van Everdingen, A. F., and W. Hurst (1949): The Application of the Laplace Transformation to Flow Problems in Reservoirs, Pet. Trans. AIME 186, 305.

Wattenbarger, R. A. (1967): Effects of Turbulence, Wellbore Damage, Wellbore Storage and Vertical Fractures on Gas Well Testing, Ph.D. Thesis, Stanford University.

Welge, H. J. (1952): A Simplified Method for Computing Oil Recovery by Gas or Water Drive, Pet. Trans. AIME 195, 91.

Wyckoff, R. D., H. G. Botset, and M. Muskat (1933): The Mechanics of Porous Flow Applied to Waterflooding Problems, Pet. Trans. AIME 103, 219.

Zana, E. T., and G. W. Thomas (1970): Some Effects of Contaminants on Real Gas Flow, J. Pet. Tech. 22(9), 1157.

8

Applications in Groundwater Hydrology

In addition to applications discussed in Chapter 7 the specific consideration of water reservoirs (aquifers) presents some special problems not associated with deeper oil reservoirs. The description and calculation of steady and transient flow include consideration of recharge from the surface and flow with a phreatic surface. Further, the problem of dispersion is complicated by interaction with surface water.

Chapters 8 and 9 are both concerned with movement of subsurface water. Figure 8.1 is a schematic of the distribution of the subsurface water. The two zones depicted are the saturated zone and the aeration zone, which are separated by the water table. The aeration zone has three subzones: soil water, an intermediate zone, and the capillary fringe. Groundwater hydrology is primarily concerned with movement of groundwater in the saturated zone. The transport of air, water, and dissolved material in the aeration zone is studied primarily in soil science.

8.1 AQUIFERS

Aquifers are porous underground formations that contain water. Aquifers are water reservoirs and may be unconsolidated or partially consolidated sand and gravel. Deeper aquifers may be permeable limestone, sandstone, or volcanic rocks. These water reservoirs are usually recharged naturally by seepage of water from above. They lose water naturally through springs or streams or artificially by being pumped. Aquifers are part of the hydrologic cycle (see Fig. 8.2).

Aquifers may be confined, partially confined, or unconfined. A confined aquifer is bounded by impervious formations. An artesian aquifer is usually confined but at such an elevation relative to the level of water in it that a well in the aquifer flows freely. The elevation is below the piezometric (pressure) surface. A phreatic aquifer, an aquifer with a zero pressure surface, is unconfined with a water table being its upper boundary. A leaky aquifer, confined or unconfined, gains or loses water through its boundary formations. These boundaries may be of very low permeability, but over large areas significant amounts of water leak through the boundary. Figure 8.3 is a sketch describing the different types of aquifers.

The reservoir description techniques discussed in Chapter 7 apply to aquifers. The same techniques (coring, logging, and pressure transient tests) are used to describe aquifers.

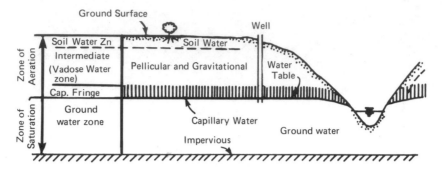

FIG. 8.1 The distribution of subsurface water. (From Bear, 1972.)

Groundwater hydrologists use a slightly different set of definitions than petro-leum engineers. Since water is the only fluid involved, Darcy's law is written as in Chapter 1 in terms of hydraulic conductivity and the head of fluid so that

$$q = \frac{K \, \Delta h}{\Delta L} \qquad\qquad (8.1)$$

where K is the hydraulic conductivity. In terms of the intrinsic permeability

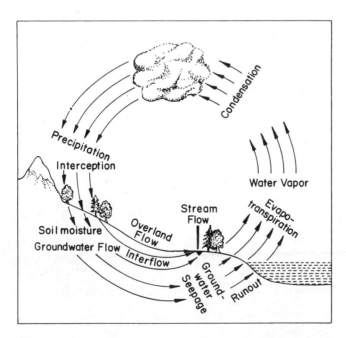

FIG. 8.2 Simplified diagram of the hydrological cycle. (From R. C. Ward, Prin-ciples of Hydrology, 2d ed. Copyright 1975 McGraw-Hill Book Co. (UK) Ltd. Reproduced by permission.)

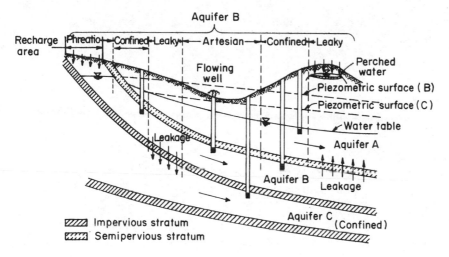

FIG. 8.3 Types of aquifers. (From Bear, 1972.)

$$K = \frac{k\rho g}{\mu} \tag{8.2}$$

The transmissivity of an aquifer is the conductance of the aquifer through its entire thickness

$$T = K\hat{h} \tag{8.3}$$

where $\hat{h}$ is aquifer thickness. The storativity describes the amount of water stored in the aquifer and is a function of the compressibility of the fluid, the matrix, the porosity of the reservoir, and its thickness. For a confined aquifer the storativity is

$$S = c_t \phi \hat{h} \tag{8.4}$$

For an unconfined aquifer the value of $\hat{h}$ depends on the water level. For an aquifer with a phreatic surface the level of fluid drops, but not all of the fluid drains by gravity—surface tension holds some back. The storativity is less by the specific retention of the water; the difference is called the specific yield. A drainage factor is applied to unconfined aquifers so that

$$B = \sqrt{\frac{T}{\alpha S_Y}} \tag{8.5}$$

where $1/\alpha$ is a delay index and S_Y is the delayed yield after a long period of time.
For leaky aquifers the leakage factor is defined as

$$L = \sqrt{Tc} \tag{8.6}$$

where c is a vertical hydraulic resistance

$$c = \frac{\hat{h}}{K_v} \tag{8.7}$$

and K_v is vertical hydraulic conductivity.

The equations of change for aquifers are those described in Chapter 3 for single-fluid flow. Jacob (1949) suggested the following form of the equation of the condition for aquifers:

$$\rho\phi\left(\beta + \frac{\alpha}{\phi}\right)\frac{\partial p}{\partial t} = -\left(\frac{\partial(\rho q_x)}{\partial x} + \frac{\partial(\rho q_y)}{\partial y} + \frac{\partial(\rho q_z)}{\partial z}\right) \tag{8.8}$$

where β is the compressibility of water and α is the vertical compressibility of the medium. For an isotropic medium with constant density

$$\frac{\phi g}{h}\left(\beta + \frac{\alpha}{\phi}\right)\frac{\partial h}{\partial t} = \frac{\partial^2 h}{\partial x^2} + \frac{\partial^2 h}{\partial y^2} + \frac{\partial^2 h}{\partial z^2} \tag{8.9}$$

For a confined aquifer of constant thickness

$$\frac{\partial h}{\partial t} = \frac{T}{S}\nabla^2 h \tag{8.10}$$

At steady state $\partial h/\partial t = 0$ and

$$\nabla^2 h = 0 \tag{8.11}$$

Solutions of Laplace's equation, Eq. (8.11) can be used to determine flow nets for aquifers.

CASE STUDY 18: Steady Flow in an Aquifer

The data in Table 8.1 are for a well in an aquifer under steady flow (USGS Water Supply Paper #2172, 1970-1974). We want to determine the steady-state flow rate for a drawdown of 30 ft at a 1500-ft radius. Well test data show the initial water level was 800 ft below mean sea level. The appropriate solution of Eq. (8.11) can be used to calculate the flow rate.

Figure 8.4 is a schematic of the well. Writing Eq. (8.11) in cylindrical coordinates

$$\frac{1}{r}\frac{\partial}{\partial r}\left(r\frac{\partial h}{\partial r}\right) + \frac{1}{r^2}\frac{\partial^2 h}{\partial \theta^2} + \frac{\partial^2 h}{\partial z^2} = 0 \tag{8.12}$$

assuming no flow as a function of angle and horizontal streamlines

$$\frac{1}{r}\frac{d}{dr}\left(r\frac{dh}{dr}\right) = 0 \tag{8.13}$$

Solving

TABLE 8.1 Aquifer Data

r_w = 16 in.

r_i = 1500 ft

h_i = 800 ft

b = 871 ft

k = 0.5 ft day^{-1}

hu = 770 ft

Source: USGS Water Supply
Paper #2172, 1970–1974.

$$h = C_1 \ln r + C_2 \tag{8.14}$$

The boundary conditions are

$$
\begin{aligned}
h &= h_w \qquad r = r_w \\
h &= h_i \qquad r = r_i
\end{aligned}
\tag{8.15}
$$

Therefore

$$C_1 = \frac{h_i - h_w}{\ln (r_i/r_w)}$$

and

$$C_2 = h_w - \frac{(h_i - h_w) \ln r_w}{\ln (r_i/r_w)} \tag{8.16}$$

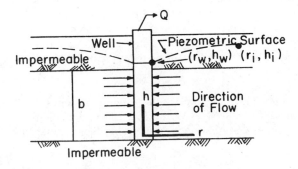

FIG. 8.4 Well schematic. (From USGS Water Supply Paper #2172, 1974.)

Substituting C_1 and C_2 into Eq. (8.14) gives

$$h = \frac{h_i - h_w}{\ln (r_i/r_w)} \ln r + h_w - \frac{h_i - h_w}{\ln (r_i/r_w)} \ln r_w \qquad (8.17)$$

The volumetric flow rate through a concentric ring is

$$q = \frac{Q}{A} = k \frac{\partial h}{\partial r} \qquad (8.18)$$

or

$$Q = 2\pi r b k \frac{\partial h}{\partial r} \qquad (8.19)$$

combining the derivatives of Eq. (8.18) and Eq. (8.19) yields at the well

$$Q_w = 2\pi b k \frac{h_i - h_w}{\ln (r_i/r_w)} \qquad (8.20)$$

The flow rate at the well is

$$Q_w = 2\pi(871)(0.5) \frac{800 - 700}{\ln (1500/1.3)}$$

$$Q_w = 11644 \text{ ft}^3 \text{ day}^{-1} = 60.6 \text{ gal min}^{-1} \blacksquare$$

8.2 AQUIFER PROPERTIES FROM PUMPING TESTS

Pumping tests in aquifers are interference tests in that one well is pumped (or allowed to recover) and the pressures at observation wells are recorded. Much of the discussion in Section 7.2 applies; however, the major differences are

> The aquifer or groundwater system is treated using the single-fluid equations. Permeability and transmissivity are much higher than in oil reservoirs; so time dependence of the pressure change characteristics may be different.
> Aquifers occur in various degrees of unconfinement, and together with discharge (or recharge) surfaces presents different boundary conditions than oil reservoirs.

The equations of change for single-fluid flow applicable to pumping tests analysis of aquifers are discussed in Sections 3.1, 3.5, and 7.2 and in hydrologic terminology in Section 8.1. The usual assumptions made in applying these equations to solution of aquifer problems are

1. The system is infinite.
2. The aquifer is homogeneous, isotropic, and of uniform thickness.

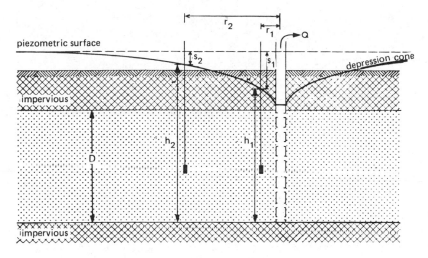

FIG. 8.5 Schematic cross section of a pumped confined aquifer. (From Kruseman and de Ridder, 1970.)

3. Prior to removal or addition of water the piezometric or phreatic surface is horizontal.
4. The pumping is at a constant rate.
5. The pumped well penetrates the aquifer.
6. The well-bore storage can be neglected.
7. Water removed from storage is discharged immediately.

Pseudo-steady-state or steady-state flow in such systems occurs if after a period of pumping the phreatic surface remains constant. Figure 8.5 is a schematic of a pumped confined aquifer which is at steady state. For such a system (Thiem, 1906),

$$T = \frac{Q}{2\pi(s_1 - s_2)} \ln \frac{r_2}{r_1} \tag{8.21}$$

Theis (1935) used the exponential integral solution of Eq. (8.10) to analyze unsteady flow in the following form

$$s = \frac{Q}{4\pi T} \int_V^\infty \frac{e^{-y} \, dy}{y} = \frac{Q}{4\pi T} W(u) \tag{8.22}$$

where

$$u = \frac{r^2 S}{4Tt} \tag{8.23}$$

The storage coefficient is

$$S = \frac{4Ttu}{r^2}$$ (8.24)

The exponential integral $W(u) = -Ei(-u)$ can be represented by the series

$$W(u) = -0.5772 - \ln u + u - \frac{u^2}{2 \cdot 2!} + \frac{u^3}{3 \cdot 3!} - \frac{u^4}{4 \cdot 4!} + \cdots$$ (8.25)

The type-curve solutions discussed in Section 7.2 follow the Theis method.
    Chow (1952) introduced the function

$$\frac{s}{\Delta s} = \frac{W(u)e^u}{2.30}$$ (8.26)

to avoid the curve matching the type-curve solution. $s/\Delta s$ is calculated from a point
on the drawdown versus time curve, where Δs is the drawdown difference in one
time cycle of the s-versus-log t plot. For large values of time or short radii the
truncated form of Eq. (8.25) is adequate, that is, keeping only the first two terms.
Cooper and Jacob (1946) showed for this region that r versus log t is a straight line
and

$$s = \frac{2.30Q}{4\pi T} \log \frac{2.25Tt}{r^2 S}$$ (8.27)

and at $t = t_0$, $s = 0$; so

$$S = \frac{2.25Tt_0}{r^2}$$ (8.28)

and at $\log t/t_0 = 1$,

$$T = \frac{2.30Q}{4\pi \, \Delta s}$$ (8.29)

where Δs is drawdown over one time cycle. If a pumping well is shut down, the
resulting pressure buildup can be analyzed to determine reservoir properties. The
pulse test approach of Johnson et al. (1967) may also be used.
    As the piezometric head is lowered in semiconfined (leaky) aquifers, a differ-
ence in head between the piezometric surface and the phreatic surface (groundwater)
causes vertical flow. Figure 8.6 is a sketch of a pumped leaky aquifer. The vertical
flow is

$$q_v = \frac{h_{phr} - h_{piez}}{c}$$ (8.30)

where c is defined in Eq. (8.7). For steady-state drawdown in a leaky aquifer
de Glee (1930) used

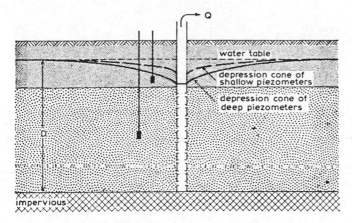

FIG. 8.6 Schematic cross section of a pumped semiunconfined aquifer. (From Kruseman and de Ridder, 1970.)

$$s_m = \frac{Q}{2\pi T} H\left(\frac{r}{L}\right) \qquad (8.31)$$

where s_m is the maximum drawdown in m at a distance r from the pumped well, L is the leakage factor of Eq. (8.6), and H is the Hankel function. The type-curve method is used with H and

$$T = \frac{Q}{2\pi s_m} H\left(\frac{r}{L}\right) \qquad (8.32)$$

$$c = \frac{L^2}{T} = \frac{1}{r/L} \frac{r^2}{T} \qquad (8.33)$$

Hantush (1956) assumed for $r/L < 0.05$

$$s_m = \frac{2.30Q}{2\pi T}\left(\log 1.12\,\frac{L}{r}\right) \qquad (8.34)$$

Using Δs_m for one cycle of log r from the straight-line portion of the curve for s_m versus log r,

$$\Delta s_m = \frac{2.30Q}{2\pi T} \qquad (8.35)$$

At $r = r_0$, $s = s_0$ and

$$1.12\,\frac{L}{r_0} = 1 \qquad (8.36)$$

For unsteady flow in a leaky aquifer Hantush and Jacob (1955) described the drawdown by the following:

$$s = \frac{Q}{4\pi T} \int_V^\infty \frac{1}{y} \exp\left(-y - \frac{r^2}{4L^2 y}\right) dy \qquad (8.37)$$

or

$$s = \frac{Q}{4\pi T} W\left(u, \frac{r}{L}\right) \qquad (8.38)$$

There are several methods due to Hantush analyzing unsteady flow in leaky aquifers depending on the data available or limiting conditions (Hantush, 1956).

If the gravity drainage is delayed, then the assumption that water removed from storage is discharged immediately is no longer valid. Boulton (1963) showed that for delayed yield the drawdown curve is in three segments. In the first, for short times, the aquifer "looks" confined. In the second, there is a decrease in the slope of the drawdown-time curve because of the gravity drainage. The third is apparently confined due to a balance between gravity drainage and withdrawal. The solution of the unsteady-state equation is complex and is of the form

$$s = \frac{Q}{4\pi T} W\left(u_{Ay}, \frac{r}{B}\right) \qquad (8.39)$$

where B is the drainage factor defined by Eq. (8.5). For early times, the first segment

$$u_{Ay} \equiv u_A = \frac{r^2 S_A}{4Tt} \qquad (8.40)$$

where S_A is the instantaneous storage coefficient. During the third segment,

$$u_{Ay} \equiv u_y = \frac{r^2 S_Y}{4Tt} \qquad (8.41)$$

where S_Y is the volume of delayed yield. In the intermediate region, the second segment if

$$r = 1 + \frac{S_Y}{S_A} \qquad (8.42)$$

is large

$$s = \frac{Q}{2\pi T} H\left(\frac{r}{B}\right) \qquad (8.43)$$

Figure 8.7 is a plot of Boulton-type curves.

For steady flow in unconfined aquifers the Dupoit assumption that the velocity of flow is proportional to the tangent of the hydraulic gradient and that flow is horizontal allows using the steady flow equation. The phreatic surface is a streamline; so along it

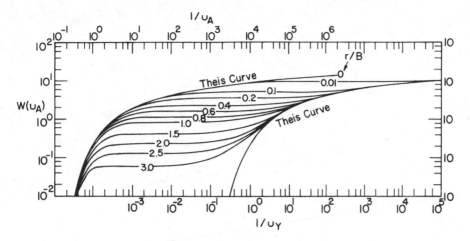

FIG. 8.7 Family of Boulton type curves: $W(u_A, r/B)$ versus $1/u_A$ and $W(u_Y, r/B)$ versus $1/u_Y$ for different values of r/B. (From Boulton, 1963.)

$$q_s = -K \frac{d\Phi}{ds} = -K \frac{dz}{ds} = -K \sin \theta \qquad (8.44)$$

The Dupoit assumption is $\Phi = \Phi(x)$ or

$$q_s = -K \tan \theta = -K \frac{dh}{dx} \qquad (8.45)$$

For unsteady flow in an unconfined aquifer the methods for unsteady-state confined aquifers are used. Table 8.2 is a summary of the methods of analysis of pumping test data for the situations discussed so far.

   If the assumptions used for the methods summarized in Table 8.2 do not apply, then more complex methods which are more specific are used. Table 8.3 is a summary of these specific methods. Most of these methods apply the super-position discussed in Section 7.2 to account for boundaries. For anisotropic systems the transformation discussed in Section 3.2 is used so that

$$s = \frac{Q}{4\pi T_e} W(u') \qquad (8.46)$$

where

$$T_e^2 = T_x^2 + T_y^2 \qquad (8.47)$$

$$u' = \frac{r^2 S}{4t T_n} \qquad (8.48)$$

$$T_n = \frac{T_x}{\cos^2(\theta + \alpha_n) + m \sin^2(\theta + \alpha_n)} \qquad (8.49)$$

TABLE 8.2

Aquifer type	Type of solution	Flow equation
Confined	Steady state	$Q = \dfrac{2\pi D(s_1 - s_2)}{\ln(r_2/r_1)}$
	Unsteady state	$s = \dfrac{Q}{4\pi kD} \displaystyle\int_u^\infty \dfrac{e^{-y}}{y}\,dy = \dfrac{Q}{4\pi kD} W(u)$
		$s = \dfrac{2.30Q}{4\pi kD} \log \dfrac{2.25kDt}{r^2 S}$
		$s'' = \dfrac{2.30Q}{4\pi kD} \log \dfrac{t}{t''}$
Semiconfined	Steady state	$s_m = \dfrac{Q}{2\pi kD} K_0\left(\dfrac{r}{L}\right)$
		$s_m = \dfrac{Q}{2\pi kD}\left(\log 1.12\dfrac{L}{r}\right)$
		$Q - Q' = \dfrac{2\pi kD(s_1 - s_2)}{\ln(r_2/r_1)}$
	Unsteady state	$s = \dfrac{Q}{4\pi kD} \displaystyle\int_u^\infty \dfrac{1}{y}\exp\left(-y - \dfrac{r^2}{4L^2 y}\right) dy$
		$= \dfrac{Q}{4\pi kD} W(u,\ r/L)$
		$s = \dfrac{Q}{4\pi kD} [2K_0(r/L) - W(q,\ r/L)]$
Unconfined with delayed yield	Unsteady state	$s = \dfrac{Q}{4\pi kD} \displaystyle\int_0^\infty 2J_0\left(\dfrac{r}{B}y\right)\dfrac{y^2}{y^2+1}\cdot$
and semiunconfined		$\cdot\,[1 - \exp\{-\alpha\gamma t(y^2 + 1)\}]\dfrac{dy}{y}$
		$= \dfrac{Q}{4\pi kD} W(u_{AY},\ r/B)$
Unconfined	Steady state	$Q = \pi k\dfrac{h_2^2 - h_1^2}{\ln(r_2/r_1)} = \dfrac{2\pi kD(s'_{m1} - s'_{m2})}{\ln(r_2/r_1)}$
	Unsteady state	As for confined aquifers

Source: Kruseman and de Ridder, 1970.

Remarks	Calculated parameters	Reference
	kD	Thiem (1906)
$u = \dfrac{r^2 S}{4kDt}$	kD and S kD and S	Jacob (1940) Chow (1952)
$\dfrac{r^2 S}{4kDt} \leq 0.01$	kD and S	Cooper and Jacob (1946)
$s'' =$ residual drawdown $t'' =$ time since pumping stopped	kD	Theis (1935)
$L \geq 3D$	kD and c	de Glee (1930)
$r/L \leq 0.05$	kD and c	Hantush and Jacob (1955)
$Q' =$ recharge rate by confining layer	kD	
$u = \dfrac{r^2 S}{4kDt}$	kD, S, and c	Walton (1962)
	kD, S, and c kD, S, and c	Hantush (1956) Hantush (1956)
$q = \dfrac{kDt}{SL^2}; \; q > \dfrac{2r}{L}; \; r > 4t_p$	kD, S, and c	Hantush (1956)
$y =$ variable of integration		
$u_A = \dfrac{r^2 S_A}{4kDt}; \; u_Y = \dfrac{r^2 S_Y}{4kDt}$	kD, S_A, S_Y, B, and $1/\alpha$	Boulton (1963)
$\gamma = \dfrac{S_A + S_Y}{S_A}; \; \gamma > 100$		
$s' = s - \dfrac{s^2}{2D}$	kD	Thiem (1906)
s is replaced by $s' = s - \dfrac{s^2}{2D}$	kD and, generally, S	Thiem (1906) Jacob (1940) Chow (1952) Cooper and Jacob (1946)

TABLE 8.3

Replaced assumption(s)	Aquifer type	Type of solution	Remarks	Calculated parameters	Reference
1. Aquifer crossed by one or more fully penetrating recharge or barrier boundaries	Confined or unconfined	Steady state	Recharge boundaries only	kD	Dietz (1943)
			Recharge and/or barrier boundaries	kD and S	Ferris et al. (1962)
			One recharge boundary	kD and S	Hantush (1959)
2. Aquifer homogeneous, anisotropic and of uniform thickness	Confined or unconfined	Unsteady state		$(kD)_x$, $(kD)_y$, and S	Hantush (1966)
			For recovery data also	$(kD)_x$, $(kD)_y$, and S	Hantush and Thomas (1966)
	Semiconfined	Unsteady state		$(kD)_x$, $(kD)_y$, S and c	Hantush (1966)
2. Aquifer homogeneous and isotropic; but thickness varies exponentially	Confined	Unsteady state	dD/dx < 0.20	kD_0 and S	Hantush (1964)
3. Prior to pumping the phreatic surface slopes in the direction of flow	Unconfined	Steady state		kD	
		Unsteady state	i < 0.20	kD and S	Hantush (1964)
4. Discharge rate variable	Confined or unconfined	Unsteady state	Step-type pumping	kD and S	Cooper and Jacob (1946)
			Continuously decreasing discharge	kD and S	Aron and Scott (1965)
			Continuously decreasing discharge	kD and S	Sternberg (1968)
			Continuously decreasing discharge	kD	Sternberg (1967)

Description	Aquifer	State	Remarks	Parameter	Reference
5. The pumped well is partially penetrating	Confined	Steady state	Screened part near top or near bottom of the aquifer	kD	Jacob (1963)
	Unconfined	Steady state		kD	Hantush (1964)
	Confined	Unsteady state	Short period of pumping	k, S and D	Hantush (1962)
		Unsteady state	Long period of pumping	kD and S	Hantush (1962)
6. Storage in the well cannot be neglected	Confined	Unsteady state	Entrance resistance is zero	kD and S	Papadopulos and Cooper (1967)
2 and 5. Two-layered aquifer with semi-pervious dividing layer	Semiconfined	Steady state	Only part is pumped	k_1D_1, k_2D_2, c_1 and c_2	Huisman and Kemperman (1951)
Added assumption: the entrance resistance of the pumped well is zero	Confined	Steady state	Upper- and lower part are pumped separately	k_1D_1, k_2D_2, c_1 and c_2	Bruggeman (1966)
		Steady state	Approximation method without piezometers	kD	Logan (1964)
			Approximation method without piezometers for very deep aquifers	kD	Gosselin (1951)
	Unconfined	Steady state	Approximation method without piezometers	kD	Logan (1964)
Added assumption as above and 5. the pumped well is partially penetrating	Confined	Steady state	Approximation method without piezometers	k	Zangar (1953)
Added assumptions: the entrance resistance of the pumped well is zero; the value of S is known.	Confined	Unsteady state	Approximation method without piezometers	kD	Hurr (1966)
	Unconfined	Unsteady state	Approximation method without piezometers	kD	Hurr (1966)
Well is free flowing: draw-down constant, discharge variable	Confined	Unsteady state		kD and S	Jacob and Lohman (1952)

Source: Kruseman and de Ridder, 1970.

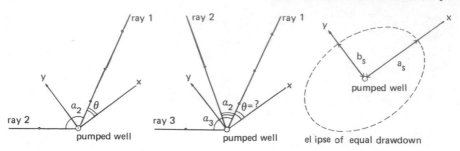

FIG. 8.8 Schematic illustration of the parameters in the Hantush and Hantush-Thomas methods for anisotropic aquifer: (a) known principal directions of anisotropy; (b) unknown principal directions of anisotropy; (c) ellipse of equal drawdown. (From Kruseman and de Ridder, 1970.)

Figure 8.8 defines θ and α_n. Numerical simulation of more complex situations can also be used [see, for example, Bredehoeft and Pinder (1970)].

CASE STUDY 19: Analysis of Pumping Test Data in an Aquifer with Delayed Yields

The standard interference test based on the exponential integral solution is used to define the reservoir parameters of confined and unconfined aquifers. The reservoir constants of concern are transmissivity T and storativity, or the coefficient of storage S. The storage coefficient is physically the volume of water released from storage under a unit decline of head across a cross-sectional area. For an unconfined aquifer it can be interpreted as the volume of water drained per total volume of water in storage. As we shall see, the coefficient of storage for an unconfined aquifer changes with time during a pump test and when equilibrium conditions are reached, s becomes constant and is known at this point as the "specific yield" of the aquifer.

For a standard pump test, the drawdown s is plotted versus time (under constant rate Q). The resulting curve is matched with the type-curve plot W(u) versus u (from the exponential integral solution). A match point is chosen and the following equations are used to obtain transmissivity T and the coefficient of storage S:

$$s = \frac{114.6Q}{T} W(u) \qquad u = \frac{1.87Sr^2}{Tt}$$

where

s = drawdown at an observation well reacting to pumped well, ft

r = distance between observation well and pumped well, ft

t = time since pumping started, days

Q = discharge at pumped well, gal min^{-1}

T = transmissivity, gal day^{-1} ft^{-1}

S = coefficient of storage, dimensionless

An example of this method of analysis has already been given in Case Study 4.

A number of assumptions underlie this analysis. Two that are of special interest here are that the coefficient of storage is constant with time and that water is released from storage instantaneously with a decline in head.

In an unconfined situation, these assumptions are invalid to the point of causing significant errors in the reservoir constants T and S. Under water-table conditions, gravity drainage is a significant part of reservoir drainage and is not taken into account by the transient flow equation (nonequilibrium method) (Walton, 1960).

This lack of instantaneous storage release is known as delayed yield. Boulton (1954, 1963) derived a method for taking into account the delayed yield in unconfined aquifers. The drainage of the reservoir consists of two parts: a volume $S \, \Delta s$ of water instantaneously released due to a lowering of the water-table Δs, between the times τ and $\tau + \Delta \tau$ and a delayed yield from storage at a time t $(t > \tau)$, from pumping $\Delta s \; \alpha S_Y e^{-\alpha(t-\tau)}$, where α is an empirical constant.

The total volume of delayed yield per unit drawdown, per unit horizontal area is

$$\alpha S_Y \int_{\tau}^{\infty} e^{-\alpha(t-\tau)} \; dt = S_Y \tag{8.50}$$

and the total effective coefficient of storage, or the specific yield is

$$SY = S_A + S_Y$$

[also, $S_A + S_Y = \eta S_A$, where $\eta = (1 + S_Y)/S_A$].

The appropriate differential equation is

$$T\left(\frac{\partial^2 s}{\partial r^2} + \frac{1}{r} \frac{\partial s}{\partial r}\right) = S \frac{\partial s}{\partial t} + \alpha S_Y \int_{0}^{t} \frac{\partial s}{\partial t} e^{-\alpha(t-\tau)} \; d\tau \tag{8.51}$$

Boulton (1963) achieved a solution of this equation $t = \tau$.

From the solution the Theis nonequilibrium solution is extended to account for delayed yield. The following dimensionless variables are introduced

$$W = \frac{s}{Q/4\pi T} \qquad u_A = \frac{r^2 S_A}{4Tt} \qquad u_Y = \frac{r^2 S_Y}{4 Tt} = \frac{(r/B)^2}{4\alpha T} = \frac{\eta - 1}{u_A} \tag{8.52}$$

where W, s, r, t, Q, T, and S_Y are defined as before and $B = (T/\alpha S_Y)^{1/2}$. W was computed and plotted versus $1/u_A$ and $1/u_Y$ for various r/B values between 0.001 and 3.0 in Fig. 8.7. Two types of curves are noted on this plot. Type A curves are those to the left of the breaks in the curves and correspond to leaky artesian (confined) aquifer conditions in the early time portion of the pump test. The type B curves

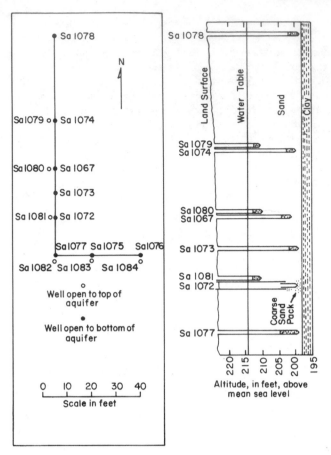

FIG. 8.9 Plan view and cross section of the aquifer-test site at Saratoga National
Historical Park, Saratoga County, N.Y.

are those to the right of the breaks in the curves and correspond to the "normal"
unconfined aquifer condition in which delayed yield storage S_Y is being released.

Field pumping data are plotted as drawdown versus time just as in the conventional analysis, and the resulting curve is matched with the type curves to ultimately
find T, S_A, and S_Y. The following situation and data illustrate Boulton's method of
analysis.

A field pumping test was conducted at the Saratoga (N.Y.) National Historical
Park. The aquifer consisted of alternately layered silty fine sand and coarse sand
from the surface to a depth of 25 ft. A plan and cross section of the well setup is
shown in Fig. 8.9. Since well SA 1072 had the most complete data, it was chosen
for this analysis. The time and drawdown data are given in Table 8.4. Well SA 1077
is the pumping well and r = 15 ft. The data from Table 8.4 are plotted on log paper
as drawdown s versus t/r^2 as in Fig. 8.10.

The plot in Fig. 8.10 is first matched with type A curve (curves to the left of
the break). When the curve is matched, the r/B value is recorded. A match point

TABLE 8.4 Drawdown and t/r^2 for Well SA 1072: r = 15.2 ft

t/r^2 (days/ft^{-2})	Drawdown s (ft)
5.10×10^{-7}	0.020
9.93×10^{-7}	0.020
1.50×10^{-6}	0.042
2.01×10^{-6}	0.073
2.49×10^{-6}	0.098
3.01×10^{-6}	0.126
3.76×10^{-6}	0.164
4.55×10^{-6}	0.198
5.24×10^{-6}	0.228
6.01×10^{-6}	0.254
9.01×10^{-6}	0.333
1.20×10^{-5}	0.383
1.80×10^{-5}	0.430
2.40×10^{-5}	0.450
3.01×10^{-5}	0.463
4.51×10^{-5}	0.476
6.02×10^{-5}	0.488
7.53×10^{-5}	0.496
9.00×10^{-5}	0.507
1.14×10^{-4}	0.512
1.44×10^{-4}	0.528
1.80×10^{-4}	0.531
2.29×10^{-4}	0.541
2.71×10^{-4}	0.548
3.76×10^{-4}	0.553
4.72×10^{-4}	0.566
6.62×10^{-4}	0.590
8.40×10^{-4}	0.605
1.02×10^{-3}	0.626
1.20×10^{-3}	0.642
1.51×10^{-3}	0.671
1.89×10^{-3}	0.691
2.43×10^{-3}	0.701
2.80×10^{-3}	0.708
3.32×10^{-3}	0.739
3.68×10^{-3}	0.764
4.08×10^{-3}	0.773
4.67×10^{-3}	0.817
5.33×10^{-3}	0.830
5.97×10^{-3}	0.839
6.62×10^{-3}	0.868
7.45×10^{-3}	0.883
7.96×10^{-3}	0.888
8.50×10^{-3}	0.898

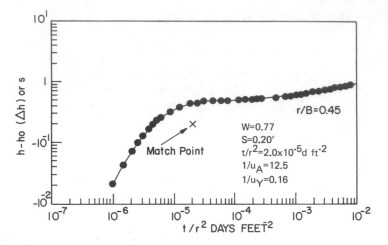

FIG. 8.10 Drawdown from Table 8.4.

is chosen and corresponding values of s, W, t/r^2, and $1/U_A$ are recorded. Figure 8.11 illustrates this procedure. The drawdown–t/r^2 curve is then transposed to the right at constant W and a match with a type B curve is found (this should be at the same r/B value). The match point chosen already is used to read off the $1/u_Y$ value (Fig. 8.12).

Now, with W, r/B, s, t/r^2, $1/u_A$ and $1/u_Y$, the reservoir constants T, S_Y, and S_A can easily be calculated from Eq. (8.52) as follows:

Given Q = 17.2 gal min^{-1}, W = 0.77, s = 0.20 ft, $t/r^2 = 2.0 \times 10^{-5}$ days ft^{-2},

$$\frac{1}{u_A} = 12.5 \qquad \frac{1}{u_Y} = 0.16$$

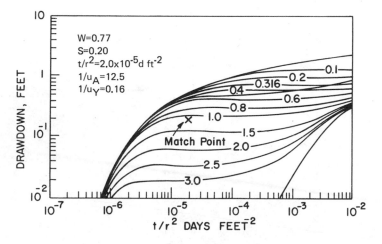

FIG. 8.11 Drawdown plot. (From Boulton, 1963.)

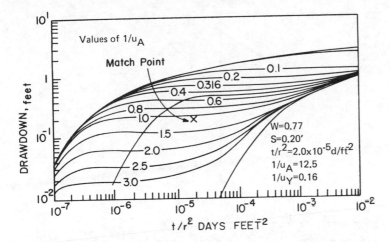

FIG. 8.12 Drawdown plot. (From Boulton, 1963.)

find T, S_A, and S_Y.

1. $T = \dfrac{WQ}{s4\pi} = \dfrac{0.77(17.2)}{0.20(4\pi)} = 5.27$ gal min^{-1} ft^{-1} = 7.59×10^3 gal day^{-1}

   ($k = 0.0142$ cm sec^{-1} = 0.00047 ft sec^{-1})

2. $S_A = \dfrac{4Tt}{u_A r^2} = \dfrac{4(7.59 \times 10^3 \text{ gal day}^{-1})(2.0 \times 10^{-5} \text{ day ft}^{-2})}{12.5} \doteq 0.0486$ gal ft^{-3}

   $\doteq 0.00649$

3. $S_Y = \dfrac{4Tt}{u_Y r^2} = \dfrac{4(7.59 \times 10^3)(2.0 \times 10^{-5})}{.16} = .51$ gal ft^{-3}

It was assumed for this theory that η is infinite. For practical cases, the method is applicable when η is large. Boulton's example resulted in a $\eta = 30$, which he stated as sufficiently large. For this analysis, $\eta = 12.5/.16 + 1 = 78$ indicating the theory to be applicable.

Conventional analysis of the test data is compared to the delayed-yield method. For the same data, one match point was chosen for the early time drawdown. Utilizing late-time drawdown from wells other than SA 1072, a second match point was chosen. The values of T and S for the early portion were, $T = 8.21 \times 10^3$ gal day^{-1} ft^{-1} and $S = 0.007$ and for the late portion, $T = 1.31 \times 10^4$ gal day^{-1} ft^{-1}, $S = 0.11$. In addition, an analysis of the change of T and S with time was also done in which T averaged 1.23×10^4 gal day^{-1} ft^{-1} and the ultimate $S = 0.15$.

In comparing these results with the results found by the delayed method, we see that the difference in transmissivity values is within tolerable limits (within an order of magnitude is considered tolerable). However, the ultimate storage coefficient, the specific yield S_Y is .51 for Boulton's method, while for the conventional analysis $S_Y = 0.15$.

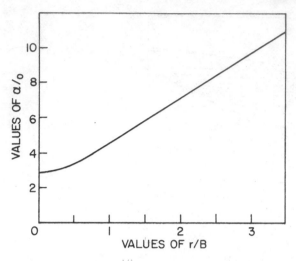

FIG. 8.13 Curve to estimate t_0.

The factor "α" in the derivation of Eq. (8.51). This "empirical constant" α is called the <u>delay index</u> and indicates the time required before delayed yield to have an effect on the drawdown curve.

$$\alpha = \frac{(r/B)^2}{4tu_Y} = \frac{(0.16)(0.45)^2}{4(0.00462 \text{ days})} = 1.75 \text{ days}^{-1}$$

Figure 8.13 is of αt_0 versus r/B. From this plot and $r/B = 0.45$,

$$\alpha t_0 = 3.3 \qquad t_0 = \frac{3.3}{1.75} \text{ days}^{-1} = 1.89 \text{ days} \quad \blacksquare$$

CASE STUDY 20: Analysis of Pumping Test Data in an Anisotropic Aquifer

In normal interference tests, the analysis assumes the reservoir is isotropic and that the transmissivity T is the same in all directions.

Preferred orientation of transmissivity (and permeability) occurs as a result of the processes of sedimentation, especially in stream-laid deposits and fracture alignments in rock units. From Eq. (3.158) the effective transmissivity is

$$T_e = (T_x T_y)^{1/2} \tag{8.53}$$

where T_x and T_y are the transmissivity in the x and y directions on an ellipse as shown in Fig. 8.14.

Hantush and Thomas (1966) devised a method for determining T_x and T_y from pump test data. The requirements for such an analysis are as follows: two rays of observation wells if the direction of anisotropy is known (one well can serve as a ray); 3 rays of observation wells if the direction of anisotropy is unknown; results of isotropic method of analysis, that is, both T and S (storage coefficient) from each ray of wells.

If after applying isotropic analysis to each ray of wells, the T's vary considerably, then application of this theory is suspect, for what is measured is the effective transmissibility T_e, and thus for rays,

$$T_{en} = T_{e1} = T_{e2} = T_{e3} = \cdots \qquad n = 1, 2, 3, \ldots$$

The drawdown s (in feet) for a completely penetrated, homogeneous, nonleaky infinite, anisotropic aquifer is

$$s = \frac{114.6Q}{T_e} W\left[\frac{2693S}{t}\left(\frac{x^2}{T_x} + \frac{y^2}{T_y}\right)\right] \qquad (8.54)$$

where

$$t = \text{time, min}$$

$$Q = \text{pumpage rate, gal min}^{-1}$$

$$s = \text{drawdown, ft}$$

$$T_e, T_x, T_y - \text{transmissibility, gal day}^{-1} \text{ ft}^{-1}$$

$$S = \text{storage coefficient}$$

$$W = \text{exponential integral for } (-u)$$

$$s = \frac{114.6Q}{T_e} W\left(\frac{2693Sr^2}{Tt}\right) \qquad (8.55)$$

$$r - \text{distance to pumped well, ft}$$

After determining T and S for each ray of wells, T_e and T_r/S are calculated. T_e is taken as the average of all T's, and T_r/S is calculated for each ray. If more

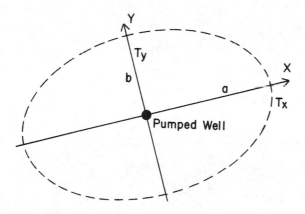

FIG. 8.14 Transmissibility ellipse.

468   Applications in Groundwater Hydrology

than one well occurs in a ray, the values of T_r/S for those wells should be roughly equal.

From Eq. (8.53) equal drawdown curves can be drawn. A time t is picked, preferably late in the test, and the drawdown value is chosen. The r value is calculated from Eq. (8.55) along each ray. Each drawdown value chosen provides an equal drawdown curve or ellipse. From each ellipse, the major axis a and minor axis b are determined.

The following equations provide T_x, Y_y, T_r, and S utilizing a and b.

$$T_x = \frac{a}{b} T_e$$

$$T_y = \frac{b}{a} T_e \qquad\qquad (8.56)$$

$$T_r = \frac{r^2}{ab} T_e$$

$$S = \frac{T_e t_f}{2693ab} W^{-1} \frac{T_e s}{114.6Q} = \frac{1}{N} \sum_{i=1}^{n} \frac{Tr_i}{\left(\frac{Tr}{s}\right)_i} \qquad\qquad (8.57)$$

In October 1963 the Illinois State Water Survey conducted a 1.5-day pumping test at Camp Grant in Rockford, Ill. The aquifer is confined and nonleaky. It consists of 62 ft of sand and gravel in a horizontal orientation at a depth of about 100 to 185 ft. The aquifer is an alluvial deposit.

The test had to be performed on existing wells which were not in a straight line, shown in Fig. 8.15. The basic data are as follows:

Match point (by well number)	W(u)	u	s (ft)	t (min)	r (ft)
M-5	1.0	0.1	0.23	1.4	300
M-4	1.0	0.1	0.23	8.0	582
M-3	1.0	0.1	0.28	28	860
M-2	1.0	0.1	0.33	60	1150
M-1	1.0	0.1	0.24	39	1420

Calculation of T's and S's (Q = 640 gal min^{-1})

Well

5    $T = \frac{114.6(640)(1)}{0.23} = 318,887$ gal day^{-1} ft^{-1} K = 0.24 cm sec^{-1}

$S = \frac{318,887(0.1)(1.4)}{(2693)(300)^2} = 0.000184$

(continued)

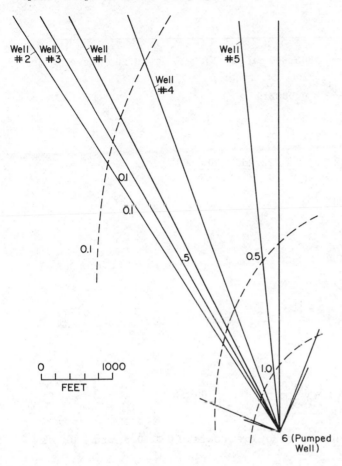

FIG. 8.15 Well layout.

Well

4    $T = \dfrac{114.6(640)(1)}{0.23} = 318,887$ gal day^{-1} ft^{-1} $K = 0.24$ cm sec^{-1}

 $S = \dfrac{318,887(0.1)(8.0)}{2693(582)^2} = 0.000280$

3    $T = \dfrac{114.6(640)(1.0)}{0.28} = 261,942$ gal day^{-1} ft^{-1} $K = 0.19$ cm sec^{-1}

 $S = \dfrac{261,942(0.1)(28)}{2693(860)^2} = 0.000368$

2    $T = \dfrac{114.6(640)(1)}{0.33} = 222,255$ gal day^{-1} ft^{-1} $K = 0.17$ cm sec^{-1}

 $S = \dfrac{222,255(0.1)(60)}{2693(1150)^2} = 0.000374$ (continued)

Well

1    $T = \dfrac{114.6(640)(1.0)}{0.24} = 305,600 \text{ gal day}^{-1} \text{ ft}^{-1}$ $K = 0.23 \text{ cm sec}^{-1}$

$S = \dfrac{305,600(0.1)(30)}{2693(1420)^2} = 0.000210$

Anisotropic Analysis

$T_e = \dfrac{(318,887)2 + 261,942 + 222,255 + 305,600}{5}$

$T_e = 285,514 \text{ gal day}^{-1} \text{ ft}^{-1}$

Well no.	$\dfrac{Tr}{S}$
5	1.7331×10^9
4	1.1389×10^9
3	7.1180×10^8
2	5.9426×10^8
1	1.3954×10^9

$s = \dfrac{(114.6)QW(u)}{T_e}$ $u = \dfrac{2693Sr^2}{Tt}$ $t = 100 \text{ min}$

Calculation of equal drawdowns for drawdowns of 0.1, 0.5, and 1.0 ft

Well no. 5

s	W(u)	u	r	$= \sqrt{\dfrac{(u)(100)(1.7331 \times 10^9)}{2693}}$
0.1	0.389	0.68	6,615	
0.5	1.95	0.084	2,325	
1.0	3.89	0.011	841	

Well no. 4

s	W(u)	u	r	$= \sqrt{\dfrac{(u)(100)1.1389 \times 10^9}{2693}}$
0.1	0.389	0.68	5,363	
0.5	1.95	0.084	1,885	
1.0	3.89	0.011	682	

Well no. 3

s	W(u)	u	r	$= \sqrt{\dfrac{(u)(100)7.1180 \times 10^8}{2693}}$
0.1	0.389	0.68	4,240	
0.5	1.95	0.084	1,490	
1.0	3.89	0.011	539	

Well no. 2

s	W(u)	u	r	$= \sqrt{\dfrac{(u)(100)5.9426 \times 10^8}{2693}}$
0.1	0.389	0.68	3,874	
0.5	1.95	0.084	1,362	
1.0	3.89	0.011	493	

Well no. 1

s	W(u)	u	r	$= \sqrt{\dfrac{(u)(100)1.3954 \times 10^9}{2693}}$
0.1	0.389	0.68	5,936	
0.5	1.95	0.084	2,086	
1.0	3.89	0.011	755	

Calculation of T_x and T_y for 1-ft curve, where a = 1430 ft and b = 400 ft:

$$T_x = \frac{a}{b} T_e = \frac{1430}{400} 285,514 \text{ gal day}^{-1} \text{ ft}^{-1}$$

$$= 1.0207 \times 10^6 \text{ gal day}^{-1} \text{ ft}^{-1}$$

$$T_y = \frac{b}{a} T_e = 7.986 \times 10^4 \text{ gal day}^{-1} \text{ ft}^{-1}$$

(or $K_x = 0.78$ cm sec^{-1} and $K_y = 0.061$ cm sec^{-1}; for 0.5-ft curve with a = 3340 and b = 960,

$$T_x = \frac{3340}{960} 285,514 = 9.93 \times 10^5 \text{ gal day}^{-1} \text{ ft}^{-1}$$

($K_x = 0.75$ cm sec^{-1})

$$T_y = \frac{960}{3340} 285,514 = 8.206 \times 10^4 \text{ gal day}^{-1} \text{ ft}^{-1}$$

($K_y = 0.063$ cm sec^{-1})

Calculation of S

$$S = \frac{T_e t_f}{(2693)ab} W^{-1} \frac{T_e s}{114.6Q}$$

$$t_f = 100 \text{ min} \qquad s = 0.5 \text{ ft} \qquad a = 3340 \text{ ft} \qquad b = 960 \text{ ft}$$

$$W^{-1}\left(\frac{285,514(0.5)}{114.6(640)}\right) = W^{-1}(1.9496) = 20.83$$

$$S = \frac{285,514(100)}{2693(3340)(960)}\, 20.83 = 0.0688 \quad \blacksquare$$

8.3 FLOW WITH PHREATIC SURFACES

Unconfined aquifers, i.e., aquifers with a phreatic surface, are treated as confined on the basis of the Dupuit approximation. This assumes that flow is proportional to dh/dx. Phreatic surface problems such as seepage through earthen dams and embankments are moving boundary value problems. These problems can be solved analytically and graphically in certain situations. Transformation into a hodograph plane is also used for certain problems. The free-surface problem is similar to the unstable flow problems discussed in Chapters 4 and 5. Analytical solutions are limited because the free surface is nonlinear. An approach is to linearize the free boundary as in treatment of the limiting cases of fingering. Numerical methods can be applied using finite element techniques [for example, see Neuman and Witherspoon (1970)].

Assume an anisotropic porous medium containing a free surface and steady flow where fluid is added to and subtracted from the system. Figure 8.16 is the sketch of an aquifer with a phreatic surface. The phreatic surface—the upper curve line in Fig. 8.16—is a potential surface and a moving boundary, so that

$$S(x, y, z, t) = 0 \tag{8.58}$$

Taking the substantial derivative of Eq. (8.58)

$$\frac{\partial S}{\partial t} + \underline{v} \cdot \underline{\nabla} S = 0 \tag{8.59}$$

If $\underline{n}$ is the accretion and $\underline{q}$ the flow,

$$\underline{v} = \frac{1}{\phi}(\underline{q} - \underline{n}) \tag{8.60}$$

where

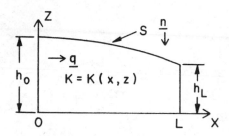

FIG. 8.16 Flow in an aquifer with a phreatic surface.

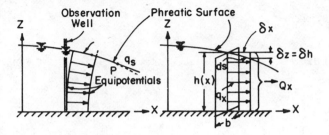

FIG. 8.17 The Dupuit assumptions. (From Bear, 1972.)

$$\underline{q} = \underline{\underline{K}}\nabla\Phi \tag{8.61}$$

Writing S in terms of Φ

$$S(x, y, z, t) = \Phi(x, y, z, t) - z = 0 \tag{8.62}$$

Combining Eqs. (8.59) to (8.62),

$$\frac{\partial\Phi}{\partial t} - \frac{1}{\phi}\left[K_x\left(\frac{\partial\Phi}{\partial x}\right) + K_y\left(\frac{\partial\Phi}{\partial y}\right) + K_z\left(\frac{\partial\Phi}{\partial z}\right) - \frac{\partial\Phi}{\partial z}(K_z + n) + n\right] = 0 \tag{8.63}$$

Equation (8.63) is the general boundary condition for a phreatic surface. Specific conditions for the phreatic surface result from Eq. (8.63). With no accretion, n = 0. For steady flow, $\partial\Phi/\partial t = 0$. For isotropic media $K_x = K_y = K_z$.

The Dupuit approximation is shown schematically in Fig. 8.17. In essence the approximation neglects the vertical flow component. The assumptions are expressed by

$$h = h(x, y) \tag{8.64}$$

$$q_x = -K_x\frac{dh}{dx} \tag{8.65}$$

$$q_y = -K_y\frac{\partial h}{\partial y} \tag{8.66}$$

Consider flow in an aquifer with an impervious horizontal bottom and a phreatic surface at the top as sketched in Fig. 8.8. The Dupuit approximation is

$$\frac{\partial\Phi}{\partial x} = \frac{\partial h}{\partial x} \tag{8.67}$$

The flow rate is

$$Q = -\int_0^{h(x)} K(x, z)\frac{\partial\Phi}{\partial x}\,dz \tag{8.68}$$

Combining Eqs. (8.67) and (8.68)

$$Q = -\frac{dh}{dx} \int_0^{h(x)} K(x, z) \, dz \tag{8.69}$$

Rearranging Eq. (8.69)

$$Q \, dx = -\left[\int_0^{h(x)} K(x, z) \, dz\right] dh \tag{8.70}$$

Defining an equivalent conductivity

$$\frac{1}{K_{eq}} = \frac{1}{L} \int_0^L \frac{h(x)}{\int_0^{h(x)} K(x, z) \, dz} \, dx \tag{8.71}$$

Combining Eqs. (8.70) and (8.71)

$$Q \, dx = -K_{eq} h(x) \, dh \tag{8.72}$$

Integrating

$$Q \int_0^L dx = -K_{eq} \int_{h_0}^{h_L} h \, dh \tag{8.73}$$

$$Q = \frac{K_{eq}}{2L} (h_0^2 - h_L^2) \tag{8.74}$$

Equation (8.74) is a general expression that can be written for homogeneous, layered, or vertically stratified aquifers by appropriate expressions for K_{eq} in Eq. (8.71).

Equation (8.64) to (8.66) contain three unknowns h, q_x, and q_y. To obtain solutions we need the continuity equation in addition. Assuming water is incompressible, ρ = constant, then

$$\phi \frac{\partial h}{\partial t} + \frac{\partial q_x}{\partial x} + \frac{\partial q_y}{\partial y} = n \tag{8.75}$$

Using the Dupuit approximation and assuming $K_x = K_y$

$$\phi \frac{\partial h}{\partial t} = \frac{K}{2} \left[\frac{\partial^2 (h^2)}{\partial x^2} + \frac{\partial^2 (h^2)}{\partial y^2}\right] + n \tag{8.76}$$

If the bottom of the aquifer is not horizontal, let h be the height of the phreatic surface and η be the height of the bottom of the reservoir relative to the datum; then

$$\phi \frac{\partial h}{\partial t} = K \left[\frac{\partial}{\partial x} (h - \eta) \frac{\partial h}{\partial x} + \frac{\partial}{\partial y} (h - \eta) \frac{\partial h}{dy} \right] + n \qquad (8.77)$$

Equations (8.76) and (8.77) are referred to as the Boussnesq equations for unsteady flow in an aquifer with accretion.

For steady flow with a horizontal bottom

$$\frac{K}{2} \left[\frac{\partial^2 (h^2)}{\partial x^2} + \frac{\partial^2 (h^2)}{\partial y^2} \right] + n = 0 \qquad (8.78)$$

If there is no accretion, n = 0, and the resulting equation is referred to as the Forchheimer equation

$$\frac{\partial^2 (h^2)}{\partial x^2} + \frac{\partial^2 (h^2)}{\partial y^2} = 0 \qquad (8.79)$$

CASE STUDY 21: Phreatic Flow with Accretion

Dupuit (1863) made two assumptions concerning unconfined flow that simplifies solutions. These assumptions are

1. Streamlines are horizontal for small inclinations of the phreatic surface.
2. The hydraulic gradient is equal to the slope of the free surface and is invariant with depth.

Though these assumptions are paradoxical, solutions based on Dupuit's assumptions compare favorably with other more rigorous methods.

For steady flow,

$$\frac{K}{2} \left[\frac{\partial^2 (h^2)}{\partial x^2} + \frac{\partial^2 (h^2)}{\partial y^2} \right] + n = 0 \qquad (8.80)$$

If one considers a two-dimensional problem with n = 0

$$\frac{\partial^2 (h^2)}{\partial x^2} = 0 \qquad (8.81)$$

Integration of this equation yields

$$h^2 = Ax + B \qquad (8.82)$$

where A and B are constants.

For steady flow on a horizontal impervious boundary Fig. 8.18, the boundary conditions are

$$x = 0 \quad h = h_1 \quad \text{and} \quad x = L \quad h = h_2$$

The elevation of the free surface along any point is

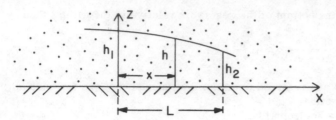

FIG. 8.18 Horizontal boundary.

$$h = \sqrt{h_1^2 - (h_1^2 - h_2^2)\frac{x}{L}} \qquad\qquad (8.83)$$

The discharge per unit width through a vertical section is

$$q = Kh\frac{dh}{dx} \qquad\qquad (8.84)$$

substituting the boundary conditions above and integrating yields

$$q = \frac{k(h_1^2 - h_2^2)}{2L} \qquad\qquad (8.85)$$

considering infiltration, n (n > 0) or evaportion (n < 0), the height at any point is

$$h = \sqrt{h_1^2 - \frac{(h_1^2 - h_2^2)}{L}x + \frac{n}{k}(L - x)x} \qquad\qquad (8.86)$$

The distance to the maximum elevation of the free surface is called a and is

$$a = \frac{L}{2} - \frac{k}{n}\frac{h_1^2 - h_2^2}{2L} \qquad\qquad (8.87)$$

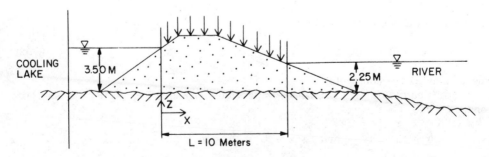

FIG. 8.19 Embankment.

A cooling lake is planned for a fossil-fuel plant. A circulation system is used so that hot water entering the lake will return to ambient temperature before being used in the plant again. However, near the hot-water discharge point, the temperature in the lake is much higher than ambient. A river runs by the lake 10 m away and an earth embankment of homogeneously placed soil ($K = 3 \times 10^{-4}$ cm sec^{-1}) separates the two bodies of water. To assess the environmental impact of the hot water discharge on the river, the quantity of seepage from the lake to the river must be known for different times of the year.

Find the quantity of seepage if the maximum infiltration rate during the year has been found to be 1 in. day^{-1}. The geometry of the situation is given in Fig. 8.19.

Maximum infiltration: n_{max} = 1 in. day^{-1} = 2.94×10^{-5} cm sec^{-1}

Find "water divide" on free surface.

$$a = \frac{L}{2} - \frac{K}{n} \frac{h_1^2 - h_2^2}{2L}$$

$$a = \frac{1000 \text{ cm}}{2} - \frac{3 \times 10^{-4} \text{ cm sec}^{-1}}{2.94 \times 10^{-5} \text{ cm sec}^{-1}} \frac{350^2 - 225^2}{2(1000)}$$

$$a = 133 \text{ cm} = 1.33 \text{ m}$$

Find maximum height of free surface

$$h_{max} = \sqrt{h_1^2 - \frac{(h_1^2 - h_2^2)}{L} x + \frac{\eta}{k} (L - x) x}$$

$$h_{max} = \sqrt{350^2 - \frac{350^2 - 225^2}{1000} 133 + \frac{2.94 \times 10^{-5}}{3 \times 10^{-4}} (100 - 133) 133}$$

$$h_{max} = 3.37 \text{ m}$$

We now compute the discharge to the <u>right</u> of the water divide by using

$$h_1 = 3.37 \text{ m} \qquad L = 10 - 1.33 = 8.67 \text{ m}$$

$$q = K \frac{h_1^2 - h_2^2}{2L}$$

$$q = 3 \times 10^{-4} \text{ cm sec}^{-1} \frac{337^2 - 225^2}{2(867)} = 0.0108 \text{ cm}^2 \text{ sec}^{-1}$$

Consider a length of embankment 50 m.

$$Q = q(50) = (0.0108 \text{ cm}^2 \text{ sec}^{-1}) 5000 \text{ cm}$$

$$= 54 \text{ cm}^3 \text{ sec}^{-1}$$

$$= 1233 \text{ gal day}^{-1}$$

If n = 0 (infiltration is zero), then

$$Q = 3 \times 10^{-4} \frac{350^2 - 225^2}{2(1000)} \; 5000 \text{ cm} = 53.91 \text{ cm}^3 \text{ sec}^{-1} = 1230 \text{ gal day}^{-1}$$

Because of the assumptions made by Dupuit, his method is <u>poor</u> for determining the actual shape of the free surface, especially at the downstream side. This is significant for stability analyses; however, Dupuit's theory is reasonably accurate for determining quantities of flow. ∎

The hodograph method of solution is applicable to steady two-dimensional flow in the vertical plane where the phreatic surface is part of the boundary. A hodograph maps the physical plane into a velocity plane as shown schematically in Fig. 8.20. The advantage gained by the mapping is that the phreatic surface is completely specified in the hodograph plane. Consider the function

$$w = q_x + iq_y \tag{8.88}$$

w describes a point (q_x, q_y) in the hodograph plane. w is not analytic; however, its complex conjugate $\tilde{w}$ is analytic

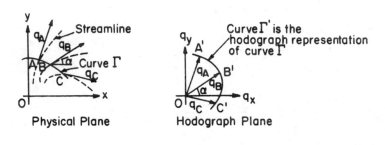

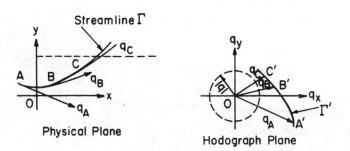

FIG. 8.20 Hodograph representation of a streamline. (From Bear, 1972.)

$$\tilde{w} = q_x - iq_y \tag{8.89}$$

and

$$\tilde{w} = -\frac{\partial \Phi}{\partial x} + i \frac{\partial \Phi}{\partial y} \tag{8.90}$$

The complex potential is

$$w = \Phi(x, \ y) + i\Psi(x, \ y) \tag{8.91}$$

where

$$z = x + iy \tag{8.92}$$

Then

$$\frac{\partial w}{\partial z} = \frac{\partial \Phi}{\partial x} + i \frac{\partial \Psi}{\partial y} = -i \frac{\partial \Phi}{\partial x} + \frac{\partial \Psi}{\partial y} \tag{8.93}$$

therefore

$$-\frac{\partial w}{\partial z} = (q_x - iq_y) = \tilde{w} \tag{8.94}$$

and

$$\left| \frac{\partial w}{\partial z} \right| = (q_x^2 + q_y^2)^{1/2} = q \tag{8.95}$$

In terms of w

$$-\frac{\partial z}{\partial w} = -\frac{1}{\partial w / \partial z} = \frac{1}{q_x - iq_y} = \frac{q_x + iq_y}{q^2} = \frac{w}{q^2} \tag{8.96}$$

In polar coordinates

$$-\frac{\partial z}{\partial w} = \frac{1}{q_x - iq_y} = \frac{e^{i\alpha}}{q} \tag{8.97}$$

and

$$\alpha = \tan^{-1} \frac{q_y}{q_x} \tag{8.98}$$

It follows from Eqs. (8.95) to (8.97) that

$$\tilde{w}(w) = -\frac{\partial w}{\partial z} \tag{8.99}$$

$$dz = -\frac{\partial w}{\tilde{w}(w)} \tag{8.100}$$

$$z = -\int_{w_0}^{w} \frac{dw}{\tilde{w}(w)} + \text{constant} \tag{8.101}$$

The procedure used with a hodograph transformation is (1) map the contour of the boundary S of the flow domain onto the $\tilde{w}$ domain, (2) reflect the contour onto the q_x axis and find the contour Γ in the $\tilde{w}$ plane, (3) map s onto the w plane, (4) determine a conformal mapping $\tilde{w} = \tilde{w}(w)$ that maps the $\tilde{w}$ plane onto the w plane, and (5) knowing $\tilde{w} = \tilde{w}(w)$, integrate Eq. (8.101).

CASE STUDY 22: Flow in an Earthen Dam with a
Phreatic Surface Using the Hodograph Method

A hodograph is a graph of the velocity throughout a cross-sectional surface. If the complex potential is

$$w = \phi - i\psi \tag{8.102}$$

then the complex velocity is

$$\frac{dw}{dz} = \frac{\partial \phi}{\partial x} - i \frac{\partial \psi}{\partial x} \tag{8.103}$$

where

$$\frac{\partial \phi}{\partial x} = q_x \tag{8.104}$$

$$\frac{\partial \psi}{\partial x} = q_y \tag{8.105}$$

are the velocity along streamlines and between potential lines, respectively; so

$$\frac{dw}{dz} = \tilde{w} = q_x - iq_y \tag{8.106}$$

The utility of the hodograph comes from the fact that the shape of the free or phreatic surface is not known in the z plane but is completely specified in the w plane.

Transform the boundaries of flow regions into the w plane by the following rules. Graphs are made in the $q_x + iq_y$ plane for simplicity.

1. Impervious boundaries: At an impervious boundary the velocity vector is in the direction of the boundary. If the boundary is at an angle α with the x axis, in the hodograph

$$\alpha = \tan^{-1} \frac{q_x}{q_y} \tag{8.107}$$

which is a straight line in the w plane passing through the origin in a direction parallel to the impervious boundary.

2. Boundary of a reservoir: At the reservoir boundary ϕ = constant (equipotential) so the velocity vector is perpendicular to the boundary. For a boundary equation

$$y = x \tan \alpha + b \tag{8.108}$$

in the hodograph plane the equation is of a straight line

$$\alpha = -\cot^{-1} \frac{q_y}{q_x} \tag{8.109}$$

which passes through the origin and is normal to the reservoir boundary.

3. Free surface: Along the free surface $\phi + ky$ = constant. Then in the hodograph plane

$$q_x^2 + q_y^2 + kq_y = 0 \tag{8.110}$$

This is the equation of a circle passing through the origin with a radius of $k/2$ and center at $(O_1, k/2)$, k is the permeability coefficient. From this graph the velocity vector at any point along the free surface is obtained from the intersection of a line from the origin to the circle at the angle equal to that between the tangent to the free surface at the point in question and the horizontal.

4. Surface of seepage: As for the free surface $\phi + ky$ = constant. Then on the hodograph

$$q_y = -q_x \cot \alpha - k \tag{8.111}$$

where α is the angle between the seepage surface and the x axis. There-

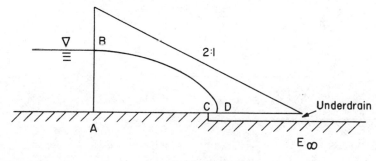

FIG. 8.21 Earthen dam with underdrain.

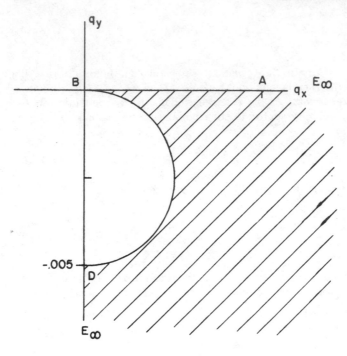

FIG. 8.22 Earthen dam with underdrain: hodograph. There is no seepage through a surface of the dam because of the underdrain: velocity at D = 0.005 cm sec^{-1} and velocity at A = 0.005 cm sec^{-1}. The flow field is shaded (k = 0.002 cm sec^{-1}).

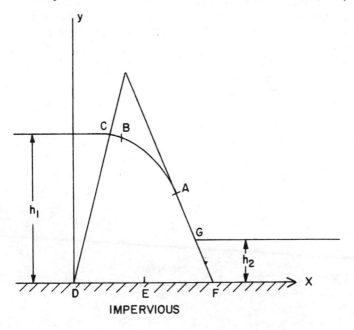

FIG. 8.23 Earthen dam without underdrain.

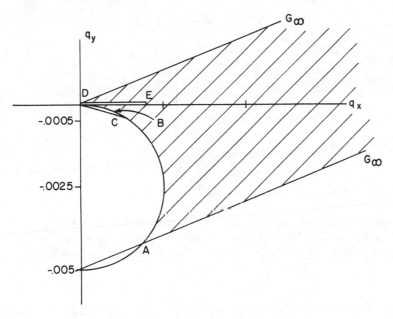

FIG. 8.24 Earthen dam without underdrain: hodograph ($v\phi A = 0.0042$ cm sec^{-1}, $k = 0.005$ cm sec^{-1}).

fore, a straight line is drawn on the w plane passing through the point (O_1, $-k$) and normal to the surface.

Figure 8.21 is a dam with an underdrain. Figure 8.22 is Fig. 8.21 transformed to the hodograph plane. Figure 8.23 is a dam without underdrain. Figure 8.24 is Fig. 8.23 transformed to the hodograph plane.

It can be seen from the figures that seepage is avoidable with proper design of of a dam, and the conditions that give no seepage are easily found from a hodograph. One of these conditions is that the seepage be perpendicular to the x axis. A drain under the dam covers this; so there is no velocity in the y direction to erode the dam. ■

8.4 DISPERSION IN AQUIFERS

Mixing occurs in aquifers when salt water or waste water comes in contact with fresh water. Examples include aquifers in contact with the sea; aquifers in contact with brackish groundwater; aquifers used for disposal of wastes; and brackish aquifers used for storage and retrieval of fresh water. Normally the density of fresh water is such that it is separated from salt water, waste water, etc. In coastal aquifers the natural gradient is toward the coast. The groundwater takes the form of a lens of fresh water floating on the salt water. The Ghyben-Herzberg hydrostatic relationship between fresh and saline water is sketched in Fig. 8.25. Saline water is found at a depth below sea level about 40 times the height of the water table

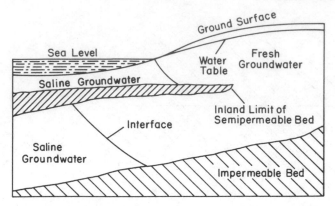

FIG. 8.25 Diagrams illustrating, in a simplified manner, the Ghyben-Herzberg hydrostatic relationship in (B) a layered coastal aquifer. [From M. A. Collins and L. W. Gelhar, Water Resources Res. 7(4), 971 (1971). Copyright by American Geophysical Union.]

above sea level. This represents the approximate hydrostatic equilibrium between fresh groundwater and the heavier seawater.

Collins and Gelhar (1971) noted that the saline wedge in the upper unconfined aquifers of Fig. 8.25 intrudes inland from the shoreline. The lower wedge may be inland or seaward depending on the fresh-water head. A sharp interface does not exist between the saline and the fresh water. Dispersion (including diffusion) is due to motion and tidal fluctuations which cause a mixed zone between the fresh and saline water.

In areas near sea coasts where aquifers are in contact with seawater the sustained reduction in the piezometric surface of the fresh groundwater results in intrusion of the seawater into the fresh groundwater. Figure 8.26 is a sketch of the equilibrium saltwater wedge in a confined coastal aquifer (Rumer and Harlemann, 1963). The position of the wedge can be calculated with the Dupuit-Forchheimer approximations discussed in Section 8.3. The specific discharge is

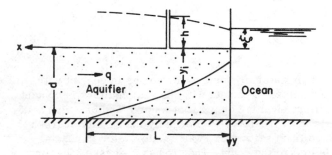

FIG. 8.26 Equilibrium wedge in a confined aquifer. [From R. R. Rumer and D. R. F. Harleman, Intruded Salt-Water Wedge in Porous Media, Proc. ASCE J. Hyd. Div. 89, HY5, 193 (1963). Copyright by the American Society of Civil Engineers.]

Dispersion in Aquifers

$$q = Ky_i \frac{dh}{dx} \qquad (8.112)$$

The condition of equal pressure along the interface yields

$$h = \frac{\Delta\rho}{\rho} y_i + \xi \qquad (8.113)$$

Combining Eqs. (8.112) and (8.113) and solving for y_i, we have

$$y_i = \left(\frac{zq\rho}{K\,\Delta\rho} + B\right)^{1/2} \qquad (8.114)$$

The constant of integration B is found at $x = 0$, $y = \sqrt{B^2}$. Henry (1959) determined the exact boundaries by hodograph. For infinite aquifers an approximation to the exact boundary condition yields

$$y_i\,(x = 0) = \frac{0.741q\rho}{K\,\Delta\rho} \qquad (8.115)$$

and

$$y_i = \left[\frac{2q\rho}{K\,\Delta\rho} + 0.55\left(\frac{q\rho}{K\,\Delta\rho}\right)^2\right]^{1/2} \qquad (8.116)$$

The results of Eq. (8.116) are very close to the solution with the exact boundary.

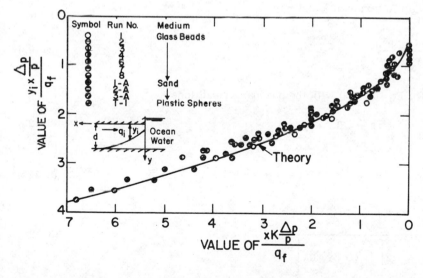

FIG. 8.27 Position of equilibrium wedge for a confined aquifer. [From R. R. Rumer and D. R. F. Harleman, Intruded Salt–Water Wedge in Porous Media, Proc. ASCE J. Hyd. Div. 89, HY5, 193 (1963). Copyright by the American Society of Civil Engineers.]

Figure 8.27 shows the results of physical model experiments compared to the Eq. (8.116).

For flow parallel to the x axis in Fig. 8.26 the mixing is expressed by

$$\frac{\partial c}{\partial t} + v_x \frac{\partial c}{\partial x} = \frac{\partial}{\partial x}\left(D_x \frac{\partial c}{\partial x}\right) + \frac{\partial}{\partial y}\left(D_y \frac{\partial c}{\partial y}\right) + \frac{\partial}{\partial z}\left(D_z \frac{\partial c}{\partial z}\right) \tag{8.117}$$

Assume transverse dispersion in the y direction is important and assume v_x independent of y and find an average $\langle v_x \rangle$.

$$\langle v_x \rangle = \frac{1}{2}\left(\frac{1}{y_i \phi} \int_0^{y_i} v_x \phi \, dy\right) = \frac{q}{2y_i \phi} \tag{8.118}$$

Then

$$\langle v_x \rangle \frac{\partial c}{\partial x} = D_y \frac{\partial^2 c}{\partial y^2} \tag{8.119}$$

Let

$$\lambda = L - x \tag{8.120}$$

and

$$\eta = y_i - y \tag{8.121}$$

Equation (8.119) in terms of λ and η is

$$\frac{\partial c}{\partial \lambda} = \frac{D_y}{\langle v_x \rangle} \frac{\partial^2 c}{\partial \eta^2} \tag{8.122}$$

The solution of Eq. (8.122) with the boundary conditions

$$c = c_0 \qquad -\infty < \eta < 0 \qquad \lambda = 0 \tag{8.123}$$

$$c = 0 \qquad 0 < \eta < \infty \qquad \lambda = 0 \tag{8.124}$$

$$\frac{\partial c}{\partial \eta} = 0 \qquad \eta = \pm\infty \tag{8.125}$$

is

$$\frac{c}{c_0} = \frac{1}{2} \operatorname{erfc} \frac{\eta}{2\left[\int_0^\lambda (D_y/y_i)\, d\lambda\right]} \tag{8.126}$$

The periodic oscillation of the ocean level due to tides produces a periodic fluctuation of the interface. The solution to the periodic fluctuations are complex and normally calculated numerically. (See Section 5.5.)

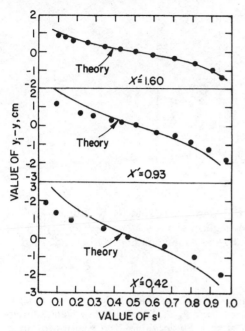

FIG. 8.28 Concentration profiles of the equilibrium wedge. [From R. R. Rumer and D. R. F. Harleman, Intruded Salt-Water Wedge in Porous Media, Proc. ASCE J. Hyd. Div. 89, HY5, 193 (1963). Copyright by the American Society of Civil Engineers.]

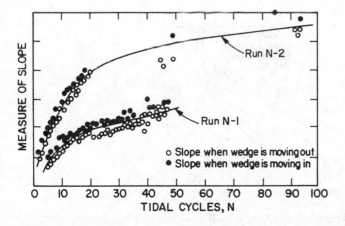

FIG. 8.29 Growth of dispersion zone measured from slopes of breakthrough and elution curves. [From R. R. Rumer and D. R. F. Harleman, Intruded Salt-Water Wedge in Porous Media, Proc. ASCE J. Hyd. Div. 89, HY5, 193 (1963). Copyright by the American Society of Civil Engineers.]

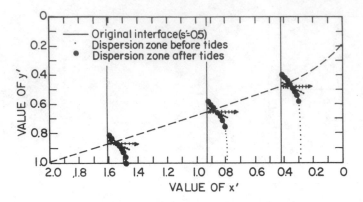

FIG. 8.30 Concentration profiles in confined aquifer. [From R. R. Rumer and
D. R. F. Harleman, Intruded Salt-Water Wedge in Porous Media, Proc. ASCE J.
Hyd. Div. 89, HY5, 193 (1963). Copyright by the American Society of Civil Engi-
neers.]

Rumer and Harlemann (1963) determined the dispersion in laboratory models
and compared the results with Eq. (8.126) on Fig. 8.28. The effect of wedge intrusion due to tides is shown on Fig. 8.29. Figure 8.30 combines the results for intrusion, tides, and dispersion.

Shamir and Dagan (1971) study interface motion numerically, without dispersion, using the Dupuit approximation to formulate the boundary. Pinder and Cooper
(1970) consider interface motion with dispersion numerically. Their results are
summarized in Fig. 8.31.

Mixing at the interface in radial flow is important in designing well systems
for disposal of waste water and in storage of fresh water in aquifers containing

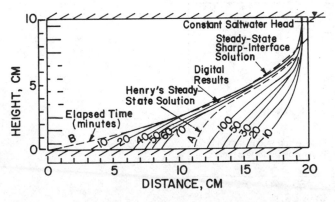

FIG. 8.31 Transient two-dimensional flow with dispersion. $Q = 0.66$ cm^2 sec^{-1},
$K = 1.0$ cm^2 sec^{-1}, $D = 0.066$ cm^2 sec^{-1}, $\rho_s = 1.025$ g cm^{-3}. [From G. F. Pinder
and H. H. Cooper, Jr., A Numerical Technique for Calculating the Transient Posi-
tion of the Saltwater Front, Water Resources Res. 6(3), 875 (1970). Copyright by
the American Geophysical Union.]

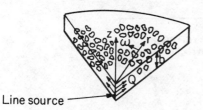

Line source

FIG. 8.32 Flow field and coordinate system for dispersion in radial flow from a line source. [From J. A. Hoopes and D. R. F. Harleman, Dispersion in Radial Flow from a Recharge Well, J. Geophys. Res. 72, 3595 (1967). Copyright by the American Geophysical Union.]

brackish water. Hoopes and Harlemann (1967) describe the symmetrical dispersion in radial flow from a single well in reference to Fig. 8.32 by the following equation where a_r is the longitudinal dispersivity.

$$\frac{\partial c}{\partial t} + \frac{A}{r}\frac{\partial c}{\partial r} = a_r\frac{A}{r}\frac{\partial^2 c}{\partial r^2} + \frac{\emptyset}{r}\frac{\partial}{\partial r}\left(r\frac{\partial c}{\partial r}\right) \tag{8.127}$$

where

$$A = \frac{Q}{2\pi bx} \tag{8.128}$$

Equation (8.127) may be transformed into a simpler form using the approximation

$$\frac{\partial}{\partial r} \doteq \frac{r}{A}\frac{\partial}{\partial t} \tag{8.129}$$

Using Eq. (8.129), Eq. (8.127) becomes

$$\frac{\partial c}{\partial t} + \frac{A}{r}\frac{\partial c}{\partial r} = \left(\frac{a_r r}{A} + \frac{\emptyset r^2}{A^2}\right)\frac{\partial^2 c}{\partial t^2} \tag{8.130}$$

For a constant step input of c_0 at $r = 0$ the solution of Eq. (8.130) is

$$\frac{c}{c_0} = \frac{1}{2}\,\mathrm{erfc}\left[\left(\frac{r^2}{2} - At\right)\left(\frac{4}{3}a_r r^3 + \frac{\emptyset r^4}{A}\right)^{-1/2}\right] \tag{8.131}$$

To reach Eq. (8.131) from Eq. (8.130) it is assumed at $\partial/\partial t = 0$ at $t = 0$. Let

$$\rho = \frac{r}{a_r} \tag{8.132}$$

and

$$\tau = \frac{At}{a_r^2} \tag{8.133}$$

Then Eq. (8.131) can be written as

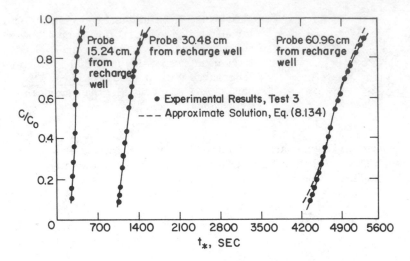

FIG. 8.33 C/C_0 versus t for three radii; radial dispersion test 3. [From J. A. Hoopes and D. R. F. Harleman, Dispersion in Radial Flow from a Recharge Well, J. Geophys. Res. 72, 3595 (1967). Copyright by the American Geophysical Union.]

$$\frac{c}{c_0} = \frac{1}{2} \, \text{erfc} \left[\frac{\rho^2/2 - \tau}{(4\rho^3/3)^{1/2}} \right] \qquad\qquad (8.134)$$

Figure 8.33 shows a comparison of physical model results with Eq. (8.134).

Kumar and Kimbler (1970) solved the radial problem including gravity segregation and layering numerically to study the effect of the various parameters. Their purpose was to consider storage in and recovery of fresh water from an aquifer containing saline water. They concluded that storage and retrieval was feasible in low-permeability, thin aquifers.

CASE STUDY 23: Numerical Solution of Flow in a
Deep Waste Injection Well

The criteria for determining the ability of a location to handle well injection of wastes are as follows:

1. A suitable reservoir exists that will take the necessary quantities of wastes under economically achievable pressures at the surface, and can be safely handled in surface equipment and the
2. A reservoir exists that preferably is sandstone and water saturated and covers an extensive area so that flow control is possible because the injected fluids travel at low velocities away from the well bone and compress and displace the formation fluid, with sufficient capacity for many years of operation.
3. The geology is such that an impervious semiplastic roof material exists in sufficient thickness to assure no migration of the waste fluid above the reservoir formation.

4. The formation fluid in the reservoir is of such a nature that the injected fluids can be made compatible with the native fluid or else such permeability exists or can be created that chemical precipitation will not plug the formation to an extent that injection pressures will become uneconomical to pump against or unsafe.

5. A stratagraphic column exists that will allow casing to be completely connected to the surface so that migration of injected fluids above the reservoir is impossible.

The Mount Simon formation in Northern Indiana was found to have these qualities after a well for waste injection was drilled in 1960. Figure 8.34 is a generalized

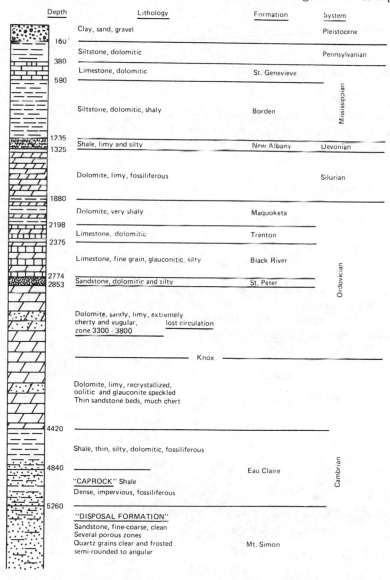

FIG. 8.34 Geologic section.

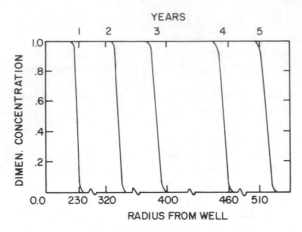

FIG. 8.35 Diffusion–only solution.

geologic section for the waste–disposal well, WD–1. The hole left for disposal below
the casing is 710 ft high, the formation itself is about 900 ft in height. The reservoir
fluid is brine with a specific gravity of 1.148. The pump purchased for the injection
was designed for 100 gal min⁻¹ at 1400 psig. The casing and head equipment were
safety rated at three times the operating pressure. The formation overlying the
Mt. Simon formation is over 600 ft of dense shale with zero vertical permeability.
The casing extends 190 ft into the Mt. Simon formation.

The governing equation for symmetrical radial dispersive flow is Eq. (8.127):

$$\frac{\partial c}{\partial t} + \frac{A}{r}\frac{\partial c}{\partial r} = a\,\frac{A}{r}\frac{\partial^2 c}{\partial r^2} + \frac{\mathcal{D}}{r}\frac{\partial}{\partial r}\left(r\frac{\partial c}{\partial r}\right) \tag{8.135}$$

This equation has been solved in an approximate form (Eq. 8.130), analytically.
The solution assumes that r is some distance from the well bore and that the dis-
persion from point to point is small compared to the dispersion already accumulated.
The initial and boundary conditions are

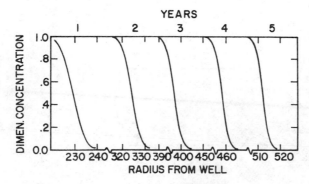

FIG. 8.36 Large dispersion solution.

$$\rho Q c_0 = dm/dt \qquad r = 0$$

$$c(oo, t) = 0 \qquad t > 0$$

$$c(r, o) = 0$$

$$\rho = \text{density}$$

$$Q = \text{flow rate} \tag{8.136}$$

$$c_0 = \text{concentration of feed}$$

$$m = \text{mass}$$

The geometry of the system is illustrated in Fig. 8.31. The solution is given by Eq. (8.131).

$$\frac{c}{c_0} = \frac{1}{2} \, \text{erfc} \left[\left(\frac{r^2}{2} - At \right) \left(\frac{4}{3} a_r r^3 + \frac{\mathcal{D} r^4}{4} \right)^{-1/2} \right] \tag{8.137}$$

The reservoir and system properties are

$$\phi = 0.06$$

$$Q = 100 \text{ gal min}^{-1} = 802 \text{ ft}^3 \text{ hr}^{-1}$$

$$h = 710 \text{ ft}$$

$$\mathcal{D} = 3.8 \times 10^{-5} \text{ ft}^2 \text{ hr}^{-1}$$

$$a_r = 0.15 \text{ cm}$$

The solutions were computed for 1, 2, 3, 4, and 5 years of continuous injection and the results are shown in Figs. 8.35 and 8.36. Figure 8.35 shows the solution with $a_r = 0$, or pure diffusion, and Fig. 8.36 shows $a_r = 15$. It can be seen that significantly more spreading of the front occurs in Fig. 8.36; however, it is still small compared to the movement of the front.

Effects such as density differences of the injected fluid from the reservoir fluid have not been accounted for here. This could cause fingering and a significantly wider radius of waste penetration. The density of the injected fluid was not known and the problem accounting for these effects is much more difficult to solve. Also not considered was the fact that the formation thickness is larger than the open well hole depth; thus some vertical dispersion and possibly convection might be possible.

REFERENCES

Aron, G., and V. H. Scott (1965): Simplified Solutions for Decreasing Flow in Wells, Proc. ASCE J. Hyds. Div. 91, HY5, 1.

Bear, J. (1972): Dynamics of Fluids in Porous Media, American Elsevier, New York.

Boulton, N. S. (1954): The Drawdown of the Water Table under Non-Steady Conditions near a Pumped Well in an Unconfined Formation, Proc. Inst. Civil Engrs. 3(3), 564.

Boulton, N. S. (1963): Analysis of Data from Non-equilibrium Pumping Tests Allowing for Delayed Yield from Storage, Proc. Inst. Civ. Eng. 26, 469.

Bredehoeft, J. D., and G. F. Pinder (1970): Digital Analysis of Areal Flow in Multi-aquifer Groundwater System: A Quasi Three-Dimensional Model, Water Resour. Res. 6, 883.

Bruggeman, G. A. (1966): Analyze van de Bodeconstanten in Een Grondpakket, Bestaande Uit Twee of Meer Watervoerende Lagen Gescheiden Door Semi-permeabele Lagan; unpublished research report.

Chow, V. T. (1952): On the Determination of Transmissivity and Storage Coefficients from Pumping Test Data, Trans. Amer. Geophys. Union 33, 397.

Collins, M. A., and L. W. Gelhar (1971): Seawater Intrusion in Layered Aquifers, Water Resour. Res. 7(4), 971.

Cooper, H. H., and C. E. Jacob (1946): A Generalized Graphical Method for Evaluating Formation Constants and Summarizing Well Field History, Trans. Amer. Geophys. Union 27, 526.

De Glee, G. J. (1930): Over Grondwaterstrommen bijwateronttrekking door middel van pullen, Thesis, J. Waltman, Delft.

Dietz, D. N. (1943): De Toepassing van Invloedsfuncties Bij het Berekenen van de Verlaging van het Grondwater Tengevolge van Wateronttrekking, Water 27(6), 51.

Dupuit, J. (1863): Etudes Theorigues et Pratigues sur le Mouvement des Eaux dans les Canaux Découverts et à Trovers les Terrains, 2d ed., Dunod, Paris.

Ferris, J. G. (1962): Theory of Aquifer Tests, U.S. Geol. Survey, Water-Supply Paper 1536-E, p. 174.

Hantush, M. S., and C. E. Jacob (1955): Non-steady Radial Flow in an Infinite Leaky Aquifer, Trans. Amer. Geophys. Union 36, 95.

Hantush, M. S. (1956): Analysis of Data from Pumping Tests in Leaky Aquifers, Trans. Amer. Geophys. Union 37, 702.

Hantush, M. S. (1959): Analysis of Data from Pumping Wells near a River, J. Geophys. Res. 64, 1921.

Hantush, M. S. (1962): Aquifer Tests on Partially Penetrating Wells, Amer. Soc. Civ. Eng. Trans. 127, 268.

Hantush, M. S. (1965): Hydraulics of Wells, in Advances in Hydroscience, V. T. Chow (ed.), Vol. I, p. 281, Academic, New York.

Hantush, M. S. (1966): Analysis of Data from Pumping Tests in Anisotropic Aquifers, J. Geophys. Res. 71, 421.

Hantush, M. S., and R. G. Thomas (1966): A Method for Analyzing a Drawdown Test in Anisotropic Aquifers, Water Resour. Res. 2, 281.

Henry, H. R. (1959): Salt Intrusion into Fresh-Water Aquifers, J. Geophys. Res. 64, 1911.

Hoopes, J. A., and D. R. F. Harleman (1967): Dispersion in Radial Flow from a Recharge Well, J. Geophys. Res. 72, 3595.

Huisman, L., and J. Kemperman (1951): Bemaling van Spanningsgrondwater, De Ingenieur 62, B29.

Hurr, T. R. (1966): A New Approach for Estimating Transmissivity from Specific Capacity, Water Resour. Res. 2, 657.

Jacob, C. E. (1974): On the Flow of Water in an Elastic Artesian Aquifer, Amer. Geophys. Union Trans. 72, 574.

Jacob, C. E. (1949): Flow of Groundwater, in Engineering Hydraulics, H. Rouse (ed.), Wiley, New York, Chapter 5.

Jacob, C. E., and S. W. Lohman (1952): Non-steady Flow to a Well of Constant Drawdown in an Extensive Aquifer, Amer. Geophys. Union Trans. 33, 559.

Jacob, C. E. (1963): Correction of Drawdowns Caused by a Pumped Well Tapping Less Than the Full Thickness of an Aquifer, in Determining Permeability, Transmissibility and Drawdown, R. Bentall (ed.), U.S. Geol. Survey Water-Supply Paper 1536-I: 272-282.

Johnson, C. R., R. A. Greenkorn, and E. G. Woods (1967): Pulse Testing: A New Method of Measuring Reservoir Continuity between Wells, Geo. Soc. Amer. Special Papers 115, 109.

Kruseman, G. P., and N. A. deRidder (1970): Analysis and Evaluation of Pumping Test Data, Bull. 11, International Institute for Land Reclamation and Improvement, Wageningen, The Netherlands.

Kumar, A., and O. K. Kimbler (1970): Effect of Dispersion, Gravitational Segregation, and Formation Stratification on the Recovery of Fresh Water Stores in Saline Aquifers, Water Resour. Res. 6(6), 1689.

Logan, J. (1964): Estimating Transmissibility from Routine Production Tests of Waterwells, Groundwater 2(1), 35.

Neuman, S. P., and P. A. Witherspoon (1970): Finite Element Method of Analyzing Steady Seepage with a Free Surface, Water Resour. Res. 6(3), 889.

Papadoupulos, I. S., and H. H. Cooper, Jr. (1967): Drawdown in a Well of Large Diameter, Water Resour. Res. 3, 241.

Pinder, G. F., and H. H. Cooper, Jr. (1970): A Numerical Technique for Calculating the Transient Position of the Saltwater Front, Water Resour. Res. 6(3), 875.

Rumer, R. R., Jr. and D. R. F. Harleman (1963): Intruded Salt-Water Wedge in Porous Media, Proc. ASCE J. Hydraul. Div. 89, HY6, 193.

Shamir, V., and G. Dagan (1971): Motion of the Seawater Interface in Coastal Aquifers: A Numerical Solution, Water Resour. Res. 7, 644.

Sternberg, Y. M. (1967): Transmissibility Determination from Variable Discharge Pumping Tests, Groundwater 5(4), 27.

Sternberg, Y. M. (1968): Simplified Solution for Variable Rate Pumping Test, Proc. Amer. Soc. Civ. Eng. 94, J. Hydraul. Div., HY1, 177.

Theis, C. V. (1935): The Relation between Lowering of the Piezometric Surface and the Rate and Duration of Discharge of a Well Using Groundwater Storage, Trans. Amer. Geophys. Union 16, 519.

Thiem, G. (1906): Hydrogrsche Methodem, Gebhart, Leipzig.

U.S. Geologic Survey (1970-1974): South Central States, Ground Water in the U.S., Water Supply Paper #2172.

Walton, W. C. (1960): Application and Limitation of Methods Used to Analyze Pumping Test Data, Water Well J., February, March.

Walton, W. C. (1962): Selected Analytical Methods for Well and Aquifer Evaluation, Illinois State Water Survey Bull., No. 49.

Ward, R. C. (1975): Principles of Hydrology, McGraw-Hill, New York.

Zangar, C. N. (1953): Theory and Problems of Water Percolation, Engineering Monographs 8, U.S. Bureau of Reclamation, Denver, Colo.

SUGGESTED READING

Eagleson, P. S. Dynamic Hydrology, McGraw-Hill, New York, 1970.

Gray, D. M. (ed.), Handbook on the Principles of Hydrology, Water Information Center, Inc., Port Washington, New York, 1970.

Kazman, R. G. Modern Hydrology, 2d ed., Harper and Row, New York, 1972.

Muskat, M., Flow of Homogeneous Fluids, McGraw-Hill, New York, 1937.

9

Applications in Soils Science

Many of the applications discussed in Chapters 7 and 8 apply in soils science. There are additional problems associated with movement of water in the unsaturated region, nonproportional flow in clayey soils, and the movement and distribution of chemicals (nutrients and pollutants) in the soil systems.

9.1 UNSATURATED FLOW

Soils scientists refer to flow in the zone of aeration of Fig. 8.1 as unsaturated flow. The flow of water at less than 100 percent saturation is considered unsaturated flow, with water the wetting fluid. The regions that do not contain liquid water contain air or water vapor. The flow is immiscible flow of a liquid and a gas.

The terminology used in soils science has evolved for the special problems of movement of soil moisture in the zone of aeration. The water content c is the volume of water in an element of soil divided by the bulk volume of the element. A soil is water-saturated when the entire pore space contains water. The water content is similar to the saturation of the wetting fluid, used in previous chapters. The pore water suction or tension is shown schematically in Fig. 9.1. The capillary pressure is

$$p_c = p_{nw} - p_w \tag{9.1}$$

If air pressure is used as the datum, assuming negligible water vapor and $p_{nw} = 0$,

$$p_c = -p_w \tag{9.2}$$

In this instance p_c is called the suction pressure or tension; in terms of pressure head,

$$h_c = \frac{p_c}{\rho_w} = -\frac{p_w}{\rho_w} \tag{9.3}$$

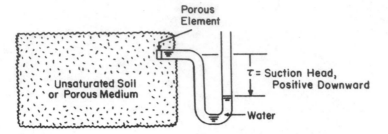

FIG. 9.1 Schematic representation of the measurement of matrix suction head τ with a tensiometer. The water in the tensiometer is at equilibrium with the water in the unsaturated soil, hydraulic communication being obtained through the porous element. (From Swartzendruber, 1969.)

The piezometric or capillary potential is

$$\Phi_c = z + \frac{p_w}{\rho_w} = z - \frac{p_c}{\rho_w} = z - h_c \tag{9.4}$$

The soil moisture characteristic curve is the suction head versus water content as in Fig. 9.2. This is similar to the capillary pressure versus saturation discussed in previous chapters. The two curves, the upper or drying curve and the lower or wetting curve, show the hysteresis discussed in Chapters 2 and 4. The specific

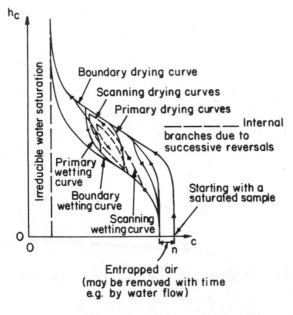

FIG. 9.2 Hysteresis in the capillary head–water content relationship for a coarse material in stable packing. (From Bear, 1972.)

water capacity is the rate of change of moisture content with decreasing suction (increasing pressure) at a given water content, the slope $dc/dn_c|_c$.

Darcy's law for soil-water systems is written as

$$\underline{q} = -K_c(c)\ \underline{\nabla}\Phi_c \tag{9.5}$$

where the effective hydraulic conductivity or capillary conductivity is a function of water content. This is Darcy's law for flow of the wetting fluid in an immiscible system (see Chapter 4). The effective hydraulic conductivity in terms of the effective permeability to water is

$$K_c(c) = \frac{k_c(c)\rho g}{\mu} \tag{9.6}$$

Soils scientists seldom use relative permeability.

If we are interested in the drainage (drying) part of the h_c versus c curve, then Darcy's law may be written as

$$\underline{q} = -K_c(c)\,[\underline{\nabla} z - \underline{\nabla} h_c(c)] \tag{9.7}$$

or

$$\underline{q} = K_c(c)\ \underline{\nabla} z - \hat{D}(c)\ \underline{\nabla} c \tag{9.8}$$

$\hat{D}(c)$ and $K_c(c)$ are functions of c

$$\hat{D}(c) = K_c(c)\left(-\frac{dh_c}{dc}\right) \tag{9.9}$$

$\hat{D}(c)$ is the soil water diffusivity. If we combine Darcy's law and the equation of continuity so that

$$\frac{\partial c}{\partial t} = -(\underline{\nabla} \cdot \underline{q}) \tag{9.10}$$

then

$$\frac{\partial c}{\partial t} = \frac{\partial K}{\partial z} + \frac{\partial}{\partial x}\left(\hat{D}\ \frac{\partial c}{\partial x}\right) + \frac{\partial}{\partial y}\left(\hat{D}\ \frac{\partial c}{\partial y}\right) + \frac{\partial}{\partial z}\left(\hat{D}\ \frac{\partial c}{\partial z}\right) \tag{9.11}$$

The distribution of static water in a homogeneous soil profile with a water table is shown in Fig. 9.3. The distribution of static water in a homogeneous soil profile without a water table is shown in Fig. 9.4. Figure 9.5 shows the saturation in the upper soil when water is applied by irrigation or due to rain on a dry soil. There is a saturation zone reaching down about 1.5 cm. Next there is a transition zone going down about 5 cm where moisture decreases rapidly. The transition zone increases in length as infiltration proceeds; the water content is approximately

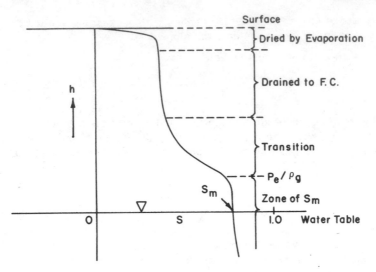

FIG. 9.3 Distribution of pseudostatic water in a homogeneous soil profile with water table. (From Corey, 1977.)

constant in this zone. The wetting zone is indicated by a rapid drop in water content. The wetting front is the visible limit of moisture penetration.

A capillary fringe exists adjacent to any free surface bounding a saturated section of soil as well as immediately above the water table. The capillary fringe may move with a water front or lag it. Figure 9.6 is a schematic of the pressure distribution in the capillary fringe at equilibrium.

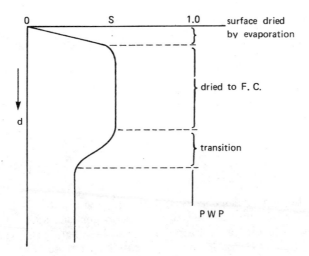

FIG. 9.4 Distribution of pseudostatic water in a homogeneous soil without a water table. (From Corey, 1977.)

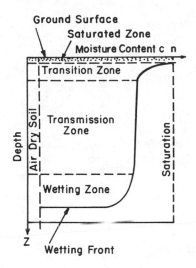

FIG. 9.5 Moisture zones during infiltration. (From Bear, 1972.)

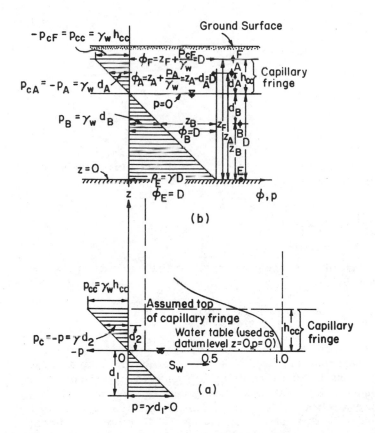

FIG. 9.6 Pressure distribution in the capillary fringe at equilibrium. (From Bear, 1972.)

During evaporation at the soil surface if the water comes from the water table, a steady moisture profile will be reached. Steady flow to or from the water table is determined using Eq. (9.8) in the z direction; so

$$\frac{dz}{dc} = \frac{\hat{D}}{q - K_c} \tag{9.12}$$

For evaporation let $q = -e$, where e is the rate of evaporation.

Water movement to plant roots is modeled by assuming the root system is an equivalent two-dimensional line sink. Equation (9.11) is written in polar coordinates as

$$\frac{\partial c}{\partial t} = \frac{1}{r} \frac{\partial}{\partial r}\left(r\hat{D} \frac{\partial c}{\partial r}\right) \tag{9.13}$$

$\hat{D}$ is assumed constant, K is constant, and assuming dc/dh_c is constant, then

$$\frac{\partial h_c}{\partial t} = \frac{\hat{D}}{r} \frac{\partial}{\partial r}\left(r \frac{\partial h_c}{\partial r}\right) \tag{9.14}$$

with conditions

$$h_c = h_0 \quad \text{at } t = t_0 \tag{9.15}$$

$$K_c \frac{dh_c}{dr} = V \quad \text{at } r = r_i \tag{9.16}$$

where V is constant water inflow velocity at a cylindrical root. An approximate solution for Eq. (9.14) with conditions (9.15) and (9.16) is

$$t - t_0 = \frac{\pi r_i V}{K_c}\left(\ln \frac{4\hat{D}t}{r^2} - 0.57722\right) \tag{9.17}$$

Numerical solutions to the soil moisture problem are available (Green et al., 1970; Hornberger and Remson, 1970).

CASE STUDY 24: Calculation of Vertically Downward Infiltrations
through Layered Soil

Infiltration is the process where water enters the surface strata of the soil and moves downward. The maximum rate at which a soil in any given condition is capable of absorbing water in this manner is called its infiltration capacity. The prevailing infiltration rate f' is equal to the infiltration capacity f only during and immediately following periods of excess rainfall. Thus in calculating infiltration capacity, it is only necessary to consider the time interval during which infiltration is occurring at maximum rate. The infiltration rate depends on many factors such as time and area. Thus the infiltration rate varies from one area to another, or the infiltration rate of a given area can change from time to time.

TABLE 9.1 Rainfall Data

Hour	Amount (in.)	Percent of total
1	0.05	1.3
2	0.05	1.3
3	0.03	0.8
4	0.02	0.5
5	0.05	1.3
6	0.05	1.3
7	0.07	1.9
8	0.08	2.2
9	0.20	5.4
10	0.20	5.4
11	0.12	3.2
12	0.12	3.2
13	0.03	0.8
14	0.02	0.5
15	0.15	4.0
16	0.15	4.0
17	0.35	9.4
18	0.35	9.4
19	0.35	9.4
20	0.35	9.4
21	0.25	6.8
22	0.25	6.8
23	0.15	4.0
24	0.15	4.0
25	0.05	1.3
26	0.05	1.3
27	0.02	0.5
28	0.01	0.3

Source: Wisler and Brater, 1954.

Infiltration often begins at a high rate and decreases to a much lower and fairly constant rate as rain continues. Other conditions that cause variations of infiltration rate for a given soil are as follows:

1. Soil moisture content
2. State of cultivation
3. Perforations of the surface soil and subsoil
4. Packing of the soil surface
5. Temperature changes
6. Depth to less permeable strata

The soil moisture content is probably the most important factor; however, the last factor is significant also since it indicates that the infiltration rate does not entirely depend on the permeability of the surface layer of the soil.

TABLE 9.2 Calculations at Various Infiltration Capacities

Hour	Total amount (in.)	Excess rainfall Infiltration capacity f (in./hr^{-1})				
		0.1	0.2	0.3	0.4	0.5
		$p = 2.0$ in.				
8	0.044	0				
9	0.108	0.008	0	0	0	0
10	0.108	0.008	0	0	0	0
11	0.070	0				
12	0.064	0				
15	0.080	0				
16	0.080	0				
17	0.188	0.088	0	0	0	0
18	0.188	0.088	0	0	0	0
19	0.188	0.088	0	0	0	0
20	0.188	0.088	0	0	0	0
21	0.136	0.036	0	0	0	0
22	0.136	0.036	0	0	0	0
23	0.080	0				
24	0.080	0				
		0.44	0	0	0	0
		$p = 3.0$ in.				
8	0.066	0				
9	0.162	0.062	0	0	0	0
10	0.162	0.062	0	0	0	0
11	0.105	0.005	0	0	0	0
12	0.096	0				
15	0.120	0.02	0	0	0	0
16	0.120	0.02	0	0	0	0
17	0.282	0.182	0.082	0	0	0
18	0.282	0.182	0.082	0	0	0
19	0.282	0.182	0.082	0	0	0
20	0.282	0.182	0.082	0	0	0
21	0.204	0.104	0.004	0	0	0
22	0.204	0.104	0.004	0	0	0
23	0.120	0.02	0	0	0	0
24	0.120	0.02	0	0	0	0
		1.145	0.336	0	0	0
		$p = 4.0$ in.				
8	0.088	0				
9	0.216	0.116	0.016	0		
10	0.216	0.216	0.116	0.016	0	
11	0.140	0.040	0			
12	0.128	0.028	0			
15	0.160	0.06	0			
16	0.160	0.06	0			

(continued)

TABLE 9.2 (continued)

Hour	Total amount (in.)	Excess rainfall Infiltration capacity f (in./hr^{-1})				
		0.1	0.2	0.3	0.4	0.5
17	0.376	0.276	0.176	0.076	0	
18	0.376	0.276	0.176	0.076	0	
19	0.376	0.276	0.176	0.076	0	
20	0.376	0.276	0.176	0.076	0	
21	0.272	0.172	0.072	0		
22	0.272	0.172	0.072	0		
23	0.160	0.06	0			
24	0.160	0.06	0			
		1.99	0.88	0.304	0	0
			p = 5.0 in.			
8	0.11	0.01	0			
9	0.27	0.17	0.07	0		
10	0.27	0.17	0.07	0		
11	0.175	0.075	0			
12	0.16	0.06	0			
15	0.20	0.10	0			
16	0.20	0.10	0			
17	0.47	0.37	0.27	0.17	0.07	0
18	0.47	0.37	0.27	0.17	0.07	0
19	0.47	0.37	0.27	0.17	0.07	0
20	0.47	0.37	0.27	0.17	0.07	0
21	0.34	0.24	0.14	0.04	0	
22	0.34	0.24	0.14	0.04	0	
23	0.20	0.10	0			
24	0.20	0.10	0			
		2.845	1.5	0.76	0.28	0
			p = 6.0 in.			
8	0.132	0.032	0			
9	0.324	0.224	0.124	0.024	0	
10	0.324	0.224	0.124	0.024	0	
11	0.21	0.11	0.01	0		
12	0.192	0.092	0			
15	0.24	0.14	0.04	0		
16	0.24	0.14	0.04	0		
17	0.564	0.464	0.364	0.264	0.164	0.064
18	0.564	0.464	0.364	0.264	0.164	0.064
19	0.564	0.464	0.364	0.264	0.164	0.064
20	0.564	0.464	0.364	0.264	0.164	0.064
21	0.408	0.308	0.208	0.108	0.008	0
22	0.408	0.308	0.208	0.108	0.008	0
23	0.24	0.14	0.04	0		
24	0.24	0.14	0.04	0		
		3.714	2.29	1.32	0.672	0.256

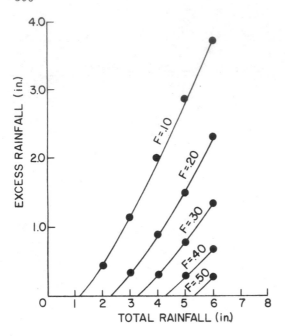

FIG. 9.7 Total rainfall.

       The actual rate and total amount of infiltration depend not only on the infiltra-
tion capacity of the soil but also on the time distribution of the rainfall. A rain of
long duration and low intensity results in more infiltration than does a short rain of
high intensity.
       There are two general methods of determining infiltration capacity. The first
is with an infiltrometer, in which water is artificially applied to a small area or ·

TABLE 9.3 Station Total Rainfall Depth Data

Station	Total rainfall (in.)	Rainfall excess (in./hr^{-1}) (from Fig. 9.1)		
		f = 0.1	f = 0.2	f = 0.3
1	5.91	3.59	2.18	1.26
2	3.81	1.80	0.74	0.21
3	3.76	1.76	0.71	0.18
4	4.13	2.10	0.93	0.34
5	4.01	1.99	0.86	0.29
6	4.09	2.08	0.91	0.32
7	4.92	2.74	1.42	0.71
8	6.05	3.72	2.31	1.33
Average:	4.58	2.47	1.26	0.58

Source: Wisler and Brater (1954).

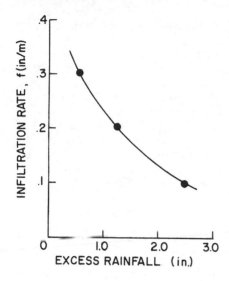

FIG. 9.8 Excess rainfall.

sample plot and the infiltration rate is more or less directly determined. The second is by analysis of the hydrograph or runoff resulting from a natural rainfall or drainage basin.

The following method was used to calculate the infiltration rate f. First, in Table 9.1, rainfall data over a time period was gathered. The fraction of the total rainfall was calculated for each hour. Then using the intensity pattern for the storm, calculations were made for a 2-, 3-, 4-, 5-, and 6-in. storm, in Table 9.2, at various infiltration capacities. The excess rainfall was then calculated, and in Fig. 9.7 the excess rainfall versus total rainfall was plotted for various rates of infiltration.

Now using the data from Table 9.3, that is, knowing the total depths of rainfall for each station, we can obtain the excess rainfall from Fig. 9.7.

Thus the average excess rainfall can be plotted against the infiltration rate as in Fig. 9.8. Therefore, if we know the excess rainfall for the sample, we can obtain the desired downward infiltration rate. ■

CASE STUDY 25: Unsteady Soil Moisture Movement in the
Vertical Direction: Numerically

A mathematical model describing water flow in a porous medium which is partially saturated with water has been developed by Green et al. (1970). The interest in flow through unsaturated soils has been motivated by the need to understand processes such as infiltration, evaporation, and gravity drainage that takes place in soils near the surface. Childs (1936) assumed the validity of a diffusion-type equation, with saturation as the dependent variable, to model soil moisture movement. Because of the nonlinear nature of the diffusivity coefficient, the diffusivity-type equation has been difficult to solve and thus requires numerical techniques.

Soil scientists researched this problem along the same lines as petroleum engineers. The petroleum reservoir-type solutions for gas-oil or water-gas flow have included an equation describing behavior of the nonwetting phase. The difference in approaches is that researchers in water resources have taken the effects of gas compressibility to be negligible.

Some experimental data taken for the purpose of verifying the various mathematical models have been reported but generally experimental checks are lacking. Very little has been reported through field tests.

In this study, the petroleum reservoir engineering approach to two-phase flow modeling has been applied to moisture movement in the unsaturated zone. The calculated expressions were compared to experimental field data collected by the U.S. Geological Survey at the Garden City Branch Experiment Station, Garden City, Kansas. The experimental procedure involved experiments in which water was fed at controlled rates to a shallow surface pond and allowed to flow downward through the underlying soil. The movement of subsurface water was monitored by radioactive tracer.

Application of the equation of continuity and Darcy's law to a porous medium in which air and water are flowing simultaneously and in one direction results in the following equations:

For the water phase,

$$\frac{\partial}{\partial x}\left[\frac{\rho_w k_{rw} k}{\mu_w}\left(\frac{\partial p_w}{\partial x} + \rho_w g \frac{\partial h}{\partial x}\right)\right] = \phi \frac{\partial(S_w \rho_w)}{\partial t} \tag{9.18}$$

For the air phase,

$$\frac{\partial}{\partial x}\left[\frac{\rho_a k_{ra} k}{\mu_a}\left(\frac{\partial p_a}{\partial x} + \rho_a g \frac{\partial h}{\partial x}\right)\right] = \phi \frac{\partial(S_a \rho_a)}{\partial t} \tag{9.19}$$

Since only air and water are assumed to be in the pores, then

$$S_a + S_w = 1 \tag{9.20}$$

In a partially saturated porous medium there is a pressure difference between water pressure and air pressure, i.e., there is capillary pressure that is a function of saturation.

$$p_c(S_w) = p_a - p_w \tag{9.21}$$

The boundary conditions applied with these equations are the constant pressure boundary.

$$\begin{aligned} p_w &= p_w \text{ constant} \\ & \qquad\qquad\qquad \text{at } x = L \\ p_a &= p_a \text{ constant} \end{aligned} \tag{9.22}$$

At the top, $x = 0$, a reflective boundary condition was specified.

The preceding equations describe flow of water from the surface to an underground aquifer if the flow can be assumed to be one-dimensional and if effects such

as osmosis and diffusion are negligible. The equation set is nonlinear, and no known
analytical solutions are available. However, these equations can be solved using
finite difference techniques.

The Taylor series was used to convert the partial differential equations to
finite difference form. An iterative procedure was used to solve the set of equations.

The output of the computer program consists of values of saturation and pressures versus depth and time. A material balance procedure was built into the program to measure how well the program solved the differential equations. Mass
balances for the water phase typically were 99 percent, while the air phase balances
were approximately 95 percent.

As mentioned previously, experimental data were obtained to verify the accuracy of the solutions to the mathematical equations. Infiltration experiments were
performed under controlled conditions on a small plot of isolated agricultural land.
A shallow circular pond with a diameter of 50 ft was constructed on the surface.
Observation wells were drilled at various radii to determine soil characteristics.
The wells ranged in depth from 8 to 30 ft. Figure 9.9 gives a gross description of
the soil characteristics. The porosity of various depths were determined from core
samples and are given in Table 9.4.

The experimental procedure was to feed water to the surface pond and allow
the water to infiltrate. Water application was continued for 59 hr. Water content in
the soil as a function of depth and time was determined from radioactive tracers.

The input data needed to compute the soil water profile using the mathematical
model are

1. Initial water content ϕS_W and pressure profile in the system
2. Rates of flow of water and air into and out of the model as functions of time
3. Viscosities of water and air

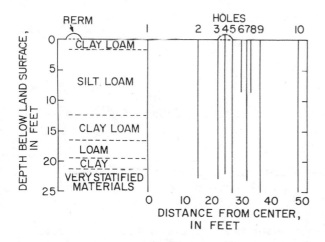

FIG. 9.9 Cross section of pond. [From D. W. Green, H. Dabiri, C. F. Weinaug,
and R. Prill, Numerical Modeling of Unsaturated Groundwater Flow and Comparison
of the Model to a Field Experiment, Water Resour. Res. 6(3), 862 (1970). Copyright by American Geophysical Union.]

TABLE 9.4 Porosity and Permeability

Depth interval (ft)	Measured porosity (%)	Depth (ft)	Final porosity used in program (%)	Final permeability used in program (darcy)
0.1–0.4	52.6	0	50	8
0.5–0.8	48.1	−1	50	6
0.9–1.2	47.0	−2	50	4
1.2–1.4	47.8	−3	50	4
1.7–1.9	46.3	−4	50	4
2.0–2.4	48.5	−5	50	4
2.5–2.9	47.8	−7	50	4
3.0–3.4	47.1	−7	50	4
3.5–3.9	48.5	−8	50	4
4.0–4.4	50.7	−9	50	4
4.5–4.9	51.5	−10	50	4
5.0–5.4	52.2	−11	50	4
5.5–5.9	54.5	−12	50	4
5.9–6.2	53.0	−13	50	4
9.0–10.0	53.4	−14	48	6
14.0–15.0	44.0	−15	48	10
18.0–19.0	47.8	−16	50	14
19.5–20.5	40.3	−17	50	8
		−18	50	2
		−19	50	4
		−20	44	4
		−21	44	4
		−22	41	4
		−23	40	4
		−24	40	4

Source: Green et al. (1970).

4. Capillary pressure as a function of water content for the various porous media in the system
5. Relative permeabilities of water and air as functions of water content for the various media in the system
6. Absolute permeability as a function of position

Little information from the preceding input data was available from the field test, so data from the literature were used. Experimental and calculated water contents as a function of depth for the center well (well 1) are shown for different times in Figs. 9.10 to 9.13.

For the initial computations no distinction in physical properties was made for the different strata in the system. Figures 9.10 and 9.11 show the initial comparison of experimental data with computed results during filtration. The computed drying rate was much too great for the drying curve. This discrepancy exists

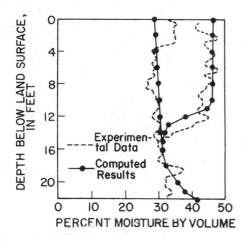

FIG. 9.10 Soil moisture: 9.5 hr infiltration. [From D. W. Green, H. Dabiri, C. F. Weinaug, and R. Prill, Numerical Modeling of Unsaturated Groundwater Flow and Comparison of the Model to a Field Experiment, Water Resour. Res. 6(3), 862 (1970). Copyright by American Geophysical Union.]

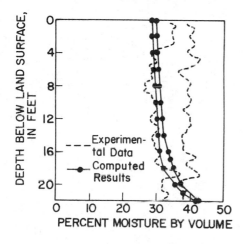

FIG. 9.11 Soil moisture: 5.25 days infiltration. [From D. W. Green, H. Dabiri, C. F. Weinaug, and R. Prill, Numerical Modeling of Unsaturated Groundwater Flow and Comparison of the Model to a Field Experiment, Water Resour. Res. 6(3), 862 (1970). Copyright by American Geophysical Union.]

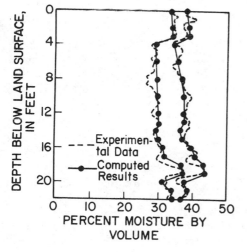

FIG. 9.12 Soil moisture: 8 days infiltration (considering hysteresis). [From D. W. Green, H. Dabiri, C. F. Weinaug, and R. Prill, Numerical Modeling of Unsaturated Groundwater Flow and Comparison of the Model to a Field Experiment, Water Resour. Res. 6(3), 862 (1970). Copyright by American Geophysical Union.]

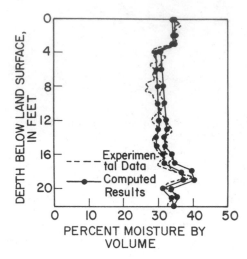

FIG. 9.13 Soil moisture: 48 days infiltration (considering hysteresis). [From D. W. Green, H. Dabiri, C. F. Weinaug, and R. Prill, Numerical Modeling of Unsaturated Groundwater Flow and Comparison of the Model to a Field Experiment, Water Resour. Res. 6(3), 862 (1970). Copyright by American Geophysical Union.]

because it was initially assumed that there was no hysteresis in the relative permeability and capillary pressure curves for the wetting and drying periods.

Action was taken to improve the fit for both wetting and drying. The initial estimates of soil properties were modified to fit the results. Thus it is not known whether the final physical data used in the model are correct.

The final comparison of the computer model with experimental data is typified by Figs. 9.12 and 9.13 for the drying period. While a good match was obtained after a modification of values for the physical properties of the porous media, more data are needed to ascertain which values are correct. The model is useful in determining where the emphasis on measurement should be placed. The results presented by Green et al. (1970) indicate that the mathematical model can be used effectively to describe the unsaturated flow behavior and that it could be used to make predictive calculations in the same area. ∎

9.2 NONPROPORTIONAL LIQUID FLOW

In Chapter 7 turbulent gas flow did not follow Darcy's law; velocity was not directly proportional to the potential gradient. In soils where hydraulic conductivity is low, deviations from Darcy's law may occur. Velocity increases more than the proportionality of the gradient. Clay and clayey soils show such behavior. Swartzendruber (1962a) represents this type of non-Darcy flow empirically as

$$q = B[i - J(1 - e^{-Ci})]$$
(9.23)

where i is the hydraulic gradient $-\Delta h/L$. B, J, and C are constants. At $q = 0$, $i = 0$ and Eq. (9.23) approaches Darcy's law and V becomes the hydraulic gradient K. If either J or C or both are zero, Eq. (9.23) is Darcy's law. Figure 9.14 shows flow velocity versus hydraulic gradient for nonproportional behavior.

There are several mechanisms postulated which individually or collectively account for the nonproportional flow behavior in soils. Low (1961) postulates that

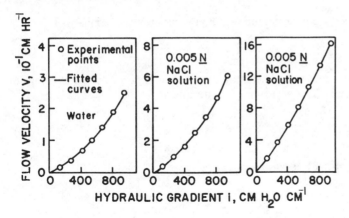

FIG. 9.14 Flow velocity versus hydraulic gradient for saturated flow through the sodium Utah bentonite-fitted curves as calculated from Eq. (9.23). [From D. Swartzendruber, Non-Darcy Flow Behavior in Liquid-Saturated Porous Media, J. Geophys. Res. 67, 5705 (1962). Copyright by the American Geophysical Union.]

soil water behaves as a non-Newtonian fluid as a result of interaction with clay particles. The nonproportionality may result from a porous medium where the particles and therefore the pores change as the velocity changes. This type of reorientation effect is possible in clay platelets (as shown schematically in Fig. 9.15). The gradient changes may cause expansion or compression of the particles, changing pore structure. Whether these effects are reversible or not is questionable.

Electrical streaming may enter the problem since there is a streaming potential

$$q = \left(K - \frac{C_e i_e}{i}\right) i \tag{9.24}$$

where C_e is the electric osmotic transport coefficient and i_e is the streaming potential gradient E/L. The ratio i_e/i is usually assumed constant for membranes and in capillary tubes. It is possible that in clayey soils i_e/i is not constant. Salt and osmotic effects may cause nonproportional behavior.

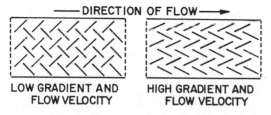

FIG. 9.15 Schematic diagram of how a highly idealized assemblage of clay platelets might reorient at higher water velocity, thus resulting in a greater hydraulic conductivity at high gradient than at low gradient. (From Swartzendruber, 1969.)

Writing Eq. (9.23) as a power series and including only the terms to i^2, then

$$q = a_3 i + a_4 i^2 \tag{9.25}$$

where a_3 and a_4 are constants. Assume steady radial flow in a spherical element of a non-Darcy porous medium. The equation of continuity is

$$\frac{1}{r^2} \frac{d(r^2 q)}{dr} = 0 \tag{9.26}$$

Integrating Eq. (9.26) twice with $q = q_1$ at $r = r_1$ yields

$$r^2 q = r_1^2 q_1 \tag{9.27}$$

Replacing i with $-dh/dr$ in Eq. (9.25) and combining with Eq. (9.27) yields

$$\frac{dh}{dr} = \left(\frac{a_3}{2a_4}\right)\left[1 - \left(r^2 - \frac{4a_4 q_1 r_1^2}{a_3^2}\right)^{1/2}\right] \tag{9.28}$$

Solving Eq. (9.28) with $h = h_1$ at $r = r_1$ and $h = h_2$ at $r = r_2$ yields

$$h_1 - h_2 = \frac{2}{a_3 G}\left[r_1 f(r_1,\, Gq_1) - r_2 f(r_2,\, Gq_1) + r_1 (Gq_1)^{1/2} \ln \frac{g(r_1,\, Gq_1)}{g(r_2,\, Gq_1)}\right] \tag{9.29}$$

where

$$G = \frac{4a_4}{a_3^2} \tag{9.30}$$

$$f(r_1,\, qr) = 1 - \left(1 + \frac{Gq_1 r_1^2}{r^2}\right)^{1/2} \tag{9.31}$$

$$q(r_1 Gq_1) = \left(\frac{Gq_1 r_1^2}{r^2}\right)^{1/2} + \left(1 + \frac{Gq_1 r_1^2}{r^2}\right)^{1/2} \tag{9.32}$$

In terms of $h - h_2$,

$$\frac{h - h_2}{h_1 - h_2} = \frac{r f_1 - r_2 f_2 + r_1 (Gq_1)^{1/2} \ln (g/g_2)}{r_1 f_1 - r_2 f_2 + r_1 (Gq_1)^{1/2} \ln (g_1/g_2)} \tag{9.33}$$

where the subscripts of f and g refer to r, r_1, r_2. If $4a_4 q_1 / a_3$ is small Eq. (9.33) reduces to

$$\frac{h - h_2}{h_1 - h_2} = \left(\frac{1}{r} - \frac{1}{r_2}\right)\left(\frac{1}{r_1} - \frac{1}{r_2}\right) \tag{9.34}$$

which is the form for Darcy flow. Figure 9.16 shows non-Darcy behavior compared to Eq. (9.29) and (9.25) on the left and Eq. (9.33) and (9.26) on the right.

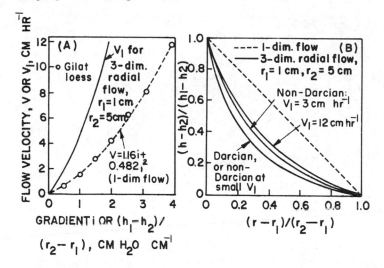

FIG. 9.16 One- and three-dimensional non-Darcy flow behavior for essentially saturated [$\tau = 10$ cm H_2O] Gilat loess: (a) velocity-gradient relationships showing q as calculated from Eq. (9.29) (solid-line curve), and Eq. (9.25) (broken-line curve) as fitted to the experimental points; (b) head dissipation patterns, solid-line curves calculated from Eqs. (9.33) (non-Darcy) and (9.34) (Darcy). (From Swartzen-druber, 1969.)

For nonproportional flow in unsaturated soils an equation similar to Eq. (9.23) is

$$q = \delta[w - \alpha(1 - e^{-\gamma w})] \qquad (9.35)$$

where

$$w = -\frac{\partial c}{\partial x} \qquad (9.36)$$

$$\delta = -B(c)\frac{\partial h_c}{\partial c} \qquad (9.37)$$

$$\alpha = -J(c)\frac{\partial c}{\partial h_c} \qquad (9.38)$$

$$\gamma = -C(c)\frac{\partial h_c}{\partial c} \qquad (9.39)$$

Therefore the hydraulic diffusivity is

$$\hat{D}(c,\ w) = \delta\left(1 - \frac{\alpha(1 - e^{-\gamma w})}{w}\right) \qquad (9.40)$$

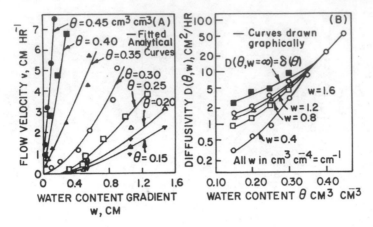

FIG. 9.17 (a) Flow velocity versus water-content gradient at selected water contents; fitted curves determined from Eq. (9.35), while experimental points for salkum silty clay loam were evaluated from data supplied by the authors and used by permission. (b) Diffusivity versus water content at specified water-content gradients; points calculated from Eq. (9.40) using δ, α, and y as fitted in part (a). (From Swartzendruber, 1969.)

Figure 9.17 shows a comparison of Eqs. (9.35) and (9.40) with nonproportional flow in unsaturated soils.

CASE STUDY 26: Infiltration into Unsaturated Soil with
Nonproportional Flow

The Darcy equation for flow may be written

$$V = Ki \tag{9.41}$$

in scalar form, where V is the steady-state flow velocity, K is the saturated hydraulic conductivity, which is considered a constant, and i is the hydraulic gradient. The hydraulic gradient i is $i = \Delta h / \ell$ where $h_2 - h_1 = \Delta h$, and h_1 and h_2 are the inlet and outlet hydraulic heads (1 is the distance between them). The Darcy equation fails at high gradients with turbulent flow. In this region, V increases less than proportionally with i.

In clay or clay containing soils over 5 percent, another deviation from Darcy's relationship has been noted. The velocity increases more than proportionally to the gradient i. Swartzendruber (1962b) represented this type of non-Darcy flow empirically as

$$V = B[i - J(1 - e^{-ci})] \tag{9.42}$$

Again, i is the hydraulic gradient while B, J, and C are constants. As i increases from zero at V = 0, equation (9.42) approaches the linear asymptote V = B(i - J), J is the gradient axis intercept of the linear asymptote. If either J or C or both are zero, Eq. (9.42) reduces to the Darcy equation.

For one-dimensional flow in the x-direction, an expression similar to Eq. (9.41) can be written

$$V = K(\tau) \frac{\partial \tau}{\partial x} \qquad (9.43)$$

The conductivity K is written as a function of τ, the suction head, because K for unsaturated flow is a function of the water content, θ and also $\tau = \tau(\theta)$; so $K(\tau)$ is written. (K for unsaturated flow is less than K for saturated flow because the air-filled pores reduce the effective cross section for liquid flow, and this increases the tortuosity of the remaining liquid flow paths.)

Similarly, for nonproportional unsaturated flow, Swartzendruber (1962b) presented this modified flow equation based on curve-fitting data.

$$V = \delta[\omega - \alpha(1 - e^{-\gamma \omega})] \qquad (9.44)$$

Equation (9.44) is derived from Eq. (9.42) by considering B, J, and C as functions of θ, along with $i = \partial h/\partial_x$, $h = z - \tau(\theta)$, and $w = -\partial \theta | \partial x$. Thus

$$\delta = B(\theta) \frac{-\partial \tau}{\partial \theta}$$

$$\alpha - J(\theta) \frac{-\partial \theta}{\partial \tau} \qquad (9.45)$$

$$\gamma = C(\theta) \frac{-\partial \tau}{\partial \theta}$$

where δ, α, and γ are constants at constant θ.

Using the method of Swartzendruber (1962b), and data published by Gardner and Hsieh (1956) a plot of the flow velocity versus water-content gradient at selected water contents was calculated.

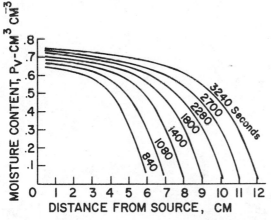

FIG. 9.18 Moisture content versus distance from source. [Data from W. Gardner and J. Hsieh, Reprinted from Soil Sci. Soc. Amer. Proc. 20, 157 (1956) by permission of Soil Science Society of America.]

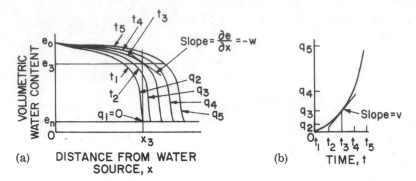

(a)   DISTANCE FROM WATER
      SOURCE, x

(b)   TIME, t

FIG. 9.19 Flow velocity and moisture gradient as functions of distance and time.
[Data from W. Gardner and J. Hsieh. Reprinted from Soil Sci. Soc. Amer. Proc.
20, 157 (1956) by permission of the Soil Science Society of America.]

    Figure 9.18 shows the data by measuring dye flows in samples of Cheshire
fine sandy loam, as the distance from the source, as a function of soil moisture
content.

    Figure 9.19a shows how the flow velocity V and moisture gradient ω for a given
value of moisture content θ are calculated. The moisture profiles show five profiles
from time t_1 to t_5. At $\theta = \theta_3$, the position coordinate for the t_3 profile is x_3. The
value of ω is the negative of the slope of the moisture profile at this point. Obtaining
v is more involved. V for θ_3 and t_3 is the volume of water per unit soil cross-
sectional area which moves per unit time across an imaginary plane at x_3. The total

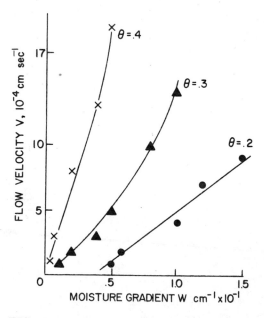

FIG. 9.20 Flow velocity versus moisture (Cheshire sandy loam).

volume of water which moved by x_3 is the area under the profile bounded by the lines $x = x_3$ and $\theta = \theta_3$. These q's are then plotted against time as in Fig. 9.19b. The slope of this curve at time t_3 is the value of v which corresponds to ω at θ_3. This calculation can be repeated to give values of v and ω for any desired θ. Figure 9.20 shows the result of these calculations using the sandy loam data. The initial moisture content was 0.1.

The plots in Fig. 9.20 can be analyzed, and fitted to Eq. (9.44) to find the values of δ, α, and γ. ■

9.3 MOVEMENT AND ADSORPTION OF CHEMICALS IN SOILS

The movement of chemicals (pollutants, nutrients, fertilizers, pesticides, etc.) in soils in liquids (solutions or gaseous states) depends on the combined effects of convection, dispersion including diffusion, and adsorption. The soil environment is complex. Soil profiles show differences in texture, structure, and chemistry with depth. Soils vary aereally in pattern and contained vegetation. Texture, structure, pore space, density, and humic nature of the soil system are interrelated in affecting movement of chemicals in soils. The microscopic character of the soils, especially the surface character of the minerals, affect adsorption. The phenomena of saturated and unsaturated flow affect chemical movement. The living matter in the soil affects the movement of the chemicals. The major mechanism of transport of chemicals in the soils is the water movement. Most of the chemicals are dissolved or dispersed in the water.

If we ignore dispersion including diffusion and consider movement down the profile—the direction parallel to gravity—only the continuity equation is necessary.

$$\frac{\partial c}{\partial t} = -v_z \frac{\partial c}{\partial z} - \frac{1}{\Phi} \frac{\partial F}{\partial t} \tag{9.46}$$

where F is the adsorption mass in the volume and $\partial F/\partial t$ is the adsorption rate. Freundlich equilibrium adsorption is normally assumed so that

$$F = kc^n \tag{9.47}$$

For most chemicals $1 \leq n \leq 1.4$. If instantaneous equilibrium is assumed between the adsorbed and solubilized component, the rate of adsorption is approximated by

$$\frac{\partial F}{\partial t} = \alpha(kc^n - F) \tag{9.48}$$

Assuming n = 1, then

$$\frac{\partial c}{\partial t} = -v_z \frac{\partial c}{\partial z} + \frac{\alpha}{\Phi}(kc - F) \tag{9.49}$$

with boundary conditions

$$c(0, t) = c_0 \qquad t \leq t_0 \tag{9.50}$$

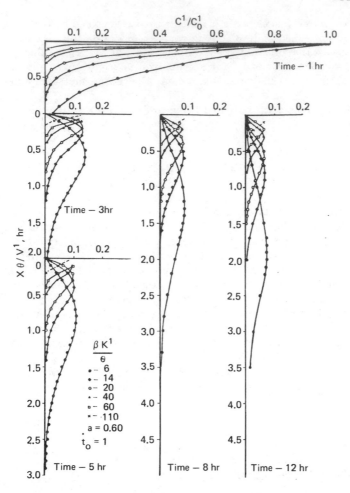

FIG. 9.21 The distribution of organic chemical in solution at different times of water application for various values of K when $\alpha = 0.60$ and $t_0 = 1$. [From J. K. Oddson, J. Letey, and L. V. Weeks, Predicted Distribution of Organic Chemicals in Solution and Adsorbed as a Function of Position and Time for Various Chemicals and Soil Properties, Soil Sci. Amer. Proc. 34, 412 (1970) by permission of the Soil Science Society of America.]

$$c(0, t) = c \qquad t > t_0 \tag{9.51}$$

Oddson et al. (1970) obtained solutions for $F(x, t)$ (complex expressions of Bessel functions) to show the effects of k, α, and t_0 on the concentration distribution in the soil profile. Figure 9.21 shows the calculated distribution of organic chemical for various values of k (β is bulk density; Θ is volumetric water content). The higher the value of k, the less the movement. Figure 9.22 shows the calculated distribution of adsorbed organic chemicals for various times and k values. In this figure the poisition of maximum adsorption follows the maximum concentration of solute.

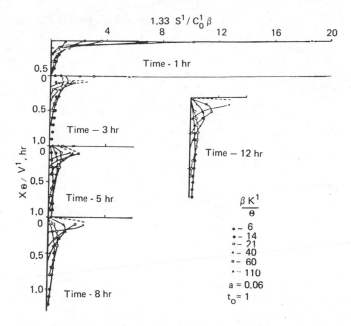

FIG. 9.22 The distribution of adsorbed organic for various times and various K values. [From J. K. Oddson, J. Letey, and L. V. Weeks, Predicted Distribution of Organic Chemicals in Solution and Adsorbed as a Function of Position and Time for Various Chemicals and Soil Properties, <u>Soil Sci. Amer. Proc.</u> <u>34</u>, 412 (1970) by permission of the Soil Society of America.]

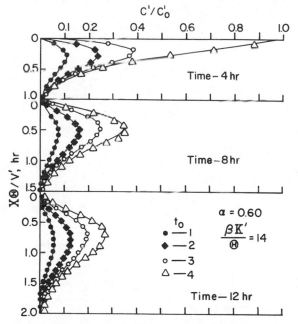

FIG. 9.23 The effect of applying different amounts of organic chemicals and their distribution in solution after different times of water application when $\alpha = 0.60$ and K = 14. [From J. K. Oddson, J. Letey, and L. V. Weeks, Predicted Distribution of Organic Chemicals in Solution and Adsorbed as a Function of Position and Time for Various Chemicals and Soil Properties, <u>Soil Sci. Amer. Proc.</u> <u>34</u>, 412 (1970) by permission of the Soil Science Society of America.]

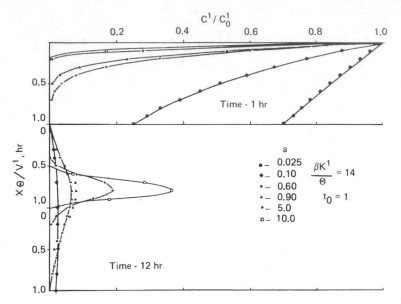

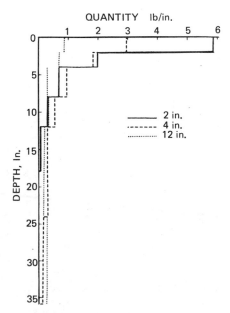

FIG. 9.24 The effect of the rate of equilibrium between adsorbed and solution chemical on the distribution in solution after 1 and 12 hr of water application when K = 14 and t_0 = 1. [From J. K. Oddson, J. Letey, and L. V. Weeks, Predicted Distribution of Organic Chemicals in Solution and Adsorbed as a Function of Position and Time for Various Chemicals and Soil Properties, Soil Sci. Amer. Proc. 34, 412 (1970) by permission of the Soil Science Society of America.]

FIG. 9.25 The distribution of monuron in soil following various applications of water. (From Upchurch and Pierce, 1957.)

Figure 9.23 shows the effect of changing t_0. The amount of chemical added and t_0 increase proportionally. Figure 9.24 shows that the rate of adsorption affects the concentration.

Figure 9.25 shows a plot of the data of Upchurch and Pierce (1957). This measured profile differs from the calculations in that the maximum concentration remains near the surface rather than moving down the profile. It appears adsorption is more severe than calculated by the model.

The value of α in Eqs. (9.48) and (9.49) can be determined from the equilibrium value of k. The mass balance is

$$\frac{\partial c}{\partial t} + \frac{\partial F}{\partial t} = 0 \qquad (9.52)$$

Integrating,

$$F(t) = c_0 - c(t) \qquad (9.53)$$

and substituting Eq. (9.53) into Eq. (9.48) yields

$$\frac{dc}{dt} = \alpha[c_0 - c(t)(k + 1)] \qquad (9.54)$$

Integrating Eq. (9.54),

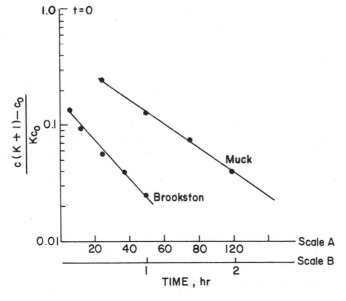

FIG. 9.26 The relationship between $[c(K + 1) = c_0]/Kc_0$ and time for two soils. [Reported by Kay and Elrick (1967) for Lindane. © 1967 The Williams and Wilkins Co.]

$$\frac{c}{c_0} = \frac{1}{k+1} - \frac{1}{k+1} e^{-\alpha(k+1)t}$$ (9.55)

Plotting $\ln\{[c(k+1) - c_0]/kc_0\}$ versus t should yield a straight line with slope $-\alpha(k+1)$. Figure 9.26 is such a plot using the data of Kay and Elrick (1967).

CASE STUDY 27: Simulation of the Distribution of Chemicals
in the Soil Profile: General Freundlich Isotherm

Chemicals such as pesticides and fertilizers are constantly being applied to soils. Wastes such as sludge and fly ash are often spread on the soil surface for disposal or are buried. Water through rainfall or irrigation causes chemicals from these materials to move through the soil profile. Most chemicals are adsorbed by the soil to a certain extent. The effectiveness of a fertilizer or pesticide is dependent, in part, on its position in the soil profile and whether it is adsorbed in solution. The potential harm of metals and other toxic chemicals leaching from waste disposal sites is also dependent on the chemicals' movement through the soil. It would be useful, therefore, to predict the concentration in solution and adsorbed as a function of position and time for various chemicals and soil types after irrigation or rainfall.

The purpose here is to illustrate one simulation technique for the distribution of chemicals in a dynamic fluid environment. The sensitivity of the parameters will be investigated. The variables were adjusted in order to observe the resulting behavior. Taking a mass balance over a differential element yields the equation

$$\frac{\partial c}{\partial t} + v_x \frac{\partial c}{\partial x} = D \frac{\partial^2 c}{\partial x^2} - \frac{1}{\phi} \frac{\partial F}{\partial t}$$ (9.56)

where

   c = concentration of species in solution
   F = amount of adsorbate on adsorbant per unit volume

This equation can be applied to various systems. The difference in each case depends on the assumptions and the boundary conditions. When this model is applied to a physical situation, the model should be a good simulation of the physical behavior. If the experimental data are different from the simulated values, it does not necessarily mean the model is incorrect. This would only indicate that boundary conditions and assumptions have not been satisfied.

Oddson et al. (1970) solved the differential equation above for organic chemicals in solution and adsorbed as a function of time and position for various chemical and soil properties.

They assumed the following:

1. The diffusion of organic chemical within the solution may be neglected in comparison with mass transport by water.
2. Assume the equilibrium condition can be expressed by the Freundlich equation

$$F = kc^n$$ (9.57)

3. Assume n is unity and therefore a linear relationship between adsorption and solution concentration at equilibrium.

4. The rate of adsorption is proportional to the difference between the amount which has already been adsorbed and the equilibrium value. Mathematically, this can be written

$$\frac{\partial F}{\partial t} = \alpha(kc - F) \tag{9.58}$$

5. The adsorption process is completely reversible; thus neglect any possible hysteresis effects.

6. The x axis will be directed vertically downward with x = 0 at the surface.

7. Assume no chemical present in the soil originally.

The organic solution with concentration c_0 was applied at the surface for $t \geq 0$. The solution is assumed to move through the soil with a velocity v_x. By symmetry, the problem is independent of y and z coordinates. Thus at any time $t > 0$ the wetting front will have moved only to a finite depth $x = v_x t$. The concentration at the surface is assumed to be constant at c_0 for a given length of time τ and zero thereafter. Thus the boundary conditions at the surface can be written

$$c(0,\ t) = c_0 \qquad 0 \leq t \leq \tau$$
$$c(0,\ t) = 0 \qquad t > \tau \tag{9.59}$$

The initial conditions are as follows:

$$F(x,\ t) = 0 \qquad 0 \leq t \leq \frac{x}{v_x}$$

$$c(x,\ t) = 0 \qquad 0 \leq t \leq \frac{x}{v_x} \tag{9.60}$$

Using the stated assumptions the mass equation becomes

$$\frac{\partial c}{\partial t} + v_x \frac{\partial c}{\partial x} + \frac{\alpha}{\phi}(kc - F) = 0 \tag{9.61}$$

Mathematically, Eq. (9.61) and the boundary conditions above constitute a "characteristic initial value problem." This can be solved by Riemann's method of integration. Omitting the details, the following expressions are obtained:

$$\text{for } 0 < t < \frac{x}{v_x}, \quad c(x,\ t) = 0 \tag{9.62}$$

$$\text{for } \frac{x}{v_x} < \frac{x}{v_x} + \tau, \quad c(x,\ t) = \frac{F(x,\ t)}{k} + c_0 \exp\left\{-\alpha[t + \frac{x}{v_x}(k - 1)]\right\} I_0(P) \tag{9.63}$$

$$\text{for } t > \frac{x}{v_x} + \tau, \quad c(x,\ t) = \frac{F(x,\ t)}{k} + c_0 \exp\left\{-\alpha[t + \frac{x}{v_x}(k - 1)]\right\} I_0(P)$$
$$- \exp(\alpha\tau) I_0(Q) \tag{9.64}$$

The evaluation of Eqs. (9.62) to (9.64) was obtained by using a program written in Fortran IV language, and using an IBM 360/50 system. The quantity adsorbed term F(x, t) was evaluated by numerical integration.

The units of F, C, v_x, and k are not the same as those traditionally used by soil scientists. The relationship between the quantities used in the equations and the traditional units are as follows:

$$F = \beta F' = \text{(bulk density of soil)} \frac{\text{adsorbed mass}}{\text{mass}}$$

$$c = \theta c' = \text{(volumetric water content)} \frac{\text{mass of solute}}{\text{cm}^3 \text{ soln}}$$

$$v = \frac{v'}{\theta} = \frac{\text{volumetric flow rate per total area}}{\text{volumetric water content}}$$

$$k = \frac{\beta k'}{\theta}$$

Concentration in the solution is a function of both time and distance the solution has traveled. There are two methods to observe the physical model. One method is to choose a position x in the soil layer and plot the concentration profile over all time. The other method is to choose a specific time t and plot the concentration profile over all positions. Oddson et al. (1970) chose the latter technique to observe the behavior of the concentration.

Figure 9.21 illustrates the distribution of organic chemical through the soil for various times. The data in Fig. 9.21 are for $\tau = 1$ hr, which means that organic chemical was available on the surface for 1 hr. After 1 hr of application, all of the organic available at the surface has entered the soil. The depth of movement for the chemical is dependent on k. The higher k, the less the movement. This is consistent with the physical situation since k is the equilibrium constant for adsorption. If more adsorption is occurring (higher k), then the chemical will move more slowly through the profile. Figure 9.21 also illustrates that maximum concentration is independent of k. The position of the maximum may vary with k, but the maximum concentration is independent of k.

Figure 9.23 illustrates the effect of applying different amounts of organic at the soil surface. In other words, the time that the organic chemical is present on the surface varies from $\tau = 1$ to 4 hr. As is seen in Fig. 9.23, the depth of movement appears to be quite independent of the amount of organic added at the soil surface. However, the concentration at a certain depth is proportional to the total amount of organic added at the soil surface. The greater the amount added, the greater the concentration.

A unique feature about this model is that it considers the rate of adsorption of organic chemical. The constant α is the rate constant for the rate of adsorption. Figure 9.24 illustrates the effect of α. During the time that the chemical is present on the surface, the depth of movement is greatly affected by α. Since α illustrates rate of adsorption, the chemical will not move as far into the soil for a large α. Once the chemical is removed from the surface, however, the opposite is true. The constant α is now for the rate of desorption. By making α large, the organic chemical moves through the soil as a wave. The depth of concentration maximum however is independent of α. Thus k influences the depth of maximum concentration for a given time, whereas α influences the magnitude of the concentration.

This model would be useful to predict the concentration profiles for chemicals in the soils. The constants k, α, and v_x would have to be determined experimentally for the type of chemical and soil present. Few data are available to estimate the value of α. Few investigators have examined the rate at which the adsorption comes to equilibrium.

Reversible adsorption was assumed. There is some evidence that this is not true for some types of soil materials. Some investigators have reported a hysteresis on the adsorption curve. The data reported are usually for adsorption rather than desorption. The desorption phase (upper part of curve) may be in error. The hysteresis effects reported would cause the predicted concentration to be in error where leaching occurs. ∎

Dispersion including diffusion affects the movement of chemicals in the soil both longitudinally and transversely. It also appears from measurements in field systems that field dispersion coefficients are several orders of magnitude larger than those measured in laboratory. Lapidus and Amundson (1952) solved the dispersion model with linear equilibrium and finite rate adsorption. The solution for this first-order rate adsorption is an integral form approximated by Ogata (1958). Lindstrom and Boersma (1970) considered a theory of chemical transport with simultaneous sorption but without convection. They considered Freundlich adsorption, first-order kinetics, and time-variant kinetics, numerically. Lai and Jurinak (1972) solved the problem of cation adsorption involving a general nonlinear exchange function, numerically. Gupta and Greenkorn (1973, 1974, 1976) consider linear and radial flow with dispersion and adsorption using Freundlich adsorption and Langmuir adsorption (bilinear adsorption) experimentally and numerically. A more complete discussion of this and other work is contained in Section 5.6.

CASE STUDY 28: Simulation of the Distribution of Chemicals in the Soils Including Dispersion, Convection, and Adsorption: Freundlich and Langmuir Adsorption

This case study involves the solution of the one-dimensional convective dispersion model:

$$\frac{\partial c}{\partial t} = D \frac{\partial^2 c}{\partial x^2} - v_x \frac{\partial c}{\partial x} - \frac{1}{\phi} \frac{\partial F}{\partial t} \tag{9.65}$$

where F is defined for Freundlich adsorption as

$$F = kc^n \tag{9.66}$$

with n usually 1, and F for Langmuir adsorption is defined as

$$F = \frac{ac}{1 + bc} \tag{9.67}$$

The formulation, boundary conditions, dedimensionalization, and numerical solution are considered in great detail in Section 5.6 and the appendix to Chapter 6. The boundary conditions used here are those for a step input, which gives the breakthrough curve for a particular solute (adsorbate).

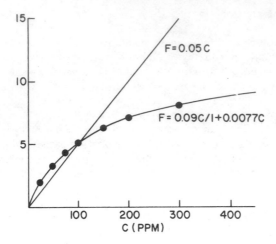

FIG. 9.27 Comparison of Freundlich and Langmuir isotherms.

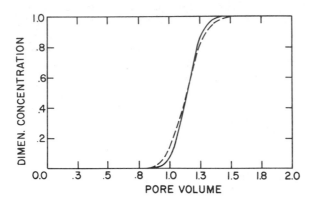

FIG. 9.28 Convective diffusion CO = 100 ppm: (———) Freundlich adsorption and (-----) Langmuir adsorption.

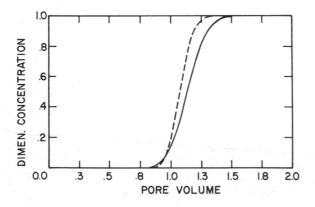

FIG. 9.29 Convective diffusion CO = 300 ppm: (———) Freundlich adsorption and (-----) Langmuir adsorption.

528

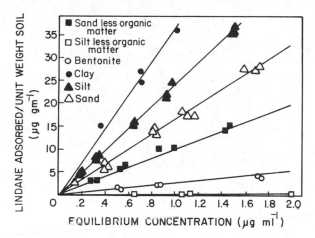

FIG. 9.30 Adsorption isotherm for lindane in soil components. [From B. D. Kay and D. E. Elrick, Soil. Sci. 104, 814 (1967). Copyright 1967 by the Williams and Wilkins Co., Baltimore.]

The dispersion coefficient and its implications have also been discussed in detail at the beginning of Chapter 5.

In this case study two particular aspects of dispersion with adsorption will be briefly discussed, and then some numerical results will be discussed for adsorption of an insecticide, lindane.

The dispersion coefficient in a convective system is a measure of the "spreading" of the concentration front. The mechanisms by which this occurs have been discussed. The appropriate measure of the significance of dispersion is the Peclet (Pe) number.* The Peclet number is the ratio of the convective to the dispersive forces of the system:

$$Pe = \frac{vL}{D} \qquad (9.68)$$

Thus for large velocities the convective forces dominate, and when the Pe number becomes large enough, the second space derivative may be ignored, and the solution is that discussed in the previous case study. The criterion for "large" depends on the particular application and accuracy desired. As the length of the system becomes sufficiently small, or the velocity becomes slow enough, the dispersive component tends to dominate. The problem with numerical solutions via divided difference schemes such as the Crank-Nicolson technique is that as the dispersion coefficient dominates, the Peclet number gets very small and the divided difference terms are manipulated as large and small numbers in basic mathematical operations, which, because they have only a limited number of digits, causes errors that are the same order of magnitude as the terms for the second space derivative. Thus these solutions fail for small enough Peclet numbers.

---

*Also called Bodenstein number.

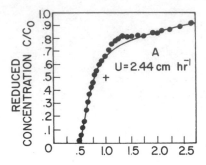 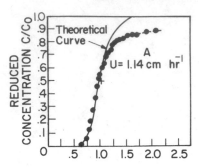

FIG. 9.31 Breakthrough curves for two different velocities of lindane from Honeywood loam. [From B. D. Kay and D. E. Elrick, Soil Sci. 104, 814 (1967). Copyright 1967 by the Williams and Wilkins Co., Baltimore.]

The functional form of the adsorption isotherm chosen is usually determined experimentally by equilibrium adsorption data. The Langmuir isotherm is used when a maximum of adsorbed species is reached, no matter what the equilibrium concentration in solution, and the Freundlich is used for linear or power dependence on the concentration. These two types cover a great many possibilities; however, some adsorbates escape description. For a large number of adsorbates a linear or power-dependent isotherm is satisfactory because the solubility limit of the compound is reached before any nonlinear effects are observed in aqueous solution. Figure 9.27 shows a Freundlich and Langmuir isotherm determined about the crossover point (point at which the curves cross); Fig. 9.28 shows the breakthrough curves for each isotherm for a c_0 of 100 ppm. Some difference is noted; however, it is relatively small and the error is evenly distributed. Figure 9.29 shows the compared breakthrough curves for $c_0 = 300$ ppm. It can readily be seen that as the concentration gets higher, the error in the curve becomes greater, as it underestimates the rate of change of the concentration. It should be pointed out that although these curves go to the limit of $c/c_0 = 1.0$, in reality this probably is not the case, as the adsorption capacity of a large number of soils is sufficient to retain a large percentage of the solute (for a step change as shown here). Some adsorbents may have a surface area for adsorption of 600 to 1000 m^2 g^{-1}. Thus for monolayer coverage, tremendous quantities of material may be adsorbed.

Kay and Elrick (1967a) solved this problem for adsorption of the insecticide lindane (1,2,3,4,5,6-hexachlorocyclohexane). Figures 9.30 and 9.31 show the adsorption isotherm and breakthrough curves, respectively. ■

REFERENCES

Bear, J. (1972): Dynamics of Fluids in Porous Media, American Elsevier, New York.

Childs, E. C. (1936): The Transport of Water through Heavy Clay Soils, I, III, J. Agric. Sci. 26, 114, 527.

Gardner, W., and J. Hsieh (1956): Water Velocity in Unsaturated Porous Media, Soil Sci. Soc. Amer. Proc. 20, 157.

Green, D. W., H. Dabiri, C. F. Weinaug, and R. Prill (1970): Numerical Modeling of Unsaturated Groundwater Flow and Comparison of the Model to a Field Experiment, Water Resour. Res. 6(3), 862.

Gupta, S. P., and R. A. Greenkorn (1973): Dispersion During Flow in Porous Media with Bilinear Adsorption, Water Resour. Res. 9(5), 1357.

Gupta, S. P., and R. A. Greenkorn (1974): Determination of Dispersion and Nonlinear Adsorption Parameters for Flow in Porous Media, Water Resour. Res. 10(4), 839.

Gupta, S. P., and R. A. Greenkorn (1976): Solution for Radial Flow with Nonlinear Adsorption, ASCE J. Env. Div. 102 EE1, 87.

Hamaker, J. W., G. A. I. Goring, and C. R. Youngsen (1966): Sorption and Movement of Lindane in Soils, Advan. Chem. Sec. 60, 23.

Hornberger, G. M., and I. Remson (1970): A Moving Boundary Model of a One-Dimensional Saturated-Unsaturated Transient Porous Flow System, Water Resour. Res. 6(3), 898.

Kay, B. D., and D. E. Elrick (1967a): Adsorption and Movement of Lindane in Soils, Soil. Sci. 104, 314.

Kay, B. D., and D. E. Elrick (1967b): Sorption and Leaching of 4-Amino-3,5,6-Trichloropicoline Acid in Soils, Soil Sci. 104, 814.

Lai, S. H., and J. J. Jurinak (1972): Cation Adsorption in One-Dimensional Flow through Soils, Water Resour. Res. 8(1), 99.

Lapidus, L., and N. R. Amundson (1952): Mathematics of Adsorption in Beds, J. Phys. Chem. 56, 984.

Lindstrom, F. T., and L. Boersma (1970): Theory of Chemical Transport with Simultaneous Sorption in Water Saturated Porous Media, Soil Sci. 110(1), 1.

Low, P. F. (1961): Physical Chemistry of Clay-Water Interaction, Advan. Agron. 13, 269.

Oddson, J. K., J. Letey, and L. V. Weeks (1970): Predicted Distribution of Organic Chemicals in Solution and Adsorbed as a Function of Position and Time for Various Chemicals and Soil Properties, Soil Sci. Soc. Amer. Proc. 34, 412.

Ogata, A. (1958): Dispersion in Porous Media, Ph.D. Thesis, Northwestern University, Evanston, Ill.

Swartzendruber, D. (1962a): Non-Darcy Flow Behavior in Liquid-Saturated Porous Media, J. Geophys. Res. 67, 5205.

Swartzendruber, D. (1962b): Modification of Darcy's Law, Soil Sci. 93, 22.

Swartzendruber, D. (1969): The Flow of Water in Unsaturated Soils, in Flow through Porous Media, R. J. M. de Wiest (ed.), Academic, New York, Chapter 6.

Upchurch, R. P., and W. C. Pierce (1957): The Leaching of Monuron from Lakdanof Sand, <u>Weeds</u> 5, 321.

Wisler, C. O., and E. F. Brater (1954): <u>Hydrology</u>, Wiley, New York.

SUGGESTED READING

Ahlrichs, J. L., The Soil Environment, in <u>Organic Chemicals in the Soil Environment</u>, Volume 1, C. A. I. Goring and J. W. Hamaker (eds.), Marcel Dekker, New York, 1972, Chapter 1.

Childs, E. C., Soil Moisture Theory, in <u>Advances in Hydroscience</u>, V. T. Chow (ed.), Academic, New York, 1967.

Corey, A. T., Mechanics of Heterogeneous Fluids in Porous Media, Water Resources Pub., Ft. Collins, Colo., 1977.

Letey, J., and J. K. Oddson, Mass Transfer, in <u>Organic Chemicals in the Soil Environment</u>, Volume 1, C. A. I. Goring and J. W. Hamaker (eds.), Marcel Dekker, New York, 1972, Chapter 6.

Nielsen, D. R., R. D. Jackson, J. W. Cary, D. D. Evens (eds.), Soil Water, American Society of Agronomy, Soil Science Society of America, Madison, Wis., 1972.

Symbols

L = length; M = mass; t = time; T = temperature; q = electric charge. (l) indicates a dimensionless quantity.

A	area (L^2)
a	connectivity, accessibility (l)
a	dispersivity (L)
B	formation volume factor (l)
Bn	Bond number (l)
B	drainage factor (l)
b	parameter in slip flow expression for K ($t^2 L/M$)
C	number of components
C	shape factor (l)
C	shape parameter (l)
$\mathscr{C}$	vertical hydraulic conductivity (L/t)
c	water content (l)
c	concentration (l)
c_B	bulk compressibility (Lt^2/M)
c_f	combined matrix fluid compressibility (Lt^2/M)
c_I	isothermal compressibility (Lt^2/M)
c_ℓ	fluid compressibility (Lt^2/M)
C_D	well-bore storage coefficient (l)
Ca	capillary number (l)
C_p	heat capacity at constant pressure ($L^2/t^2 T$)
C_v	heat capacity at constant volume ($L^2/t^2 T$)
c_p	pore compressibility (Lt^2/M)
c_s	matrix, solid compressibility (Lt^2/M)
D	diameter (L^2)
D	dispersion (L^2/t)
D_L	longitudinal dispersion (L^2/t)

D_p	particle diameter (L)
D_T	transverse dispersion (L^2/t)
$\hat{D}$	soil water "diffusion" (L^2/T)
$\mathscr{D}$	binary diffusion coefficient (L^2/t)
d	characteristic microscopic length (L)
d	pore diameter (L)
E	emf ($L^2 M/t^2 q$)
E	Young's modulus (M/Lt^2)
e	volume strain (l)
$e^{(0)}$	initial volume strain (l)
F	amount of absorbate on absorbent (l)
F	degrees of freedom (l)
F	dimensionless capillary ratio (l)
F	electrical resistivity factor (l)
F	Helmholtz free energy ($L^2 M/t^2$)
$\underline{F}$	body forces (LM/t^2)
f	friction factor (l)
f_n	fraction of nonwetting fluid flowing (l)
f_p	friction factor based on particle diameter (l)
f_s	superficial surface free energy ($L^2 M/t^2$)
ΔG_d	differential gas production (L^3/t)
g	acceleration due to gravity (L^2/t^2)
H	Hankel function (l)
H	non-Newtonian bed factor (l)
h	energy per unit weight of fluid (hydraulic head for water) (L)
h	thickness (L)
h_c	suction head (L)
$\underline{I}$	unit matrix (l)
i	current (q)
i	$\sqrt{-1}$
i	negative hydraulic gradient (L)
i_e	electrical streaming potential (q)
J	Leverett J factor, reduced capillary pressure (l)
j	mass flux ($M/L^2 t$)
K	conductance parameter (L/t)
K	hydraulic conductivity (L/t)
K	scale factor (l)
$\tilde{K}$	slip conductivity (L/t)
K_c	effective hydraulic conductivity of water (L/t)
K_i	vaporization equilibrium ratio of component i (l)
K_L	variance of marked particle in direction of flow, longitudinal dispersion (L^2/t)
K_L	variance of marked particle perpendicular to flow, transverse dispersion (L^2/t)

K^{ad}	adsorption equilibrium constant (1)
k	intrinsic permeability (L^2)
k	kinetic rate constant
$\hat{k}$	thermal conductivity (LM/t^3T)
k_L	effective permeability to water as a function of water content (L^2)
k_r	effective permeability of nonwetting fluid (L^2)
k_{rg}	relative permeability of gas (1)
k_{ri}	relative permeability (1)
k_{ro}	relative permeability of oil (1)
k_{rog}	two-phase gas–oil relative permeability (1)
k_{row}	two-phase oil–water relative permeability (1)
k_{rw}	relative permeability of water (1)
L	characteristic macroscopic length (L)
L	leakage factor (1)
L	moles of liquid (M)
l	length (L)
ℓ	pore length (L)
M	molecular weight (1)
m	parameter in power law model (M/t^2)
m	mass (M)
m	mobility ratio (1)
N	number of particles (1)
N	number of pores (1)
$\underline{N}$	number flux ($1/L^2$)
n	coordination number, ratio of void to total volume (1)
n	shape factor (1)
n	parameter in power law model (1)
$\underline{n}$	outward directed normal (1)
n_i	moles of component i (M)
ΔN_d	differential oil production (L^3/t)
P	number of phases (1)
$\mathscr{P}$	$p - p_0 + \rho\phi$ (M/t^2L)
Pe	Peclet number (1)
Pe_L	Peclet number based on length (1)
Pr	Prandtl number (1)
p	pressure (M/t^2L)
p_c	capillary pressure (M/t^2L)
p_c	critical pressure (M/t^2L)
p_D	dimensionless pressure (1)
p_m	arithmetic average pressure $(p_1 + p_2)/2$ (M/t^2)
p_0	reference pressure (M/t^2L)
p_r	reduced pressure (1)

$p^{(0)}$ initial pressure ($M/t^2 L$)

p^* lowered vapor pressure in pores (Kelvin) ($M/t^2 L$)

p° vapor pressure ($M/t^2 L$)

Q ultimate capacity of dry adsorbent in gram equivalents adsorbed per gram of dry medium (l)

Q volumetric flow rate (L^3/t)

Q_m volumetric flow rate at average pressure p_m (L^3/t)

q amount of gram equivalents adsorbed per gram of dry medium (l)

$\underline{q}$ seepage velocity (L/t)

q_∞ capacity of adsorbent at saturation (l)

Re Reynolds number (l)

$\hat{Re}$ Reynolds number in terms of inertial parameter (l)

Re_p Reynolds number based on particle diameter (l)

R capillary radius (L)

R electrical resistance ($L^2 M/q^2 t$)

R gas constant ($L^2 M/t^2 T$)

R_H hydraulic radius (L)

R_{nw} interface curvature (l)

R_s solution gas/oil ratio (l)

r radius

r_d effective drainage radius (L)

r_e drainage radius (L)

r_w radius of investigation (L)

r_w well radius (L)

r_1, r_2 radii of curvature in Laplace equation (L)

S entropy ($L^2/t^2 T$)

S saturation (l)

S storativity ($L^2 t^2/M$)

S surface area (L^2)

S specific surface (l)

S_A surface area of pores (L^2)

S_e effective saturation (l)

S_g gas saturation (l)

S_L total liquid saturation (l)

S_n nonwetting fluid saturation

S_o oil saturation (l)

S_p particle surface area (L^2)

S_w surface between solid and fluid

S_w water saturation (l)

S_w wetting fluid saturation (l)

S_{wi}	residual saturation (1)
S_{wr}	irreducible saturation (1)
s	distance along an arc (L)
s	drawdown (L)
s	skin effect (1)
T	temperature (T)
T	transmissivity ($L^2 t/M$)
T_c	critical temperature (T)
T_r	reduced temperature (1)
t	time (t)
t_L	time lag (t)
t_0	characteristic time of Noll simple fluid (t)
U	velocity of approach at $x = -\infty$ for Hele-Shaw model (L/t)
u	displacement (L)
V	moles of vapor (M)
V	velocity at $x - \infty$ for Hele-Shaw model (L/x)
V	volume (L^3)
V*	V/V_p pore volume (1)
$\tilde{V}$	velocity of particle following most probable path (L/t)
$\tilde{V}$	volume per mole (L^3/M)
V_b	bulk volume of a porous medium (L^3)
V_{b_1}	bulk volume at 1 atm (L^3)
V_c	critical volume (L^3)
V_f	fluid volume (L^3)
V_p	particle volume (L^3)
V_{p_1}	pore volume at 1 atm (L^3)
V_r	reduced volume (1)
V_s	solid, matrix volume (L^3)
$\underline{V}_{DF}$	Dupuit-Forchheimer velocity (L/t)
V_v	void volume (L^3)
v	voltage, emf ($L^2 M/t^2 q$)
v	velocity (L/t)
v	interstitial velocity (L/t)
$\bar{v}$	time-averaged velocity (L/t)
$\langle v \rangle$	area-averaged velocity (L/t)
v_∞	velocity of approach (L/t)
w	complex function
w	mass rate of flow (M/t)
x	coordinate (L)
x_i	mole fraction of component i in liquid (1)
y	coordinate direction (L)

y_i mole fraction of component i in vapor (1)

z compressibility factor (1)

z coordinate (direction of gravity) (L)

z_i mole factor of component i in a system (1)

α angle between velocity vector and the outward directed normal

α attenuation (1/L)

α parameter in Ellis model (1)

α vertical compressibility (Lt^2/M)

$\alpha(\delta)$ void distribution factor (1)

α_0 parameter in non-Newtonian model of porous medium (1)

β coefficient of pressure expansion (Lt^2/M)

β compressibility of mole (Lt^2)

β phase factor (1/L)

β turbulence factor (1)

Δ difference

∂ partial differential

δ flow correction factor (1)

δ increment (L)

δ pore diameter (L)

δ_{ek} effective frequency parameter (1)

δ_{et} dynamic frequency parameter (1)

δ_k frequency parameter (1)

δ_t frequency parameter (1)

δ_{ij} Kronecker delta (1)

δ_z unit vector in z direction (1)

$\underline{\nabla}$ vector partial differential operator (1/L)

$$\underline{i}\ \partial/\partial x + \underline{j}\ \partial/\partial y + \underline{k}\ \partial/\partial z$$

∇^2 $\partial/\partial x^2 + \partial/\partial y^2 + \partial/\partial z^2$ ($1/L^2$)

ϵ incremental volume strain (1)

ϵ_y incremental strain (1)

Γ propagation constant (1/L)

γ heat capacity ratio (1)

γ wave number (1/L)

γ_{12} specific free energy of interface between fluids 1 and 2 (L^2M/t^2)

γ_{s1} specific free energy of interface between the solid and fluid 1 (L^2M/t^2)

λ coefficient of elasticity (1)

λ finger width parameter (1)

λ Lamé constant (M/Lt^2)

λ inertial parameter (1)

λ mobility ratio (1)

λ pore size distribution index (1)

μ shear modulus (M/Lt^2)

μ viscosity (LM/t)

μ_i chemical potential of component i (L^2t^2/M)

μ_0	apparent viscosity (LM/t)
μ_0	characteristic viscosity of a Noll simple fluid (LM/t)
μ_p	plastic viscosity (LM/t)
ν	kinematic viscosity (L²/t)
ν	Poisson's ratio (1)
ν_t	thermal conductivity parameter (1)
Ω	wavelength (L)
ω	frequency (1/t)
ω_i	mass fraction of component i (1)
Φ	potential (L²/t)
ϕ	porosity (1)
Ψ	stream function (L²/t)
ψ	pseudopressure (M/Lt²)
θ	contact angle of liquid-solid interface)
ρ	density (M/L³)
ρ_b	bulk density of matter (M/L³)
ρ_0	initial density at p = 1 atm (M/L³)
ρ_0	steady-state density (M/L³)
σ	electrical conductivity (q²T/L²)
σ	incremental fluid pressure (M/Lt²)
σ	interfacial tension $\sigma \equiv \gamma_{12}$ (M/t²)
σ	pore fluid tension (M/Lt²)
σ	shear stress (M/Lt²)
$\bar{\sigma}$	hydrostatic tension (M/Lt²)
σ^*	effective interfacial tension (M/Lt²)
σ'	incremental effective stress (M/Lt²)
σ_k	shear stress for porous medium (M/Lt²)
σ_R	shear stress at capillary wall (M/Lt²)
τ	fluid stress (M/Lt²)
τ	tortuosity (1)
$\tau_{\frac{1}{2}}$	shear stress at $\frac{1}{2}\tau_0$ (M/Lt²)
τ_0	yield stress (M/Lt²)
τ_y	initial stress (M/Lt²)
$\tau_y^{(0)}$	total stress (M/Lt²)
τ_y'	effective stress (M/Lt²)
τ_w	tortuosity of wetting fluid (1)
η_{eff}	effective viscosity in Ellis models (M/Lt)
η_0	parameter in Ellis model (1)
$\langle\,\rangle$	space average

Overscores

^   identifying mark
~   per mole, perturbation
-          time average

Subscripts

f   fractured
g   gas
i   coordinates
l   liquid
n   nonwetting
o   oil
r   residual
s   standard conditions
w   water
w   wetting

Note: The Society of Petroleum Engineers has endorsed a Final SPE Metric
Standard. The International System of Units (SI) is reported in the Journal of
Petroleum Technology, September 1982, pp. 2019-2056.

Index

541